Commercial Building Construction

About the Author

David A. Madsen

MADSEN DESIGNS INC.

David A. Madsen is the president of Madsen Designs Inc. (www.madsendesigns.com). David is Faculty Emeritus of Drafting Technology and the Autodesk Premier Training Center at Clackamas Community College in Oregon City, Oregon. David was an instructor and department chairperson at Clackamas Community College for nearly 30 years. In addition to community college experience, David was a Drafting Technology instructor at Centennial High School in Gresham, Oregon. David is a former member of the American Design and Drafting Association (ADDA) Board of Directors and was honored by the ADDA with Director Emeritus status at the annual conference in Louisville, Kentucky. David has extensive experience in mechanical drafting, architectural design and drafting, and building construction as a commercial construction worker, architectural designer, and general contractor in Oregon for over 30 years. David holds a Master of Education degree in Vocational Administration and a Bachelor of Science degree in Industrial Education. David is the coauthor of Cengage Learning *Architectural Drafting and Design*, *Engineering Drawing and Design*, *Print Reading for Architecture and Construction Technology*, and *Print Reading for Engineering and Manufacturing Technology*; and Goodheart-Willcox *AutoCAD and Its Applications: Basics, Advanced, and Comprehensive,* and *Geometric Dimensioning and Tolerancing.* Also the coauthor of Pearson *Civil Drafting Technology*, and Routledge *Modern Residential Construction Practices.*

Commercial Building Construction

Materials and Methods

DAVID A. MADSEN

President, Madsen Designs Inc.
www.madsendesigns.com
Faculty Emeritus
Former Department Chairperson Drafting Technology
Autodesk Premier Training Center
Clackamas Community College, Oregon City, Oregon
Director Emeritus
American Design Drafting Association

New York   Chicago   San Francisco
Athens   London   Madrid
Mexico City   Milan   New Delhi
Singapore   Sydney   Toronto

Library of Congress Cataloging-in-Publication Data

Names: Madsen, David A., author.
Title: Commercial building construction : materials and methods / David A.
   Madsen.
Description: [New York] : McGraw Hill Education, [2021] | Includes index. |
   Summary: "This textbook introduces students to building construction
   methods and systems, with real-world examples and modern 3D
   illustrations that reflect current industry practices better than other
   books"— Provided by publisher.
Identifiers: LCCN 2020038954 | ISBN 9781260460407 (hardcover) | ISBN
   9781260460414 (ebook)
Subjects: LCSH: Commercial buildings—Design and construction—Textbooks.
Classification: LCC TH4311 .M32 2020 | DDC 690/.52—dc23
LC record available at https://lccn.loc.gov/2020038954

McGraw Hill books are available at special quantity discounts to use as premiums and sales promotions, or for use in corporate training programs. To contact a representative please visit the Contact Us page at www.mhprofessional.com.

Commercial Building Construction: Materials and Methods

1 2 3 4 5 6 7 8 9   CCD   26 25 24 23 22 21

ISBN   978-1-260-46040-7
MHID   1-260-46040-1

The pages within this book were printed on acid-free paper.

Sponsoring Editor
   Ania Levinson

Editing Supervisor
   Donna M. Martone

Production Supervisor
   Pamela A. Pelton

Acquisitions Coordinator
   Elizabeth Houde

Project Manager
   Warishree Pant,
   KnowledgeWorks Global Ltd.

Copy Editor
   KnowledgeWorks Global Ltd.

Proofreader
   KnowledgeWorks Global Ltd.

Indexer
   ARC Indexing Inc.

Art Director, Cover
   Jeff Weeks

Composition
   KnowledgeWorks Global Ltd.

Contents

Preface

This textbook provides easy-to-read, comprehensive, and highly illustrated coverage of commercial building construction practices that conform to industry standards. Each chapter provides comprehensive descriptions, real-world practices, realistic examples, actual two-dimensional (2D) commercial construction drawings, three-dimensional (3D) illustrations, and related tests, research problems, and print reading problems. The examples illustrate typical construction practices used throughout the United States and Canada. Chapters are organized in a format that is consistent with the process used to take commercial construction projects from preliminary concept through all phases of commercial building construction.

The following describes several special features found in this textbook.

Special Note Feature

> **NOTE**
> A special Note box feature is used where appropriate to provide reference to other chapters, additional explanation, helpful tips, professional information, or alternate practice.

Commercial Construction Applications

Actual real-world commercial construction applications are found throughout this textbook to reinforce what you learn with current practices used in the commercial construction industry. There is no substitute for observing the construction of actual projects and hearing accounts from architects, engineers, contractors, and material suppliers related to actual construction projects.

Chapter Tests

Each chapter ends with a content-related test used for examination, review, or other purposes described in your course syllabus. Each chapter test is organized into the following test groups:

Know and Understand. This test group is designed for reinforcement and digital platform auto-grading. This group contains true and false, and multiple-choice questions.

Analyze and Apply. This test group requires you to apply concepts learned in the chapter in order to answer the question. This section consists of short answer and calculation questions.

Go to the Downloads & Resources tab at **www.mhprofessional.com/Commercial Building Construction-Ancillaries** to access the tests.

Chapter Problems

Critical Thinking Problems. These problems are generally short essay or calculation problems that require you to analyze content in the chapter. These problems are intended to evaluate comprehension and application of higher-level learning objectives. Creative thinking problems are provided to help expand your knowledge by researching given subjects.

Go to the Downloads & Resources tab at **www.mhprofessional.com/Commercial Building Construction-Ancillaries** to access the creative thinking problems.

Image Bank

All figures found in this textbook are provided in an image bank where you can access and look at the chapter figures on your computer screen where you can zoom in and out and pan for better observation, because full drawings reduced to fit on a textbook page are often difficult to read.

Go to the Downloads & Resources tab at **www.mhprofessional.com/Commercial Building Construction-Ancillaries** to access the images provided in this textbook.

Glossary

The most comprehensive building construction glossary available anywhere is located in this textbook. Glossary terms are **bold** in chapter content, where they are defined immediately and also placed in the glossary for additional alphabetical reference. Each term is clearly defined using descriptions that are related to construction industry practices. In addition to terms found throughout this textbook, the glossary contains many more building construction-related terms.

Complete Set of Commercial Working Drawings

As you use this textbook to study commercial construction, you can discover the importance of being able to read actual architectural drawings that are used in commercial construction. A complete set of plans for the Brookings South Main Fire Station are used as examples throughout this textbook.

Go to the Downloads & Resources tab at **www.mhprofessional.com/Commercial Building Construction-Ancillaries** to access the images provided in this textbook.

Print reading problems have questions that ask you to find information on the Brookings South Main Fire Station plans. The print reading problems can help you become familiar with the format of construction documents and reinforce your ability to seek specific information. This is an important skill for you to master as you prepare to enter the construction industry. Go to **Downloads & Resources tab at www.mhprofessional.com/Commercial Building Construction-Ancillaries** to access the complete set of plans for the Brookings South Main Fire Station.

LEARNING COMMERCIAL BUILDING CONSTRUCTION:
MATERIALS AND METHODS

Commercial Building Construction: Materials and Methods is designed for you. Chapters are presented in individual learning segments that begin with basic concepts and build until each chapter provides complete coverage of each topic. The content of each chapter is divided into logical learning segments, providing you with an opportunity to see two-dimensional and three-dimensional illustrations of construction applications and practices. Content is based on commercial building construction practices and standards, and trends in the commercial construction industry. Use the text as a learning tool while in school, and take it along as a desk reference when you enter the profession. The amount of written information is comprehensive but kept as concise as possible. Examples and illustrations are used extensively. The following are a few helpful hints to use as you learn commercial building construction materials and methods using this textbook:

1. *Read the text.* The text content is designed for easy reading. Descriptions, practices, and applications are given in as few, easy-to-understand words as possible. Do not pass up the reading, because the content helps you to understand construction concepts and how to build every aspect of a new project.

2. *Look carefully at the examples.* The figure examples are presented in a manner that is consistent with *Commercial Building Construction: Materials and Methods*. Look at the examples carefully in an attempt to understand the intent of specific applications. If you are able to understand why something is done a certain way, it is easier for you to apply the concepts to construction applications in school and on the job.

3. *Use the text as a reference.* Construction practices can be complex and varied, so always be ready to use the reference if you need to verify how a specific application is handled. Become familiar with the definitions and use of technical terms. It is difficult to memorize everything in this textbook, but after hands-on use of the concepts and field experience, construction applications should become second nature.

4. *Learn each concept and skill before you continue to the next.* The text is presented in a logical learning sequence. Each chapter is designed for learning development, and chapters are sequenced for your construction knowledge to grow from one chapter to the next. Chapter tests and problem assignments are presented in the same learning sequence as the chapter content.

5. *Practice.* Development of good construction skills depends to a large extent on experience and practice. It can take several months of training and experience to become proficient with construction practices. You should take building construction courses at your school and continue with on-the-job training with a commercial contractor or in an apprenticeship program.

6. *Read actual commercial plans.* Reading as many actual commercial plans as possible gives you a chance to organize thoughts about how the project is built in relationship to what you learn in school and on the job. Your ability to effectively read commercial plans can help you to communicate in the construction industry.

7. *Attend building construction or related schooling.* Find education courses in building construction, commercial construction, building codes, or surveying at your

high school, trade school, community college, or apprenticeship program. Most construction technology programs in schools have quality professional training, equipment, and practices used on actual construction projects. Compare programs and evaluate the equipment, facilities, instructorsand projects when selecting a school for your education.

INSTRUCTOR RESOURCES

The *Instructor's Manual* for *Commercial Building Construction: Materials and Methods* contains the following features to help you develop and teach commercial building construction courses.

Chapter Test Answers

The *Instructor's Manual* has each chapter test provided with the questions repeated and the answer to each question given.

Chapter Problem Solutions

The *Instructor's Manual* provides written or graphic solutions to end-of-chapter problems. Actual student answers can vary but should include the fundamental content provided.

Image Bank

All figures found in this textbook are provided in an image bank where you can access and look at the chapter figures on your computer screen, and pan in and out for better observation, because full drawings reduced to fit on a textbook page are often difficult to read. The image bank also includes the complete set of plans for the Brookings South Main Fire Station used as examples throughout this textbook. Use the image bank as desired to create your own teaching presentations and print reading problems.

Go to the Downloads & Resources tab at **www.mhprofessional.com/Commercial Building Construction-Ancillaries** to access the images provided in this textbook.

ACKNOWLEDGMENTS

Special thanks to all industry professionals who provided images that correlate with content throughout this textbook. A credit is provided with each image acknowledging the contribution.

Cover Photo

The cover photo displays the University of Florida Health Hospital project showcased in Chapter 2, *Sustainable Technology*. Commercial construction application provided by Green Globes® and administered by the Green Building Initiative®.

Title Page Photo

The title page photo displays the Central Academic Building, Texas A&M University in San Antonio, Texas showcased in Chapter 5, *Masonry Construction*. Commerical construction application provided by Acme Brick.

Modern Residential Construction Practices

Many of the drawings in this book were previously published in *Modern Residential Construction Practices*, Routledge, 2017, and are reproduced with permission of Taylor & Francis (ISBN: 978-1-138-28489-0). *Modern Residential Construction Practices* provides easy-to-read, comprehensive, and highly illustrated coverage of residential building construction practices that conform to industry standards in the United States and Canada. The book is organized in a format that is consistent with the process used to take residential construction projects from preliminary concept through all phases of residential building construction. An ideal textbook for college level construction programs, the book is packed with useful features such as problems, summaries, key notes, a detailed glossary, and online materials for both students and educators. See: https://www.routledge.com/9781138284890.

GRAPHISOFT graphisoft.com

Special thanks is given to the following for their help in providing content and graphics for the building information modeling (BIM) content in Chapter 1, *Plans, Specifications, and Construction Management*.

Julianna Gulden, Public Relations Manager, GRAPHISOFT

Tibor Szolnoki, Team Leader, Training and Certification team, GRAPHISOFT

Graphisoft

GRAPHISOFT® ignited the BIM revolution in 1984 with ARCHICAD®, the industry's first BIM software for architecture. GRAPHISOFT continues to lead the industry with innovative solutions such as its revolutionary BIMcloud®, the world's first real-time BIM collaboration environment; and BIMx®, the world's leading mobile app for lightweight access to BIM for nonprofessionals. GRAPHISOFT is part of the Nemetschek Group. For more information, visit www.graphisoft.com.

ARCHICAD

GRAPHISOFT ARCHICAD is the architects' BIM software of choice. Natively available for both Windows® and Macintosh®, ARCHICAD offers a complete end-to-end design and documentation workflow for architectural and design practices of any size. According to user and analyst reviews, ARCHICAD is the easiest to learn and most fun to use. ARCHICAD's unique multithreaded software architecture offers best-in-class software performance for projects of any size.

BIMx

GRAPHISOFT BIMx is the most popular BIM presentation and coordination app on both the iPad and iPhone and on Android phones and tablets. BIMx features the

"BIM Hyper-model"—a unique concept that helps nonprofessionals easily explore the BIM project by superimposing 2D plans on 3D model views. Real-time model cut-throughs, in-context measuring and project markups in the model context make BIMx your best on-site BIM companion.

BIMcloud

GRAPHISOFT BIMcloud is the industry's first and most technologically advanced team collaboration solution. Featuring its unique, patented "Delta Server" technology, BIMcloud enables secure, real-time teamwork between members of the project team, regardless of the size of the design project, the location of the offices, or the speed of the Internet connection. BIMcloud is available both in private and public cloud configurations.

Green Building Initiative

Special thanks to Shaina Weinstein, Vice President of Engagement, Green Building Initiative for content and a commercial sustainable construction application. Green Building Initiative® (GBI) is a nonprofit organization that owns and administers the Green Globes® green building assessment and certification in the United States and Canada. GBI is dedicated to accelerating the adoption of building practices that result in energy-efficient, healthier, and environmentally sustainable buildings. The GBI seeks to be innovative and provide responsive customer service by collectively moving toward a sustainable built environment. GBI recognizes that credible and practical green building approaches for commercial and governmental construction are critical in this effort. Responding to the reality that one size does not fit all in sustainable improvements, GBI seeks to create a tailored approach to sustainability that takes into account building type, purpose, and occupants.

Green Globes®

Green Globes is a science-based building rating system that supports a wide range of new construction and existing building project types. Designed to allow building owners and managers to select which sustainability features best fit their building and occupants. The Green Globes custom-tailored approach helps produce the most sustainable outcomes based on building type and purpose.

Commercial Construction Green Building Application

This University of Florida Health Hospitals commercial construction application uses Green Globes to realize the full potential of quality sustainable design and construction practices that can be achieved with the adoption of building practices that result in energy-efficient, healthier, and environmentally sustainable buildings.

Commercial Construction Solar Application

This commercial construction application focuses on the solar installation at the Prodeo Academy and Metro Schools College Prep of Minneapolis, Minnesota.

Commercial Construction Erosion Control Application

The commercial construction erosion control application, provided by ECBVerdyol, is an example of improved technology being used to provide erosion control in extreme conditions and location.

Commercial Construction Precast Concrete Application

The commercial construction precast concrete application demonstrates how precast concrete enhances the beauty of the Academy Museum of Motion Pictures.

Commercial Construction Masonry Application

This commercial construction application is provided, in part, by Acme Brick and developed by Ray Don Tilley provides a masonry project that demonstrates the beauty, creativity, durability, and historic significance of using brick masonry construction. The project is the Central Academic Building, Texas A&M University in San Antonio, Texas.

Commercial Construction Steel Construction Application

This commercial construction application is provided by the Steel Framing Industry Association. The project is SOVA ON GRANT, a 12-story, 211 unit luxury apartment home building in Denver, Colorado.

Commercial Construction Mass Timber Application

This commercial construction application taken, in part, from a Nordic Structures project is an example of the universal characteristics of mass timber construction systems, combining glulams, CLT, timber concrete-composite systems, steel-concrete, and steel-wood all in one project. The project is the John W. Olver Design Building at the University of Massachusetts.

Commercial Building Construction Stair Application

BUILD recently completed this project in the Ballard neighborhood of Seattle. The upper grade is a full 14 ft above the street, while the hillside slopes parallel to the street. In order to produce an efficient and usable stair project, the design process focused heavily on the stair design strategy, both inside and out.

Special Appreciation

Special thanks is given to David P. Madsen, my son and best friend, for his creation of most 3D illustrations found throughout this textbook, and for the preparation of all figures and correlated captions. The highly detailed construction 3D models found in this textbook are unprecedented and allow students and instructors the opportunity to easily visualize construction practices described in the content.

CHAPTER 1

Plans, Specifications, and Construction Management

INTRODUCTION

This chapter covers the types of drawing found in a set of commercial construction plans and written information provided in specifications based on commercial construction industry practices and standards. Drawings and specifications from a small commercial building are used throughout this chapter and also in other chapters to demonstrate these applications. The building is the Brookings South Main Fire Station designed by JLG Architects, Alexandria, MN. Actual commercial construction drawings can be difficult to read on textbook pages because of the scale when reduced on a printed page. For this reason and for you to clearly view drawings and images, there is an image bank provided.

> Go to Downloads & Resources tab at **www.mhprofessional.com/Commercial Building Construction-Ancillaries** to access the images provided in this chapter and correlate with the content of this chapter. Here you can look at the chapter figures on your computer screen, pan as desired and zoom in and out for better observation. You can also use the images to create your own PowerPoint presentations.

COMMERCIAL CONSTRUCTION DOCUMENTS

Documents is a general term that refers to all drawings and written information related to a project. **Construction documents** are drawings and written specifications prepared and assembled by architects and engineers for communicating the design of the project and administering the construction contract. The two major groups of construction documents are **bidding requirements** and **contract documents**. Bidding requirements are used to attract bidders and provide the procedures to be used for submitting bids.

Bidding requirements are the construction documents issued to bidders for the purpose of providing construction bids. The word **bid,** also called **estimate,** is an estimate for the construction cost of a project or portion of a project. Contract documents are the legal requirements that become part of the construction contract. Contract documents are where the construction drawings and specifications are found.

Construction Drawings

The drawings described in this chapter are two-dimensional (2D) drawings that are commonly used when creating commercial construction drawings. Some drawings are three-dimensional (3D) used to provide a realistic appearance. The drawings are prepared using a computer-aided design and drafting (CADD) software program. There are several companies that provide design software for commercial construction-related disciplines, such as architecture, structural, and civil applications. Professional CADD programs allow the designer to perform a variety of tasks in addition to creating 2D and 3D drawings. CADD software can also be used to create **3D models.** A 3D model is a mathematical representation of any 3D surface of an object developed using specialized software. Professional CADD software programs can create 2D drawings and 3D models of a building at the same time. These CADD programs can also perform building information modeling activities. **Building information modeling (BIM)** is a computer technology 3D modeling process that gives architecture, engineering, and construction (AEC) professionals the creativity and tools to efficiently plan, design, construct, and manage buildings and infrastructure throughout the entire process from concept to completion. BIM applications go beyond planning the project and proceed throughout the building life cycle. **Building life cycle** refers to the observation and analysis of a building over the course of its entire life. The supporting processes of building life cycle management include cost management, construction management, project management, facility operation, and **green building** applications. Green building refers to a structure and construction processes that are environmentally responsible and resource efficient throughout a building life cycle. BIM is described in detail later in this chapter and green building is covered in Chapter 2, *Sustainable Technology*. The term **architecture, engineering, and construction (AEC)** used together and collectively referred to by AEC is an division of the construction industry that provides architectural design, engineering design, and building construction services.

Construction drawings are a principal part of the set of construction documents. The individual drawings needed depend on the specific requirements of the construction project. The drawings for a small **addition** might fit on one or two pages, whereas the drawings for a commercial building might be on over a hundred pages. An addition is construction that adds a new portion of building to an existing structure. Drawings vary in how much information they show, depending on the use, the project phase, and the desired representation. In addition to plan views, elevations, sections, and details, drawings can have **schedules** that have a complete list of components, items, or parts to be furnished in the project. The term **plan view** is a 2D view looking down on the feature being represented as shown in Figure 1.1. The term **elevation** refers to a 2D exterior view of a building, as shown in Figure 1.2. The terms **section** and **details** are described later in this chapter.

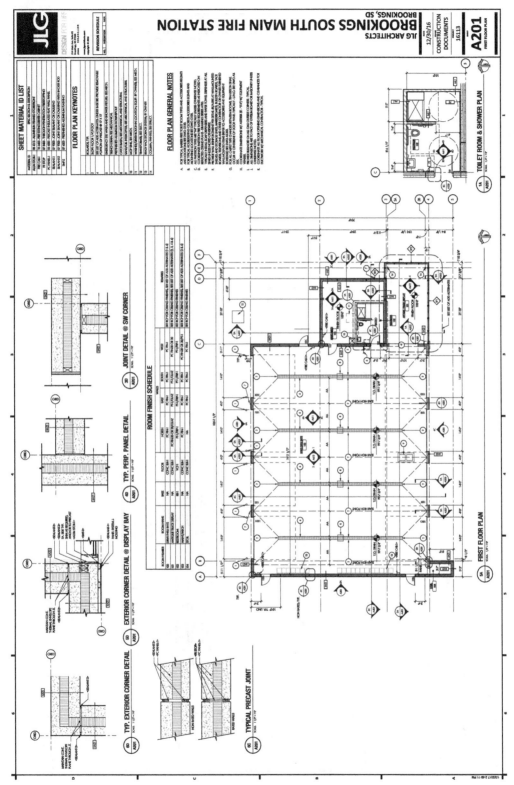

Figure 1.1 A plan view is a two-dimensional (2D) view looking down on the feature being represented. (*Courtesy of JLG Architects, Alexandria, MN 56308.*)

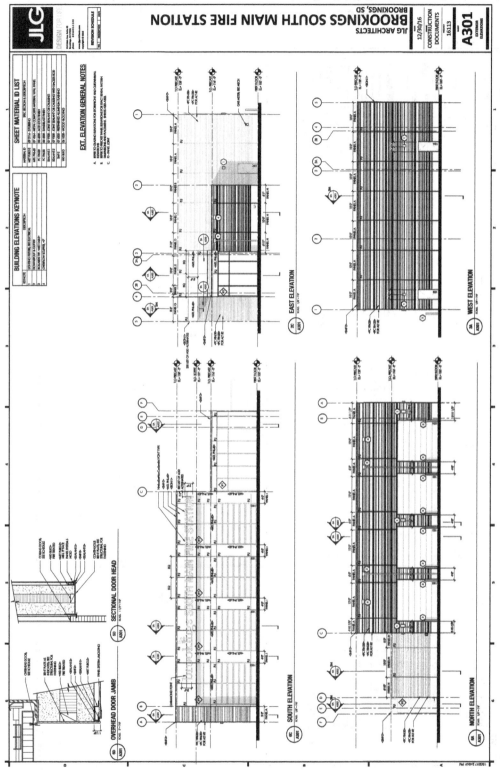

Figure 1.2 An elevation is a two-dimensional (2D) exterior view of a building. (*Courtesy of JLG Architects, Alexandria, MN 56308.*)

Coordinating Drawings and Specifications

A complete set of construction documents contains **drawings** and **specifications.** The combined drawings and specifications are often referred to as **plans.** Drawings provide quantity, capacity, location, and general written information in the form of images and notes, whereas the **specifications** clearly define items, such as minimum requirements, physical properties, chemical composition, and installation procedures, in written and table form. Specifications are described in detail later in this chapter. One person or a unified team of people should coordinate drawings and specifications for the construction project. The elements used in the drawings and specifications, such as symbols, abbreviations, and terminology, are standardized to help avoid confusion. After the construction documents are prepared in a professional manner, the construction coordination must be conducted with effective communication. Drawings locate and identify materials and include the assembly of components, dimensions, details, and diagrams. Drawings have **notes** that are written information on the drawings used to describe and identify material or processes. There are **general notes** and **specific notes.** General notes relate to the entire drawing, as shown in Figure 1.3. Specific notes describe individual features, as shown in Figure 1.4. Additional notes include **keynotes.** A keynote is a note found on the drawings, with each keynote identified with a letter or number, or combination of letters and

ROOF PLAN GENERAL NOTES

A. COORDINATE FINAL SIZE AND FINAL LOCATION OF ALL EQUIPMENT SUPPORT WITH THE APPROPRIATE EQUIPMENT MANUFACTURER.
B. PROVIDE POSITIVE SLOPE TO ALL ROOF DRAINS. ROOF SLOPE TO BE 1/4" PER FOOT MINIMUM.
C. SEE DRAWING A203 FOR ROOF DETAILS. ROOFING CONTRACTOR TO VERIFY ALL ROOF DETAILS COMPLY WITH APPROVED ROOFING SYSTEM. IT IS THE RESPONSIBILITY OF THE ROOFING CONTRACTOR TO SUPPLY ANY ADDITIONAL DETAILS AND/OR MATERIALS TO ENSURE COMPLIANCE WITH WARRANTY.
D. REFER TO MECHANICAL AND ELECTRICAL DOCUMENTS FOR ALL PIPES, CURBS, VENTS, DUCTS, CONDUITS, AND OTHER FEATURES EXTENDING THROUGH ROOF SURFACES WHICH REQUIRE FLASHING AND COORDINATE SIZE AND LOCATION OF SAME.
E. ALL ROOF MANUFACTURERS TO SUPPLY ANY ADDITIONAL MATERIALS FOR THEIR SYSTEM AS REQUIRED.
F. FOR ROOF ASSEMBLY TYPES SEE SHEET G102.
G. INLET OF OVERFLOW ROOF SUMPS SHALL BE LOCATED 2" ABOVE ROOF SUMP INLET (TYP) U.N.O.
H. SEE MECHANICAL DRAWINGS FOR FINAL QUANTITIES & FINAL LOCATIONS OF VENTS THRU ROOF.
I. PROVIDE SPLASHBLOCKS AT ALL PRIMARY AND SECONDARY RAIN CONDUCTOR TERMINATIONS ONTO SINGLE-PLY MEMBRANE ROOFING
J. INSTALL ROOF FRAMES PRIOR TO CUTTING & REMOVING OF ANY METAL DECK.

R.S. - ROOF SUMP
H.P. - HIGH POINT
O.S. - OVERFLOW ROOF SUMP
S.C. - SCUPPER
V.T.R. - VENT THRU ROOF

Figure 1.3   General notes relate to the entire plan. (*Courtesy of JLG Architects, Alexandria, MN 56308.*)

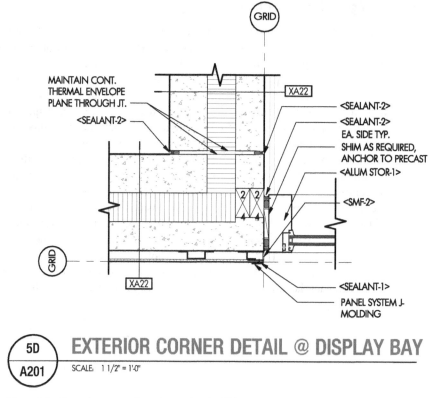

Figure 1.4 Specific notes describe individual features. (*Courtesy of JLG Architects, Alexandria, MN 56308.*)

numbers or symbols next to or pointing at a specific feature on the drawing that correspond to the description of the keynote in a list, legend, or in a general note. Detailed written information, beyond the scope of notes, is placed in the specifications that are described later. Abbreviations and symbols used in the set of drawings are represented as an approved standard and can be shown and labeled in a legend for reference purposes. A **legend** is a feature on a drawing that shows and names symbols used on the drawing. See Figure 1.5. An **abbreviation** is a shortened form of a word or phrase that takes the place of the whole word or phrase, such as mm for millimeter. **Symbols** are images that represent a feature on a drawing or in specifications; for example, the symbol for feet is ', the symbol for inch is ", and the symbol for diameter is Ø. The specifications are used to define the specific quality and type of material, equipment, and installation.

Schedules

A **schedule** is a grouping of related items with corresponding distinguishing features, with a heading and a minimum of three columns of related information. A schedule formats information into rows and columns in order to more easily present design information. Schedules are used on drawings to help simplify communication by providing certain items in a table format. See Figure 1.6. Schedules can be placed on the drawings or in the specifications. The information found in schedules or on the drawings generally do not

ELECTRICAL SYMBOLS

THESE SYMBOLS COMPRISE A STANDARD LIST; NOT ALL SYMBOLS MAY APPEAR ON THIS PROJECT.

ALL MOUNTING HEIGHTS ARE TO CENTER OF DEVICE ABOVE FINISHED FLOOR. MOUNTING HEIGHTS INDICATED ON ARCH. WALL ELEVATIONS ORAS NOTED SPECIFICALLY ON THE DRAWINGS OR IN THE SPECIFICATIONS SHALL TAKE PRECEDENCE OVER MOUNTING HEIGHTS LISTED BELOW.

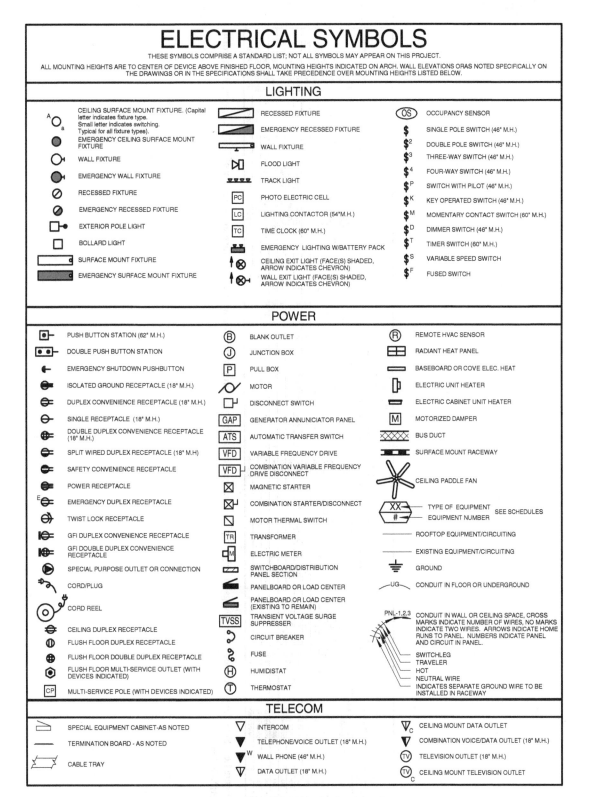

Figure 1.5 A legend is a feature on a drawing that shows and names symbols used on the drawing. (*Courtesy of JLG Architects, Alexandria, MN 56308.*)

DOOR SCHEDULE

DOOR NUMBER	ROOM NAME	SIZE			DOORS				FRAME				DETAILS			RATING	HW SET	NOTES
		WIDTH	HEIGHT	THK	MAT'L	TYPE	FINISH	GLZ	MAT'L	TYPE	FINISH	GLZ	JAMB	HEAD	SILL			
FIRST FLOOR																		
100A	DRIVE-THRU BAYS	3'-0"	7'-0"	1 3/4"	ALUM	F	PREFIN.	-	ALUM	2	PREFIN.	-	5C/A220	5C/A220	4C/A220	-	SEE SPEC	
100B	DRIVE-THRU BAYS	3'-0"	7'-0"	1 3/4"	ALUM	F	PREFIN.	-	ALUM	2	PREFIN.	-	5C/A220	5C/A220	4C/A220	-	SEE SPEC	
100C	DRIVE-THRU BAYS	3'-0"	7'-0"	1 3/4"	ALUM	F	PREFIN.	-	ALUM	2	PREFIN.	-	5C/A220	5C/A220	4C/A220	-	SEE SPEC	
100D	DRIVE-THRU BAYS	14'-0"	14'-0"	2"	ALUM	OH Glass Door - Panel	PREFIN.	INSULATED TEMP/INSULATED	SL	SL	-	-	6D/A301	5D/A301		-	SEE SPEC	GLASS
100E	DRIVE-THRU BAYS	14'-0"	14'-0"	2"	ALUM	OH Glass Door - Panel	PREFIN.	INSULATED TEMP/INSULATED	SL	SL	-	-	6D/A301	5D/A301		-	SEE SPEC	GLASS
100F	DRIVE-THRU BAYS	14'-0"	14'-0"	2"	ALUM	OH Glass Door - Panel	PREFIN.	INSULATED TEMP/INSULATED	SL	SL	-	-	6D/A301	5D/A301		-	SEE SPEC	
100G	DRIVE-THRU BAYS	14'-0"	14'-0"	2"	ALUM	OH Glass Door - Panel	PREFIN.	INSULATED TEMP/INSULATED	SL	SL	-	-	6D/A301	5D/A301		-	SEE SPEC	GLASS
100H	DRIVE-THRU BAYS	14'-0"	14'-0"	2"	ALUM	OHS	PREFIN.	-	SL	SL	-	-	6D/A301 (SIM.)	5D/A301 (SIM.)		-	SEE SPEC	SOLID
100J	DRIVE-THRU BAYS	14'-0"	14'-0"	2"	ALUM	OHS	PREFIN.	-	SL	SL	-	-	6D/A301 (SIM.)	5D/A301 (SIM.)		-	SEE SPEC	SOLID
100K	DRIVE-THRU BAYS	14'-0"	14'-0"	2"	ALUM	OHS	PREFIN.	-	SL	SL	-	-	6D/A301 (SIM.)	5D/A301 (SIM.)		-	SEE SPEC	SOLID
102	DRIVE-THRU BAYS	3'-0"	7'-0"	1 3/4"	HM	F	FIELD PAINT	-	HM	1	PT	-	1C/A220	1C/A220	-	90MIN	SEE SPEC	SIGN 'B'
103	MAINT/MECH	3'-0"	7'-0"	1 3/4"	HM	F	FIELD PAINT	-	HM	1	PT	-	1C/A220	1C/A220	-	90MIN	SEE SPEC	SIGN 'D'

Figure 1.6 Schedules are used on drawings to help simplify communication by providing certain items in a table format. (*Courtesy of JLG Architects, Alexandria, MN 56308.*)

8

repeat information found in the specifications. The drawings, schedules, and specifications must be carefully coordinated so the information is consistent.

MasterFormat™

MasterFormat: Master List of Numbers and Titles for the Construction Industry is a master list of numbers and subject titles for organizing information about construction work results, requirements, products, and activities that are divided into a standard sequence. The MasterFormat divisions and sublevels can be used as a checklist to ensure that every required specification is included. MasterFormat is an effective system for indexing project specifications. The MasterFormat is described in detail later in this chapter.

THE ARCHITECT AND STRUCTURAL ENGINEER

Commercial buildings require the entire set of plans be designed and submitted to the building official by a state-licensed **registered architect (RA)**, or **professional engineer (PE)**. An RA is a person trained in the planning, design, and oversight of the construction of buildings and is licensed to practice architecture in a specific state or states where they pass rigorous qualifications. A PE is an engineer who is registered or licensed within a specific state or states where they pass rigorous qualifications needed to offer professional services. A PE is generally licensed in one or more specific disciplines, such as a structural engineer who performs the engineering on the building structure, or civil engineer who does earth-related engineering. An architectural firm can have one or more RAs involved in building design and can also have discipline-related PEs responsible for specific portions of engineering for a project.

A **structural engineer** works with architects to engineer the structural components of a building. Structural engineering is generally associated with commercial buildings and for the structural design of residential buildings. The structural engineer also works with civil engineers to design bridges and other structures related to road and highway construction. Mechanical engineers can also be involved in the structural project by designing machinery supports and foundation components.

Structural engineers or architects produce **design drawings.** Design drawings contain all the details required to prepare structural drawings. Design drawings provide data on loads, **axial forces, moments,** and **shear forces.** Axial forces are forces working along the axis of a structure such as a column. Moments are a measure of resistance to changes in the rotation of an object, also referred to as **moment of inertia.** Shear forces are forces caused when two construction pieces move over each other. Design drawings also contain information of each framing member, precise dimensions, the location of each beam and column, and general notes for reference.

Structural engineering drawings and **detail drawings** show the results of structural designing in a condensed form. Drawings, general notes, schedules, and specifications serve as instructions to the contractor. The drawings must be complete and have sufficient detail so no misinterpretation can be made. Structural drawings can be with or independent of architectural drawings and other drawings, such as electrical, plumbing, or heating, ventilating, and air-conditioning (HVAC). When necessary, structural drawings are clearly cross-referenced to architectural drawings. See Figure 1.7.

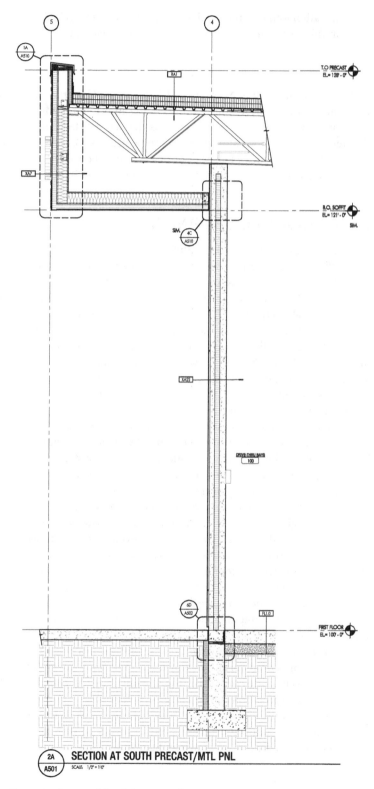

Figure 1.7   Structural detail. (*Courtesy of JLG Architects, Alexandria, MN 56308.*)

ANALYZING A SET OF COMMERCIAL CONSTRUCTION DOCUMENTS

Commercial construction documents are the complete set of drawing and specifications for a project. A set of construction documents for a commercial project can have more than a hundred pages. Because of the complex nature of the drawings and the number of disciplines involved, the set of documents is often divided into several major groups. Common major groups include architectural, civil, structural, mechanical (HVAC), plumbing, electrical, and fixture drawings. An architect normally prepares the architectural drawings. The architect then coordinates with engineering staff, or consulting engineering firms that prepare the drawings for their specific disciplines. A **consulting engineer** is a professional expert engineer in a specific discipline who works with the architect on a certain phase of a project. An architect plans, designs, and supervises the construction of buildings, and has specialized advanced education and practical experience resulting in licensing by a state or multiple states to professionally practice architecture. The **structural engineering** firm generally prepares the related structural drawings, which include the steel and concrete parts of the project. Structural engineering is a branch of **civil engineering** dealing primarily with the design and construction of structures, such as bridges, buildings, and dams. **Civil engineering** is a professional engineering discipline that deals with the design, construction, and maintenance of the physical and naturally built environment, including projects like roads, bridges, canals, dams, and buildings.

Drawing Sheets

Drawing sheets used for commercial construction projects are commonly the large sheet format used for architectural, mechanical, and piping drawings. The American Society of Mechanical Engineers (ASME) establishes drawing sheet sizes in ASME Y14.1 *Decimal Inch Drawing Sheet Size and Formats* and ASME Y14.1M *Metric Drawing Sheet Size and Format*. ASME inch sheet sizes are classified with letters A through F, and metric sheet sizes are sized A0 through A4. The American Institute of Architects (AIA) adopted the U.S. National CAD Standard where drawing sheet sizes are specified as ARCH A through ARCH F. The ASME E-size 44 × 34-in. (A0 size 1189 × 841 mm) and the ARCH E 36 × 48-in. sheet sizes are common for architectural drawings, but other sheet sizes are also used.

Most drawing sheets have a border and sheet blocks. The ASME standards have very specific sheet blocks and sheet block locations. **Sheet blocks** are a group of informational areas normally surrounded by boarder lines and grouped in one consistent location on the drawings. Architectural drawings generally have less restrictive sheet block requirements than those found in mechanical drafting for manufacturing. Architectural drawings commonly have a title block and a revision history block. The **title block** provides a variety of information about the drawing, as shown in Figure 1.8. Title block information can include content such as company name and logo, company address and phone number, revision history block, engineering stamp, customer name and project name, project number, approval signatures, predominant drawing scale, and sheet size. **Revision history block,** also called the **revision block,** is used to record changes to the drawing and is generally located in or next to the title block. A typical architectural revision block is shown in Figure 1.9.

Numbering Sheets

A set of architectural drawings requires a specific page-numbering system because of the large number of pages. When a set of drawings has a number of pages that are easy to manage, it is common to see pages numbered in consecutive order, such as 1, 2, 3, 4. If a sheet

Figure 1.8   Architectural drawing title block. (*Courtesy of JLG Architects, Alexandria, MN 56308.*)

REVISION SCHEDULE		
NO.	**DESCRIPTION**	**DATE**
1	Revision 1	Date 1

Figure 1.9   Architectural revision block. (*Courtesy of JLG Architects, Alexandria, MN 56308.*)

has subpages, then letters might follow the sheet number alphabetically, such as 4A, 4B, 4C, 4D. The Brookings South Main Fire Station drawing sheets correlate drawings by disciplines, such as architectural (A), civil (C), electrical (E), mechanical (M), and structural (S). Each discipline has its own sheet numbers, such as A201, A301, A401, and A501. The actual numbering of pages, and page descriptions, varies between regions of the country, architectural and structural engineering companies, and client preference. *The Architect's Handbook of Professional Practice* recommends a decimal page-numbering system. The following general numbering system is an example of how sheets are categorized by disciplines and activities, but this can change according to each project.

ARCHITECTURAL
T1   Title sheet, site demolition, and survey
A1   Site plan and details
A2   Grading plan and details
A3   First-floor plan and details
A4   Second-floor plan, sections, and details
A5   Enlarged plans, interior elevations, and details
A6   Exterior elevations and details
A7   Building and wall sections and details
A8   Roof plan and details
A9   Details
A10 Reflected ceiling plan and details

CIVIL
U1   Site utilities
U2   Erosion control plans and details
U3   Public utility plan
U4   Utility details

LANDSCAPE
L1   Irrigation system plan and legend
L2   Irrigation system details and notes
L3   Planting plan, details, and notes

STRUCTURAL
S1   Foundation plans
S2   Floor plans
S3   Roof framing plans
S4   Typical details, elevations, schedules
S5   Foundation details
S6   Framing details
S7   General notes (either here or on the first sheet)

MECHANICAL
M1   First-floor plumbing plan, legends
M2   Second-floor plumbing plan, notes

M3   Details and schedules
M4   First-floor HVAC plans and legends
M5   Second-floor HVAC plan
M6   Roof-mounted HVAC equipment plan
M7   HVAC details and schedules

ELECTRICAL
E1   Notes, legend, riser
E2   First-floor lighting plan
E3   Second-floor lighting plan
E4   First-floor power plan
E5   Second-floor power plan
E6   Roof-mounted equipment power plan
E7   Details and schedules
E8   First-floor communications plan and legend
E9   Second-floor communications plan

FIXTURES
F1   Fixture plan and schedules
F2   Details

The general groups and elements within the groups can differ, depending on the company practice and the complexity of the building being designed. For example, the Index of Drawings for the Brookings South Main Fire Station project demonstrated throughout this chapter represents a sheet-numbering system for a small commercial building, as shown in Figure 1.10.

Each sheet numbering category within a general group can have additional pages. These pages are numbered with the sequential decimals of .1, .2, .3. Therefore, the architectural drawings might have pages such as

A1.1, A1.2, A1.3

A2.1, A2.2, A2.3, A2.4

Structural drawings can have a series of sheets that are numbered such as

S1.1, S1.2

S2.1, S2.2, S2.3

S3.1, S3.2, S3.3, S3.4, S3.5

The decimal sheet-numbering system can be divided even further by adding .01, .02, .03 to the existing numbers as needed. For example, additional pages in the series of S3.1 are numbered S3.1.01, S3.1.02, S3.1.03.

Sheet Zoning

Zoning on drawing sheets is a system of numbers along the top and bottom margins and letters along the left and right margins of a sheet. Zoning allows the drawing to be read like a road map. For example, to find information located at zone 4C, you find 4 on the top or bottom margin and C along a side margin and look vertically and horizontally to where these intersect to find the desired information.

INDEX OF DRAWINGS

SHEET NUMBER	SHEET NAME
GENERAL	
G100	COVER SHEET
G101	TITLE SHEET
G102	TYPICAL ASSEMBLIES
G110	LIFE SAFETY PLANS
G120	MOUNTING HEIGHTS
CIVIL	
C101	GENERAL NOTES & SPECIFICATIONS
C102	GENERAL NOTES & SPECIFICATIONS
C103	STORM WATER POLLUTION PREVENTION PLAN
C201	EXISTING CONDITIONS
C202	EROSION CONTROL & REMOVAL
C203	SITE LAYOUT
C204	SITE GRADING
C205	POND LAYOUT & DETAILS
C206	SITE UTILITIES
C301	MISCELLANEOUS SITE DETAILS
C302	STANDARD PLATES & DETAILS
C304	STANDARD PLATES & DETAILS
STRUCTURAL	
S101	GENERAL STRUCTURAL NOTES
S102	FOOTING & FOUNDATION PLAN
S201	ROOF FRAMING PLAN
S301	STRUCTURAL DETAILS
S302	STRUCTURAL DETAILS
S303	STRUCTURAL DETAILS
S304	STRUCTURAL DETAILS
S305	STRUCTURAL DETAILS
S306	STRUCTURAL DETAILS

INDEX OF DRAWINGS

SHEET NUMBER	SHEET NAME
ARCHITECTURE	
A201	FIRST FLOOR PLAN
A203	ROOF PLAN
A220	DOOR SCHEDULE & GLAZING
A301	EXTERIOR ELEVATIONS
A302	PRECAST PANEL ELEVATIONS
A303	PRECAST PANEL ELEVATIONS
A401	BUILDING SECTIONS
A501	EXTERIOR WALL SECTIONS
A502	EXTERIOR WALL SECTIONS
A510	EXTERIOR ASSEMBLY DETAILS
A611	INTERIOR ELEVATIONS
A612	INTERIOR ELEVATIONS
A701	FIRST FLOOR REFLECTED CEILING PLAN
MECHANICAL	
M200	UNDERFLOOR PLUMBING PLAN
M201	FIRST FLOOR PLUMBING PLAN
M202	FIRST FLOOR HVAC VENTILATION PLAN
M203	FIRST FLOOR HVAC PIPING PLAN
M300	MECHANICAL DETAILS
M301	MECHANICAL DETAILS
M400	MECHANICAL SCHEDULES
M401	MECHANICAL SCHEDULES
ELECTRICAL	
E200	FIRST FLOOR POWER & COMM. PLAN
E201	FIRST FLOOR LIGHTING PLAN
E202	ENLARGED ELECTRICAL PLAN
E300	ELECTRICAL DETAILS
E400	ELECTRICAL SCHEDULES

Figure 1.10 The Index of Drawings, found on sheet G100, for the Brookings South Main Fire Station project demonstrated throughout this chapter represent a sheet numbering system for a small commercial building. (*Courtesy of JLG Architects, Alexandria, MN 56308.*)

Drawing Scales

Drawing scale, also called **scale,** is a measurement unit representing a proportional relationship between a reduced-size drawing and the actual full-sized feature. Drawings are scaled so the objects represented can be illustrated on standard sheet sizes. The selected scale depends on

- The actual size of the objects being drawn
- The amount of detail to show

- The number of views to be drawn
- The selected sheet size
- The amount of dimensioning and notes required
- Company and industry standards

Architectural scales are based on each inch representing a specific increment of feet. Each foot is subdivided into 12 parts to represent inches and fractions of an inch, and the degree of precision depends on the specific scale. A sample architectural scale is $1/2'' = 1'-0''$. The drawing scale is commonly placed below the drawing title where the scale applies.

Metric scales are based on a ratio that is the relationship of one measurement to another. A common metric scale is 1:50, where 1 m on paper equals 50 m on the drawing.

The following shows common architectural scales with approximate metric scale equivalents and the drawings that typically use these scales, other scales are also available:

Architectural Scales	Metric Scales	Typical Drawings
$1/8'' = 1'-0''$	1:100	Site plans, large floor plans
$1/4'' = 1'-0''$	1:50	Floor plans, foundation plans, and elevations
$1/2'' = 1'-0''$	1:20	Sections
$3/4'' = 1'-0''$	1:10	Details

Civil engineer **scales** are commonly used to draw large projects, such as site plans, roads, utilities, and bridges. On a civil engineer scale, 1 in. represents 10 ft increments, such as $1'' = 10'$, $1'' = 20'$, and $1'' = 100'$. The civil engineer scale divides 1 in. into equal decimal units of 10, 20, 30, 40, 50, 60, and 80.

Some drawings such as diagrams and schematics are drawn without scale for convenience, because they do not represent a physical size object. When this is done, the scale is noted below the drawing title as NTS or NOT TO SCALE.

COMPONENTS ON A SET OF COMMERCIAL ARCHITECTURAL PLANS

A set of commercial architectural plans can have a large number of sheets with specific elements of the building provided in a general format or with specific construction elements shown in sections or in details. The term **section,** also called **cross section** and **sectional view,** is a type of drawing that shows a cut through the building to display the construction practices being used along with construction materials and principal dimensions found along the cut. A building cross section is shown in Figure 1.11. Sections range in scale from $1/4'' = 1'-0''$, $1/2'' = 1'-0''$, $3/4'' = 1'-0''$, or $1'' = 1'-0''$. The term **details** refers to enlarged sections used to show exact construction requirements at a specific location, such as a footing, connection, or building application, and range in scale from $3/4'' = 1'-0''$, $1'' = 1'-0''$, $1-1/2'' = 1'-0''$, or $3'' = 1'-0''$. A detail is shown in Figure 1.12. The use of sections and details is described and shown, in the following content, together with various other types of drawings found on the plans.

Although not all complete sets of structural drawings contain the same number of sheets or type of information, some of the representative drawings include a floor plan, a foundation

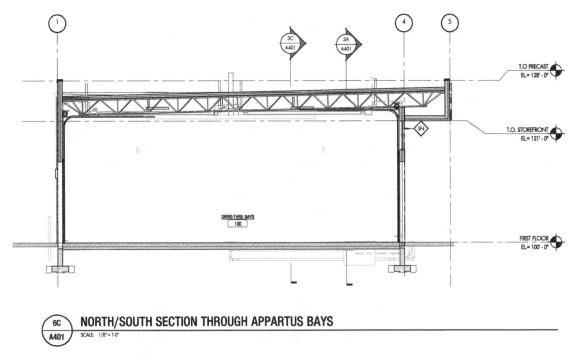

6C / **A401** — NORTH/SOUTH SECTION THROUGH APPARTUS BAYS
SCALE: 1/8" = 1'-0"

Figure 1.11 A building cross section. (*Courtesy of JLG Architects, Alexandria, MN 56308.*)

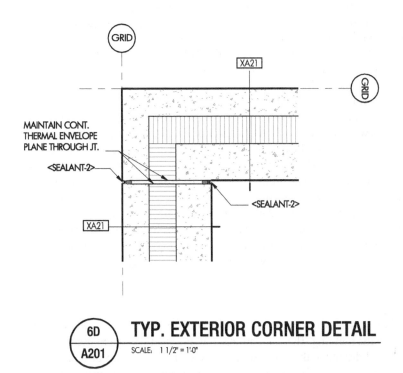

6D / **A201** — TYP. EXTERIOR CORNER DETAIL
SCALE: 1 1/2" = 1'-0"

Figure 1.12 A construction detail. (*Courtesy of JLG Architects, Alexandria, MN 56308.*)

plan and details, a concrete slab plan and details, a roof framing plan and details, a roof drainage plan, building section(s), exterior elevations, a panel plan, elevations, and wall details.

Floor Plan

The **floor plan** is a scale drawing showing the arrangement, sizes, and location of rooms and other features in each separate story of a building. A floor plan is a representation provided by an imaginary horizontal cut made through the building at approximately 4′ (1220 mm) above the floor line. The floor plan is generally designed and drawn by the architect. The floor plan drawing is then used as the layout for the other associated drawings from consulting engineers, for example, the mechanical (HVAC), plumbing, electrical, and structural engineers. The structural engineering firm draws the floor plan in some situations and depending on the business practices and project. The floor plan for a small commercial building and part of the set of plans is shown in Figure 1.13.

Correlating Details and Sections

Detail and section drawings are labeled with identifying symbols and a title below the detail or section drawing. The symbol correlates the detail or section between the view and the location on the sheet where the detail or section is taken. Detail drawings are keyed to the plan view using **detail markers.** Detail markers are usually drawn as a circle of about 3/8″ to 1/2″ (9.5 to 13 mm) in diameter on the plan view and a coordinating circle of about 3/4″ (19 mm) in diameter under the associated detail. Each detail marker is divided in half. The top half contains the detail number and the bottom identifies the page number on which the detail is drawn. A typical detail marker is shown in Figure 1.14a. Notice the detail markers associated with the example drawings in this chapter. Detail marker style can vary depending on the company standards and practices.

Symbols used for section identification are called cutting-plane lines. A cutting-plane line is placed in a view identifying the location where a sectional view is cut through at that location. A cutting-plane line symbol is drawn using a circle the same size as the detail marker circle with a vertical, horizontal, or angle line through the center. The line through the cutting-plane line orients the symbol in the same direction where the cut is made through the view for the section. The top half of the cutting-plane line circle has a letter identifying the section, and the bottom half has a number or letter and number combination, identifying the sheet where the section view is located. A solid filled triangle is generally attached to the cutting-plane line circle. The triangle is like an arrow where the point identifies the direction of sight for the cutting plane. A cutting-plane line symbol can be drawn entirely through the view or symbols place on each side of the view, or partially through a view for a partial section. Cutting-plane line symbols can vary between companies. Figure 1.14b shows options and labels the parts of cutting-plane line symbols. Notice the cutting-plane line symbols associated with the example drawings in this chapter.

Foundation Plan and Details

The **foundation plan** is a scale drawing used to display construction features and dimensions for the **foundation** of the building. The foundation is the construction system used to support the structure loads and distribute the loads to the ground. The foundation plan scale is normally 1/4″ = 1′-0″. The purpose of the foundation plan is to show the supporting system for the walls, floor, and roof. The foundation can consist of continuous perimeter footings that

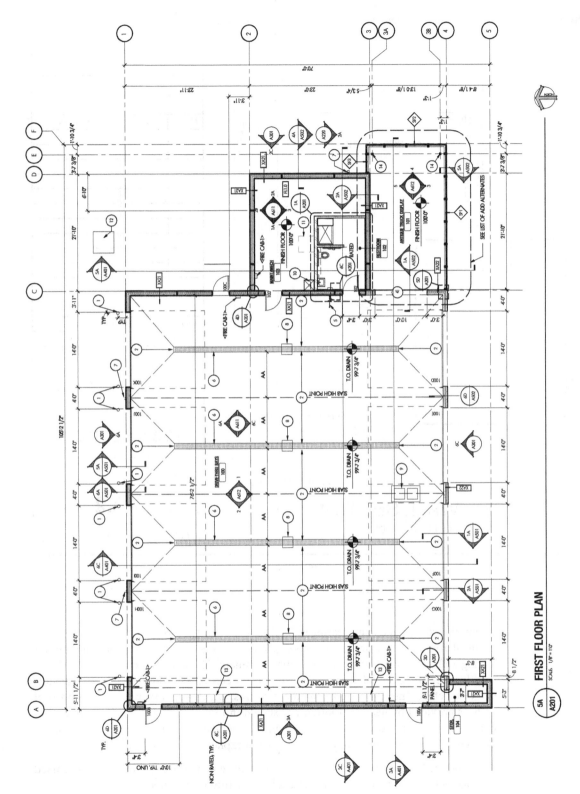

Figure 1.13 A floor plan. (*Courtesy of JLG Architects, Alexandria, MN 56308.*)

5A **FIRST FLOOR PLAN**
A201   SCALE 1/8" = 1'-0"

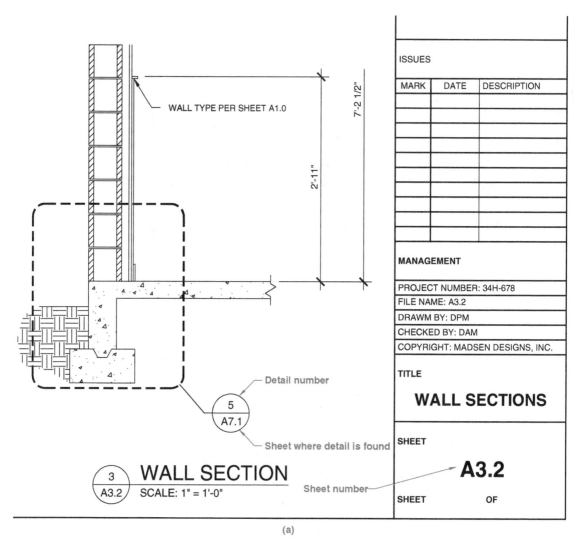

Figure 1.14   (a) Typical detail marker. (*Continued*)

support the exterior and main bearing walls of the building. The footings in these locations are rectangular in shape and are continuous concrete supports centered at exterior and interior bearing walls. The foundation plan for a small commercial building is shown in Figure 1.15.

There are also pedestal footings that support the concentrated loads of specific elements of the structure. Interior support for the columns, which support upper floor and roof loads, is provided by pedestal footings, as shown in Figure 1.16. Anchor bolts and other metal connectors are shown and located on the foundation plan. The terms used in the previous content are defined in the following and explained further in other chapters:

Concrete is a construction material made from cement, sand, gravel, and water mixed together and set in a form to make a solid structure when cured.

Continuous perimeter footings are the lowest member of the foundation system used to spread the loads of the structure on the supporting ground, and the continuous perimeter footings form the boundary of a building.

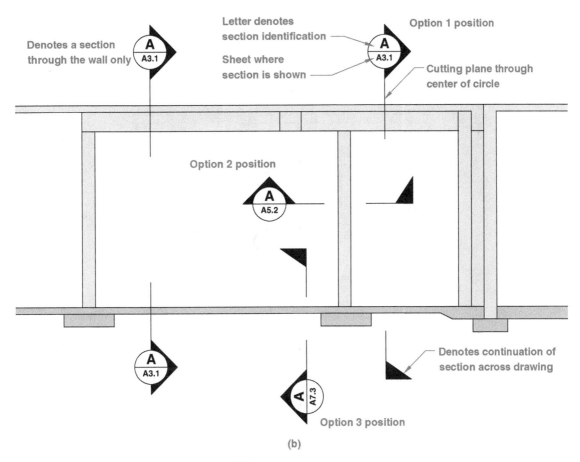

Figure 1.14   (b) Typical cutting-plane line symbols.

Bearing walls, also called **load-bearing wall,** is a wall that supports the weight of the structure resting on it from above, and by transferring the weight to the foundation structure.

Pedestal footing is a concrete structure provided to carry the loads from supported elements like columns to a footing below the ground, and the pedestal width is usually greater than the height.

Column is a vertical structural member designed to transmit a compressive load, and it is typically made of reinforced concrete but can be constructed from wood, steel, fiber-reinforced polymer, cellular PVC, or aluminum.

Anchor bolts are steel L-shaped bolts that are imbedded in the top of the foundation wall or other locations and extend out far enough to fasten the sill or other member with a washer and nut at each bolt.

Sill is a continuous pressure treated wood member that provides a barrier between the foundation wall and the framing above.

Foundation details provide information on the concrete foundation at the perimeter walls, and the foundation and pedestals at the center support columns. Possible details

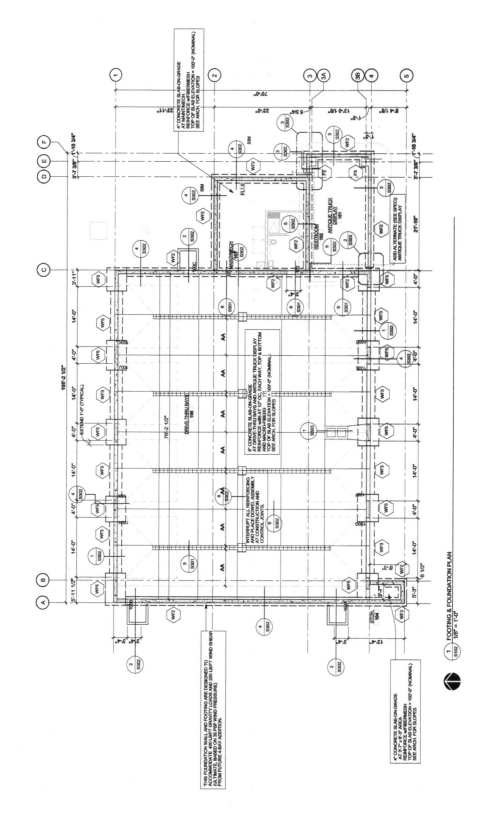

Figure 1.15 The foundation plan for a small commercial building. (*Courtesy of JLG Architects, Alexandria, MN 56308.*)

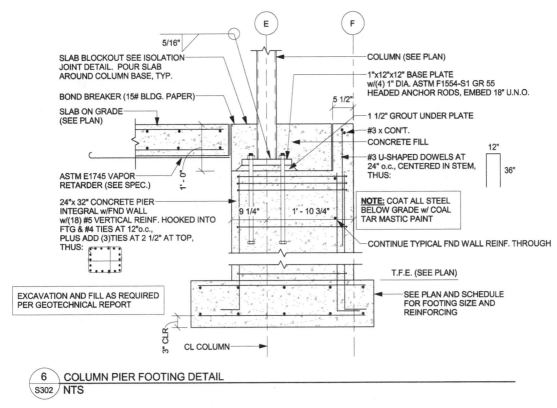

SLAB BLOCKOUT SEE ISOLATION
JOINT DETAIL. POUR SLAB
AROUND COLUMN BASE, TYP.

BOND BREAKER (15# BLDG. PAPER)

SLAB ON GRADE
(SEE PLAN)

ASTM E1745 VAPOR
RETARDER (SEE SPEC.)

24"x 32" CONCRETE PIER
INTEGRAL w/FND WALL
w/(18) #5 VERTICAL REINF. HOOKED INTO
FTG & #4 TIES AT 12"o.c.,
PLUS ADD (3)TIES AT 2 1/2" AT TOP,
THUS:

EXCAVATION AND FILL AS REQUIRED
PER GEOTECHNICAL REPORT

5/16"

COLUMN (SEE PLAN)

1"x12"x12" BASE PLATE
w/(4) 1" DIA. ASTM F1554-S1 GR 55
HEADED ANCHOR RODS, EMBED 18" U.N.O.

5 1/2"

1 1/2" GROUT UNDER PLATE

#3 x CON'T.

CONCRETE FILL

#3 U-SHAPED DOWELS AT
24" o.c., CENTERED IN STEM,
THUS:

NOTE: COAT ALL STEEL
BELOW GRADE w/ COAL
TAR MASTIC PAINT

CONTINUE TYPICAL FND WALL REINF. THROUGH

T.F.E. (SEE PLAN)

SEE PLAN AND SCHEDULE
FOR FOOTING SIZE AND
REINFORCING

9 1/4"   1' - 10 3/4"

12"

36"

1' - 0"

3" CLR

CL COLUMN

6
S302

COLUMN PIER FOOTING DETAIL
NTS

Figure 1.16 Foundation column pier footing detail. (*Courtesy of JLG Architects, Alexandria, MN 56308.*)

include retaining walls, typical exterior foundation and rebar schedule, typical interior bearing wall, and typical pedestal and rebar schedule. The **rebar schedules** are charts placed next to the detail that key to the drawing information about the rebar used in the detail, as shown in Figure 1.17. **Rebar** is the term used to identify steel reinforcing bars in the construction industry and described in detail in Chapter 4, *Concrete Construction and Foundation Systems*.

Concrete Slab Foundation Plan and Details

A **concrete slab foundation system** uses minimum 3-1/2″ (90 mm) reinforced concrete that is referred to as a slab that is a flat concrete pad poured directly on the ground or other engineered material. The **concrete slab foundation plan** is drawn to outline the concrete

FOOTING SCHEDULE			
MARK	**SIZE**	**REINFORCING**	**COMMENTS**
F3	3'-0"x3'-0"x1'-0"	(5) #4 EACH WAY, BOTTOM	
WF2	2'-0"x1'-0" CON'T.	(3) #5 x CON'T.	
WF3	3'-0"x1'-0" CON'T.	(4) #5 x CON'T.	
WF5	5'-0"x1'-2" CON'T.	(6) #5 x CON'T. #6 @ 12" oc TRANSVERSE	

Figure 1.17 Basic footing schedule with rebar. (*Courtesy of JLG Architects, Alexandria, MN 56308.*)

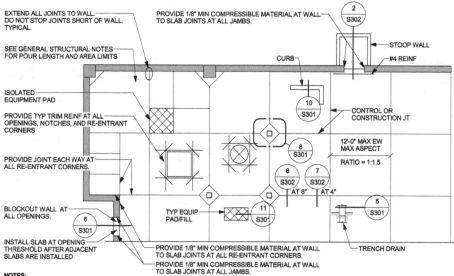

Figure 1.18   Concrete slab on grade foundation plan. (*Courtesy of JLG Architects, Alexandria, MN 56308.*)

used to construct the concrete floor or floors. Items often found on concrete slab plans include floor slabs, slab reinforcing, expansion joints, pedestal footings, metal connectors, anchor bolts, and any foundation cuts for doors or other openings. The openings and other items are located and labeled on the plan, as shown in Figure 1.18.

The **concrete slab details** are drawn to provide information of the intersections of the concrete slabs. These details include interior and perimeter slab joints, as in Figure 1.19, and other slab details. The drawings include information such as slab thicknesses, reinforcing sizes and locations, and slab elevations (heights). The term **elevation,** related to surveying, site planning, and disciplines such as plumbing and HVAC, refers to the height of a feature from a known base, which is usually given as 0 (zero elevation).

Roof Framing Plan and Details

The purpose of the **roof framing plan** is to show the plan view of major structural components that occur at the roof level. Figure 1.20 shows a roof framing plan using truss framing. Figure 1.21 shows a panelized roof framing plan.

Roof framing details are required to show the construction methods used at various member intersections in the building roof structure. Roof details can include the following intersections: wall to beam, beam to column, beam splices and connections, truss details, bottom chord bracing plan and details, purlin clips, cantilever locations, and roof drains, as shown in Figure 1.22. The following are brief definitions of the previously used terms, and these terms have additional descriptions later in this textbook:

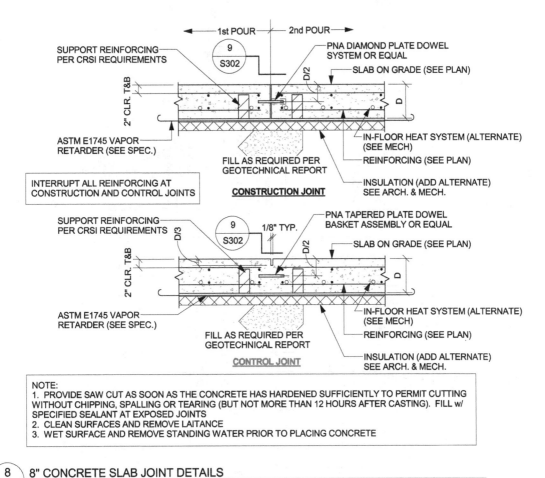

Figure 1.19 Concrete slab detail. (*Courtesy of JLG Architects, Alexandria, MN 56308.*)

Beams are horizontal construction members that are used to support floor systems, wall or roof loads.

A **truss** is a prefabricated or job-built construction member formed of triangular shapes used to support roof or floor loads over long spans. The term **span** is the horizontal dimension of a construction member between supports.

A **chord** is the outer members of a truss that defines the truss shape.

Cantilever is a structure that is supported at one end and is self-supporting on the other end where it projects into space.

A **roof drain** means a fitting or device that is installed in the roof to allow storm water to discharge into a downspout. The term **downspout** is a vertical pipe that is connected to the roof drain or gutters for the purpose of moving rainwater from the roof or gutter to the ground or to a rain drain pipe in the ground. **Gutter** is a channel at the edge of the roof eave for moving rainwater from the roof to the downspout.

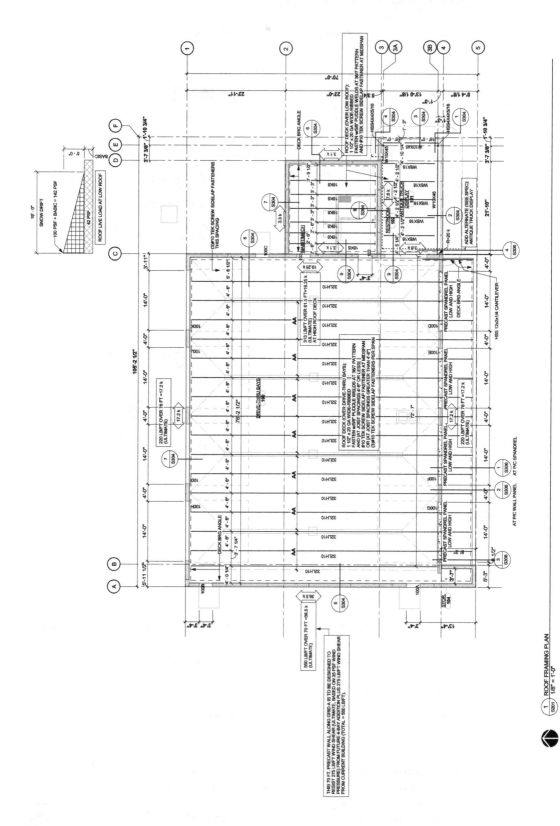

Figure 1.20 Roof framing plan. (*Courtesy of JLG Architects, Alexandria, MN 56308.*)

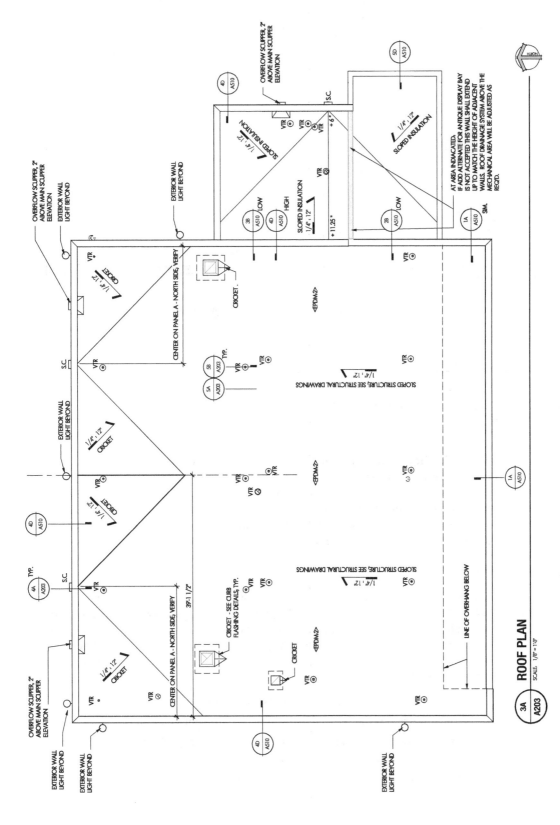

Figure 1.21 A roof panel plan. (*Courtesy of JLG Architects, Alexandria, MN 56308.*)

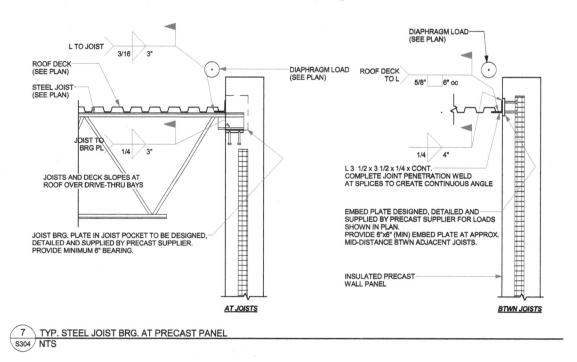

Figure 1.22 Typical roof detail. (*Courtesy of JLG Architects, Alexandria, MN 56308.*)

Notice the **elevation symbol** shown in Figure 1.23. The elevation symbol is commonly used on construction drawings to give the elevation of locations from a known zero elevation. The zero elevation might be at the first floor or other good reference point such as the top of a foundation wall. Elevation symbols are used together with standard dimensioning practice as needed on construction drawings and are often found on sections, details, and elevations but can be found in other places.

Roof Drainage Plan

The **roof drainage plan** can be part of a set of structural drawings for some buildings, although it can be considered part of the plumbing or piping drawings, depending on the particular company use and interpretation. The purpose of this drawing is to show the elevations of the roof and provide for adequate water drainage required on low slope and flat roofs, as shown in Figure 1.24. Some of the terminology associated with roof drainage plans includes the following:

Roof drain (RD): A screened opening to allow for drainage

Overflow drain (OD): A backup drain in case the roof drains fail

Down spout (DS): Usually a vertical pipe used to transport water from the roof

Gutter or **scupper:** A water collector usually on the outside of a wall at the roof level to funnel water from the roof drains to the downspouts

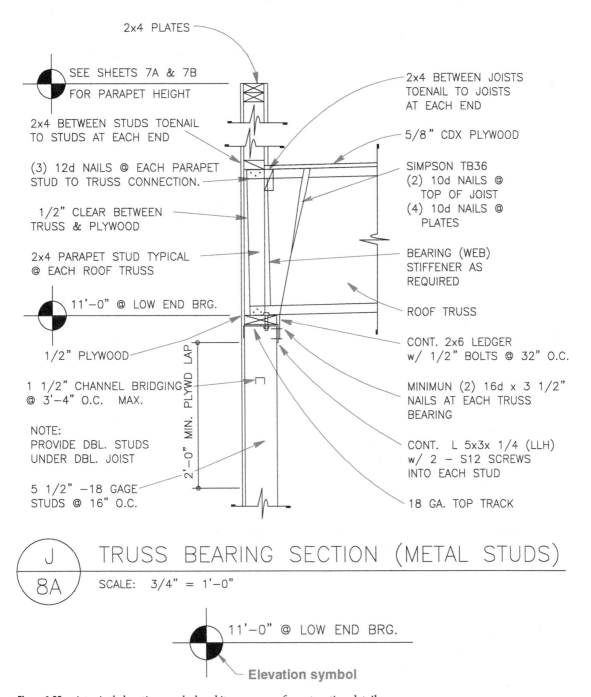

Figure 1.23   A typical elevation symbol and its use on roof construction details.

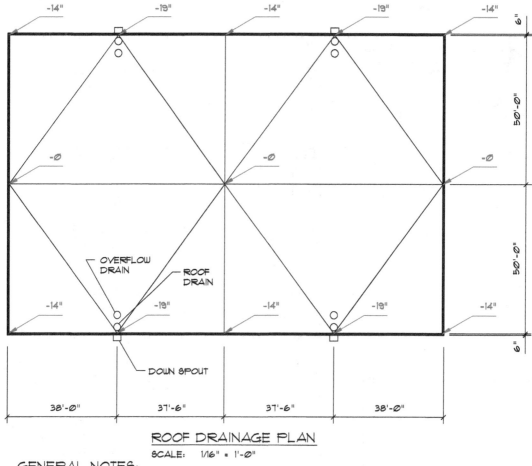

ROOF DRAINAGE PLAN
SCALE: 1/16" = 1'-0"

GENERAL NOTES:

1. ROOF AND OVERFLOW DRAINS SHALL BE LARGE GENERAL PURPOSE TYPE W/ NON-FERROUS DOMES AND 4"∅ OUTLETS.
2. OVERFLOW DRAINS SHALL BE SET W/ INLET 2" ABOVE ROOF DRAIN INLET AND SHALL BE CONNECTED TO DRAINS LINES INDEPENDENT FROM ROOF DRAINS.
3. USE A 4"H × 7W SCUPPER W/5" NOMINAL (3 3/4 × 5) RECTANGULAR CORRUGATED DOWNSPOUT. PROVIDE A 6" × 9" CONDUCTOR HEAD @ TOP OF DOWNSPOUT.

Figure 1.24   Roof drainage plan.

Building Sections

A **building section** is usually a section through the entire building used to show the relationship between the plans and details. The building section is considered a general arrangement or construction reference and is often drawn at a small scale such as 1/4″ = 1′-0″. However, some detailed information is provided with regard to building elevations (heights) and general dimensions, material size, and name. The overall section is not intended to provide explanation of building materials, but building materials can be shown and labeled. The details more clearly show building materials, because they are drawn at a larger scale. Some less-complex buildings, however, can show a great deal of construction information on the building section and use fewer details. Overall building sections, are also commonly called **typical cross sections,** show the general arrangement of construction through the entire building and often have details correlated to them, as shown in Figure 1.25. In some situations, partial sections are useful in describing portions of the construction that are not effectively described with the building section and are larger areas than normally identified with a detail, as shown in Figure 1.26.

Exterior Elevations

Exterior elevations are drawings that show the external appearance of the building. An exterior **elevation** is drawn at each side of the building to show the relationship of the building to the final grade, height dimensions and elevations, length dimensions, the location of openings, wall heights, roof slopes, exterior building materials, and other exterior features. A front elevation is generally the main entry view and is drawn at a 1/4″ = 1′-0″ scale, depending on the size of the structure. Other elevations can be drawn at 1/4″ = 1′-0″ scale or a smaller scale such as 1/8″ = 1′-0″ scale. The elevation scale depends on the size of the building, the amount of detail shown, and the sheet size used. Many companies prefer to draw all elevations at the same scale showing an equal amount of detail in all views. An elevation can be omitted if it is the same as another. Elevations can be labeled as FRONT ELEVATION, REAR ELEVATION, RIGHT ELEVATION, and LEFT ELEVATION. Elevations are also often labeled by compass orientation, such as NORTH ELEVATION and SOUTH ELEVATION.

Elevations can provide a great deal of detail at a scale of 1/4″ = 1′-0″. Often elevations for commercial buildings are drawn at a small scale representing minimal detail, as shown in Figure 1.27.

Interior elevations or details such as an interior finish elevation are drawings that show the inside appearance of specific characteristics such as cabinets, architectural details, and other inside features that need to be represented for construction, as shown in Figure 1.28.

PICTORIAL DRAWINGS

The drawings described previously in this chapter are 2D drawings that are commonly used when creating commercial construction plans. 2D drawings are prepared using a computer-aided design and drafting (CADD) software program. There are several companies that provide design software for commercial construction-related disciplines, such as architecture, structural, and civil applications. **Pictorial drawings** show a view of a building or construction detail similar to the way it would look in the real world. Pictorial drawings, such as 3D models, are sometimes used in the set of architectural and structural drawings when it is necessary to represent something more clearly than in a 2D drawing. Pictorial drawings are not required, but they are used when there is a need to aid in the visualization of specific construction applications. Figure 1.29 shows the use of a pictorial drawing. **CADD renderings** are also used to create realistic 3D images. Many architectural CADD software programs automatically allow the creation of 2D plans, 3D models, and **3D renderings** at the same time. A 3D rendering is a view of a 3D model that has been converted into a realistic image using materials, textures, colors, shading, and other special effects. A 3D rendering is shown in Figure 1.30.

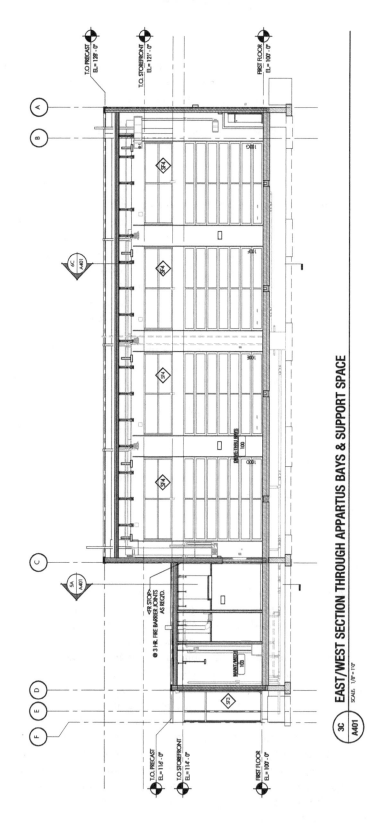

Figure 1.25 Typical cross sections show the general arrangement of construction through the entire building. (*Courtesy of JLG Architects, Alexandria, MN 56308.*)

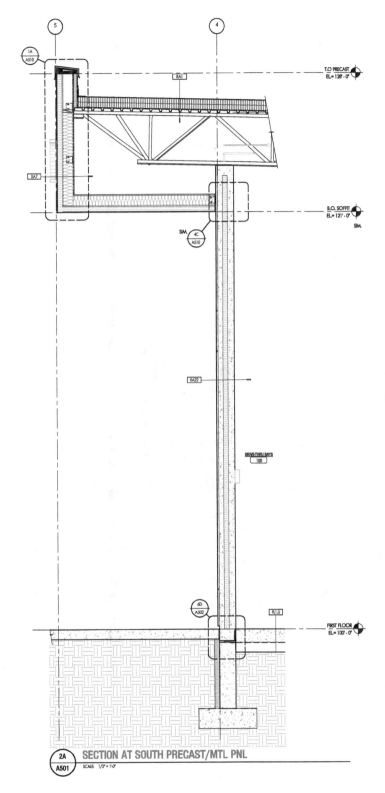

Figure 1.26 Typical wall section. (*Courtesy of JLG Architects, Alexandria, MN 56308.*)

33

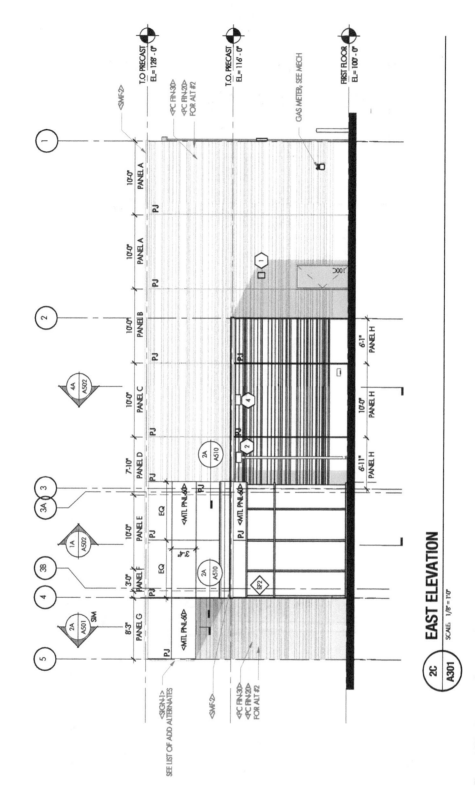

Figure 1.27 Exterior elevation. (*Courtesy of JLG Architects, Alexandria, MN 56308.*)

34

Figure 1.28   Interior elevation. (*©Deskcube–Can Stock Photo Inc.*)

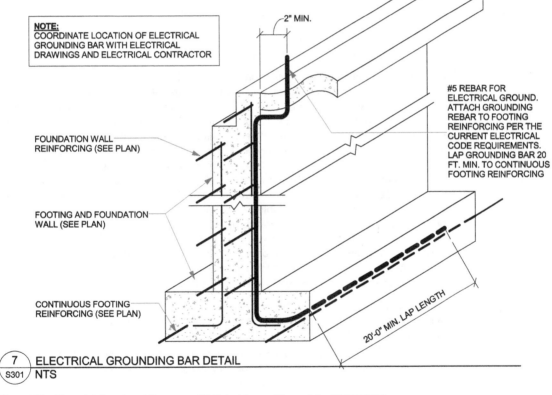

NOTE:
COORDINATE LOCATION OF ELECTRICAL
GROUNDING BAR WITH ELECTRICAL
DRAWINGS AND ELECTRICAL CONTRACTOR

2" MIN.

#5 REBAR FOR
ELECTRICAL GROUND.
ATTACH GROUNDING
REBAR TO FOOTING
REINFORCING PER THE
CURRENT ELECTRICAL
CODE REQUIREMENTS.
LAP GROUNDING BAR 20
FT. MIN. TO CONTINUOUS
FOOTING REINFORCING

FOUNDATION WALL
REINFORCING (SEE PLAN)

FOOTING AND FOUNDATION
WALL (SEE PLAN)

CONTINUOUS FOOTING
REINFORCING (SEE PLAN)

20'-0" MIN. LAP LENGTH

7   ELECTRICAL GROUNDING BAR DETAIL
S301   NTS

Figure 1.29   Pictorial drawing. (*Courtesy of JLG Architects, Alexandria, MN 56308.*)

Figure 1.30   Architectural rendering. (*Courtesy of JLG Architects, Alexandria, MN 56308.*)

COMMERCIAL CONSTRUCTION CADD APPLICATION

CADD is used in commercial construction architecture to design and draw plans. There are also many other capabilities found in CADD systems. For example, **Autodesk Seek** (seek.autodesk .com) free Web service allows architects, engineers, and other design professionals to discover, preview, and download branded and generic BIM files, models, drawings, and product specifications directly into active design sessions in **Autodesk Revit** or **AutoCAD** software. Autodesk transferred the operations and customer support related to the Autodesk Seek customers with high-quality BIM content services on the **BIMobject Cloud.** BIM is described in detail later in this chapter. Autodesk Revit is BIM software for architects, structural engineers, engineers, and contractors. AutoCAD is a computer-aided design (CAD) program used for 2D and 3D design and drafting. The BIMobject Cloud is a platform for manufacturer-specific BIM content. Features let you browse thousands of high-quality objects quickly and find exactly what your project needs. Download is free of charge and includes format options to match your preferred software. For building product manufacturers, BIMobject Cloud offers a unique way to connect with the professional designers responsible for specifying and recommending their products for purchase. BIMobject Cloud provides the following features:

- *Manufacturer-supplied product information:* Access building information models, drawings, and product specifications for more than 35,000 commercial and residential building products from nearly 1000 manufacturers.

- *Powerful parametric search technology:* Search by key attributes, including dimensions, materials, performance, sustainability, or manufacturer name using industry-standard classifications.

- *Preview and explore models before downloading:* View, rotate, zoom, and slice product models and then download the accurate files directly into your design session. For Revit models, preview the family parameters and associated type catalogs before downloading.

- *Multiple formats:* Select the formats that work for you, such as Revit, DWG™, DGN, and SKP files; Microsoft® Word documents; three-part specifications; and PDFs.

- *Share designs:* You can share designs with peers by uploading them directly from your AutoCAD files. You can choose to share the current drawing or select a block definition within the drawing. Thumbnails, title, and metadata are automatically extracted and indexed. Shared designs can be searched and downloaded by anyone. A complete set of structural applications are provided. You can view an enlarged image and description of a product, and use viewing options such as DWG, RFA, DXF, PDF, and Word files.

Similar products are available from other CADD software developers and from product manufacturers.

DRAWING REVISIONS

Drawing revisions are common in the architectural, structural, and construction industry. Revisions can be caused for a number of reasons, for example, changes requested by the owner, job-site corrections, correcting errors, or code changes. Changes are done in a formal manner by submitting an **addendum** to the contract, which is a written notification of the change or changes and is accompanied by a marked-up or red-lined drawing or print that represents the change.

Revision Clouds

A **revision cloud** is placed around the drawing area that is changed by a revision. The revision cloud is a cloudlike circle around the change, as shown in Figure 1.31. There is also a triangle with a revision number inside that is placed next to the revision cloud or along the revision cloud line, as shown in Figure 1.31. The triangle is commonly called a **delta note** or **flag note.** The number in the flag note triangle is then correlated to a revision note

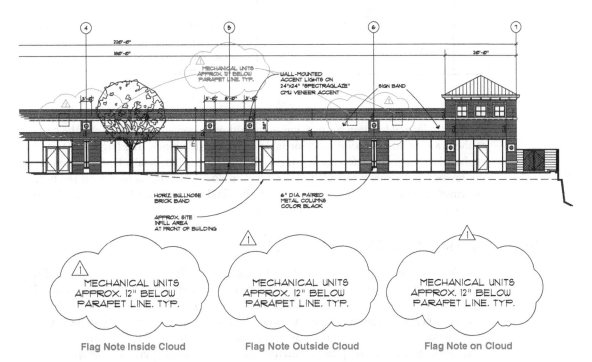

Figure 1.31   A revision cloud is a cloudlike circle around a drawing change. There is also a triangle with a revision number inside placed next to the revision cloud or along the revision cloud line.

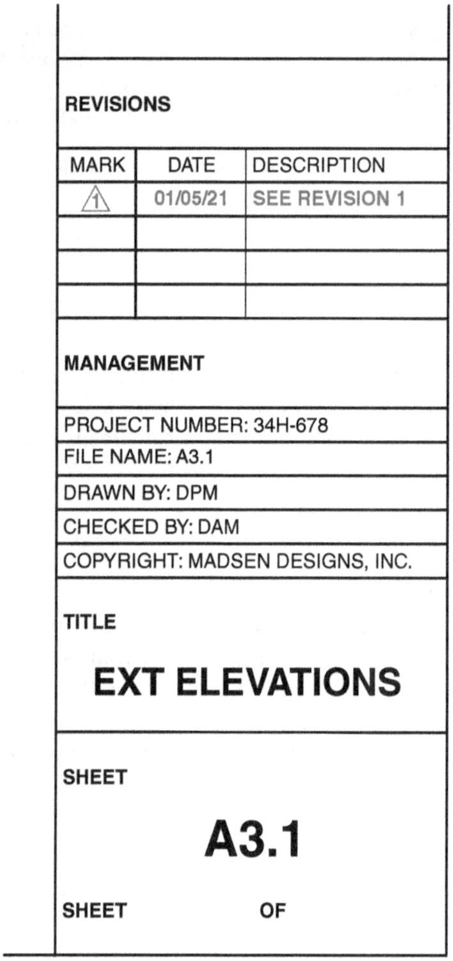

REVISIONS		
MARK	DATE	DESCRIPTION
⚠	01/05/21	SEE REVISION 1

MANAGEMENT

PROJECT NUMBER: 34H-678

FILE NAME: A3.1

DRAWN BY: DPM

CHECKED BY: DAM

COPYRIGHT: MADSEN DESIGNS, INC.

TITLE

EXT ELEVATIONS

SHEET

A3.1

SHEET   **OF**

Figure 1.32   The number in the revision cloud flag note triangle is correlated to a revision note placed somewhere on the drawing or in the title block.

placed somewhere on the drawing or in the title block, as shown in Figure 1.32. Each company has a desired location for revision notes, although common places are in the corners of the drawing, in a revision block or table, or in the title block. The flag note is used to explain the change. If a reference is given in the title block, then detailed information about the revision is normally provided in the revision document that is filed with the project information. The revision document is typically filled out and filed for reference, because changes can cause increased costs in a project.

GENERAL CONSTRUCTION SPECIFICATIONS

Specifications are written documents that describe detailed requirements for products, materials, and workmanship on which the construction project is based. A specification is an exact statement describing the characteristics of a particular aspect of the project. Specifications communicate information about required products to be used in construction, as a basis for competitive construction bidding, and to measure compliance with contracts. Proprietary product, method, and end result are performance specification methods commonly used in the construction industry.

Proprietary product specification provides specific product names and models for desired applications. When this type of specification is used, a named product can be followed by or equivalent, which allows for equal alternatives and helps promote competition in providing the product. Proprietary product specifications can limit competition, increase cost, and decrease flexibility.

Method specifications outline material selection and construction operation process to be followed in providing construction materials and practices. Method specifications provide the final desired structure, such as the concrete thickness and strength, or the lumber dimensions, spacing, species, and grade. Method specifications allow for more flexibility, but the owner is responsible for the performance.

End-result specifications provide final characteristics of the products and methods used in construction, and the contractor can use a desired method for meeting the requirements. End-result specifications often provide minimum and maximum as a range of acceptable completion. For example, under concrete slab gravel can be specified between 4 and 8 in. thick with specific compaction given. End-result specifications can use statistical methods to estimate overall material quality based on a limited number of random samples. End-result specifications place construction quality on the contractor by defining the desired final product. Such specifications can allow the contractor freedom in achieving that final product, which can lead to innovation, efficiency, and lower costs.

SPECIFICATIONS FOR COMMERCIAL CONSTRUCTION

Specifications for commercial construction projects are used for business, commercial, or industry applications with documents that are often more complex and comprehensive than specifications for residential construction. Commercial project specifications can provide highly detailed instructions for each phase of construction. Specifications can establish time schedules for the completion of the project. Also, in certain situations, the specifications include inspections in conjunction with or in addition to those required by a local jurisdiction. Construction specifications often follow the guidelines of the individual architect or engineering firm, although a common format has been established, titled: *MasterFormat: Master List of Numbers and Titles for the Construction Industry,* which is published by the Construction Specifications Institute (CSI) and the Construction Specifications Canada (CSC). The following is contact information for the CSI and CSC:

The Construction Specifications Institute (CSI)
99 Canal Center Plaza, Suite 300
Alexandria, VA 22314 (www.csinet.org)

Construction Specifications Canada (CSC)
120 Carlton Street, Suite 312
Toronto, ON, M5A 4K2 (www.csc-dcc.ca)

The MasterFormat numbers and titles offer a master list of numbers and subject titles for organizing information about construction work results, requirements, products, and activities into a standard sequence. Construction projects use many different delivery methods, products, and installation methods. Successful completion of projects requires effective communication among the people involved. Information retrieval is nearly impossible without a standard filing system familiar to each user and based on industry standards. The MasterFormat numbers and titles document facilitate standard filing and retrieval systems throughout the construction industry. MasterFormat is suitable for use in project manuals for organizing cost data, for reference keynotes on drawings, for filing product information and other technical data, for identifying drawing objects, and for presenting construction market data.

Each MasterFormat number and title defines a section, which is arranged in *levels*, depending on its depth of coverage. The broadest collections of related construction products and activities are level one titles, otherwise known as **divisions.** Each division in the *MasterFormat: Master List of Numbers and Titles for the Construction Industry* is made up of level 2, level 3, and occasionally level 4 numbers and titles assigned by MasterFormat, each of which defines a gradually more detailed area of **work results** to be specified. Work results are traditional construction practices that typically result from an application of skills to construction products or resources.

The *MasterFormat* numbers are established using a six-digit system. The following is an example showing how the list of numbers and titles are used in the numbering system:

Division 04—Masonry

The first two numbers, such as 04 in this example, represent the division and are also called **level 1.** The complete list of divisions is given in the next section of this textbook.

04 05 Common Work Results for Masonry

The second pair of numbers—05 in this example—is referred to as **level 2.** In this case, Common Work Results for Masonry is a subcategory of Masonry.

04 05 19 Masonry Anchorage and Reinforcing

The third pair of numbers—19 in this example—is called **level 3.** In this case, Masonry Anchorage and Reinforcing is a subcategory of Common Work Results for Masonry.

04 05 19.26 Masonry Reinforcing Bars

Occasionally, level 4 numbers are provided, such as .26 in this example. When level 4 numbers are used, they follow level 3 numbers and are separated from level 3 numbers with a dot. Level 4 numbers are used when the amount of detail requires additional level of classification. In this case, Masonry Reinforcing Bars is a subcategory of Masonry Anchorage and Reinforcing.

An example of the six-digit numbering system with levels **1** through **4** is shown in Figure 1.33.

MasterFormat Division Numbers and Titles

The MasterFormat has these two main groups:

1. Procurement and contracting requirements
2. Specifications

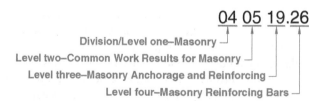

Figure 1.33 An example of the six-digit numbering system with levels one through four for the MasterFormat: Master List of Numbers and Titles for the Construction Industry.

Procurement and Contracting Requirements are referred to as series zero in the MasterFormat, because they begin with a 00 level one numbering system prefix. These documents are not specifications. They establish relationships, processes, and responsibilities for projects.

The **Specifications group** in the MasterFormat contains the construction specifications subgroups and their related divisions. Some divisions are identified as reserved for future additions or specific user applications. The following is an outline of the divisions found in the Procurement and Contracting Requirements and Specifications groups in the MasterFormat:

PROCUREMENT AND CONTRACTING REQUIREMENTS GROUP

    Division 00   Procurement and Contracting Requirements

SPECIFICATIONS GROUP

    General Requirements Subgroup
    Division 01   General Requirements
    Facility Construction Subgroup
    Division 02   Existing Conditions
    Division 03   Concrete
    Division 04   Masonry
    Division 05   Metals
    Division 06   Wood, Plastics, and Composites
    Division 07   Thermal and Moisture Protection
    Division 08   Openings
    Division 09   Finishes
    Division 10   Specialties
    Division 11   Equipment
    Division 12   Furnishings
    Division 13   Special Construction
    Division 14   Conveying Equipment
    Divisions 15 through 19 Reserved

FACILITY SERVICES SUBGROUP

    Division 20   Reserved
    Division 21   Fire Suppression
    Division 22   Plumbing
    Division 23   Heating, Ventilating, and Air-Conditioning
    Division 24   Reserved
    Division 25   Integrated Automation
    Division 26   Electrical
    Division 27   Communications
    Division 28   Electronic Safety and Security
    Division 29   Reserved

SITE AND INFRASTRUCTURE SUBGROUP

    Division 30   Reserved
    Division 31   Earthwork
    Division 32   Exterior Improvements
    Division 33   Utilities

Division 34   Transportation
Division 35   Waterway and Marine Construction
Division 36 through 39 Reserved

PROCESS EQUIPMENT SUBGROUP

Division 40   Process Integration
Division 41   Material Processing and Handling Equipment
Division 42   Process Heating, Cooling, and Drying Equipment
Division 43   Process Gas and Liquid Handling, Purification, and Storage
Equipment
Division 44   Pollution Control Equipment
Division 45   Industry-Specific Manufacturing Equipment
Division 46 and 47 Reserved
Division 48   Electrical Power Generation
Division 49   Reserved

The UniFormat Uniform Classification System

The **UniFormat** is a uniform classification system for organizing preliminary construction information into a standard order or sequence on the basis of **functional elements.** Functional elements, also referred to as **systems** or **assemblies,** are common major components in buildings that perform a known function regardless of the design specification, construction method, or materials used. The use of UniFormat can provide consistent comparable data across an entire **building life cycle.** Building life cycle refers to the observation and examination of a building over the course of its entire life. The life cycle of a building considers everything about the building from design, commissioning, operation, and decommissioning.

The purpose of UniFormat is to achieve consistency in economic evaluation of projects, enhance reporting of design program information, and promote consistency in filing information for facility management, drawing details, and construction market data. UniFormat classifies information into nine level 1 categories that can be used to arrange brief project descriptions and preliminary cost information. The first level 1 category is Project Description, which includes information about the project, through cost estimating and funding. The last eight level 1 categories are referred to as Construction Systems and Assemblies, which include the construction applications and practices, such as foundation, roofing, exteriors, electrical, and plumbing. Each of the Construction Systems and Assembly categories are identified with a letter and title as follows:

A—Substructure

B—Shell

C—Interiors

D—Services

E—Equipment and Finishings

F—Special Construction and Demolition

G—Building Sitework

Z—General

UniFormat has a numbering system that divides each level 1 category into levels 2, 3, 4, and 5 titles with set alphanumeric labels. The following is an example showing the first three levels of a UniFormat alphanumeric system for a specific category:

Level 1: D Services

Level 2: D20 Plumbing

Level 3: D2010 Plumbing Fixtures

> **NOTE**
>
> The difference between MasterFormat and UniFormat is that MasterFormat is a material-based organization of building content, and UniFormat is a systems-based organization of building content.

BUILDING INFORMATION MODELING

Building information modeling (BIM) is often discussed in the construction industry but with varying definitions. Some people say BIM is a type of software or the 3D virtual model of buildings. Others say BIM is a process or a collection of all building data organized in a database, easy to search visually and numerically. BIM is all of these capabilities and more.

BIM starts with a 3D digital model of a building, as shown in Figure 1.34. This model is more than just geometry and textures. A true BIM model consists of the **virtual** equivalents of the actual parts that make up the constructed building. These elements have all the required physical and logical characteristics of their real counterparts. These intelligent

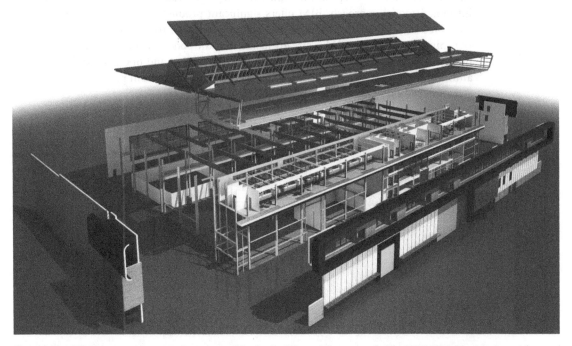

Figure 1.34   BIM starts with a 3D digital model of a building. (*Courtesy of GRAPHISOFT. The Nuclear Advanced Manufacturing Research Centre | UK Bond Bryan Architects www.bondbryan.com.*)

elements are the digital prototypes of the physical building elements, such as walls, columns, windows, doors and stairs. BIM simulates the building and understands its behavior in a digital environment before construction begins. The term virtual refers to something that appears to have the properties of a real or actual object or experience.

BIM Is More Than Just 3D Modeling

The **BIM model** contains all the necessary information to describe the building for construction and management purposes. The model is made of real architectural elements with non-graphical information attached, such as materials, cost, and manufacturer details. As a result, the **BIM modeling software** coordinates the 3D model, the 2D documentation, and the attached BIM data. In this sense, a typical BIM modeling software adds much more value to the project than a traditional CAD tool.

The purpose of using CAD is to create 2D and 3D project documentation in a digital environment. BIM goes far beyond CAD by providing a new communication platform for all participants of the building project. BIM requires a completely new way of thinking and the acceptance of new methodologies and principles by all involved in the project.

BIM Applications Imitate the Real Building Process

Instead of creating drawings from 2D line-work, buildings are virtually modeled from digital equivalents of real construction elements, such as walls, windows, slabs, and roofs, as shown in Figure 1.35. This allows architects to design buildings in the same way they are built. Since all data is stored in the central virtual building model, design changes are automatically followed-up on individual drawings generated from the model. With this integrated model approach, BIM offers significant productivity gains and serves as the basis for better-coordinated designs and a computer model-based building process. While switching from CAD to BIM is already justified by the benefits achieved during the design phase, BIM offers further benefits during the construction and operation of buildings by allowing the reuse of the already existing digital data for different purposes such as construction or facility management.

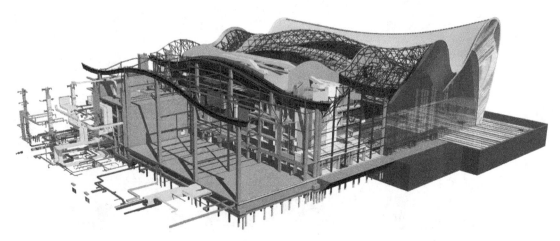

Figure 1.35   With BIM, buildings are virtually modeled from digital equivalents of real construction elements, such as walls, windows, slabs, and roofs. (*Courtesy of GRAPHISOFT. Irina Viner-Usmanova Rhythmic Gymnastics Center in the Luzhniki Complex, Moscow, Russia | CPU PRIDE www.prideproject.pro.*)

BIM Is a Platform for Communication

Architects must continuously share their design with many project team members during the **project life cycle.** A project life cycle is the sequence of phases that a project goes through from its initiation to its completion. The BIM model provides the ideal platform for sharing the building data inside and outside of the office. **Industry Foundation Classes (IFC)** and other file protocols allow the BIM program to communicate with diverse applications, such as structural, energy analysis, and collision detection programs. The IFC data model is intended to describe architectural, building, and construction industry data. IFC is a platform-neutral, open file format specification that is not controlled by a single vendor or group of vendors. Developers of BIM applications are one of the leading forces behind IFC standards. IFC allows seamless sharing and exchange of the building models with all kinds of applications related to building design, construction, and management. Because BIM information is preserved during the data transfer, IFC is gaining acceptance worldwide as the standard for BIM data exchange. **Extensible Markup Language (XML),** also used by BIM, is a general-purpose description language that can describe different kinds of data. XML is ideal for connecting different systems through the Internet. Specialized XML languages allow computers to exchange and make use of information between building design models and a wide variety of engineering analysis tools. The BIM model can also be shared among the team of architects by using GRAPHISOFT **Teamwork,** for example, and project file **hotlinking** methods. Teamwork is based on a client-server architecture and is designed to ensure maximum flexibility, speed, and data safety to enable teams—even those spread out around the world—to collaborate on large projects. Hotlinking is a process of linking directly to other project files or predefined typical project components. **Integrated project delivery (IPD)** is a new approach to the design and construction of buildings that is based on a cooperative working relationship, shared risk and reward, and open exchange of data. This collaborative project delivery method depends largely on the latest technologies including BIM, **4D modeling** and **5D modeling.** 4D modeling is a term often used in the construction industry as a time dimension to a 3D CAD model that allows teams to analyze the sequence of events on a timeline and visualize the time it takes to complete tasks within the construction process. 5D modeling is a term used in the CAD and construction industries, referring to the intelligent linking of individual 3D CAD components or assemblies with time-schedule constraints and then with cost-related information. See Figure 1.36.

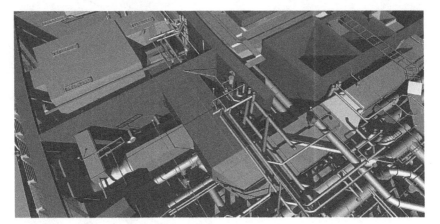

Figure 1.36   Architects must continuously share their design with many project team members during the project lifecycle, allowing seamless sharing and exchange of the building models with all kinds of applications related to building design, construction, and management. (*Courtesy of GRAPHISOFT. © BIMES, Dubai www.bimes.com.*)

BIM as the Foundation for Construction Coordination

Adding time and cost information to the BIM model opens the way for automated construction solutions. The 5D modeling concept, as this approach is often called, helps to solve many problems of the construction industry that largely result from inefficient communication among architects, designers, engineers, and constructors. It also helps to achieve more accurate cost estimations even in the early phases of project design.

BIM Is an Open System

Project participants often require different data types and formats. BIM programs allow architects to export projects in several file formats. This way the BIM model can be opened in various programs with continuity and without data loss. Open BIM is a universal approach to the collaborative design, construction, and operation of buildings based on standards and workflows, as shown in Figure 1.37. Open BIM is an initiative of several leading software vendors using the open **buildingSMART Data Model,** the software-neutral IFC data model and represents a modern approach to interdisciplinary collaboration for all members of the AEC industry. The following are open BIM characteristics:

- Supports a transparent, open workflow, allowing project members to participate regardless of the software tools they use

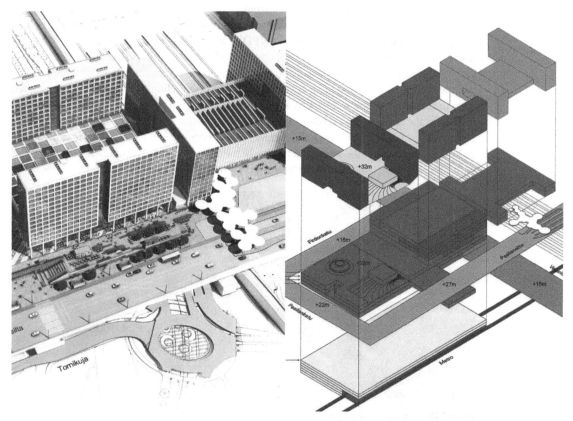

Figure 1.37   Open BIM is a universal approach to the collaborative design, construction, and operation of buildings based on standards and workflows. (*Courtesy of GRAPHISOFT. © YIT Group TRIPLA axonometric view.*)

- Creates a common language for widely referenced processes, allowing industry and government to procure projects with transparent commercial engagement, comparable service evaluation and assured data quality
- Provides enduring project data for use throughout the building lifecycle, avoiding multiple input of the same data and consequential errors
- Small and large (platform) software vendors can participate and compete on system-independent solutions
- Strengthens the online product supply side with more exact user demand searches and delivers the product data directly into the BIM

3D Visualization

The most basic use of a BIM model is for visualizations of the proposed building, as shown in Figure 1.38. This helps in the design decision-making process by comparing design alternatives and for promoting the design to a client and others who have a deciding role in construction of the project. The latest technological advances allow the quick, real-time visualization of BIM projects even on mobile devices without the need for complex rendering applications.

Project Collaboration

Designing, constructing, and maintaining a building is a highly complex process that requires the close cooperation of many people. The architect or the general contractor must be in the center of this communication process, and the BIM model provides the ideal platform for communicating the building information. Internal communication within the office is just as important as external relationships. BIM supports the cooperation of

Figure 1.38 The most basic use of a BIM model is for visualizations of the proposed building. (*Courtesy of GRAPHISOFT. Architect: Svetlana Kravchenko | Rendered image.*)

project team members with automated collaboration techniques to increase productivity and reduce the risk of coordination errors. Most BIM applications have built-in collaboration solutions that help the architectural team in sharing and coordinating the projects.

Data Management

BIM contains management information that supports scheduling, manpower needs, and construction cost coordination that allows the budget and estimated costs of a project to be monitored at any given time.

Change Management

BIM data is centrally stored, automatically reproducing building design modifications throughout the project. For example, a floor plan change is updated in sections, elevations, specifications, and all other locations. This process works to create documentation faster and provides strict quality assurance through coordination in the entire project. See Figure 1.39.

Building Simulation

BIM models contain architectural data with the full depth of the building information related to the different engineering disciplines, such as structural design, plumbing, electrical, mechanical, and sustainability information that can easily be simulated well in advance to observe the design results. See Figure 1.40.

Collision Detection. A collision occurs when two or more construction elements physically intersect. Collision detection is a BIM feature that works by checking the relationships between any two features or groups of elements. The groups are defined by criteria selected by the architects. For example, an architect might run collision detection between construction and mechanical, electrical, and plumbing (MEP) elements; and between concrete and steel construction elements; or to check clearance for exit routes. See Figure 1.41.

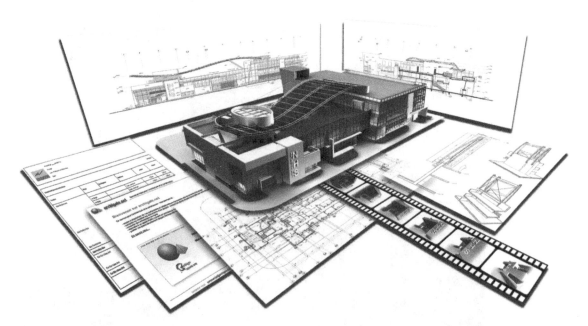

Figure 1.39   BIM data is centrally stored, automatically reproducing building design modifications throughout the project. For example, a floor plan change is updated in sections, elevations, specifications and all other locations. (*Courtesy of GRAPHISOFT. paastudio, NHS Building.*)

Figure 1.40 BIM models contain architectural data with the full depth of the building information related to the different engineering disciplines, such as structural design, plumbing, electrical, mechanical, and sustainability information that can easily be simulated well in advance to observe the design results. (*Courtesy of GRAPHISOFT. Charles Perkins Centre, Australia | fjmt – fjmtstudio. com Photo © Keira Yang Zhang.*)

Figure 1.41 Collision detection is a BIM feature that works by checking the relationships between any two features or groups of elements. (*Courtesy of GRAPHISOFT. MEP model shown on ARCHICAD using IFC communication | ©YIT Group.*)

Energy Evaluation

Energy use evaluation is integrated into the BIM program, offering an easy-to-use workflow for performing dynamic building energy calculations on projects, as shown in Figure 1.42. Energy evaluation is a tool that allows architects to monitor and control all architectural design parameters that influence building energy performance. The energy evaluation tool performs reliable dynamic energy evaluation at all stages of the design process, so architects can make informed decisions regarding the building's energy efficiency. Including energy evaluation in the architectural design process makes it easy to create projects that conform to or exceed energy efficiency standards and model programs described in this chapter.

Building Operation

Data entered into a BIM model can be reused throughout the entire building life cycle. This helps reduce the operation, management, and maintenance costs of buildings. Asset management, tenant management, maintenance scheduling, and many more services are now based on the digital database stored in the BIM project model file. See Figure 1.43.

Professional BIM Program Characteristics

There are several model-based design products available on the market today, and not all can equally fulfill the requirements for being a true BIM solution. The following set of features can help determine if a BIM program has the necessary characteristics:

- The BIM model should be able to serve all deliverables during the entire building lifecycle.

- Real BIM models contain geometry and an abundance of additional information necessary to coordinate, document, list, and manage the building based on its intelligent BIM model.

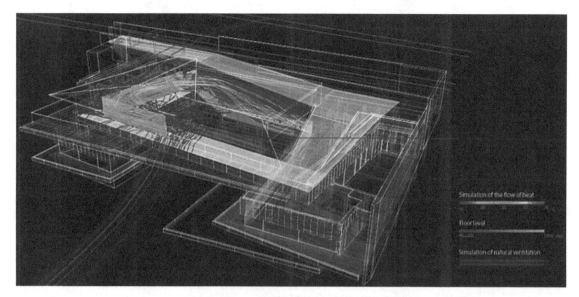

Figure 1.42   Energy use evaluation is integrated into the BIM program, offering an easy-to-use workflow for performing dynamic building energy calculations on projects. (*Courtesy of GRAPHISOFT. "On the Water Guest House" project air flow analysis | Nikken Sekkei.*)

Figure 1.43   Data entered into a BIM model can be reused throughout the entire building lifecycle. (*Courtesy of GRAPHISOFT. Project by Enzyme APD | www.weareenzyme.com | and PLA group.*)

- The BIM authoring tool should cover the complete BIM workflow without the need for changing tools and/or workflows in the middle of the project.
- Design, documentation, and building operation are supported throughout the building life cycle.
- BIM tools must be compatible with one another, so interdisciplinary design teams can collaborate on the different design aspects of the same building.
- The BIM tool supports open standards and open workflows that enable coordination with consultants regardless of the type or version of BIM tool used.
- The BIM model is designed to support real-time sharing for teams and projects of any size.
- The BIM solution can handle that growth, including its ability to use the latest hardware enhancements with the model.

- Determine if the maximum possible project size is supported at the same time throughout the entire application.
- The BIM solution is compliant with local design standards and includes local content, such as intelligent building objects, local project, and listing templates, and attribute sets that follow local standards.

New BIM Technologies

Technological advances are happening in every industry, including the AEC industry. There are many advances being made in BIM applications. The following sections highlight some of the latest innovations that influence the future of the AEC industry.

Virtual Reality
Virtual reality (VR) technology is becoming an essential part of the building design and construction process in the AEC industry. VR refers to a world that appears to be a real or actual world, having many of the properties of the real world. VR allows all audiences to easily perceive spaces. VR can improve design communication with clients and collaboration with team members during the design process. VR can offer a solution for efficient timesaving in online design discussions.

Artificial Intelligence
Artificial intelligence (AI) trends are already changing the way buildings are being constructed. Artificial intelligence is the theory and development of computer systems able to perform tasks that normally require human intelligence, such as visual perception, speech recognition, decision making, and translation between languages. Roles traditionally carried out by humans are now being performed by machines. There are three areas in construction where AI has the most potential. These are building design and analysis, construction technologies, and building operation. GRAPHISOFT has embedded AI into ARCHICAD, introducing **predictive design technology** to automate stair design. Predictive design technology works by analyzing historical and current data and generating a model to help predict future outcomes.

Algorithmic Design
Algorithms are widespread in many areas of everyday life, encouraged by unprecedented data-processing power. Algorithms are a process or set of rules to be followed in calculations or other problem-solving operations. The role of the architect in algorithmic design is not to design the building, but to design the system that designs the building. Algorithmic design is often thought of as a schematic design tool used only at the beginning of the design process. Ideally, algorithmic design and the constant exchange of BIM information are linked throughout the design process. See Figure 1.44. GRAPHISOFT provides a variety of ways to communicate between 3D CAD applications and ARCHICAD, reflecting the variety of algorithmic design workflows.

Go to Downloads & Resources tab at **www.mhprofessional.com/Commercial Building Construction-Ancillaries** to access the images provided in this chapter and correlate with the content of this chapter. Here you can look at the chapter figures on your computer screen to pan and zoom in and out for better observation, because full drawings reduced to fit on a textbook page are often difficult to read. The image bank also includes the complete set of plans for the Brookings South Main Fire Station used as examples throughout this textbook.

Figure 1.44   Algorithmic design and the constant exchange of BIM information are linked throughout the design process. (*Courtesy of GRAPHISOFT.*)

Go to **Downloads & Resources tab at www.mhprofessional.com/Commercial Building Construction-Ancillaries** to access the test and creative thinking problems for this chapter. The chapter test can be used for review or to evaluate content knowledge depending on your course objectives. Creative thinking problems are provided to help expand your knowledge by researching given subjects.

Print reading problems have questions that ask you to find information on the Brookings South Main Fire Station plans. The print reading problems can help you become familiar with the format of construction documents and reinforce your ability to seek specific information. This is an important skill for you to master as you prepare to enter the construction industry. Go to **Downloads & Resources tab at www.mhprofessional.com/Commercial Building Construction-Ancillaries** to access the complete set of plans for the Brookings South Main Fire Station.

CHAPTER 2

Sustainable Technology

INTRODUCTION TO SUSTAINABLE TECHNOLOGY

Building science is the collection of scientific knowledge that focuses on the analysis and control of a buildings' physical environment. Building science typically includes the analysis of construction materials used to create the building, the effects of geographical and geological observations on the building, and the effects of the buildings on its occupants. Building science applications help to advance construction best practices and offer alternate construction methods and energy sources that improve the building. Reducing energy consumption and required maintenance are often at the top of lists for sustainable design and construction priorities, while protecting the quality of future life on earth is also an important consideration when designing, building, or buying a commercial property today. Improving energy efficiency is a good way to reduce oil and fossil fuels dependency, and address environmental concerns connected with commercial construction. Private organizations, local and state governments, and national agencies sponsor projects that provide incentives and education for architects, designers, and contractors in an effort to promote building practices that are good today and protect the future. The phrases or terms **sustainable, environmentally friendly,** and **environmentally sound** refer to design and construction practices using renewable materials, such as wood from certified forest farms, earthen or natural materials, and recycled products. These buildings are also healthier for the occupants because of the extensive use of low- and nontoxic materials. Using construction materials and collecting energy from renewable resources helps protect the environment and reduce the **carbon footprint** of the building. Renewable energy resources can include geothermal, solar, water, and wind energy. The carbon footprint of a building is a measure of the impact the building has on the environment over its lifetime by producing **greenhouse gases,** which are measured in units of **carbon dioxide,** during the construction, maintenance cycle, and deconstruction of the building. The term **footprint** is used to identify the foundation dimensions upon which the building is constructed. Carbon dioxide has the chemical formula CO_2, and is a colorless, odorless gas present in the atmosphere and formed when any fuel containing carbon is burned. CO_2 is produced by many events that occur on earth. CO_2 is breathed out of the lungs during respiration, is produced by the decay of organic matter, is removed during volcano eruptions, and is also used by plants in **photosynthesis.** Carbon dioxide is also used in refrigeration, fire extinguishers, and carbonated drinks. Photosynthesis is the process by which green plants and some other organisms use sunlight to synthesize foods from carbon dioxide and water. The carbon footprint measurement of a building is a measure of its theoretical influence in contributing to global warming. **Carbon offsets** are abstract tools used to negate the impact of a building's carbon footprint through the development of alternative projects, such as solar or wind energy, or reforestation. Greenhouse gases are the gases that absorb global radiation and contribute to the **greenhouse effect.** The main greenhouse gases are water vapor, methane, carbon dioxide, and ozone. The greenhouse effect is atmospheric heating caused by solar radiation being transmitted inward through the earth's atmosphere with less radiation transmitted outward due to absorption by gases in the atmosphere.

A **carbon-neutral** building can be created through the use of natural, low footprint materials, increased energy efficiency, renewable energy purchases, and carbon offset purchases, or all of these applications. Carbon neutral is a state in which the net amount of carbon dioxide or other carbon compounds emitted into the atmosphere is reduced to zero by carbon offsets. Construction practices that help create carbon-neutral buildings often use renewable energy for heating and cooling, implement highly energy- and water-efficient appliances and mechanical systems, and use construction materials that have a low carbon footprint from responsible manufacturers. In addition to the materials and energy sources used in the construction of a commercial building, the carbon footprint can be reduced in other ways. For example, people can do their part by carpooling, use of mass transit, or use airline and train travel for trips that reduce automobile driving. People can also be involved in or donate to programs that offset carbon dioxide emissions, such as renewable energy or tree-planting programs.

SUSTAINABLE TECHNOLOGY SUPPORTS A HEALTHY ARCHITECTURAL ENVIRONMENT

Throughout this textbook you can learn how constructing an energy-efficient building involves issues related to construction techniques, construction materials, and heating, ventilating, and air-conditioning (HVAC) design. Techniques that result in an airtight structure are important when the goal is to design and build for energy efficiency. Controlling the movement of air into and out of the structure is an important part of controlling the temperature. Methods used to reduce outside air infiltration range from the selection and proper installation of airtight building materials and insulation to the use of caulking products and continuous vapor retarders. These products and construction methods are described in detail in Chapter 10, *Insulation and Barriers, and Indoor Air Quality and Safety.*

Controlling indoor air quality and maintaining a healthy living environment is also a very important consideration when designing and constructing for energy efficiency. The system needs to be properly balanced to remove the risk of problems associated with poor indoor air quality. Air quality issues can range from stale air to a buildup of harmful contaminants, including radon, carbon monoxide, or mold spores. These issues can be easily addressed with a well-designed and properly installed heat recovery and ventilation system. Sources of pollutants and methods used to deal with these problems are discussed in Chapter 13, *Mechanical, Plumbing, and Electrical Systems*, with complete coverage of HVAC systems.

The selection of environmentally friendly building materials and products is equally important. Avoid the use of manufactured wood products, such as fiberboard, particleboard, and plywood, unless they are manufactured using environmentally safe chemical processes and adhesives. Use water-based caulking, paint, and finishes. Keep synthetic carpeting to a minimum, or eliminate it completely by using natural fiber products, such as wool and cotton. Combustion must be complete in appliances that use gas fuel, and they must be properly ventilated to the exterior. Acceptable alternatives include electric appliances or sealed combustion appliances. Avoid using treated wood products in the construction of the **building envelope.** Building envelope consists of the roof, exterior walls, and floor of a structure, forming a barrier that separates the interior of the building from the outdoor environment. Consider steel-frame construction in regions where treated wood is commonly used for termite and rot control, although there is an increased carbon footprint associated with the production of steel.

The following is a checklist of features, products, and chapters within this textbook that can be considered for designing and building an energy-efficient and healthy environment:

- Design and build a high-performance building envelope using quality materials and efficient construction practices. See Chapter 4, *Concrete Construction and Foundation Systems*, Chapter 5, *Masonry Construction*, Chapter 6, *Steel Construction*, Chapter 7, *Roof Construction and Materials*, Chapter 8, *Wood Construction*, and Chapter 10, *Insulation and Barriers, and Indoor Air Quality and Safety*. Design and build with high-quality environmentally friendly insulation, caulking, and vapor barriers. See Chapter 10, *Insulation and Barrier, and Indoor Air Quality and Safety*.

- Use properly installed high-performance windows and exterior doors. See Chapter 9, *Doors, Windows, and Installations*.

- Use unbleached plasterboard and plaster rather than chemically produced sheetrock and materials. See Chapter 12, *Finish Work and Materials*.

- Install a properly designed heat-recovery ventilation system, also called an air-to-air heat exchanger. See Chapter 13, *Mechanical, Plumbing, and Electrical Systems*.

- Install a high-performance particle air filtration system. See Chapter 13, *Mechanical, Plumbing, and Electrical Systems*.

- Use low-toxicity ceramic tiles and hardwood floors or natural fiber carpets rather than synthetic carpets. See Chapter 12, *Finish Work and Materials*.

- Use wood sheathing and low-formaldehyde decking. See Chapter 8, *Wood Construction*.

- Consider low-toxicity foam insulation or other nonchemical insulation products. See Chapter 10, *Insulation and Barrier, and Indoor Air Quality and Safety*.

- Use water-base, solvent-free, low-toxicity paints and finishes. See Chapter 12, *Finish Work and Materials*.

- Construct a tightly sealed foundation to help keep out radon gas and moisture. See Chapter 4, *Concrete Construction and Foundation Systems*.

- Use an energy-efficient heat pump such as a geothermal unit that produces no indoor pollutants. See Chapter 13, *Mechanical, Plumbing, and Electrical Systems*.

- Install gas appliances and fireplaces that use only outdoor air and exhaust to the outside without any downdraft. See Chapter 13, *Mechanical, Plumbing, and Electrical Systems*.

- UV lights mounted inside the ductwork can neutralize up to 98 percent of microorganisms. See Chapter 13, *Mechanical, Plumbing, and Electrical Systems*.

- Install an electrical system that is designed to **reduce stray voltage.** Stray voltage is a low voltage present on grounded metal objects, and can be controlled with quality equipment, fixtures, and proper grounding. See Chapter 13, *Mechanical, Plumbing, and Electrical Systems.*

- Use a sealed and properly sized metal duct system with ducts insulated on the outside or preferably located in conditioned spaces. Avoid fiberglass insulation inside ducts. See Chapter 13, *Mechanical, Plumbing, and Electrical Systems.*

- Install a central vacuum system that vents particles out of the house into a filtered container in the garage. See Chapter 13, *Mechanical, Plumbing, and Electrical Systems.*

- Install a garage ventilation system. See Chapter 13, *Mechanical, Plumbing, and Electrical Systems.*

- Use natural roofing products. See Chapter 7, *Roof Construction and Materials.*

INTERNATIONAL ENERGY CONSERVATION CODE

States and local jurisdictions added energy-saving measures to building codes in an effort to achieve better energy efficiency. Energy codes have been successful in reducing energy consumption through the use of energy-efficient windows, added insulation, and improved air-infiltration barriers. States and local jurisdictions normally adopt national model energy codes as a basis for their own codes. In 1998, the first International Energy Conservation Code (IECC) was published by the International Code Council (ICC). The IECC introduced the most significant efficiency increase in the history of the national model energy code by specifying practices that make new and renovated buildings more energy efficient than past practices. Most local jurisdictions have adopted the IECC as the foundation for their energy codes. Some states and local jurisdictions have modified the IECC to deal with specific regional or local concerns.

> **NOTE**
>
> Building codes are comprehensive and detailed, and local codes can vary from the international codes. A complete examination of codes is beyond the scope of this textbook. You should consider studying local and national building codes, or taking codes courses that can be offered at your local trade schools, community colleges, or local jurisdictions.

ENERGY-EFFICIENT CONSTRUCTION PROGRAMS

Architects today often use design and construction options that provide sustainable high-quality insulation, air-infiltration barriers, proper caulking, energy-rated windows, quality materials, and energy-efficient mechanical equipment. The following describes energy conservation architectural design and construction programs and practices.

USGBC Leed Program

The United States Green Building Council (USGBC) members came together to create a standard for **green building,** called the Leadership in Energy and Environmental Design (LEED) Green Building Rating System. This rating system is a national standard for developing energy-efficient **sustainable buildings.** Green building refers to a structure and construction processes that are environmentally responsible and resource-efficient throughout a **building life cycle.** Building life cycle refers to the observation and examination of a building over the course of its entire life, including design, construction, operation, maintenance, renovation, and demolition. Sustainable buildings are buildings capable of maintaining their desired function into the future. The terms **green building** and **sustainable building** can be used interchangeably. The following LEED goals represent the approach to sustainability, as taken from the USGBC website:

- Define green building by establishing a common standard of measurement.
- Promote integrated, **whole-building** design practices. The term **whole building** refers to the building assembly designed and built to maximize sustainable and economic function through the use of energy and other resources, building materials, site preservation, and indoor air quality for a structure to run at its maximum efficiency, provide a comfortable and healthy environment, and have the minimum impact on the environment.
- Recognize environmental leadership in the building industry.
- Stimulate green competition.
- Raise consumer awareness of green building benefits.
- Transform the building market.

The LEED building certification program recognizes commercial construction projects that demonstrate the highest sustainability performance standards. The LEED certification requires that the project meet some mandatory requirements, and earn additional points for implementing high-performance, healthy and sustainable building practices. The points scored are measured using the LEED rating system. In the LEED program, various levels of certification are awarded to new buildings and rehabilitated structures, based on their point value. When completing their construction and operational objectives, a building can earn enough points to be awarded certified, silver, gold, or platinum levels. For more information about the LEED program, visit the U.S. Green Building Council website at www.usgbc.org/leed.

The Construction Product Sustainability Information Reporting Guide

The Construction Specifications Institute (CSI) and Construction Specifications Canada (CSC) publish detailed construction **specification** formats and procedures that are described in Chapter 1, *Commercial Plans, Specifications, and Construction Management.* A specification is an exact statement describing the characteristics of a particular feature of the project. The CSI developed the *GreenFormat: The Construction Product Sustainability Information Reporting Guide.* GreenFormat is a CSI system that allows manufacturers to accurately report the sustainability measuring properties of their products and provides designers, contractors, and building operators with basic information to help meet green requirements. When using GreenFormat, construction product manufacturers complete an online GreenFormat reporting questionnaire that collects the sustainable information about their product. Data collected is presented in a system for access when making sustainability design decisions. Access to the GreenFormat report is available through www.greenformat.com. Users can print reports on specific products with information grouped in categories covering product sustainability. New topics are added and outdated topics are removed as sustainability issues evolve. The structure can be applied to all construction products and product categories.

Green Building Initiative

Green Building Initiative® (GBI) is a nonprofit organization that owns and administers the Green Globes® green building assessment and certification in the United States and Canada. GBI was established in 2004 and is headquartered in Portland, Oregon. The GBI is dedicated to accelerating the adoption of building practices that result in energy-efficient, healthier, and environmentally sustainable buildings. The GBI seeks to be innovative and provide responsive customer service by collectively moving toward a sustainable built environment. GBI recognizes that credible and practical green building approaches for commercial and governmental construction are critical in this effort. Responding to the reality that one size does not fit all in sustainable improvements, GBI seeks to create a tailored approach to sustainability that takes into account building type, purpose, and occupants.

GBI became the first Building Rating Organization to become an American National Standards Institute (ANSI) Developer. GBI used the ANSI process, recognized for being open, balanced, and consensus-based, to create ANSI/GBI 01-2010: Green Building Assessment Protocol for Commercial Buildings out of the Green Globes environmental design and assessment rating system for new construction. In 2014, GBI began the process of updating the ANSI/GBI 01-2010 standard to keep the Green Globes New Construction Rating System timely and consensus based. Green Globes identifies opportunities and provides effective tools to achieve success. A nationally recognized green rating assessment, guidance and certification program, Green Globes works with the construction industry to realize sustainability goals for new construction projects, existing buildings and interiors.

Green Globes is a science-based building rating system that supports a wide range of new construction and existing building project types. Designed to allow building owners and managers to select which sustainability features best fit their building and occupants, Green Globes recognizes projects that meet at least 35 percent of 1000 available evaluation points. The Green Globes custom-tailored approach helps produce the most sustainable outcomes based on building type and purpose. The Green Globes cloud-based program provides project team members personal access and upload information. This minimizes coordination time and shares documentation tasks.

Green Globes provides support from an expert dedicated to each project. Green Globes third-party assessors interact with project teams and building owners in real time to create partnerships throughout the process. The assessors are skilled in green building design best practices, engineering, construction, and facility operations. Assessors review documentation, answer questions, conduct on-site building assessments, and offer insightful suggestions to enhance sustainable practices. The assessment concludes in providing a project with a score of up to 1000 points. The percentage of 1000 points given can grant one of the following Green Globes ratings:

Four Green Globes—85 to 100 percent: Demonstrated national leadership and excellence in the practice of water, energy, and environmental efficiency to reduce environmental impacts

Three Green Globes—70 to 84 percent: Demonstrated leadership in applying best practices regarding energy, water, and environmental efficiency

Two Green Globes—55 to 69 percent: Demonstrates excellent progress in achieving reduction of environmental impacts and use of environmental efficiency practices

One Green Globes—35 to 54 percent: Demonstrates a commitment to environmental efficiency practices

COMMERCIAL CONSTRUCTION GREEN BUILDING APPLICATION

This commercial construction application is an example of quality sustainable design and construction practices that can be achieved with the adoption of building practices that result in energy-efficient, healthier, and environmentally sustainable buildings.

University of Florida Health Hospitals Rely on Green Globes to Realize Their Full Potential

Recent years have seen tremendous growth at University of Florida Health (UF Health). To respond to increasing needs, the organization recently expanded through the addition of the new, state-of-the-art Heart & Vascular and Neuromedicine Hospitals on the UF Health campus.

"UF Health is committed to being the best academic health center in the Southeast and beyond," explains Patrick Spoden, the project's architect and project manager at Flad & Associates. Combined, the two towers' 540,000 ft^2 includes 15 operating rooms, 216 private rooms, and an adjacent parking structure.

Green Globes certification was a great fit for the organization's goals says John Chyz of AEI Affiliated Engineers, who was the project's Lead Sustainability Consultant. "Green Globes is a nationally recognized framework with a robust rating system, and we used its quantifiable data to help guide the project's sustainable strategies," Chyz recalls. The process encouraged the team to push themselves in several areas. For example, while the contractor had a construction waste management plan, Green Globes criteria inspired them to find a drywall recycler for leftover drywall. "We also decided to specify **fly ash** content in our concrete mix designs for the building core and shell, which was partially driven by the life cycle performance path in Green Globes," Chyz says. Fly ash is a fine powder that is a by-product of burning pulverized coal in electric generation power plants. Fly ash forms a compound similar to Portland cement when mixed with lime and water. Fly ash improves the strength and segregation of the concrete and makes concrete easier to pump. Fly ash is an environmentally friendly construction material, because it is a by-product. Replacing Portland cement with fly ash could significantly reduce carbon emissions associated with construction Portland cement has a very high embodied energy because its production requires a great deal of heat. Fly ash is also recognized as an environmentally friendly material because it is a by-product. Replacing Portland cement with fly ash reduces the greenhouse gas emissions when compared with Portland cement production. The production of 1 ton of Portland cement generates approximately 1 ton of CO_2, compared with no CO_2 generated with fly ash production.

The hospitals' award of Four Green Globes confirms the success of UF Health and the team's efforts. The project claims the following intelligent solutions:

- An extensive green roof area helps mitigate storm runoff.
- A municipally reclaimed water source is used for all irrigation and supplements cooling tower makeup water.
- A closed-loop chilled water system cools the sterilizer equipment, resulting in huge process-water savings.
- In-depth acoustic studies led to a sound masking system that reduced the need for full-height gypsum walls in certain areas.
- A combined heat and power system provides power and hot water for the hospital, and the building equipment runs on high-efficiency motors and variable frequency drives.

High Performance

According to Chyz the Green Globes assessor delivered insightful feedback to the project. During the Stage I Assessment, the assessor assisted the team through the energy-modeling process to ensure the team properly modeled and accounted for the combined heat and power system. The assessor also encouraged the team to explore the **CO_2 equivalent emissions path** in the energy performance criteria, which helped raise its Green Globes standing. Carbon dioxide equivalent emissions path is a standard for measuring carbon footprints by express the impact of greenhouse gas in terms of the amount of CO_2 created using a ratio to convert the various gases into equivalent amounts of CO_2.

"Our Assessor truly understood the complexities involved in healthcare projects," Chyz remarks. "Having the opportunity to share a dialog with an expert in our industry and draw upon him as a resource to get where we wanted to go proved to be a refreshing and rewarding experience."

Chyz kept the assessor's suggestions handy as the project moved toward its Stage II Assessment. "Whenever the owner wanted to push the envelope a bit further on the project, we had those suggestions readily available to discuss additional options with them," Chyz says.

Patient welfare and positive outcomes were the top priority, and the team also included sustainable measures with proven health benefits like **low-VOC materials** and patient-room access to daylight and views. Patients also have technology that allows them to manage their personal environments. "All these things improve the patient experience and ideally lead to faster recovery times," Spoden notes. "With our Four Green Globes rating, UF Health can clearly demonstrate that these hospitals have the best interests of the environment, community and patients in mind." Low volatile organic compounds (VOC) materials refers to paints, sealants, adhesives, and cleaners and other products that have a very low or zero VOC that are not harmful to the environment and humans. The completed UF Health project is shown in **Figure 2.1**.

Energy Star

Energy Star is a joint program of the U.S. Environmental Protection Agency (EPA) and the U.S. Department of Energy created to help save money and protect the environment through energy-efficient products and practices and through superior product energy efficiency. Energy Star-qualified construction must meet strict guidelines for energy efficiency set by the EPA. These projects include additional energy-saving features that typically make them 20 to 30 percent more efficient than standard construction. Energy Star is independently verified to meet the requirements. Energy savings are based on heating, cooling, and hot water energy use and are typically achieved through a combination of

- **Building envelope** upgrades: The building envelope consists of the roof, exterior walls, and floor of a structure, forming a barrier that separates the interior of the building from the outdoor environment.
- High-performance windows.
- Controlled air infiltration.
- Upgraded heating and air-conditioning systems.
- Tight duct systems.
- Upgraded water heating equipment.

These features contribute to improved quality, comfort, lower energy demand, and reduced air pollution. Energy Star also encourages the use of energy-efficient lighting,

Figure 2.1 The completed UF Health project. (*Courtesy of the Green Building Initiative and Mark Herboth Photography LLC.*)

appliances, and features designed to improve indoor air quality. Verification of energy efficiency by a third party is an essential step in getting the Energy Star label and certificate. Verification is generally dependent on the construction method using a **Builder Option Package (BOP)**. A BOP represents a set of prescriptive construction specifications for a specific climate zone, based on performance levels for the thermal envelope, insulation, windows, orientation, HVAC system, and water heating efficiency for the climate zone.

NET-ZERO ENERGY OR NET METERING

Net-zero energy can be achieved with the help of energy-producing systems, such as solar collectors, photovoltaic modules, geothermal systems, wind generators, small-scale hydroelectric generators, or biopower. **Net-zero energy** means that the installation generates as much or more energy than the facility consumes through the course of a year. To accomplish this, the installation is connected to the utility grid. The energy is fed into the power grid when more energy is produced than needed, and the facility draws power back from the grid when not producing enough electricity. The first step to reaching net-zero is consumption reduction. The lesser power used, the lower the amount of power needs to be generated. The following provides a variety of commercial construction applications and systems that are used to help achieve net-zero energy.

SOLAR ENERGY DESIGN

Solar energy is produced by sunlight that can be captured when it is transferred to something that has the ability to store heat or energy. Solar energy can be captured and regulated in order to heat water, air, the **thermal mass** of a building, or create electricity. Thermal mass is a dense material that can effectively absorb and store heat, and release the heat as the building cools at night. Most areas of the earth receive about 60 percent direct sunlight each year, and in very clear areas, up to 80 percent of the annual sunlight is available for use as solar energy. When the sun's rays reach the earth, the air and the features on the earth become heated through **thermal radiation.** Thermal radiation is the heat that comes from the sun to the earth or emitted by other heat sources. Certain dense materials such as concrete can absorb more heat than less dense materials such as wood. During the day the dense materials absorb and store solar energy. Then at night, they release the energy in the form of heat. Some substances, such as glass, absorb thermal radiation while transmitting light. This is the basic concept that makes solar heating possible. Solar radiation enters a structure through a glass panel and warms the surfaces of the interior areas. The glass along with insulating window coverings keeps the heat inside by absorbing the radiation.

Two basic commercial uses for solar energy are heating spaces and hot water. Solar space heating systems are passive, active, or a combination of passive and active. **Passive solar systems,** also called **architectural solar systems,** use no mechanical devices to retain, store, or radiate solar heat. **Active solar systems,** also called **mechanical solar systems,** use mechanical devices to absorb, store, and use solar heat. A combination of passive and active solar systems retain, store, or radiate solar heat with the aid of some mechanical devices.

The potential reduction of fossil fuel energy consumption and free availability of natural sunlight can make solar heat an economical alternative. A number of factors contribute to the effective use of solar energy, including energy-efficient construction techniques and full insulation of a building to reduce heat loss and air infiltration.

An auxiliary heating system is often used as a backup or to supplement solar heat. The amount of heat needed from the auxiliary system depends on the effectiveness of the solar system and the building design. The primary and the supplemental systems should be professionally engineered for optimum efficiency and comfort.

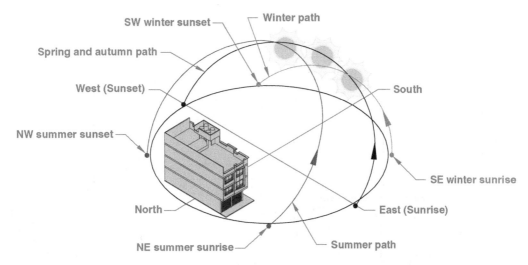

Figure 2.2 A southern exposure provides the best site orientation for solar construction.

A southern exposure provides the best site orientation for solar construction. A perfect solar site allows the structure to have an unobstructed southern exposure. Figure 2.2 shows the sun's path in relationship to the southern exposure of a building on a solar site.

Interior space planning is an important factor to consider for taking advantage of solar heat. It is best to place most commonly used areas that are frequently occupied during the daytime along south-facing walls. In the northern hemisphere, buildings with major glass walls facing south receive daytime sunlight. Communal spaces, such as the meeting spaces, offices, and lobbies, should be on the south side of the building, while inactive areas, such as bath rooms, copy rooms, computer hardware, and laundries, should be located on the north side where a cooler environment is normally desirable. Also, the parking garage should be placed on the north, northeast, northwest side, or under the building, where it acts as a barrier insulating the high use areas from cold exterior elements. The inactive areas provide shelter from the colder northern exposure. Concrete and masonry walls also allow the building to be built into a slope on its northern side, or berms constructed on the north side as shelter from the elements. Figure 2.3 shows a floor plan that has good interior design for solar orientation.

Living with Solar Energy Systems

Solar energy and energy-efficient construction require a commitment to conserving energy. Architects and clients must evaluate cost against potential savings and become aware of the responsibility of living with energy conservation. Solar systems can be designed that provide some heat from the sun and require little or no involvement from the owner to assist the process. Such a minimal energy-saving design is worth the effort in many cases. Active solar space heating systems are available that can provide a substantial amount of needed electrical energy. These systems are automatic and also do not generally require involvement from the owner, although there are operation procedures and maintenance schedules that need to be conducted. Alternately, some passive solar heating systems require participation by the owner. For example, a mechanical shade can be used by the owner to block the

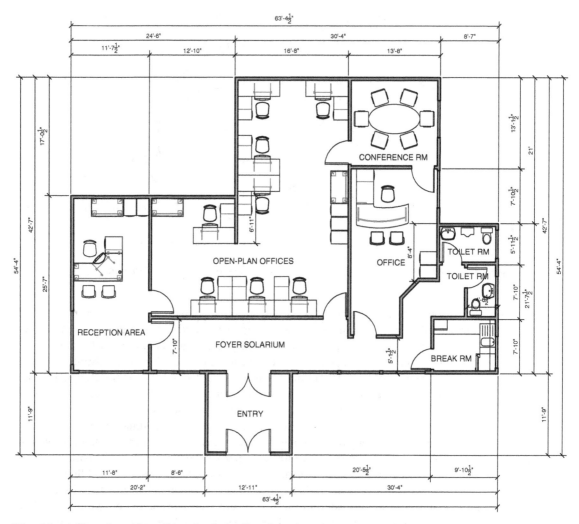

Figure 2.3 A floor plan with good interior design for solar orientation.

summer sun rays from entering southern-exposure windows. The interior heat should be retained as long as possible during the winter months when the sun is heating the high use areas. In the morning the owner needs to open window coverings to allow sunlight to enter and heat the rooms. In the early evening, before the heat begins to radiate out, the window coverings are closed to keep the heat in longer. These functions can be automatic and set on appropriate timers.

Codes and Solar Rights

Building permits are generally required for the installation of active solar systems or the construction of passive solar systems. Some installations also require plumbing and electrical permits. Verify the exact requirements for solar installations with local building officials. During the initial planning process, and check the local zoning ordinances. For example, some areas restrict building height. If the planned solar system encroaches on this zoning rule, then a different approach or a **variance** to the restriction must be considered. A variance is a request to depart from zoning requirements. An approved variance permits the owner to use the land in a manner not otherwise permitted by zoning ordinances. A variance is not a change in zoning but is a waiver of zoning requirements for the specific application.

Access to sunlight is not always guaranteed. A solar project can be built in an area that has excellent solar orientation, and then a few years later, a tall structure that blocks the sun may be built on adjacent property. Adjacent property trees can also grow tall and reduce sunlight. Determine the possibility of such problems during initial design stages. Some local zoning ordinances, laws, or even deed restrictions can protect the right to light. In the past, laws generally provided the right to receive light from above the property but not from across neighboring land. This situation is changing in many areas of the country.

Passive Solar Systems

In passive solar architecture, the structure is designed so the sun directly and indirectly warms the interior. A passive solar system allows sunlight to enter the structure and be absorbed into a structural mass. The stored heat then warms the inside space during the day and also when the sun has receded. Window coverings are used to control the amount of sunlight entering, and vents help control temperature. The structure is the storage system in passive solar construction. The amount of material needed to store heat depends on the amount of sunlight, the desired temperature within the structure, and the ability of the material to store heat. Materials, such as water, steel, concrete, and masonry, have good heat capacity. Several passive solar architectural methods have been used individually and together, including south-facing glass, thermal storage walls, roof ponds, solariums, and efficient envelope construction.

South-Facing Glass, Solariums, and Thermal Mass

Direct solar gain is heat created by the sun, which is captured in a structure through south-facing windows. Large window areas facing south can provide up to 60 percent of a building's heating needs when the windows are insulated at night with tight-fitting insulated coverings. Heat gain during the day is quickly lost when window insulation is not used at night. It is important for energy from the sun to be retained in floors constructed of or covered with a high-heat-absorbing material such as tile, brick, or concrete, and special walls made of concrete or masonry or containing water tubes. Figure 2.4 shows a typical application of direct-gain south-facing glass.

Active Solar Systems

The future looks bright for solar technologies. The costs and efficiencies of high-technology systems are improving. Solar alternatives should be considered as conventional heating and cooling costs rise and become more concern about oil and gas shortages increase. A qualified solar engineer should evaluate the site and recommend solar design alternatives when preparing a preliminary design. Additional assistance is usually available from the state or the national Department of Energy.

Active solar systems are part of a group of **green power** systems. Green power is electricity generated from resources, such as solar, wind, geothermal, biomass, and low-impact hydro facilities.

Active solar systems for space heating use collectors to gather heat from the sun. The heat in the collectors is transferred to a fluid, and moved to an area of heat storage.

Figure 2.4 Typical application of direct-gain south-facing glass on a small office building. (*Courtesy of Chief Architect Software.*)

Heat stored in the system is transported to the interior space by transferring the heat to air and transferring it though insulated ducts, or by transferring the heat to water, and pumping the water through plumbing fixtures. **Ducts** are the pipes used to move hot or cold air in a heating, air-conditioning, and ventilation system. An active solar heating system generally requires a backup heating system that is capable of handling the entire heating needs of the building when solar gain is reduced to a minimum. A suitable backup system can include a forced-air system that uses the same ducts as the solar system, or a ductless heat pump system. Heating and cooling systems are described in Chapter 13, *Mechanical, Plumbing, and Electrical Systems.* The backup system is generally fueled by electricity, natural gas, or oil.

Solar Collectors

Solar collectors convert sunlight to heat and can be nearly 100 percent efficient. The number of solar collectors needed to provide heat depends on the size of the building and the volume of heat needed. These are the same requirements for sizing a conventional heating system. Collectors are commonly placed in rows on a roof or on the ground near the structure, as shown in Figure 2.5. The best solar collector placement requires an unobstructed southern exposure. The angle of the solar collectors should be in proper relationship to the angle of the sun during winter months, when the demand for heat is greatest. The best angle for the solar collector is when the winter sun hits at about 90° to the collector. In general, the solar collector angle should roughly match the **latitude** of your location. Latitude is an angle measured from the point at the center of the earth, and lines of latitude are imaginary lines running east to west around the earth. The equator has a latitude of 0° and the North Pole has a latitude of 90°, so your building has a latitude of 45° if you are half way between the equator and the North Pole. The degree symbol is (°). There are several websites that can help you determine the best angle for solar collector installation in your area by searching the Internet using

Figure 2.5   Solar collectors are commonly placed in rows on a roof or on the ground near the structure.

words such as *determine solar collector angle*. Professional solar collector installers also know the correct angle of installation in your area. Some solar collectors are designed to get solar heat by direct and reflected sunlight. Reflecting, or focusing, lenses can help concentrate sunlight on the collector surface. Figure 2.6 shows an example of solar collector angle.

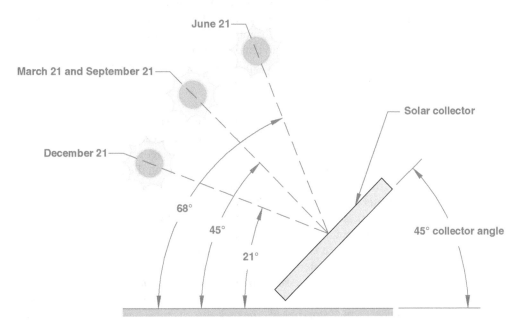

Figure 2.6   Preferred solar collector angle. (*Previously published in Modern Residential Construction Practices, Routledge, 2017 and reproduced here with permission.*)

Photovoltaic Modules

Photovoltaic (PV) cells turn light into electricity. The word **photovoltaic** comes from the Greek word *photo,* meaning light, and *voltaic,* meaning to produce electricity by chemical action. In the photovoltaic cells, photons strike the surface of a **silicon wafer,** which is a semiconductor diode that stimulates the release of electric charges that are guided into a circuit where they become a useful electric current. Photovoltaic modules produce direct current (DC) electricity. This type of power is useful for many applications and for charging storage batteries. An **inverter** is used to change DC current to alternating current (AC). AC is the type of current that powers electrical systems. Some photovoltaic systems use the DC current immediately to power DC motors as in hot water-pumping systems. The energy produced from photovoltaic cells must be stored when solar electric systems are not used immediately, or if an energy reserve is required for use when sunlight is not available. Batteries are the most common storage devices, allowing the stored electricity to be used when needed. DC and AC systems can be used to supply the electrical needs, and AC systems can be connected to a **utility grid** where electricity can be shared. **Utility grid,** also referred to as **grid,** is the transmission system for electricity that is a network of coordinated power providers and consumers that are connected by transmission and distribution lines and operated by one or more control centers. The system can draw on the grid for extra electricity if needed during times of peak power usage. The system can also return extra unused power to the grid where the electricity can be purchased by the utility, providing income or electricity credits for the owner. In most areas of the country, the grid is required to purchase excess power from private sources. Because battery storage is expensive and space consuming, a municipally connected solar electric system is the most popular and least expensive way to take advantage of solar power.

The solar modules need to be located and positioned where they receive maximum exposure to direct sunlight for the longest period of time every day. It is also important to keep distances to **electrical loads** to a minimum. The electrical loads are the circuits, appliances, and equipment that use electricity. It is also important to confirm that shade from buildings and trees do not block the sunlight. Photovoltaic solar collectors are constructed by placing individual photovoltaic cells in groups called modules, and the modules are combined in groups of six to create the photovoltaic collector, as shown in Figure 2.7.

Storage

Active solar collectors transfer heat energy to a storage area and then to the interior space during periods of sunlight. After the demand for heat is met, the storage facilities allow the heat to be contained for use when solar activity is reduced, such as at night or during cloud cover. The kind of storage facilities used depends on the type of collector system. There are water, rock, and chemical storage systems.

Water storage is excellent because water has a high capacity for storing heat. Water in a storage tank is used to absorb heat from a collector. When the demand for heat exceeds the collector output, the hot water from the storage tank is pumped into a radiator, or through a water-to-air heat exchanger for dispersal to a forced-air system. Domestic hot water can be provided by a water-to-water heat exchanger in the storage tank. A **heat exchanger** is a device for transferring heat from one medium to another.

Rock storage is often used when the collectors contain air rather than water. Heated air flows over a rock storage bed. The rock absorbs some heat from the air while the rest of the heat is distributed to the interior space. When the solar gain is minimal, cooler air passes over the rock storage, where it absorbs heat and is distributed by fans to the living space.

Figure 2.7 Photovoltaic solar collectors are constructed by placing individual photovoltaic cells in groups called modules, and the modules are combined in groups of six to create the photovoltaic collector.

Domestic hot water can also be provided by an air-to-water heat exchanger located in the rock storage or in the hot air duct leading from the collector.

Chemicals used in collectors and in storage facilities absorb large amounts of heat at low temperatures. Chemicals with very low freezing temperatures can absorb heat when the outside temperature is low during winter months.

Solar Thermal Water Heating

Solar water heating (SWH) or **solar hot water (SHW)** systems comprise several innovations and many renewable energy technologies that have been well established for many years. SWH uses solar radiation to heat water or air in buildings. SWH systems are **close-coupled** or **pump-circulated**, and active or passive. A close-coupled system has a horizontally mounted storage tank directly above the solar collectors. No pumping is required because the hot water naturally rises into the tank. A pump-circulated system has a ground- or floor-mounted storage tank below the collectors. In this system, a circulating pump moves water or heat transfer fluid between the collectors and the tank.

SWH systems can deliver hot water over most of the year. There may not be enough solar heat gain to deliver sufficient hot water during winter months in some locations. A gas or electric booster is normally used to heat the water during these periods. An active SWH system heats water by using a collector, on the roof, on the ground, or on a wall facing the sun that heats working fluid that is pumped. A passive SWH system moves the product by natural convection. Heat is stored in a hot water storage tank with either system. The solar hot water storage tank is normally larger than in conventional water heating because of the need to store more hot water during inactive times. The heat transfer fluid (HTF) for the absorber can be the hot water from the tank, but is commonly a separate loop of fluid containing antifreeze and a corrosion inhibitor that delivers heat to the tank through a heat exchanger made of a coil of copper tubing within the tank.

SWH systems commonly have a backup electric, gas, or fuel oil central water heating system that is activated when the water in the tank falls below a minimum temperature setting. The backup water heat system allows the hot water system to work all year in cooler climates.

Architectural Features Associated with Solar Systems

Solar systems can be enhanced by incorporating specific design characteristics and construction materials into the project. Some of these design characteristics and construction materials can be added to the design without increasing the cost of the project, while other options can be expensive. The enhancements can include an air-lock entry, roof overhang considerations, clerestory windows, skylights, light shelves, thermal storage walls, roof ponds, green roofs, solariums, and solar architectural concrete products.

Air-Lock Entry

An energy-efficient design element is an **air-lock entry,** known as a **vestibule.** This is an entry that provides a chamber between an exterior and interior door to the building. Commercial buildings are required to have a vestibules on primary entrance doors leading to and from spaces greater than or equal to 3000 ft² (298 m²). Office buildings with lobbies connected to corridors with a combined total area of 3000 ft² require only those areas that cannot be closed off from the entrance door to be included when a vestibule is required. The vestibule is designed so the interior and exterior doors are not open at the same time, by a distance between doors of at least 7 ft (2100 mm). This distance forces the occupants to close one door before reaching the other door. Vestibule design variables should be researched for these designs. The main idea behind an air-lock entry is to provide a chamber that is always closed to the interior area by a door. When the exterior door is opened, the air lock loses heat, but the heat loss is confined to the small space of the vestibule and the warm air of the interior area is exposed to minimum heat loss. Figure 2.8 shows the portion of a floor plan demonstrating an air-lock entry. A revolving door system can be designed as a commercial entrance option in addition to the vestibule. **Revolving doors** provide an entry system that is always open, an always closed at the same time. Revolving door systems eliminate drafts, while reducing noise and air pollution.

Roof Overhang

The sun angle changes from season to season. The sun is lower on the horizon in the winter and higher in the summer. An overhang can shield a major glass or even wall area from the heat of the summer sun and also allow the lower winter sun to help warm the occupants.

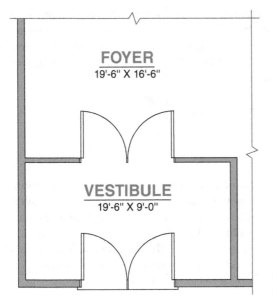

Figure 2.8 Portion of a floor plan demonstrating an air-lock entry.

Figure 2.9 shows an example of how a properly designed overhang can aid in the effective use or protection from sun heat. An overhang that provides nearly 100 percent window shading at noon on the longest day of the year can be calculated with a formula that divides the window height by a factor determined in relationship to **latitude:**

$$\text{Overhang} = \frac{\text{Windowsill height}}{\text{F}}$$

North Latitude	F		North Latitude	F
28°	8.4		44°	2.4
32°	5.2		48°	2.0
36°	3.8		52°	1.7
40°	3.0		56°	1.4

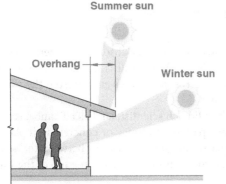

Figure 2.9 An example of how a properly designed overhang can aid in the effective use or protection from sun heat. (*Previously published in Modern Residential Construction Practices, Routledge, 2017 and reproduced here with permission.*)

The following is an example used to calculate the recommended southern overhang for a location at 36° latitude and provide for a standard window height of 6'-8". Latitude is the angular distance between an imaginary line around the earth parallel to the equator and including the equator. F is a fixed value established at various degrees of north latitude shown in the previous table.

$$\text{Overhang (OH)} = \frac{6'\text{-}8'' \text{ (Window height)}}{3.8 \text{ (F at 36° latitude)}}$$

$$= \frac{6.6}{3.8} = 1.8 \approx 1'\text{-}10''$$

Overhang protection can be constructed in ways other than a continuation of the roof structure. For example, an awning, porch cover, or trellis can be built to serve the same function as an overhang. Alternative methods of shading from summer heat and exposing window areas to winter sun can be achieved with mechanical devices. These movable devices require the occupant to be aware of the need for shade or heat at different times of the year. Figure 2.10 shows mechanical shading options used to protect windows from summer heat and exposing window areas to winter sun.

Clerestory Windows

Clerestory windows can be used to provide light and direct solar gain to a second or higher floor area and increase the total solar heating capacity. Clerestory windows are a row of windows set along the upper part of a wall. Clerestory windows can also be used to help ventilate a structure during the summer months, when the need for cooling is greater than the need for heating. Clerestory window applications are often found on contemporary architectural styles.

Skylights

Skylights can be used effectively for direct solar gain. A skylight is a window in a roof used to allow sunlight to enter. Skylights placed on a south-sloping roof provide direct solar gain during the winter. Skylights can cause the area to overheat in the summer unless ventilation or a shade cover is provided. Some manufacturers have skylights that open and can provide sufficient ventilation. Figure 2.11 shows an application of an operable skylight.

Light Shelves

Light shelves are often placed underneath clerestory windows to redirect light and solar energy upwards. **Reflected light** is light bounced off a ceiling that has a uniform quality to help reduce the need for artificial lighting. In solar systems, light shelves can be used much in the same manner as roof overhangs, to protect the interior from overheating during summer months, and allowing sunlight during winter.

Thermal Storage Walls

Thermal storage walls are constructed of heat-absorbing material, such as concrete, masonry, or water-filled cylinders. The storage wall is usually constructed across south-facing windows. The thermal wall receives and stores energy from the sun during the day and releases the heat slowly at night.

The **Trombe wall**, designed by a French scientist, Dr. Felix Trombe, is a thermal storage wall constructed as a massive dark-painted masonry or concrete wall placed a few inches inside and next to south-facing glass. The sun heats the air between the wall and the glass. The heated air rises and enters the room through vents at the top of the wall. At the same

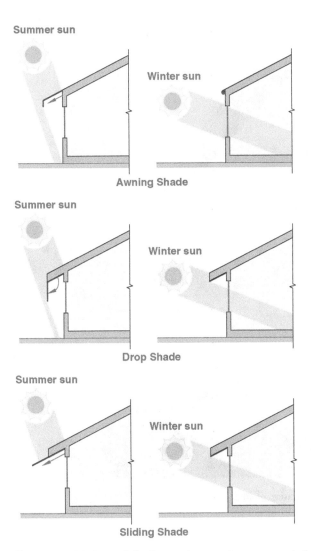

Summer sun

Winter sun

Awning Shade

Summer sun

Winter sun

Drop Shade

Summer sun

Winter sun

Sliding Shade

Figure 2.10 Mechanical shading options used to protect windows from summer heat and exposing window areas to winter sun. (*Previously published in Modern Residential Construction Practices, Routledge, 2017 and reproduced here with permission.*)

time, cool air from the floor level of adjacent areas is pulled in through vents at the bottom of the wall. The vents in the Trombe wall must be closable to avoid losing the warm air at night. The heat absorbed in the wall during the day radiates back into the room during the night hours. The Trombe wall also acts to cool the structure during summer. This happens when warm air rises between the wall and glass and is vented to the outside and air currents work to pull cooler air from an open north-side window or vent. Figure 2.12 shows how the thermal storage wall system functions to help heat the home during winter months and cool during the summer. Large vertical water-filled tubes or drums painted dark to absorb heat can also be installed as the Trombe wall. The water tubes store heat during the day and release heat at night.

Figure 2.11   An application of an operable skylight.

Roof Ponds

A **roof pond** is usually constructed of containers filled with antifreeze and water on a flat roof. The water is heated during the winter days, and then at night the structure is covered with an insulated blanket, which allows the absorbed heat to radiate into the interior space. This process functions in reverse during the summer, when the water-filled units are covered with insulation during the day and uncovered at night to allow stored heat to escape. It is best for the building to be constructed of a good thermal-conducting material such as steel to assist the radiation of heat at night. Figure 2.13 shows an example of the roof pond system.

Green Roofs

A **green roof** is also known as a **rooftop garden**. The green roof has plants over the roof structure to help reduce building temperatures, filter pollution, and lessen water runoff. A green roof reduces the **urban heat island effect** and helps temper heating and cooling loads in the building. The urban heat island effect means that city areas are warmer than suburbs or rural areas due to less vegetation, more land coverage, and other infrastructure. The green roof also slows stormwater runoff from the roof and lessens the load on the city's wastewater system. A green roof installation is shown in Figure 2.14.

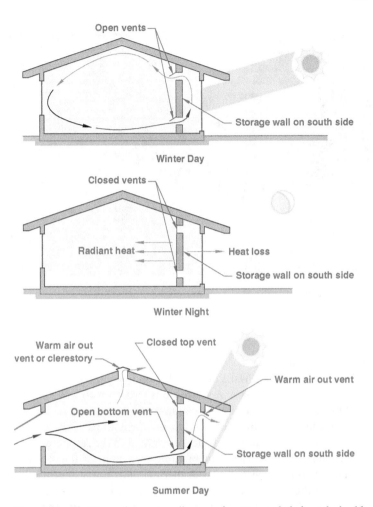

Open vents

Storage wall on south side

Winter Day

Closed vents

Radiant heat

Heat loss

Storage wall on south side

Winter Night

Warm air out vent or clerestory

Closed top vent

Warm air out vent

Open bottom vent

Storage wall on south side

Summer Day

Figure 2.12 The thermal storage wall system functions to help heat the building during winter months and cool during the summer. (*Previously published in Modern Residential Construction Practices, Routledge, 2017 and reproduced here with permission.*)

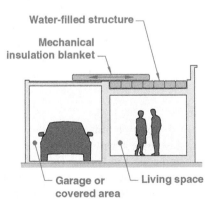

Water-filled structure

Mechanical insulation blanket

Garage or covered area

Living space

Figure 2.13 Roof pond system. (*Previously published in Modern Residential Construction Practices, Routledge, 2017 and reproduced here with permission.*)

Figure 2.14   A green roof.

Solariums and Sunrooms

A **solarium** is a room with walls of glass that go beyond the eaves. Solariums often have roofs made entirely out of glass. A **sunroom** is a room with walls of glass that stop at the eaves and normally have traditional looking roofs. A solarium or sunroom is constructed on the south side of a building next to high-use areas and an excellent place for a cafeteria or lounge. Solar energy is absorbed and transmitted to the rest of the building through the solarium. A solarium can extend the building area by providing an outdoor-style atmosphere where plants grow well all year. A thermal mass can also be used in the solarium to store heat from the sun. Heat from the solarium can be circulated throughout the entire interior by natural convection or through a forced-air system. The circulation of hot air during the winter and cool air during the summer allows the solarium to function like a Trombe wall.

The possibility of the solarium overheating during summer days can be reduced by mechanical ventilators or a mechanical humidifier. Exterior shading devices can also be rolled down over the solarium glass area to help reduce overheating. Landscaping with southern **deciduous trees** shading, the solarium can also provide for necessary summer cooling. Deciduous trees lose their leaves in the winter. Figure 2.15 shows how a solarium can provide an excellent environment that combines the comfort of the indoors with the atmosphere of the outdoors and provides solar heat and circulates summer cooling.

Solar Architectural Concrete Products

Solar collectors can be built into a patio, driveway, tennis court, pool deck, side surface of a building wall or fence, or tile roof using **solar architectural concrete products.** Solar architectural concrete products absorb and collect heat from the sun and outside air and transfer the heat into water, glycol, or another heat-transferring fluid passing through embedded tubes. The finished surface can resemble products such as cobblestone, brick,

Figure 2.15 A solarium can provide an excellent environment that combines the comfort of the indoors with the atmosphere of the outdoors and provides solar heat and circulates summer cooling.

or roof tiles. They are manufactured from a specially formulated mixture that is strong, dense, very conductive, and waterproof. **Glycol** is ethylene glycol or ethylene alcohol that is a colorless, sweet liquid used as antifreeze and solvent.

GEOTHERMAL SYSTEMS

Geothermal heating and cooling systems are mechanically similar to a conventional **air source heat pump,** except that they use available water to cool the refrigerant or to extract heat, rather than using outside air for cooling or heating as with an air source heat pump. The conventional heat pump is described in Chapter 13, *Mechanical, Plumbing, and Electrical Systems*. The difference between air source and ground or water source temperature is that air temperature can change much more than ground or water temperatures. This gives the ground or water source heat pump more capacity and makes it more efficient.

Water can store great amounts of geothermal energy because of its high **specific heat,** which is the amount of energy required to raise the temperature of any substance 1°F (−17°C). The specific heat of air is only 0.018, so it can absorb and release only 1/50 of the energy that can be released by water. To produce a given amount of heat, 50 times more air volume than water is required to pass through the unit. Geothermal heating and cooling equipment is designed to use the constant, moderate temperature of the ground (50 to 55°F) to provide space heating and cooling, or domestic hot water, by placing heat exchangers in the ground, or in water wells, lakes, rivers, or streams. The basic types of geothermal systems are water and refrigerant based. The **water-based system** has **closed-loop** and **open-loop** options. The closed-loop system has water or water-antifreeze fluid pumped through polyethylene tubes. The open-loop system has water pumped from a well or reservoir through a heat exchanger and then discharged into a drainage ditch, a field tile, a reservoir, or another well. The **refrigerant-based system** is also known as direct exchange where refrigerant flows in copper tubing around a heat exchanger. Direct exchange represents only a small number of geothermal systems. A geothermal system operates by pumping groundwater from a supply well and then circulating it through a heat exchanger, where either heat or cold is transferred by a refrigerant. The water, which has undergone only a temperature change, is then returned through a discharge well back to the strata. Lakes, rivers, ponds, streams, and swimming pools can be alternate sources of water. A **pond-loop system** geothermal exchanger is inserted into a lake, river, or other natural body of water to extract heat by pumping the water up to the heat exchanger. To cool, the geothermal heat pump extracts heat from within the structure and transfers it to the fluid, where the heat dissipates into the ground as it circulates. Another alternative is to insert a geothermal **Freon** exchanger directly into a well, where the natural convection of the groundwater temperature is used to transfer hot or cold temperature as needed. Freon, also called R-22, is a trademark used for a variety of nonflammable gaseous or liquid fluorinated hydrocarbons employed primarily as working fluids in refrigeration and air-conditioning and as aerosol propellants. Sometimes the thermal transfer must be assisted by forcing the Freon to circulate through tubes within the well by a low-horsepower pump. In 1997, fluorocarbons were added to the **Kyoto Protocol,** and are known to deplete the ozone layer. Environmentally friendly building practices avoid fluorocarbon use. The Kyoto Protocol, also known as the Kyoto Accord, is an international treaty among industrialized nations that sets mandatory limits on greenhouse gas emissions. New Freon standards in the United States are part of the Clean Air Act, which is enforced by the U.S. Environmental Protection Agency (EPA). R22 Freon has been discontinued for use in new air conditioners and will no longer be manufactured by year 2020. Government regulations require replacement of R22 Freon with a less environmentally damaging substance **R410a,** referred to by a common brand name Puron®. The transition from R22 to R410a Freon requires Freon R-22 air conditioners and heat exchangers to be replaced. R410a has been approved for use in new air conditioners. R410a is a hydrofluorocarbon (HFC) which does not contribute to ozone depletion.

A geothermal system can assist a solar system that uses water as a heat storage and transfer medium. The water becomes the source for operating the geothermal system after being heated by solar collectors and has given up enough heat to reduce its temperature to below 100°F (38°C).

The earth can also be used to extract either heat or cold. This application is possible, because the ground maintains a constant temperature below the **frost line.** The frost line is the depth of frost penetrating the soil, and this depth varies with geographic area. Thermal extraction is done through the use of a closed-loop system consisting of polyethylene tubing filled with an antifreeze solution and circulated through the geothermal system. Another method is to use a vertical dry-hole well, which is sealed to allow it to function as a closed-loop system. Figure 2.16 shows the installation of a horizontal earth closed loop system, a vertical earth closed loop system, a pond or lake closed loop system, and a water well open loop system.

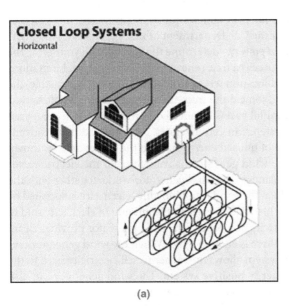

(a)

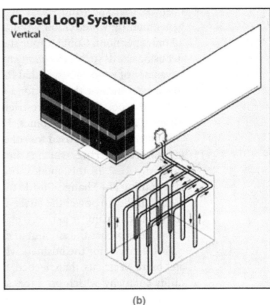

(b)

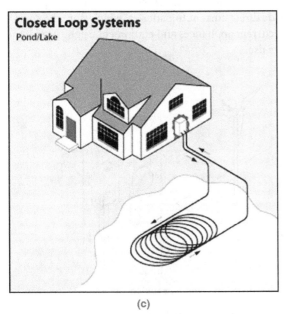

(c)

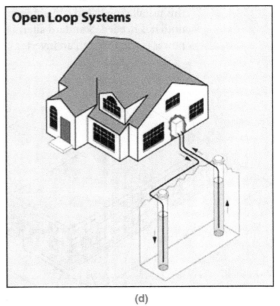

(d)

Figure 2.16 The installation of a horizontal earth closed loop system, a vertical earth closed loop system, a pond or lake closed loop system, and a water well open loop system. (*Courtesy of the US Department of Energy.*)

WIND ENERGY

Wind turbines capture energy from the wind. A wind turbine is turned by wind passing over propellers that power an electric generator, creating a supply of electricity. Large-scale applications use several technologically advanced turbines grouped together in what is referred to as a **wind farm.** This wind power plant generates electricity that is fed into a local utility for distribution to customers. Small-scale wind turbines are also available for business use. The wind speed in an area determines the effectiveness of possible power generation by wind. Good wind areas have an annual minimum average wind speed of 13 miles per hour. Contact your state or the U.S. Department of Energy, or search the Internet using words such as *location for wind power* to determine the feasibility in your area. The advantage of wind energy is that the source is a free, renewable, clean resource. Limitations to consider include the need for a good location where unobstructed wind is available, the initial cost, possible zoning restrictions, some noise from the turbines, and possible hazard for birds flying into the turbines. The initial cost is balanced out over time when compared with the cost of fuels used for other systems. In certain states, consumers can take advantage of **net-metering,** which is the sale of unused energy back to the power grid, as shown in Figure 2.17. In this application, the wind generator is connected to the electric meter and nothing else changes inside the business. The wind generator works together with the electric utility to power the facility. When the wind is not blowing, electricity is supplied by the utility company. A **net-zero** energy installation is when the amount of electricity sold to the utility matches the amount of electricity purchased. The wind generator provides clean, quiet electricity for the business when there is enough wind. When the wind generator creates more electricity than needed, the system allows the owner to sell electricity back to the utility company, which produces an **energy positive** system. The wind generator can also provide electric energy for remote locations, telecommunications sites, water pumping, and other rural applications, or when it is not possible to connect with the utility company. In this installation, wind generators provide direct current to batteries, which store the energy until it is needed. Standard alternating current appliances and equipment can be used if the power is run through an inverter before use.

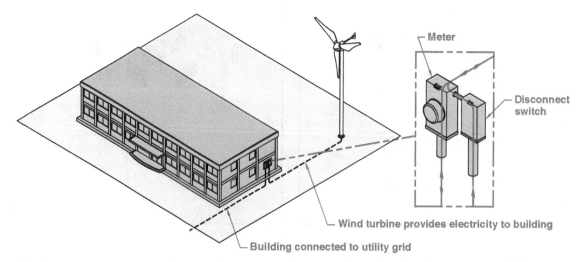

Figure 2.17 In some states, consumers can take advantage of net-metering, which is the sale of unused energy back to the power grid.

HYDROELECTRIC POWER

Hydroelectric generators convert the energy from falling water into electricity. Hydro-electric power for commercial use depends on a good source of flowing water with specific characteristics. Two key factors for a good source of flowing water are **flow rate** and **head.** Flow rate is calculated in gallons per minute (GPM). Head is the vertical distance water falls, measured in feet. The site should be evaluated and tested by a hydrology engineer to determine its feasibility for generating hydroelectric power. Examples of sites where hydro-electric power generation is possible include 100 GPM water flow falling 10 ft through a pipe, or 5 GPM falling 200 ft through a pipe, can supply enough power to run a small business. A dam can be constructed to form a reservoir to help increase the flow rate and head, but dam construction is normally very expensive and can cause additional environmental concerns. Local codes and environmental regulations must be considered before planning a hydroelectric project. This can be confirmed with the local planning department, Department of Environmental Quality, or the Department of Natural Resources. Power produced from hydroelectric sources is clean and free. Hydroelectric power limitations include ini-tial cost and environmental impact, such as blocking fish passage. There are several types of small-scale hydroelectric power generators, with the most common being the Pelton wheel turbine and the submersible propeller. Pelton wheel turbine work best with head of at least 50 ft, but works well with low flow rates. This system operates when water is channeled through a pipe and is forced across a series of cups that spin a wheel. The wheel is connected to a turbine that generates electricity. The submersible propeller can operate in water as little as 1 ft deep, and best with 3 to 30 ft head in 50 GPM or more stream flows. The water flow turns the propellers that operate the turbine that generates electricity. Figure 2.18 shows how the small-scale hydroelectric generator provides electricity. Some hydroelectric gen-erators can be placed directly in the stream, but funneling the stream into an intake pipeline at a higher elevation than the turbine can increase the flow rate and the head.

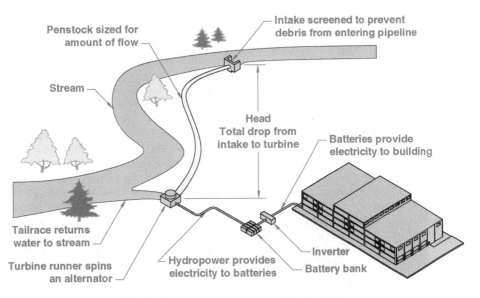

Figure 2.18   A small-scale hydroelectric generator can provide electricity.

BIOPOWER

Biopower, also referred to as **biomass energy,** uses **biomass** to generate electricity in a way that is cleaner and more efficient than most other electricity generation techniques. Biomass generating facilities typically use natural **biofuels** to produce steam, which drives a turbine that turns a generator to produce electricity. Biofuel is biomass used as fuel. Biomass means natural material, such as trees, plants, agricultural waste, and other **organic material.** Organic material is any material that originated as a living organism. Biomass refers here to plant matter grown to generate electricity or produce biofuel, but it also includes plant or animal matter used for production of fibers, chemicals, or heat. Biomass can also include biodegradable wastes that can be burned as fuel. Biomass excludes organic material such as coal or petroleum, because coal and petroleum combustion increases the carbon dioxide content in the atmosphere. In addition to providing clean energy, biopower protects the environment by preventing scrap lumber, agricultural cuttings, forest debris, and other organic waste from being open-burned or dumped in landfills. Biopower is twice as effective at reducing greenhouse gases as other forms of renewable energy and is carbon neutral. Biopower also reduces carbon dioxide emissions, while diverting organic materials that would otherwise decompose, be open-burned, or accumulate in the forest as overgrowth material. Biopower improves the forest health and reduces firefighting costs by clearing forest waste, which otherwise would be fuel for fires.

SUSTAINABLE DEVELOPMENT AND GREEN BUILDING ON BROWNFIELD SITES

Brownfield sites are parcels of land that have the presence or potential presence of a hazardous substance, pollutant, or contaminant. Brownfield sites can be found in every state and complicate the expansion, redevelopment, or reuse of the land. Common examples are abandoned gas stations and dry cleaners, railroad properties, factories, and closed military bases. Brownfields can also include properties that are under-utilized for various socioeconomic reasons, such as abandonment, obsolescence, tax delinquency, and disease.

COMMERCIAL CONSTRUCTION SOLAR APPLICATION

This commercial construction application focuses on the solar installation at the Prodeo Academy and Metro Schools College Prep of Minneapolis, Minnesota, shown in **Figure 2.19**.

A high-tech rooftop solar array added to a recently redeveloped Minneapolis North Loop property near Target Field has become a prominent feature on the Minneapolis skyline. The owners of the North Loop Campus Building, a multi-tenant charter school campus, recently made the decision to convert the building's rooftops to a sprawling solar array with the help of All Energy Solar. The project was one of the first and largest participants in the 2018 Minneapolis Green Business Solar Cost Share program, which helps business owners overcome financial obstacles to implementing solar systems. The 300 **kilowatt (kW)** flat roof system on the building was installed in August of 2018 and accounts for over 850 individual solar panels across seven different rooftop sections. A **kilowatt** is a measure of power. Kilowatt produced by this array is projected to offset nearly two-thirds of the entire facility's electricity need, based on historical electrical load data. The solar array is projected to offset nearly 5500 metric tons of atmospheric carbon emissions. This is equivalent to planting over 140,000 trees or displacing nearly 6,000,000 lb of coal burned for fossil fuel energy.

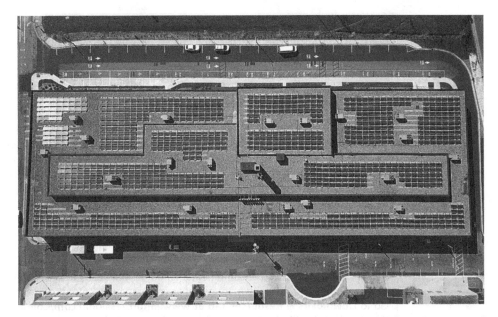

Figure 2.19 The solar installation at the Prodeo Academy and Metro Schools College Prep of Minneapolis, Minnesota. (*Courtesy of All Energy Solar.*)

Sending a High Energy Message

"The optics of a big rooftop solar array in a major metropolitan downtown sends a message about how versatile and financially feasible solar has become," said Michael Allen, president and co-owner of All Energy Solar. "This North Loop Campus now sports a solar installation that will catch the attention of hundreds of thousands of people from the highways and high-rises nearby. 'If solar works in snowy Minneapolis, in a location where real estate is in high demand, it must be living up to the hype,' is the takeaway message."

"We know the students will be curious and hopefully inspired to see what human ingenuity can accomplish," Allen said. "After all, it is their generation that will take our clean energy technologies to a whole new level. With this project, there's an impactful story of environmental responsibility through reduced emissions and a lower carbon footprint in this school building, and that story will come home to more than a thousand students, families, staff and visitors each day."

"Our charter school tenants and their students are quite literally the 'next generation,'" said Michael Johander, President, North Loop Campus LLC. "We thought it would be a great opportunity to not only harvest all of the environmental benefits of solar energy, but to do so in the laboratory-like setting provided by our schools."

"All Energy Solar was able to pipe real-time monitoring of our solar electricity output to TV screens in the building so that students can see how much electricity is being produced, the amount of carbon dioxide emissions our system is saving, and the equivalent number of trees that would need to be planted to process those emissions," said Johander. "We hope that by weaving this information into students' daily lives, we can help build an awareness of the importance of renewable energy and maybe even inspire some revolutionary thinking. After all, you never know when the next Albert Einstein will be walking through the halls of our building!"

Go to Downloads & Resources tab at **www.mhprofessional.com/Commercial Building Construction-Ancillaries** to access the images provided in this chapter and correlate with the content of this chapter. Here you can look at the chapter figures on your computer screen to pan and zoom in and out for better observation, because full drawings reduced to fit on a textbook page are often difficult to read. The image bank also includes the complete set of plans for the Brookings South Main Fire Station used as examples throughout this textbook.

Go to **Downloads & Resources tab at www.mhprofessional.com/Commercial Building Construction-Ancillaries** to access the test and creative thinking problems for this chapter. The chapter test can be used for review or to evaluate content knowledge depending on your course objectives. Creative thinking problems are provided to help expand your knowledge by researching given subjects.

Print reading problems have questions that ask you to find information on the Brookings South Main Fire Station plans. The print reading problems can help you become familiar with the format of construction documents and reinforce your ability to seek specific information. This is an important skill for you to master as you prepare to enter the construction industry. Go to **Downloads & Resources tab at www.mhprofessional.com/Commercial Building Construction-Ancillaries** to access the complete set of plans for the Brookings South Main Fire Station.

CHAPTER 3

Construction Site and Excavation

INTRODUCTION

This chapter describes site planning characteristics and site preparation practices. Construction site surveying basics are provided to help you understand practices used to locate commercial construction-related features on a building site. Types of survey stakes are identified and survey stake labeling is discussed for your use on a construction project. Erosion and sediment control practices are important for you know how to protect the environment prior to during and after construction.

SITE ORIENTATION

Site orientation is one of the preliminary factors that an architect takes into consideration when starting the design process. Site orientation is the placement of a structure on the property with certain environmental and physical factors taken into consideration, to help maximize energy efficiency, durability, and comfort of the occupants. The specific needs of the occupants, such as business characteristics and habits, perceptions, and aesthetic values, are important and need to be considered. Site orientation combines the values of the owner with other factors that influence the location of the building. Site orientation is predetermined in some cases. For example, the street frontage often controls the front of the building. In many cases, the property line **setback** requirements do not allow much construction placement flexibility. Setback is the minimum distance required between the structure and the property line. In such a case, site planning has a minimal influence on site orientation. This chapter presents factors that can influence site orientation, such as terrain, view, sunlight, wind, and sound.

Terrain Orientation

Terrain, also called **topography,** is the characteristics of land, especially as considered with reference to its natural features such as flat or sloping. Topography is represented as lines at given heights of the geographical landscape. **Figure 3.1** shows terrain from flat in the

Figure 3.1   Terrain from flat in the foreground to sloping upward to the hills beyond can determine the architectural design. (*Courtesy of the National Park Service.*)

foreground to sloping upward to the hills beyond. Terrain affects the type of structure to be built and the amount of **excavation** or **infill** required on the site. Excavation is any man-made cut, cavity, trench, or depression in the earth surface formed by earth removal. This includes the excavation for a building that is referred to as a **dig-out** in the construction industry. **Infill** is where extra ground or other material is brought to the site to build up a sloping area, cavity, or depression in the existing terrain.

Construction on any site requires that a **slope** be **graded** away from the building or other drainage systems provided. This is to help ensure that water drains away from the building. The slope, or **grade** of land, or grade, refers to the amount of incline of the earth surface. Grading refers to the modification of natural topography using earth removal practices. Grading is used to provide suitable topography for construction and control runoff and erosion.

A level construction site is a natural location for a single-level or multistory building, as shown in Figure 3.2. Sloped sites are a natural location for multilevel or **daylight basement** building or parking garage. A daylight basement structure is typically built on a front to rear slope where the amount of slope allows for a fully exposed basement wall at one side and a completely underground basement at the opposite side. Figure 3.3 shows a building on a slope site. A single-level project is a poor choice on a sloped site because of the extra con-struction cost for excavation or building up the foundation. An architect can take advantage of slope construction sites by designing on columns. This requires careful geological and structural engineering to ensure safe construction. A **column** is a vertical structural mem-ber designed to transmit a compressive load and is typically made of reinforced concrete or concrete block, but it can be constructed from wood, steel, fiber-reinforced polymer, cellular PVC, or aluminum.

View Orientation

View orientation provides optimum exposure to a view and can be a major factor in the purchase of property for commercial construction. The view can be of mountains, city lights,

Figure 3.2   A level construction site is a natural location for a single-level or multistory building. (*Courtesy of the National Aeronautics and Space Administration.*)

Figure 3.3   Buildings on a slope site.

Figure 3.4   An example of view orientation. (*©bonzodog—Can Stock Photo Inc.*)

a lake, a river, or the ocean, for example. Figure 3.4 shows windows placed to take advantage of a view. These view sites are usually more expensive than comparable sites without a view. The design typically has large windows taking advantage of the view from as many areas as possible. It is best to provide an environment that allows the occupants to feel as though they are part of the view.

View orientation can conflict with the advantages of other orientation factors, such as solar or wind orientation. Such a trade-off can be necessary when a buyer pays a substantial amount to purchase a view site. When the view requires that a large glass surface face a wind-exposed nonsolar orientation, some energy-saving alternatives should be considered as described throughout this textbook.

Solar Orientation

The sun can be an important factor in site orientation. Sites with correct **solar orientation** allow for excellent exposure to the sun. There should not be obstacles such as tall buildings, evergreen trees, or hills that have the potential to block the sun. Generally, a site located on a south slope can have these characteristics.

When a site has southern exposure that allows for correct solar orientation, a little basic astronomy can contribute to proper building placement. Figure 3.5 shows how a southern orientation relative to the sun path provides the maximum solar exposure.

Establishing South

A perfect solar site allows the structure to have unobstructed southern exposure. When a site has this potential, true south should be determined. This determination should be established in the preliminary planning stages. Other factors that contribute to orientation, such as view, can also be taken into consideration at this time. If view orientation requires a structure be turned slightly away from south, it is possible that the solar potential will not be significantly reduced.

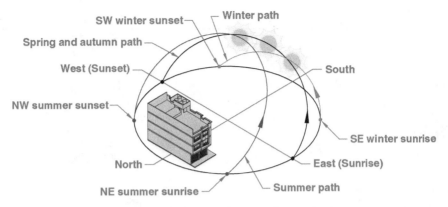

Figure 3.5   A southern orientation relative to the sun path provides the maximum solar exposure.

True north is determined by a line from the North Pole to the South Pole. When a compass is used to establish north-south, the compass points to **magnetic north.** Magnetic north is not the same as true north. True north is the same as geographic north, which is the North Pole. The difference between true north and magnetic north is known as the **magnetic declination.** Magnetic north is the direction of the earth magnetic pole and is the direction where the north-seeking pole of a compass points when free from **local magnetic influence.** Local magnetic influence is also called local magnetic disturbance and local attraction. Local magnetic influence is an abnormality of the magnetic field of the earth, extending over a relatively small area, due to local magnetic influences. Figure 3.6 shows the compass relationship between true north-south and magnetic north. Magnetic declination differs between locations on the earth. Figure 3.7 shows a map with lines that represent the

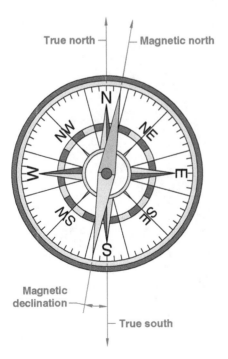

Figure 3.6   The compass relationship between true north-south and magnetic north.

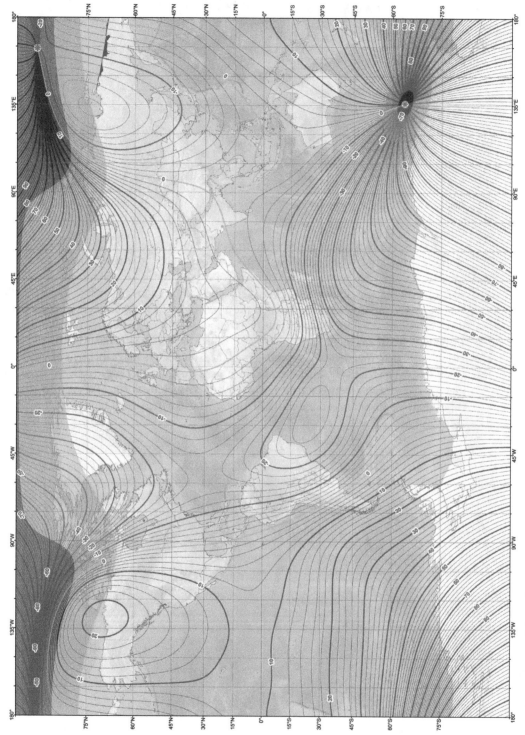

Figure 3.7 A map with lines that represent the magnetic declination across the earth. (*Courtesy of the National Centers for Environmental Information.*)

magnetic declination at different locations throughout the United States. Magnetic declination can also be determined using a calculator provided on the National Geographic Data Center website. An example is a magnetic declination of 16.5° east, which occurs in northern California, means that the compass needle points 16.5° to the east of true north, or 16.5° to the west of true south. True south is 16.5° to the left at this location, as you face toward magnetic south.

Solar Site-Planning Tools

Instruments are available that calculate solar access and demonstrate shading patterns for any given site throughout the year. **Solar access** refers to the availability of direct sunlight to a construction site. Some instruments provide accurate readings for the entire year at any time of day, in clear or cloudy weather. Search the Internet with the entry *Solar Site-Planning Tools* to get solar site-planning information and learn more about available tools.

Solar Site Location

A solar site in a rural or suburban location where there can be space to take advantage of a southern solar exposure allows for site placement flexibility. Some other factors can also be considered when selecting an urban solar site. Select a site where zoning restrictions have maximum height requirements. This prevents future development from blocking the sun from an otherwise good solar orientation. Avoid a site where large **coniferous trees** hinder the full potential of the sun exposure. This factor is more important for a one-story office building than for a multistory complex. Coniferous trees are cone-bearing trees that are normally evergreen. Local planning restrictions can require that trees be identified and remain on the site. In this situated, southern exposure can be achieved with coniferous trees to the north and **deciduous trees** on the south side. The coniferous trees can effectively block wind without interfering with the solar orientation. Deciduous trees lose their leaves in the winter. The deciduous trees provide shade from the hot summer sun, the winter sun exposure is not substantially reduced in the winter when these trees lose their leaves.

WIND ORIENTATION

The term **prevailing winds** refers to the direction from which the wind most frequently blows in a given area of the country. For example, if the prevailing winds are said to be southwesterly, that means the winds in the area generally flow from the southwest. There are some locations that have southwesterly prevailing winds, but during certain times of the year, severe winds can blow from the northeast, for example. The factors that influence these conditions can be mountains, large bodies of water, valleys, canyons, or river basins. The prevailing winds in the United States are from west to east, although some areas have wind patterns that differ from this because of geological and environmental factors.

Wind conditions should be taken into consideration when deciding site orientation. In many cases, different site orientation characteristics can present a conflict. For example, the best solar orientation can be in conflict with the best wind orientation. The best view could be of the ocean, but the winter winds can also come from that direction. An excellent view can be on top of a hill where the site is exposed to wind, while placing the building behind the hill protects from wind but eliminates the view. One of the factors used to evaluate orientation can outweigh another. Personal judgment can be the final ruling factor. A good combination can be achieved if careful planning is used to take all of the environmental factors into consideration.

Wind conditions that influence site location can be found in almanacs, in the local library, and on the Internet. Look for subjects such as climate, microclimate, prevailing winds, and wind conditions. Evaluate the direction of prevailing winds in an area by calling the local weather bureau, by discussing it with local businesses, or by searching the Internet. Select an area where winter winds are at a minimum or where there is protection from these winds. Commercial architectural site planning often takes place regardless of prevailing wind conditions, but factors such as placing large windows away from the wind and structural engineering factors must be evaluated.

SOUND ORIENTATION

If the construction site is for a one-story office building, then that is level with, or slightly below a road can have less noise than a site that is above and overlooking the sound source. A few landscape designs can also contribute to a quieter environment. Berms, trees, hedges, and fences can all be helpful. Some landscape materials deflect sounds, while others absorb sounds. The density of the sound barrier has an influence on sound reduction, although even a single hedge can help reduce a sound problem. A mixture of materials can most effectively reduce sound. Keep in mind that deciduous trees and plants help reduce noise problems during the summer, but they are poor sound insulation in the winter when the leaves have fallen.

Construction methods and materials can also reduce sound transmission into the building. For example, heavy multistory commercial buildings construction often uses materials such as concrete and masonry that are effective in minimizing sound inside the building. Some insulation products reduce sound transmission. Triple-glazed high-quality windows can also help reduce outside sound.

A number of differing factors can influence site orientation. The priority of these factors can have a large impact on final site selection. Solar orientation can be important to one builder, but another can consider view orientation as the main priority. A perfect construction site has elements and design features of each orientation characteristic. When a perfect site is not available, the challenge is to orient for the best advantage of each potential design feature. It is possible to achieve some elements of good site orientation with excavation and landscaping techniques. Always take advantage of natural conditions whenever possible.

TOPOGRAPHY

Topography is a physical description of land surface showing its variation in **elevation,** known as **relief,** and locating other natural features. Elevation is the vertical difference between two points. Surface relief can be represented with graphic symbols that use shading methods to accent the character of land, or the differences in elevations can be shown with **contour lines.** Site plans that require surface relief identification generally use contour lines. Contour lines, also called contours, are lines that join points of equal **elevation** (height) above an established zero level, and contours help demonstrate the general lay of the land.

A good way to visualize the meaning of contour lines is to look at a lake or ocean shoreline. When the water is high during the winter or at high tide, a high-water line establishes a contour line at that level. As the water recedes during the summer or at low tide, a new lower-level line is established. This new line represents another contour line. The high-water line goes all around the lake at one level, and the low-water line goes all around the lake at another level. These two lines represent contours, or lines of equal elevation. The horizontal distance between contour lines is known as **contour interval.** When the contour lines are far apart, the contour interval shows relatively flat or gently sloping land. When the contour lines are close together, the contour interval shows land that is steeper. Contour lines shown on a map or site plan have a space placed periodically along the length, and the numerical value of the contour elevation, above sea level, is inserted in the space. Figure 3.8 shows sample contour lines. The contour lines that show the elevation values are called **index contours.** Generally, every fifth contour line is used as an index contour line. The other contour lines are referred to as **intermediate contours.** Index contours are generally drawn thicker than intermediate contours, much like the numbered lines on a ruler or scale. Figure 3.9 shows a graphic example of land relief in pictorial form and contour lines of the same area in plan view. The term **elevation** refers to the height of a feature from a known base, which is usually given as 0 (zero elevation).

Figure 3.10 shows a site with contour lines. Commercial site plans typically require contour lines showing topography along with property corner elevations, street elevation at a driveway, and the elevation of the finished floor levels of the building. In some applications, slope can be identified and labeled with an arrow showing the direction of the slope.

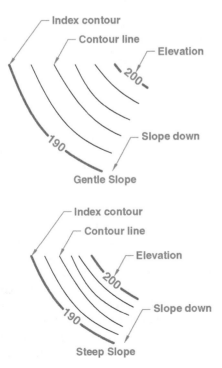

Figure 3.8   Contour lines.

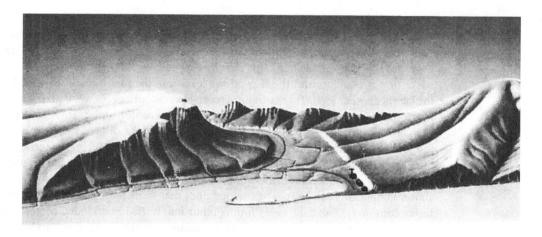

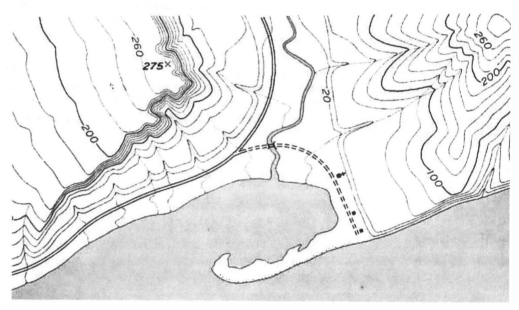

Figure 3.9   A graphic example of land relief in pictorial form and contour lines of the same area in plan view. (*Courtesy of the U.S. Geological Survey.*)

LAND CLEARING

Commercial construction sites often need to have the land cleared of undesirable trees, brush, rocks, and other materials. Trees can be selectively cut and harvested and specific desirable trees can be protected and remain on the site to enhance future landscaping and to provide shade. Land clearing should be performed in a manner that reduces destructive impact on the land. Brush, stumps, and unwanted trees can be mulched using a forestry mulching machine, also called a **forestry mulcher.** The forest mulcher shreds all types of vegetation, leaving behind natural mulch, which is broken-down vegetation that is reintroduced into the earth, making the soil richer and more fertile. Heavy-duty forestry mulchers can clear up to 15 acres of vegetation a day, depending on terrain, density, and type of material.

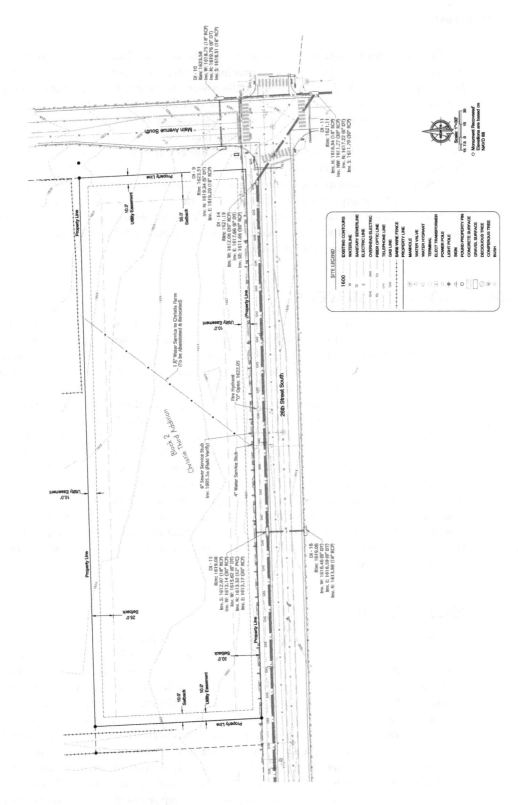

Figure 3.10 A site with contour lines. *(ILG Architects, Alexandria, MN.)*

Trees and brush from land clearing can also be used for biomass. **Biomass** means natural material, such as trees, plants, agricultural waste, and other organic material. The biomass is then used to develop biopower. **Biopower,** also referred to as **biomass energy,** uses biomass to generate electricity in a way that is cleaner and more efficient than most other electricity generation techniques. Biomass generating facilities typically use natural biofuels to produce steam, which drives a turbine that turns a generator to produce electricity. Biomass can also be converted directly into **biofuels.** The two most common types of **biofuels** are ethanol and **biodiesel.**

Excavators and dump trucks can be used to remove unwanted materials from the construction site, or large quantities of rock can be crushed into gravel using a rock crushing machine, called a **crusher.** A crusher is a machine designed to reduce large rocks into smaller rocks, gravel, or rock dust. Crushers are used to reduce the size, or change the form, of waste materials for disposal or recycling.

INTRODUCTION TO BASIC SURVEYING

In construction, a **survey** is the result of establishing the exact corners, boundaries, and topography of a piece of land using **surveying** techniques. Surveying has been used to establish land boundaries, maps, and for construction projects for about 5000 years. Surveying for construction is the measurement of dimensional relationships, horizontal distances, elevations, directions, and angles, on the earth surface, especially for use in locating property boundaries, construction layout, and site plan drafting. **Land surveying** is the technique, profession, and science of accurately determining the three-dimensional location of points and the distances and angles between points on the earth. The location of points, distances, and angle between points are used to establish land maps and boundaries for site plans, construction site locations, ownership, property legal descriptions, and government documentation. Surveyors are the people who do surveying using special equipment, geometry, trigonometry, physics, engineering, and land law to accomplish their task. Surveying is required in the planning and completing of all construction projects, mapping, and the definition of legal boundaries for land ownership. The following leads you through the basics of a site survey and how to read a site plan and survey stakes found on a construction site.

CONSTRUCTION SURVEYING

A general definition for **surveying** is a branch of applied mathematics that is a science concerned with determining the area of any portion of the earth surface by measuring distances, angles, and directions of characteristics such as boundary lines and the contour of the surface by accurately defining the features in notes and on drawings. **Surveying for construction** is the measurement of dimensional relationships, horizontal distances, elevations, directions, and angles, on the earth surface especially for use in locating property boundaries, construction layout, and site plan drawings. A **survey** is the result of establishing the exact corners, boundaries, and elevations of a piece of land using surveying techniques. Construction site surveying, described in this chapter, is performed by professional surveyors. A **surveyor** is a professional who uses current technology surveying equipment and practices for surveying as a specialist who provides a map file of the job site from which civil designers and drafters develop drawings. Commercial construction workers generally do not do site surveying activities, but construction workers often do on site measuring and leveling related to the construction project. **Leveling** is a process of determining the height of one level relative to another. Leveling is used in surveying and in construction to establish the elevation of a point relative to a **datum,** or to establish a point at a given elevation relative to a datum. A datum is theoretically exact reference point, axis, or plane. The term **location**, as related to surveying, specifies a geographic point or area. Identifying locations and measuring the distance between locations are fundamental concepts associated with most civil engineering applications. This chapter explains methods used to make measurements, establish elevations, and determine the location, including measuring horizontal and vertical distances between points. Refer to the textbook *Civil Drafting Technology*, eighth edition, published by Pearson (ISBN-13: 978-0-13-443604-3) for detailed information about professional surveying.

MEASURING HORIZONTAL DISTANCES

Measuring the horizontal distance between two points is a fundamental surveying and civil engineering design and construction process. The length of a property line, distance between structures, and width of a road are basic examples of horizontal measurements. Surveyors and construction workers use a variety of methods to measure horizontal distances depending on project requirements. The following information introduces methods of measuring horizontal distances.

Electronic Distance Measurement

Surveyors use **electronic distance measurement (EDM)** to measure distances whenever possible. An EDM device measures distance using electromagnetic or microwave radio waves, or infrared or laser light waves. Current EDM devices primarily use infrared and laser technology. An EDM device transmits light waves to a reflector, which reflects the light back to the transmitting instrument. The time required for the signal to return to the device is measured and translated into a distance. A second approach uses the laser wavelength and phase shift of the reflected beam to determine distance. Both approaches are common and produce similar precision. Distances are recorded and stored onboard the instrument or in a separate device. Distances can be measured on a level line of sight or on a slope.

Lasers in EDM devices can be used to measure distances between the surveying instrument and a prism or prism array mounted on a tripod. Greater distances can require larger prism arrays. A single reflector uses what is called a **corner cube prism,** which is a corner cut off of a cube of glass. A corner cube prism reflects the beam of light back onto the incoming path. It can be slightly out of alignment with the EDM device and still provide an accurate measurement. A **total station** is a tripod-mounted instrument that contains an EDM device. Horizontal and vertical angles between the total station and the measured points, when combined with a distance measurement, can provide location and elevation (XYZ) data for any point.

Some laser EDM devices can also operate in reflectorless mode, in which a prism is not required to measure the distance. The laser beam can be aimed at an object, such as a wall, curb, or street intersection, and the distance is measured. Reflectorless EDM technology in total stations is generally accurate for distances less than 0.75 mi (1200 m). A handheld EDM that uses a laser beam reflected off of objects can also be used to measure distances such as measuring buildings for construction projects.

Traditional Methods of Measuring Distances

Even with the advent and refinement of EDM and **GPS,** traditional methods of measuring distance, such as **pacing** and **taping,** continue to be useful for some purposes and serve as a valuable reference. Pacing is a way to measure distance by walking and counting the number of steps taken. Pacing is inaccurate compared to other forms of distance measurement but is useful for quickly taking rough measurements without using equipment. For example, a pace is your average step when walking, so if your step measures 3 ft, then 30 paces is about 90 ft (3 × 30 = 90). A **pedometer** can be used to pace long distances, general over 200 ft. A pedometer is an instrument that estimates the distance traveled by walking and records the number of steps taken. GPS is the acronym for Global Positioning System, which is the United States global navigation satellite system (GNSS) that uses a network of satellites in combination with earth-based receivers to determine locations on the earth. GNSS is a generic term for any system of satellites that provide global coverage for geographic positioning. A detailed discussion covering GPS is beyond the scope of this content.

The **measuring wheel** shown in Figure 3.11 is a device that measures and records distance as a wheel is rolled along the ground. An **odometer,** much like the device that

Figure 3.11 The measuring wheel is a device that measures and records distance as a wheel is rolled along the ground. (*Kara-Fotolia.*)

records the distance you travel in your vehicle, is similar to a measuring wheel and is also used to measure and record distance. Measuring wheels and odometers offer accuracy similar to or slightly better than pacing, and are generally used to take rough measurements over long distances.

Taping is the process of using a tape measure, or tape, to measure distances. Taping is a slower process than using EDM, especially over longer distances or on steep slopes with distances generally less than 100 ft (30 m) such as measuring buildings and staking out small construction sites. Figure 3.12 shows an example of a steel tape. There are many different types of tapes for different applications. A 100-ft-long steel tape graduated in feet and hundredths of feet is common. A 30-m-long steel tape graduated in meters and mm is a common metric tape. Fiberglass, or woven, tapes are generally less expensive and stronger than steel tapes and come in a variety of lengths. Taping is also known as **chaining,** which is a term leftover from the use of the old 66-ft-long Gunter chain.

Figure 3.13*a* shows how taping is used to measure distances. Hubs or markers of some sort are placed at each point where a reading is to be made. A **plumb bob** is a weight hanging from a string used to establish a vertical line to a point. Additional equipment used for taping include range poles for sighting points and guiding the workers, taping pins for marking locations on the tape, hand levels to help establish a horizontal tape, and tools to apply tension to the tape. When tape measurements must be made on a slope, the process is often referred to as **breaking chain.** See Figure 3.13*b*.

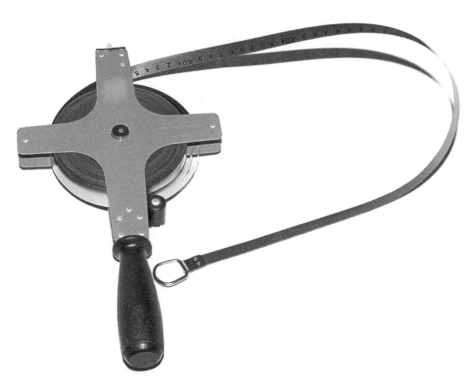

Figure 3.12   An example of a steel tape. (*lnzyx–Fotolia.*)

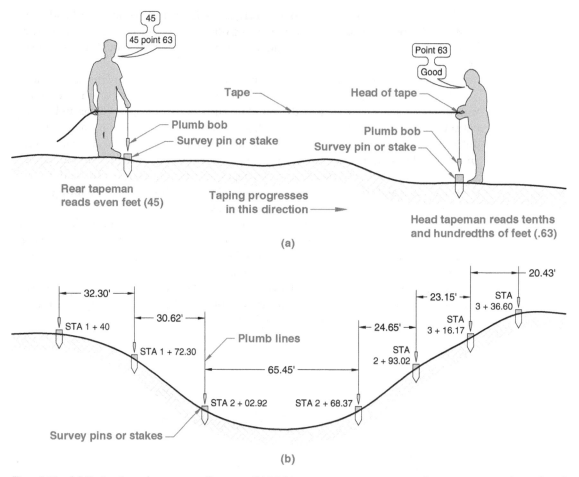

Figure 3.13 (*a*) Taping is used to measure distances. (*b*) Making tape measurements on a slope is a process often referred to as breaking chain.

MEASURING VERTICAL DISTANCES

Elevation is the height of a geographic location above or below a theoretically exact reference point, axis, or plane. A **vertical datum** provides the zero elevation reference for elevations of surfaces and features on the earth. A **geodetic vertical datum** is the principal datum used to express land elevations and to provide vertical survey control. A **reference datum** used for a survey can be any solid object with a known elevation from the vertical datum, such as a **bench mark (BM)** or a temporary control point established by a surveyor. A BM is a survey mark intended to be permanent that specifies the latitude, longitude, and elevation of a point. Measuring the elevation of points is an essential surveying and civil engineering design and construction process. The existing and finished elevations of a building site, depth of an underground utility line, and height of roadway features such as curbs are basic examples of elevation measurements. Surveyors use a variety of methods to measure vertical distances depending on project requirements. The following section provides valuable information about the process of determining elevation by differential leveling and gives you important background knowledge.

> **NOTE**
>
> **Mean sea level (MSL)** is the average of high and low tides taken by a tide gage over an extended period of time. Geodetic vertical datums provide elevation data relative to a reference surface defined by MSL and can have a datum point at a tide gage.

Leveling

Leveling is the process of determining the difference in elevations between points. There are different systems of leveling. **Direct leveling,** also known as **spirit leveling,** is the theory of measuring vertical distances relative to a horizontal line, and using the measurements to calculate the differences between the elevations of points. Direct leveling is the most common system of leveling associated with surveying and construction site layout work. The term **differential leveling** refers to the method used to determine the difference in elevation between two points. Differential leveling begins at a point of known elevation and works toward points with unknown elevations.

Field work and preliminary site work often require approximations before accurate leveling or surveying can begin. Rough leveling estimates can be accomplished with the use of a **hand level,** shown in Figure 3.14. A hand level is also used in taping to help establish a horizontal tape. Accurate differential leveling is normally performed using a level mounted on a tripod, and a level rod. A **surveyor level** is an instrument with a telescope and spirit level or compensator used to establish a horizontal line for measuring elevations. A **level rod,** also known as a **level pole, level staff,** or **Philadelphia rod,** is a rod graduated to fractions of feet or meters used to measure elevations and distances when viewed through a level, **transit,** or **theodolite.** A leveling rod has two sliding sections graduated in hundredths of a foot. The front of the rod has the graduation increasing from zero at the bottom. The back of the rod has graduations decreasing from 13.09 ft at the bottom to 7 ft. The rod is made in two sliding sections to reduce the length for carrying and transporting. Readings of 7 ft (2.1 m) or less, and up to 13 ft (4.0 m), can be measured. The rod must be fully extended, when higher measurements are needed to avoid reading errors. The transit and theodolite, shown in Figure 3.15, are used by the surveyor to measure both horizontal and vertical angles. The purpose of these instruments is similar, but the theodolite is more accurate than a transit. These instruments have a minimum accuracy of 1 minute of angle and quality theodolites can measure angles to an accuracy of one-tenth of a second of angle.

Figure 3.16*a* shows an example of an **automatic level.** An automatic level maintains a horizontal line of sight once leveled. The **digital level,** shown in Figure 3.16*b*, includes a graphical display that provides a readout of elevations. Field measurements can be stored in digital level internal memory or transferred to a USB storage device or field computer. The data files can be transferred to an office computer. Figure 3.16*c* shows a tripod on which a

Figure 3.14   Rough leveling estimates can be accomplished with the use of a hand level. (*Courtesy of Trimble.*)

Figure 3.15 The transit and theodolite are used by the surveyor to measure both horizontal and vertical angles. (*©kadmy–Can Stock Photo Inc.*)

(a)

(b)

(c)

Figure 3.16 (*a*) An automatic level maintains a horizontal line of sight once leveled. (*b*) The digital level includes a graphical display that provides a readout of elevations. Field measurements can be stored in digital level internal memory or transferred to a USB storage device or field computer. The data files can be transferred to an office computer. (*c*) A tripod on which a level is mounted. (*Courtesy of Topcon.*)

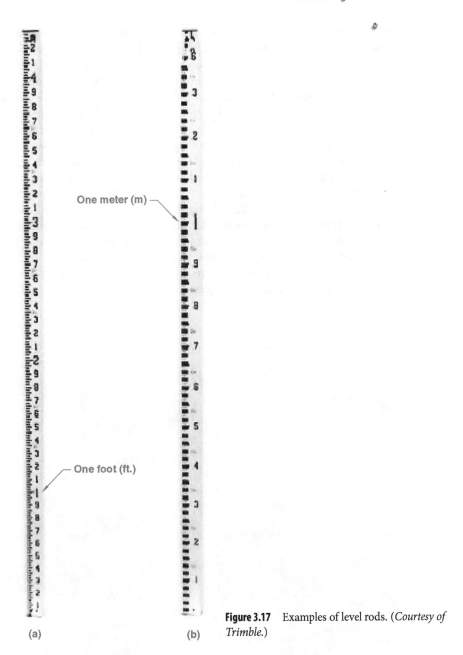

One meter (m)

One foot (ft.)

(a)   (b)

Figure 3.17   Examples of level rods. (*Courtesy of Trimble.*)

level is mounted. Figure 3.17 shows examples of level rods. A footplate used with some level rods provides a stable, well-defined foundation for the rod. A **footplate** is a cone-shaped device that has pointed feet anchored in the ground.

Leveling Process

Leveling begins by anchoring the tripod firmly in the ground at a location approximately halfway to the point with an unknown elevation to be measured. The instrument is attached to the tripod and leveled with three leveling screws on the base of the instrument. A **spirit level** or electronic display indicates when the instrument is level. A spirit level, also called a **bubble tube,** is a glass tube containing liquid and a bubble that is level when the bubble is adjusted between two marks on the tube. The rod is then placed on a point with a known elevation. Accurate recording of elevations must begin at known elevation points. A survey mark, or **monument (MON),** such as the BM shown in Figure 3.18, is an example of a monument point with a known elevation. A monument is a fixed point such as a section corner, a rock, a tree, an iron rod driven in the ground, or an intersection of streets.

When running levels, it is best to begin and end on a MON, such as a BM, because these markers provide checks to detect any errors in the leveling work.

A reading is now taken on the rod at the point of known elevation, identified as the **backsight (BS).** The BS is the rod or target location in surveying from which a reference measurement is made. The BS reading is added to the elevation of the known point to give the **height of the instrument (HI).** The rod is then placed on a point with an unknown elevation to be measured, identified as the **foresight (FS).** The FS is the rod or target location in surveying to which an elevation and/or location reading is taken. The reading taken at the FS is subtracted from the HI to give the unknown elevation. See Figure 3.19. An **intermediate foresight (IFS)** is a separate elevation taken in addition to the run of levels for information purposes.

Figure 3.18   An example of a monument point with a known elevation. (*Courtesy of Trimble.*)

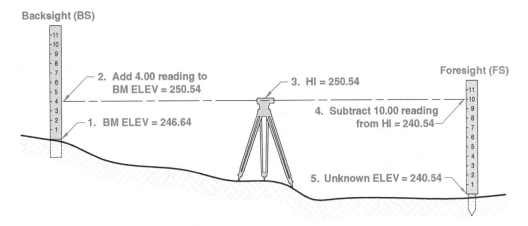

Figure 3.19   Using differential leveling to find an elevation with a level and rod.

Surveyors often establish semipermanent points, referred to generically as **local control points.** Local control points are called **temporary bench marks (TBM)** because they are temporary, but serve the same purpose as a BM. The most temporary TBM, known as a **turning point (TP),** is used when level measurements must be made over long distances or large changes in elevation. ATP is a TBM such as a large rock, fire hydrant (FH), concrete curb or foundation, wooden stake, or metal spike, on which the rod is rested and used as a pivot for the rod. The TP acts as both a BS and an FS. An FS measurement to a TP is often taken using the top of (TO) an object such as an FH or curb as the TP. Once the TP elevation is determined, the instrument is physically moved to the next station point (STA). The rod is then physically turned to face the instrument. The instrument reading as an FS is subtracted from the HI to find the level of the TBM, and then the instrument and tripod are physically moved ahead of the rod and reset. The next reading to the TP is a BS, which is added to the TBM elevation to become the HI. Now, the rod can be moved to the next FS position. Figure 3.20 illustrates leveling with several TPs required to measure elevations over long distances.

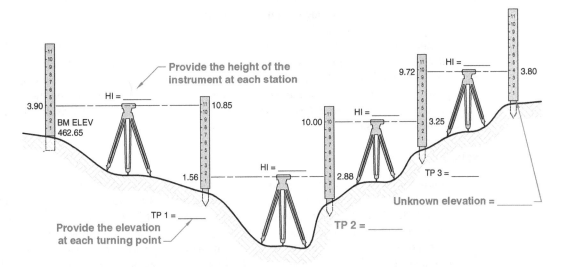

Figure 3.20   Using turning points to find an unknown elevation. The unknown elevation should equal 467.05′.

Leveling Example

For a better understanding of the leveling calculation process, take time to work through the example, shown in Figure 3.20. Before beginning this exercise, remember the basic rule for calculating elevation when running levels: add the BS and subtract the FS. In this example, the beginning point elevation of the BM is 462.65′. The level rod is placed on the BM. The level instrument is set up an appropriate distance away, in the direction of the unknown elevation. The rod reading of 3.90′ is taken at the BS. The BS reading of 3.9′ is added to the BM elevation of 462.65′ to get the HI of 466.55′ (3.9′ + 462.65′ = 466.55′).

Now that the HI is known, the rod can be moved ahead to the next location, which is determined by terrain and distance. This rod location becomes TP 1. The rod is physically placed on a stable object and is not moved again until both FS and BS readings are made. The instrument reading is taken on the FS. In this case, the level rod reading is 10.85′, which is subtracted from the HI of 466.55′ to get the TP 1 elevation of 455.70′ (466.55′ − 10.85 = 455.70′).

The instrument can now be moved to the next location, and the process is repeated. Notice in Figure 3.20 that the second position of the instrument in the valley is the same situation as the first instrument setup. Therefore, the next reading to be taken, the BS is read and calculated the same as the first calculation in this exercise. The rod reading of 1.56′ is taken at the BS. The BS reading of 1.56′ is added to the TP 1 elevation of 455.70′ to get the HI of 457.26′ (1.56′ + 455.70′ = 457.26′).

The rod is moved ahead to TP 2, and an FS reading of 2.88′ is taken. The FS reading of 2.88′ is subtracted from the HI of 457.26′ to get the TP 2 elevation. Complete this exercise as described above to determine the unknown elevation. The answer is provided at the end of this section.

Reading a Level Rod

A typical U.S. customary (foot) level rod is divided into whole feet (ft), tenths, and one-hundredths of a foot. See Figure 3.21. Metric (meter) level rods are also available. The whole feet values on the level rod are shown in large red numerals. The large black numerals indicate tenths of a foot and are located at a pointed black band on the rod. Each black band is one-hundredth of a foot wide. Notice the small numbers located every three-tenths of a foot on the rod face. These small red numbers indicate the whole feet value and are references in case the view through the instrument shows only a small portion of the rod, and no large red number is visible.

Rods that can be read by the person operating the instrument are called **self-reading rods.** A rod with a **target vernier** is often used to measure readings over distances longer than 200 to 300 ft (60 to 90 m). Once the rod person precisely locates the vernier as directed by the person running the instrument, the rod person reads the graduations on the vernier and records the elevation. A target vernier is a secondary scale that has 10 divisions equal in distance to 9 divisions on the main scale that allows measurements of 0.001′.

Precision Leveling

A **digital level** can be used to process elevations and distances electronically when used with a special rod. The face of the rod is graduated with a bar code. See Figure 3.22. The reverse side of the digital rod contains standard numerical increments. The digital level captures the image of the bar code and compares it with the programmed image of the entire rod, which is stored in memory. The comparison of the captured image and the programmed image allows the height and distance to be calculated. These readings can be stored in memory or transferred to another location for processing.

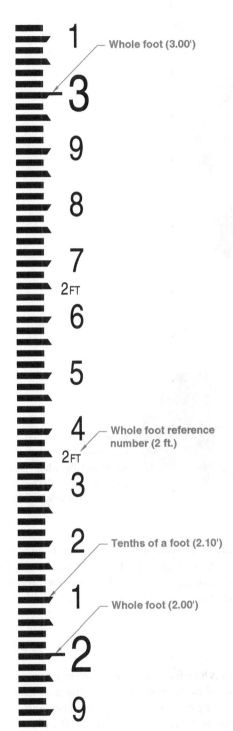

Whole foot (3.00')

1

3

9

8

7

2FT

6

5

4 — Whole foot reference number (2 ft.)

2FT

3

2 — Tenths of a foot (2.10')

1 — Whole foot (2.00')

2

9

Figure 3.21 A typical U.S. customary (foot) level rod is divided into whole feet (ft), tenths, and one-hundredths of a foot.

Digital Level

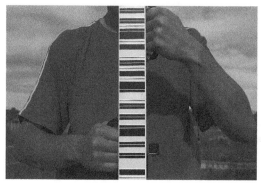

Digital Level Rod

Figure 3.22 A digital level can be used to process elevations and distances electronically when used with a special rod. The face of the rod is graduated with a bar code. (*Courtesy of Trimble.*)

An **optical-micrometer level** uses a special lens that can be rotated to vertically deflect the incoming light ray from the line of sight. The level instrument can then subdivide the level rod graduations to achieve an accuracy of ±0.02 of the level rod graduation. The highest order of leveling and surveying, called first order, can be achieved with optical-micrometer levels.

Leveling Field Data

Field data based on leveling contains location and elevation information. Figure 3.23 shows an example of leveling field notes and calculations. The left column contains the location of the rod. The BS reading is taken looking back toward the point of beginning. HI is the height of instrument calculated by adding the reading taken from the BS. IFS is the intermediate foresight, with one measurement taken at the top of a fire hydrant (TO FH). FS is the foresight taken and subtracted from the HI to give the elevation (ELEV) of the new TP, which is listed in the last column. The leveling process continues until the final elevation is determined.

The difference between the totals of the BSs and the totals of the FSs equals the difference in elevation between the BM and the final elevation. This value can be used to check for error

Leveling Computation					
STA	BS(+)	HI	IFS(-)	FS(-)	ELEV
BM E635	4.62				735.62
		740.24			
TP 1	6.81			3.78	736.46
		743.27			
TP 2	1.04			11.36	731.91
		732.95			
TO FH			6.77		726.18
TO CURB				10.19	722.76
	BS total = 12.47			FS total = 25.33	735.62 - 722.76 ‾‾12.86
				Level check FS 25.33 - BS 12.47 ‾‾12.86	

Figure 3.23   Leveling field notes and calculations.

in the intermediate calculations. Notice at the bottom of the right column that the difference between the BM and final elevation is calculated to be 12.86. The last row of the field notes, labeled Level check, shows the totals of the BS and FS columns. The difference between these two is also 12.86, which means there have been no errors in the intermediate calculations.

Construction Laser Levels

The **electronic** or **laser level** is commonly used for construction work such as leveling to set elevation grade stakes for road, parking lot, and driveway grades, and elevations for excavations, fills, concrete work, plumbing, and floor level verification. There are several different types of laser levels and accessories for specific applications. The **rotating beam laser level** or **rotating laser level** projects a 360° rotating laser beam in a plane. The beam appears on a surface as a line that can be detected anywhere around the instrument. Rotating lasers are used for a variety of purposes from basic surveying and site layout to establishing grades.

A **pipe laser level** projects a laser beam in front of the instrument that appears on a surface as a dot. Pipe lasers are typically used to align pipe and are often mounted directly to pipes and features associated with pipelines. A **line laser level** projects a laser beam in front of the instrument that appears on a surface as a line or crosshairs. Line lasers are used for applications such as leveling and plumbing floors, walls, and ceilings. A **plumb laser level** projects a beam in front of the instrument that appears on a surface as a dot. Plumb lasers are used for leveling and plumbing applications when a reference line is not necessary. Most laser levels can produce a laser beam or beams horizontally, vertically, or as a specified angle as needed. Laser beams can be effective to a radius of more than 2600 ft (800 m). A **laser detector,** also known as a **laser receiver,** is used to detect the beam from a laser level when the beam cannot be seen, such as over far distances or in bright light.

> **NOTE**
>
> The answer for the unknown value in the leveling example exercise earlier in this section is 467.05′.

SITE SURVEY

A **site survey** identifies property corners, border lines, and elevations and can include the location of construction corners, building outlines, and corner elevations. Figure 3.24 shows the survey of a small subdivision of land with six separate lots or sites. Lot number 3 is highlighted showing the corners and boundaries of one site. The terms **lot** and **site** can be used

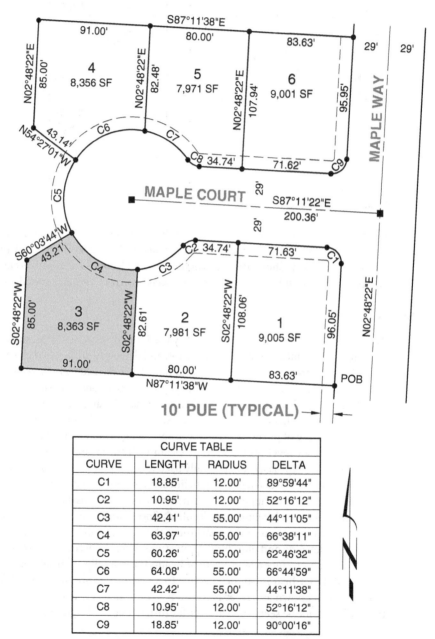

CURVE TABLE			
CURVE	LENGTH	RADIUS	DELTA
C1	18.85'	12.00'	89°59'44"
C2	10.95'	12.00'	52°16'12"
C3	42.41'	55.00'	44°11'05"
C4	63.97'	55.00'	66°38'11"
C5	60.26'	55.00'	62°46'32"
C6	64.08'	55.00'	66°44'59"
C7	42.42'	55.00'	44°11'38"
C8	10.95'	12.00'	52°16'12"
C9	18.85'	12.00'	90°00'16"

Figure 3.24   The survey of a small subdivision of land with six separate lots or sites. Lot number 3 is highlighted showing the corners and boundaries of one site. (*Previously published in Modern Residential Construction Practices, Routledge, 2017 and reproduced here with permission.*)

interchangeably, except that a site can also show the location of the construction project. The property lines of each lot are commonly labeled with distances and bearings. The property lines are generally displayed as a line with long and two short dashes or as solid lines, as shown in Figure 3.24, although other line types can be used. The distances between corners are generally displayed above each property line as feet and tenths of a foot using the foot (′) symbol, such as 112.93′. The **bearing** of each property line is usually placed below the line. Bearings are directions with reference to one quadrant of the compass. There are 360° in a circle or compass, and each quadrant has 90°. Degrees are divided into minutes and seconds. There are 60 minutes (60′) in 1° and 60 seconds (60″) in 1 minute. Bearings are measured clockwise or counterclockwise from north or south. For example, a reading 45° from north to west is labeled N 45° W, as shown in Figure 3.25. Fractions of a degree are used when a bearing reading requires additional accuracy with minutes and seconds. For example, S 30°20′10″ E reads from south 30 degrees 20 minutes 10 seconds to east.

The site survey begins with a **monument,** known as the Point of Beginning (POB). This point is a fixed location that is generally an iron rod driven into the ground. Find the POB in the lower right corner of Figure 3.24. As you look at Figure 3.24, see the example of property line labeling on the west property line of Lot 3, which is 85.00 for the length and S 02°48′22″ W for the bearing.

Many sites have property lines that are arc shaped, such as the lines around the **cul-de-sac** in Figure 3.24. A cul-de-sac is a street that is closed at one end and usually has a large radius turning area. Property line arcs are often labeled directly on the property line, such as $R = 50.00$ $L = 48.39$, where R is the **radius** of the arc, and L is the **length of the arc,** as shown in Figure 3.26a. Another method is to label the arc property line of each lot with characters such as C1, C2, C3 through C9, as shown in Figure 3.24. These characters correlate with the information found in a **curve table**, as in Figure 3.24. Figure 3.26a shows how the radius and arc lengths are measured. Site surveys also typically show a **delta angle** for an arc, which is represented by the symbol Δ. The delta angle is the **included angle** of the arc. The included angle is the angle formed between the center and the end points of the arc, as shown in Figure 3.26b. Figure 3.26c shows the final labeling of the curve.

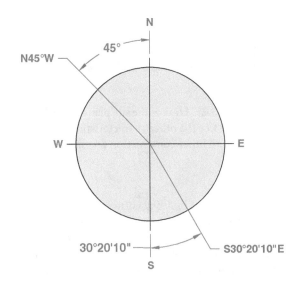

Figure 3.25 Bearings are measured clockwise or counterclockwise from north or south. For example, a reading 45° from north to west is labeled N 45° W, as shown here. (*Previously published in Modern Residential Construction Practices, Routledge, 2017 and reproduced here with permission.*)

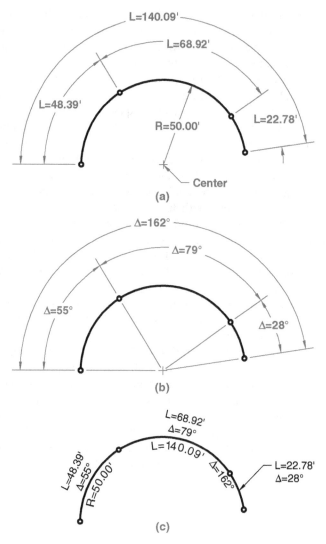

Figure 3.26   (*a*) Measuring radius and arc lengths. (*b*) The included angle is the angle formed between the center and the end points of the arc. (*c*) The final labeling of the curve. (*Previously published in Modern Residential Construction Practices, Routledge, 2017 and reproduced here with permission.*)

A complete site plan drawing is shown in Figure 3.27. Look at the site plan and read the distances and bearings of each property line. Also notice the other characteristics found on the site plan, such as building location, elevations, and other construction features.

Reading Survey Stakes

Survey stakes are used to control alignment and grade of building corners, roads, erosion control, and other features during construction. Survey stakes are markers surveyors use in surveying projects to prepare job sites, mark property boundaries, and provide information about claims on natural resources such as timber and minerals. Survey stakes are generally made of natural wood such as fir and hemlock in different sizes and lengths, but they can be

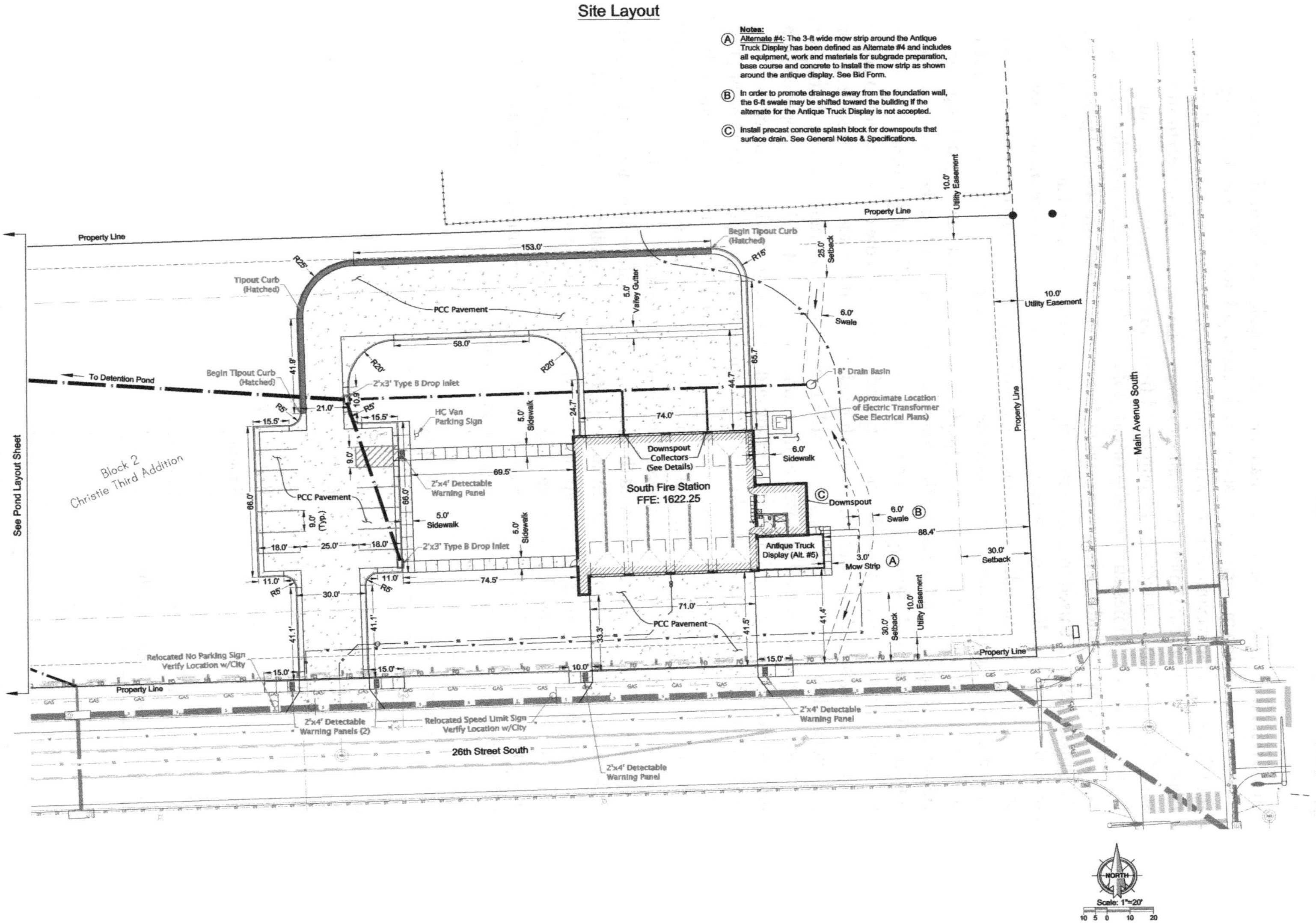

Site Layout

Figure 3.27 A portion of a complete site plan. (*ILG Architects, Alexandria, MN.*)

made from metal, plastic, and other materials. Survey stakes can be made in a range of sizes and colors for different purposes. A survey stake has a pointed end to make it easy to drive into the ground. Surveyors mark or write information on survey stakes to identify property boundaries and provide information about cutting, filling, and grading earth to required specifications. Survey stakes can be purchased from survey supply vendors. Survey stakes also have specific names such as **alignment stakes, offset stakes, centerline stakes, cut stakes, fill stakes, grade stakes, slope stakes, hub stakes, marker stakes,** and **guard stakes.** Alignment stakes are used to align a roadway. Offset stakes are placed at a desired distance from the actual building corner stakes to prevent the loss of reference information. Offset stakes are placed far enough away so they are not damaged or destroyed by equipment being operated in the construction area. Centerline stakes represent the centerline of a construction feature such as a driveway, road, or pipe. Cut stakes indicate a lowering of the ground or elevation. Fill stakes indicate raising the ground or elevation. Grade stakes provide elevation information. **On-grade stakes** indicate the ground, cut, or fill is at the desired grade and does not need additional cut or fill. The term **grade** can refer to the current condition of the land but can have multiple meanings. **Finish grade** means the grade after all required work is performed to cut or fill the area. **Subgrade** is a preliminary grade prior to doing work to finish grade. **Slope stakes** are used to determine the point at which the proposed **slope** intersects the existing ground. The terms **grade** and **slope** can be used interchangeably. **Slope** is described by the ratio of the rise divided by the run between two points on a line or plane, as shown in Figure 3.28. **Hub stakes** are generally 2 × 2 × 12 in. with red and blue painted tops, used mainly to show specific well-defined survey reference points such as finish grade. A **marker stake,** also called a **reference stake,** is usually next to a hub stake, and is a short **lath** stake with a marker card attached that provides survey information for a hub. A **guard stake** is usually a 48-in. lath with paint and flagging used as visibility marker and protection for the guard stake. Lath is a thin strip of wood, typically available in 1/2″ × 2″ × 48″.

Measurements provided on survey stakes are in feet, tenths, and hundredths of a foot using the foot (′) symbol, such as 26.85′, or in meters and centimeters, such as 2.57. The following are abbreviations and symbols found on survey stakes:

Abbreviation or Symbol	Meaning
O or OF	Offset
C	Cut
F	Fill
FG	Finish grade
RP	Reference point
SG	Subgrade
UC	Undercut
CL (Art, set with L over C)	Centerline
OO (Art, set with line through)	Grade

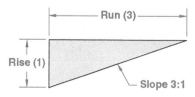

Figure 3.28 Slope is described by the ratio of the rise divided by the run between two points on a line or plane.

Most survey stakes are raw wood with color paint or ribbon to designate a specific location or operation on the construction site. Common stakes are 1 × 2 × 16 in., or lath stakes, which are 1/2 × 2 × 48 in. The top of a band of blue paint or a blue ribbon near the bottom of the survey stake represents the finish grade location. The top of the survey stake can be painted or can have a ribbon with a tail streamer. The color designates a specific location or operation on the construction site. For example, blue ribbon stakes can represent the foundation line. Follow the layout of the blue ribbon stakes to see the foundation outline. Yellow ribbon stakes can be used to locate a road centerline. Follow the layout of the yellow ribbon stakes to see the road centerline, for example. The following examples show a variety of survey stake colors and information. Figure 3.29 shows a survey stake with color code ribbon, survey log reference number, finish grade elevation, and grade ribbon. Figure 3.30 shows a survey stake with location from hub, centerline symbol, and grade symbol.

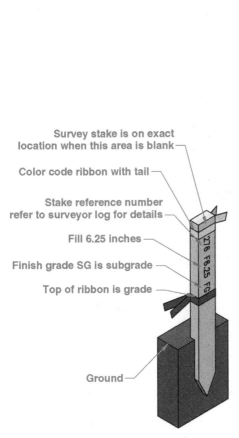

Figure 3.29 Survey stake with color code ribbon, survey log reference number, finish grade elevation, and grade ribbon. (*Previously published in Modern Residential Construction Practices, Routledge, 2017 and reproduced here with permission.*)

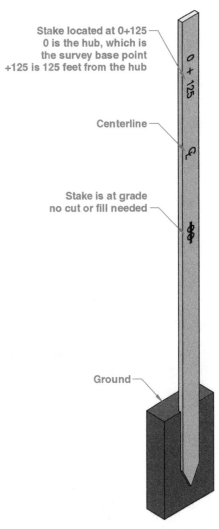

Figure 3.30 Survey stake with location from hub, centerline symbol, and grade symbol.

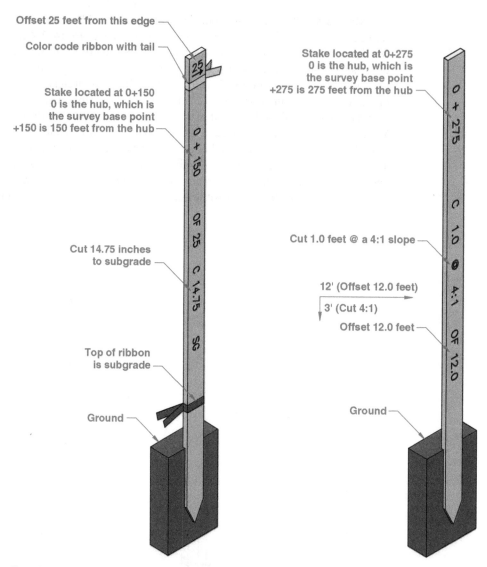

Figure 3.31 Survey stake with offset distance, color code ribbon, location from hub, cut depth to subgrade, and grade ribbon.

Figure 3.32 Survey stake with location from hub, cut depth at 4:1 slope, and offset distance and depth.

Figure 3.31 shows a survey stake with offset distance, color code ribbon, location from hub, cut depth to subgrade, and grade ribbon. Figure 3.32 shows a survey stake with location from hub, cut depth at 4:1 slope, and offset distance and depth.

Hub and guard stakes were briefly defined earlier. 2 × 2 × 9-, 12-, or 18-in hard-wooden stakes are used as **hubs.** Hubs are important survey starting points or they mark important survey points, work points, or reference points, which are to remain in place for future use. Hub stakes are normally driven flush with the ground and a **survey tack** is set in the top of the hub to mark the exact survey point. A survey tack is a small, sharp, broad-headed nail. The hub is driven flush to make it less at risk to damage and its location is usually marked by a guard stake. Guard stakes are generally $1/2 \times 2 \times 18$ in. and are used to mark and describe hub locations. Guard stakes are driven in the ground at an angle over the hub stake and a flagged lath stake is often placed near to help clearly identify the location to help keep the stakes safe from damage. Guard stakes usually have information describing the point, as shown in Figure 3.33.

A **grade control station** provides a 2×2-in. stake offset from the construction site and identified with three flagged lath stakes protecting it from damage. The grade control station provides grade and offset information to the specific grade at construction.

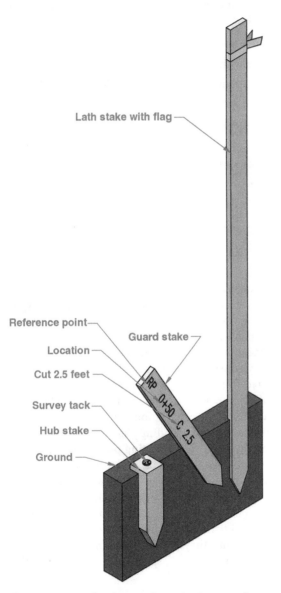

Lath stake with flag

Reference point
Location
Cut 2.5 feet
Survey tack
Hub stake
Ground

Guard stake

RP
0+50   C 2.5

Figure 3.33   Guard stakes are driven in the ground at an angle over the hub stake and a flagged lath stake is often placed near to help clearly identify the location to help keep the stakes safe from damage. Guard stakes usually have information describing the point.

GRADING PLAN

A **grading plan** shows the existing and proposed contours, elevations, and grades of the construction site before and after **excavation.** A grading plan includes the excavation for the building that is referred to as a **dig-out** in the construction industry. The grading plan shows where areas need to be cut and filled and shows structures such as retaining walls. This is referred to as **cut and fill.** Cut and fill is the excavation process involving the removal of earth, which is the cut, and moving earth to another location, which is the fill. The grading plan for a specific construction site can be included on the site plan. Figure 3.34 shows the site plan with grading information for the excavation. Information found on the site plan and grading plan can include

- Property boundaries with dimensions and bearings
- Structure perimeter with dimensions
- Roads, driveways, parking area patios, and walkways with dimensions
- Retaining walls with specifications and dimensions
- Sanitary sewers, storm sewers, and utilities such as electrical, water, and cables
- Contour lines
- Cut and fill specifications
- **Cutting plane line** for profile
- **Profile** drawing
- Roads and streets
- Title and scale
- North symbol

The grading plan usually has a separate **profile** drawing, which is a section view cut through the construction site. A profile can also be called a **site section.** A profile is a vertical section of the surface of the earth, and underlying earth that is taken along any desired fixed cutting-plane line. The profile of a construction site is usually through the building excavation location, but more than one profile can be drawn as needed. The profile for road and utility construction is normally placed along the centerline of the feature. Profiles are created from the contour lines at the section location. The contour map and its related profile are commonly referred to as the **plan and profile.** Profiles can be used to show road grades, site excavation, sanitary and storm sewer, and utilities. Figure 3.35 shows the profile through a grading plan of a pond. The location of the profile is established on the grading plan using a **cutting plane** symbol as shown with detailed options in Figure 3.36. Cutting plane symbols are represented as an arrow and line. The line represents where the cut is made through the site and the arrow shows the line of sight when looking at the site plan-related profile section. The cutting plane symbol is generally a circle divided in half with a letter in the top half and a number in the bottom half. The letter is the section identification, such as A used in this example. If there are other sections in this set of plans, they continue in alphabetical order, such as B and C. The number in the bottom half, such as 7, specifies the page where the section is found in the set of plans.

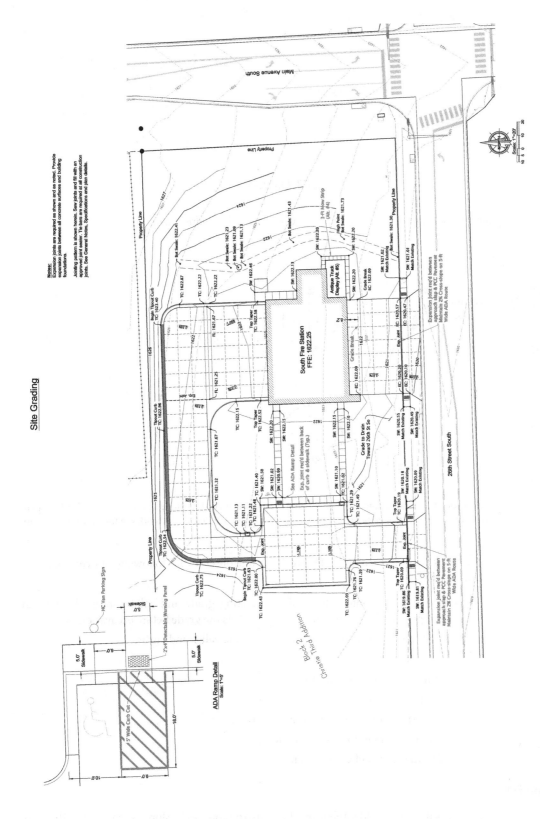

Figure 3.34 Site plan with grading information for the excavation. *(JLG Architects, Alexandria, MN.)*

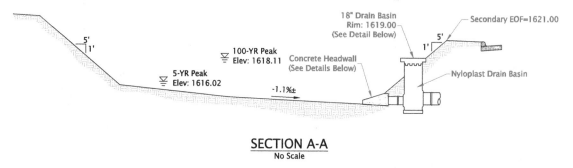

SECTION A-A
No Scale

Figure 3.35   A profile through a grading plan of a pond. (*JLG Architects, Alexandria, MN.*)

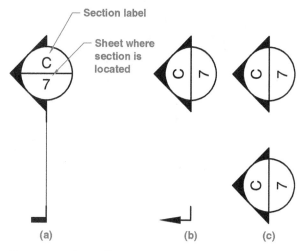

Figure 3.36   The profile location is established on the grading plan using a cutting-plane line symbol. (*a*) Cutting-plane labeled vertically. (*b*) Cutting-plane labeled horizontally. (*c*) Cutting-plane labeled on both sides of the cutting-plane line.

Look at the site plan in Figure 3.34 and observe the following drawing and note information provided.

- The dashed contour lines of the existing grade and solid contour lines of the proposed excavation. The existing grade is the grade before excavation.
- Site features such as roads, driveways, sidewalks, and curbs.
- Elevations and related notes.
- Building outline and location.
- Property lines.
- Erosion control swales and slope arrows.
- Expansion joint locations and notes.
- North arrow and graphic scale.

Cut and Fill

A ratio value can be identified on the construction grade slope, such as 2:1. This ratio is the slopes of cut and fill from the excavation site measured in feet of horizontal run to feet of vertical rise called the **angle of repose.** Two units of run to 1 unit of rise is specified as 2:1. The actual angle of repose for cuts and fills is normally determined by approved soils engineering or engineering geology reports. The slope of cut surfaces can be no steeper than is safe for the intended use and cannot exceed 1 unit vertical in 2 units horizontal (1:2). Alternative designs can be allowed if soils engineering and engineering geology reports state that the site has been investigated and give an opinion that a cut at a steeper slope is stable and does not create a hazard to property. Fill slopes cannot be constructed on natural slopes steeper than 1 unit vertical in 2 units horizontal (1:2). In order to provide a bond with the new fill, the ground surface must be prepared to receive fill by removing vegetation, previous unstable fill material, topsoil, and other unsuitable materials. Other requirements include soil engineering where stability, steeper slopes, and heights are issues. Soil engineering can require **benching** the fill into sound material and specific drainage and construction methods. A **bench** is a fairly level step excavated into the earth material on which fill is placed. Any grading plan with a cut and fill design must be professionally engineered. A civil engineer is responsible for the proper design and construction.

Site Grading

Site grading is the construction process of changing the elevation and slope of the land to civil engineering specifications and to site survey requirements at and near the proposed construction site. A grading plan and related drawings is prepared to establish areas of the site to be graded, identify drainage patterns, and establish runoff velocities into existing streams, rivers, and lakes. Also included in the grading plan is information about when earthwork will start and stop, establish the degree and length of finished slopes, and describe where and how excess material is disposed, or where **borrow** materials is obtained if needed. The term **borrow** or **borrow pit,** also known as a **sand box,** is an area where material such as soil, gravel, or sand has been dug for use at another location. Land grading should be a key consideration for **construction sequencing.** Construction sequencing is a specified work schedule that manages the coordination and timing of construction activities and stages from the start of land development to final occupancy. Each stage of construction can also have a sequence. Site preparation, for example, has the sequence of surveying, excavation activities, and the installation of erosion and **sediment** control measures. Sediment is earth particles and other material that drop out of the stormwater runoff. One goal of a construction sequence schedule is to reduce on-site erosion and off-site sedimentation and minimize exposed soils at any given time during construction. Incorporate in the plan any berms, diversions, and other stormwater practices that require excavation and filling. Part of the site grading process involves cut and fill described earlier. Site grading removes existing earth down to firm undisturbed soil to provide for a solid base upon which to build the foundation. Earth removal to firm undisturbed soil is common on most sites, but it is sometimes necessary to bring in material to fill or build up a construction site. When this is done, the new material must be of a quality specified by a civil engineer and be compacted to specifications required by civil engineering. Site grading is also used to level and shape the land for landscaping and to provide proper drainage to control the flow of stormwater. Proper site grading is important to help reduce erosion and keep stormwater from entering the foundation and from adversely impacting adjacent properties or pollute downstream waters. Site grading can also include moving earth to create berms, channels, ditches, or vegetated swales to capture or filter stormwater and keep it away from the building and adjacent properties. **Detention basins** can also be created to temporarily detain stormwater and release it slowly; and prevent sediment-laden water from leaving the site, or both. Another method used to control water runoff is **site benching,** where each adjacent site or lot is graded separately toward the adjoining property line. This provides a slight elevation reduction at the property line allowing water to runoff away from structures. Figure 3.37 shows a variety of site grading applications.

It is best to use surface runoff control methods when possible, because these techniques allow the water runoff to soak into the soil instead of running away from the site. Alternately, **French drains** can be used to capture and drain water away from the structure and direct it to local stormwater channels. French drains are most often used to drain surface water when the topography is not favorable to surface flow, such as a flat or level site. French drains are common drainage systems that can be used to prevent ground and surface water from penetrating or damaging building foundations. Adjacent topsoil is graded

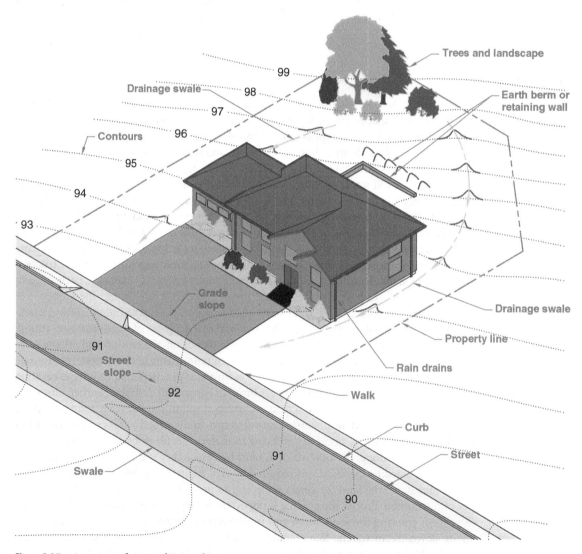

Figure 3.37   A variety of site grading applications.

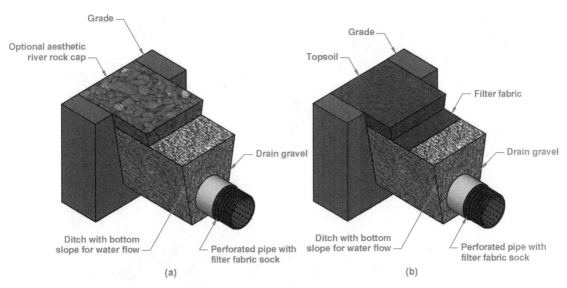

Figure 3.38   French drain system. (*Previously published in Modern Residential Construction Practices, Routledge, 2017 and reproduced here with permission.*) (*a*) Small drain gravel layer at the bottom of the ditch followed by a perforated pipe with filter fabric sock and additional drain gravel to the surface. (*b*) Drain gravel capped with filter fabric and then covered with soil.

toward the French drain to direct water into the drain, as shown in Figure 3.38. A **French drain** is a piping system in a ditch with the bottom sloped in the desired direction of water flow. There is normally a small drain gravel layer at the bottom of the ditch followed by a perforated pipe with **filter fabric sock** and additional drain gravel to the surface, with an optional top layer of **river rock** for landscaping appearance. River rock is available in many sizes and colors, but small naturally gray rock is common for this application. A **filter fabric** sock is a filter fabric product made to fit over the drain pipe to keep material from plugging the drain pipe. Filter fabric is a water-filtering porous material and soil stabilization product made to allow water to pass through, but keep dirt and other materials from passing through. River rock is smooth and circular, because it comes from river and creek bottoms where water force and friction has removed sharp edges. River rock maintains spaces between the rocks allowing water to flow through, compared with crushed rock, which is angular rock that packs tightly and can minimize water flow. French drains can also be capped with drain fabric and then covered with soil and planted with grass. The drain fabric keeps the soil from entering and clogging the drain system below. This type of French drain system collects subsurface water but may not collect surface water unless the cover material is porous, which is the common use of a French drain. A **catch basin** can be installed at one end of the French drain system if it is important for the French drain to collect surface water. However, it is generally referred to as a regular drain when a catch basin is used. A catch basin is inlet structure for a drain or drain system and is designed to drain excess rain and groundwater from the adjacent area, as shown in Figure 3.39. The grade is sloped toward the catch basin so water naturally flows into the structure. Catch basins are usually rectangular concrete boxes with a steel **grate** on top to help keep debris from entering the pipe below. A grate is a framework of metal bars in the form of a grille set into a catch basin, wall, pavement, or other location, serving as a cover or guard allowing air, light,

Figure 3.39 A catch basin is inlet structure for a drain or drain system and is designed to drain excess rain and groundwater from the adjacent area.

and water to pass through. The most common use of catch basins is next to **curbs** in road construction where they are placed at specific intervals to collect and drain rainwater from the roadway and adjacent properties. A curb is a raised concrete edging to a road, street, or path. Curbs can also be constructed with brick, stone, or other decorative materials when used in landscaping applications.

Site grading is performed with heavy equipment, such as dozers, backhoes, graders, and excavators, shown in Figure 3.40. Earth can also be loaded into dump trucks for removal from the site. Graders and compactors are also used to level and compact the earth to create a smooth, even surface, or to cover drainage systems. If a site has poor soil quality, it can be necessary to remove existing soil and bring in a new layer of topsoil for final landscaping needs.

Figure 3.40   Site construction equipment. (*a*) Dozer. (*b*) Backhoe. (*c*) Excavator. (*d*) Compactor. (*e*) Grader.

CORNER STAKING, EXCAVATION LAYOUT, AND FOOTING EXCAVATION

Corner staking is placing survey stakes at each corner of the proposed building location. Corner stakes are set by a professional surveyor, because they must be accurate. The surveyor refers to the site plan to establish the exact building corners. A building perimeter line is established by running string lines between each corner stake and marking the lines with paint on the ground or by survey alignment of intermediate building line stakes for large structures. For concrete slab construction, the footing excavation is done to the

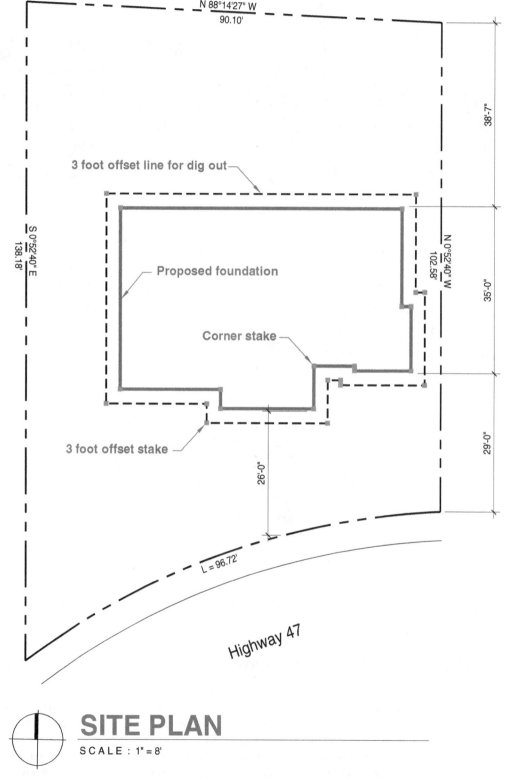

N 88°14'27" W
90.10'

38'-7"

3 foot offset line for dig out

S 0°52'40" E
138.18'

N 0°52'40" W
102.58'

35'-0"

Proposed foundation

Corner stake

29'-0"

3 foot offset stake

26'-0"

L = 96.72'

Highway 47

SITE PLAN

SCALE : 1" = 8'

Figure 3.41   Corner stakes and offset stakes for a site plan.

corner stakes. When the building has a concrete foundation construction, the surveyor sets 2 to 3 ft (590 to 880 mm) offset **stakes,** which provides for the additional space needed to set concrete forms for footings and foundation walls. The offset stakes can be set further from the corner stakes and building line for a large structure with expanded concrete footings. The excavator removes the earth to the desired depth everywhere inside the offset lines. This is called the dig-out. The dig-out depth for a foundation is determined by type of structure, geological and civil engineering as specified on the plans. Figure 3.41 shows the corner stakes and offset stakes for a site plan. The exact building corner stakes are placed again and a new outline is established after the excavation is complete. Now, the concrete crew can set forms for footings and foundation walls.

ENVIRONMENTAL PROTECTION

Protecting existing trees and other vegetation, streams, rivers, lakes, and neighboring green spaces is extremely important when preparing a construction site and throughout construction to final landscaping. Environmental protection is part of the design process that continues through the entire construction process. Figure 3.42 shows an erosion control plan. As you look at the erosion control plan in Figure 3.42, notice these erosion control features:

- The X-X-X line is a **silt fence,** which is an important part of erosion control. The silt fence stops rainwater runoff from the property. The use of silt fences is described later in this chapter.

- The 15′ **PSDE** is a **stormwater easement,** which is a low area or swale between properties that allows stormwater to drain away from buildings. PSDE stands for Public or Private Storm Drain Easement.

- The covered **stockpiles** are noted for a couple of reasons. One is to provide space on the site for the stockpiling of materials, and the other is to note that stockpiles are covered to minimize erosion of the pile. In this application, the term **stockpile** refers to the earth material that is piled and stored during excavation for later use on the site, such as backfill. Stockpile can also refer to the on-site storage of other construction material, such as lumber and steel, or the piling of construction waste for later removal.

- A gravel exit is required to reduce mud and dirt tracking from vehicle tires onto streets. No construction or work is allowed outside of the silt fence, even on land that is part of the lot.

Contractors can be required to have a **Stormwater Pollution Prevention Plan (SWPPP)** as part of their operating procedures. SWPPPs are a requirement of the National Pollutant Discharge Elimination System (NPDES) that regulates water quality when associated with construction or industrial activities. The SWPPP addresses all pollutants and their sources, including sources of sediment associated with construction, construction site erosion, and all other activities associated with construction activity and controlled through the implementation of **Best Management Practices (BMPs).** BMPs are procedures that provide effective and practical means of achieving construction goals while making the best use of available resources. Many local jurisdictions and land development projects have environmental protection requirements. The civil engineer and contractor must determine these requirements and include them during the design process and update them

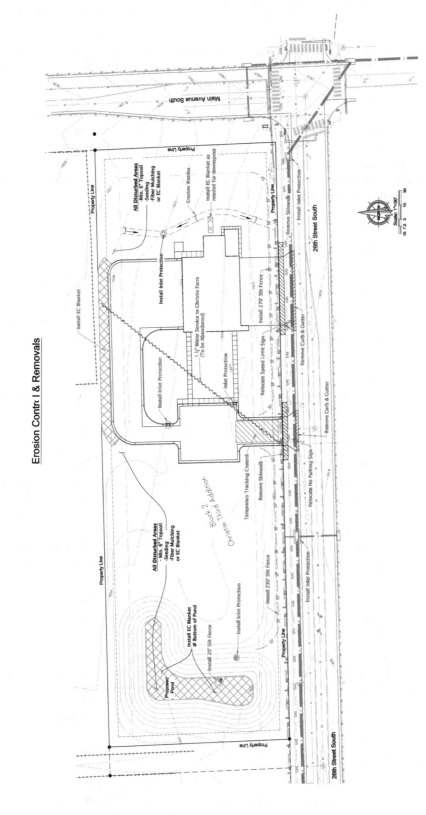

Figure 3.42 Erosion control plan. (*JLG Architects, Alexandria, MN.*)

daily throughout the project construction. A SWPPP is not limited to the following but can include

- Protection of natural resources such as topsoil, natural green areas, and existing trees, ground cover, and plants
- Stormwater management and control
- Air quality control, including exhaust emissions and dust control
- Construction noise control and restricted operation hours as needed
- Prevention of hazardous materials and wastes releases into the environment
- Construction waste disposal and recycling
- Protection of historical and archeological sites

Some sites have good **topsoil** and some have poor or no topsoil. Topsoil is the upper part of the soil rich in degraded vegetation, which is usually rich in nutrients and most favorable for landscaping lawn and plant growth. Good topsoil is valuable and should be saved on the site and re-spread after construction. After the footings and foundation are completed and all groundwork is done, the excavated area is **backfilled** with earth and topsoil as shown in Figure 3.43. Backfill is the refilling of an excavated area with material from the excavation or trucked in as needed. When a site has poor or no topsoil, trucked in quality topsoil is recommended for landscaping.

Natural areas of vegetation should be protected, because they give natural erosion control, hold stormwater, provide **biofiltration,** and aesthetic values to a site during and after construction. A natural area is a geographical area that has a physical and cultural originality developed through natural growth rather than design or planning. Biofiltration is a pollution control technique using living material to capture and biologically degrade process pollutants. Natural areas process higher quantities of stormwater runoff and have higher filtering capacity, require less maintenance than newly planted areas, do not require

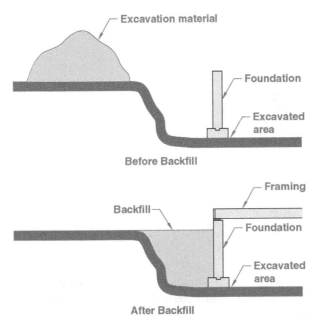

Figure 3.43 After the footings and foundation are completed and all groundwork is done, the excavated area is backfilled with earth and topsoil.

time to establish, existing root structures are denser, and living root systems help hold soil in place. Natural areas also reduce stormwater runoff by catching rainfall, hold soil particles in place, maintaining the ability of soil to absorb water, and as a result it helps biofiltration, provides buffers and screens against noise. Natural areas also provide immediate habitat for wildlife, enhance local aesthetics, and can increase property value.

Existing trees, especially mature trees, should be protected by designing them into the landscape plan. Mature trees can also increase property value. Trees need to be protected during construction, because any damage to the trunk, roots, and branches can cause a tree to die even months after construction is complete. It is best to install a protective fence around trees to help keep equipment a safe distance away. The protective fence should be as far away from the tree as possible, but no closer than the **drip line.** The drip line is the area directly located under the outer circumference of the tree branches, as shown in Figure 3.44. Tree roots generally extend to and often beyond the drip line, and trees get most of their water at the roots around the drip line. It is also important to keep machinery from compacting the soil near the tree. Water runs off of compacted soils and may not provide adequate nutrients for the tree.

Dust control is required on some projects, which means that the contractor must water the site to keep dust from occurring during construction and until the site has landscaping that eliminates future dust problems.

Construction fuel and other hazardous materials must be properly contained and stored so they do not spill onto the earth and harm the environment. This kind of soil and water pollution can result in a serious fine, and cleanup can be very expensive.

Many locations have noise restrictions. For example, construction cannot start before 7 a.m. and must end by 6 p.m. These requirements need to be determined before construction begins, because an abuse can result in a violation. Additionally, all machinery should be properly muffled and other noise kept to a minimum.

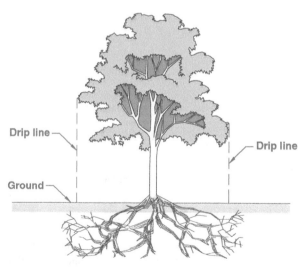

Figure 3.44   The tree drip line is the area directly located under the outer circumference of the tree branches. (*Previously published in Modern Residential Construction Practices, Routledge, 2017 and reproduced here with permission.*)

Construction material waste should be recycled when possible. Some materials can be recycled directly into the same product for reuse. The easiest construction materials to recycle are lumber, drywall, concrete, plywood, metal, wire, and roofing. Other materials can be reconstituted into other usable products. Recycling construction materials has the following advantages:

- Conserves natural resources by reducing the demand for raw materials.
- Conserves energy and water, because manufacturing with recycled materials requires less processing than using raw materials.
- Reduces air and water pollution, because manufacturing from recycled materials is generally a cleaner process.
- Reduces material sent to landfills.
- Protects the environment and health by reclaiming hazardous waste.
- Reduces the cost of construction. Reduced waste equals reduced raw material consumption.

A first step for construction waste recycling is waste reduction. Stockpile partly used materials alongside new material, so tradesmen can easily see what is available. Separate remaining waste on-site. Establish a material sorting plan and have all workers involved. On-site separation can be done at little or no additional cost, and there can be companies or nonprofit organizations in your area that schedule construction material pickup. Nonprofit organizations collect and distribute unused construction materials to needed projects.

Protection of historical and archeological sites is required in sensitive areas. Native American and Hawaiian archeological sites, and other archeological and historic sites must be protected and specific local regulations are critical to follow.

CONSTRUCTION SITE STORMWATER RUNOFF CONTROL

According to the U.S. Environmental Protection Agency (EPA), uncontrolled stormwater runoff from construction sites can significantly impact rivers, lakes, and estuaries. Sediment in water bodies from construction sites can reduce the amount of sunlight reaching aquatic plants, clog fish gills, smother aquatic habitat and spawning areas, and obstruct navigation. The most obvious stormwater runoff control that you have probably seen at or near construction sites is a silt fence, which is described in detail later in this chapter. There are two main categories of control, which are erosion control and sediment control. Many different control methods found in each category are taken in part from the EPA website and described in the following discussion.

Erosion Control

Erosion is the process that causes soil and rock to be removed from the earth surface and deposited in other locations by natural processes such as wind or water flow. Erosion is a natural process, but human activity has dramatically increased global erosion and has caused degradation of certain sensitive areas. Erosion is often accelerated during the construction process due to the removal of vegetation, ground cover, and other naturally occurring erosion control measures. There are many ways to help control, lessen, and prevent erosion. The following briefly describes these methods. Some techniques involve products where manufacturer instructions and specifications should be followed.

Chemical Stabilizers

Chemical stabilizers or soil binders provide temporary soil stabilization. These are typically sprayed on the ground to hold soil in place and minimize erosion from runoff and wind. These materials are easily applied to the surface of the soil, can stabilize areas where vegetation cannot be established, and provide immediate protection.

Compost Blankets

Compost blankets are a layer of compost material placed on the soil in disturbed areas to reduce erosion. The compost absorbs, filters, and slows water flow. Additionally, seeds can be mixed into the compost before it is applied to promote revegetation.

Geotextiles

Geotextiles are porous fabrics also known as **filter fabrics, road rugs,** or **construction fabrics.** Synthetic construction fabrics are used for a variety of purposes such as separators, turf reinforcement, filtration and drainage, and erosion control. For example, turf reinforcement mats are used to stabilize soil and aid in plant growth by holding seeds, fertilizers, and topsoil in place.

Mulching

Mulching is an erosion control practice that uses materials, such as grass, hay, wood chips, wood fibers, or straw, to stabilize exposed or recently planted soil surfaces and is similar to the use of compost described earlier. Mulching is highly recommended and is most effective to prevent erosion from starting and to help establish vegetation growth. In addition to stabilizing soils, mulching can reduce stormwater velocity and improve water absorption. Mulching can also aid plant growth by holding seeds, fertilizers, and topsoil in place, preventing birds from eating seeds, retaining moisture, and insulating plant roots against extreme temperatures.

Riprap

Riprap is a layer of large rocks used to protect soil from erosion on steep slopes or slopes that are unstable because of seepage problems. Riprap can be any large rocks, including boulders, down to 4 to 6 in (100 to 150 mm) rock. Riprap is commonly used to stabilize stream and river side slopes, inlets and outlets for culverts, bridges, slope drains, grade stabilization structures, and storm drains.

Seeding and Sodding

Seeding or **sodding** can be used to provide long-term, short-term, or immediate grass growth on a desired area. **Seeding** is used to control and reduce erosion and sediment loss by establishing permanent vegetative cover, and typically much more effective when used in conjunction with a cover like mulch or blankets. Seeding is economical, adaptable to different site conditions, and allows selection of a variety of plant materials. **Sodding** is an immediate and permanent erosion control practice where grass **sod** is installed on exposed soils. Sod, also called **turf grass,** is grass that has already been planted in sections that are grass and soil held together by roots and fine mesh or other materials. Sod is available in pieces 16 to 24 × 48 in., or 12 to 18 in. wide × 40 to 60 in. long rolls. Sod is more expensive than seed, but necessary when immediate grass and erosion control is required.

Soil Retention Structures

Soil retention structures, while not for erosion or sediment control, are used to hold soil in place on a sloped site. Soil retention structures include wood, concrete, or steel structures. Reinforced soil-retaining structures can be effective when sites have very steep slopes or loose, highly erodible soils. Design of soil retention structures often requires civil or structural engineering for use at a specific site application. A fairly inexpensive soil bracing system uses lumber to support the excavated face of a solid soil slope. Continuous sheeting material, such as steel, concrete, or wood, can be used to continuously cover the entire slope. Concrete, masonry, or wood **retaining walls** can be used to provide permanent slope support. Retaining walls are concrete, masonry, or wood structures designed to restrain soil on a slope between two different elevations. Retaining walls can be purely structural or structural and designed into the finish landscaping, as shown in Figure 3.45.

Soil Roughening

Soil roughening is a temporary erosion control practice often used while grading. Soil roughening involves increasing the relief of the soil surface with horizontal grooves stair-stepping or ripping parallel to the contour of the slope using construction equipment. Slopes that are not fine graded and left in a roughened condition reduce erosion. Soil roughening reduces runoff velocity, increases water penetration, reduces erosion, traps sediment, and prepares the soil for seeding and planting.

Temporary Slope Drain

A **temporary slope drain** can be used to divert stormwater from one elevation to another with a corrugated metal, plastic, or concrete pipe extending from the upper to the lower

Figure 3.45   Retaining walls can be purely structural or structural and designed into the finish landscaping. (©*Redi-Rock International – flickr.*)

Figure 3.46   A temporary slope drain can be used to divert stormwater from one elevation to another with a corrugated metal, plastic, or concrete pipe extending from the upper to the lower elevation. (*Courtesy of Tennessee Department of Environment & Conservation.*)

elevation, as shown in Figure 3.46. A berm is constructed to divert the water into the pipe at the inlet and the outlet is flared and reinforced with rock and riprap to minimize erosion. The pipe has watertight seams and is anchored throughout its length. Temporary slope drains, also called **pipe slope drains,** carry runoff without causing erosion on or at the bottom of the slope. This technique is temporary and commonly used for less than 2 years, or until permanent drainage structures are installed and slopes are permanently stabilized.

Sand Fence

Sand fences, also called **wind fences,** are barriers made of small, evenly spaced wooden slats or fabric used to reduce wind velocity and to trap blowing sand, as shown in Figure 3.47. Sand fences can be used as perimeter controls around open construction sites to keep sediments from blowing off-site on to roads, streams, and adjacent properties. The spaces between the fence slats allow wind and sediment to pass through but reduce wind speed, and deposit sediment along the fence. Wind fences are used on construction sites with large areas of cleared land with loose, fine-textured soils and in arid or coastal regions where sand and dust that commonly blows off-site.

Check Dam

Check dams are small, temporary structures constructed across a swale or channel. Check dams slow the speed of concentrated flow in a water channel. Multiple check dams can be placed along the channel to retain sediment as stormwater runoff flows through the channel. Check dams can be constructed out of gravel, rock, sandbags, logs, or treated lumber. Rocks used in check dams should be 2 to 15 in. The elevation at the center of the check dam should be at least 6 in lower than its two ends. See Figure 3.48.

Figure 3.47   Sand fences, also called wind fences are barriers made of small, evenly spaced wooden slats or fabric used to reduce wind velocity and to trap blowing sand.

Figure 3.48   Check dams slow the speed of concentrated flow in a water channel. Multiple check dams can be placed along the channel to retain sediment as stormwater runoff flows through the channel. (*Courtesy of Tennessee Department of Environment & Conservation.*)

Grass-Lined Channel

A **grass-lined channel** can be an attractive part of the landscaping and used to carry storm-water runoff. Grass or other vegetation lining the channel slows down water flow. Grass-lined channels are excellent for controlling and distributing runoff. Grass channels can include check dams to improve runoff storage, decrease flow rates, and help sediment removal.

Slope Diversion

Slope diversions are constructed by creating channels laterally across slopes to intercept the down-slope flow of runoff and reduce the possibility of erosion. When placed correctly, slope diversions slow potential flow velocity and decrease the overall water volume moving down the slope within a given time frame. Slope diversions can be permanent structures like grass-lined channels, or they can be built for temporary runoff control above and below a construction site. A slope diversion can be a berm or a combination berm and channel built along the perimeter of and within the disturbed part of a site. A slope diversion can be an earthen perimeter control built as a ridge of compacted soil, often accompanied by a ditch or swale with a vegetated lining constructed at the top or base of a sloping site. The slope diversion at the top of a site controls surface runoff from entering the site, and can improve working conditions on the construction site. The slope diversion at the base of a construction site helps control runoff from entering adjacent properties.

Sediment Control

Sediment control is any practice used to detain sediment-loaded stormwater on a construction site so that it does not flow on to other properties or cause water pollution in a stream, river, lake, or ocean. Sediment controls are often used with one or more of the previously described erosion control methods. The following describes the most common sediment control applications.

Dust Control

Some locations require dust control, especially in arid or semiarid regions. Dust can be a threat to the environment and to human health. Heavy construction equipment used in land clearing, excavation, demolition, and construction traffic can cause dust. Dust control measures often depend on topography, existing cover, soil characteristics, and expected rainfall. Dust can be controlled by watering the ground with sprinklers or spraying the ground with a water truck. Vegetation such as grass can be planted to stabilize soil in areas of the site where there is no construction activity. Applying compost to an area can help control dust, but watering can still be necessary to prevent the compost from drying out and becoming dust. Protecting construction areas from wind with barriers such as existing trees and shrubs, or by using hay bales or wind fences can help. **Gravel** placed on roadways can help reduce dust problems caused by trucks kicking up dirt. Gravel is a mixture of three sizes or types of material: stone, sand, and fines. Gravel can also establish a road base and a road surface. Chemical adhesives can be used to control dust. Products include asphalt, latex emulsion, or resin-water emulsions, and calcium chloride. Only use chemical applications that are biodegradable or water soluble and determine their possible effect on the environment before use.

Brush Barrier

Brush barriers are perimeter sediment control structures constructed of material such as small tree branches, root mats, stone, or other debris left over from site clearing. Brush barriers can be covered with a filter fabric to stabilize the structure and improve barrier efficiency.

The edge of the filter fabric cover should be buried in a trench 4 in. deep and 6 in. wide on the drainage side of the barrier. Brush barriers are only possible on sites where brush, tree roots, and rocks are cleared from the site. Brush barriers are temporary, because the material decays over time, or is removed after construction because it is normally unsightly and undesirable. To be more efficient, contractors often prefer to remove the material immediately and use other sediment controls.

Compost Filter Berm

A **compost filter berm** is a long raised bed of compost placed along the site contours to slow runoff and control erosion. The compost filter berm can effectively retain and block sediment. Compost filter berms are usually placed along the perimeter of a site, or at intervals along a slope. Berms can be planted with grass or left unplanted. Grass filter berms are normally left in place and provide long-term control and filtration of stormwater runoff. When construction is finished, unplanted berms are often spread around the site as a soil improvement.

Compost Filter Sock

A **compost filter sock** is compost in a round or oval-shaped mesh tube. The compost filter socks can effectively retain water, and block sediment. Compost filter socks are usually placed along the perimeter of a site, or at intervals along a slope or site contours to slow runoff. Staked compost filter socks can also be used as check dams in small drainage ditches and above a catch basin to slow water flow and trap sediment. Compost filter socks can be planted with grass or left unplanted. Grass filter socks can be left in place and provide long-term control and filtration of stormwater runoff. Filter socks are flexible and can be filled in place or filled and moved into position.

Construction Exit Stabilization

Construction exit stabilization provides gravel on the exit driveway and road to stabilize the ground to keep mud and dirt from vehicle tires leaving the area as mud attached to vehicles. Filter fabric can be placed under the gravel to provide additional stabilization. The gravel used in this application is usually 3 to 4 in. (74 to 98 mm) clean, angular rock depending on the ground conditions. Another advantage of this practice is the gravel can be used as the road and driveway base for final pavement or concrete to complete the driveway. Good surface gravel needs a percentage of crushed rock, which gives strength to support loads, especially in wet weather. Surface gravel also needs a percentage of fines to fill the voids between the gravel for stability. Fines also bind the material together, allowing a gravel road to form a crust and shed water. An additional graveled area can be used as a vehicle washing station at the site entrance. The wash stations allow workers to remove sediment from vehicles before they leave the site.

Filter Rolls

Fiber rolls are tube-shaped erosion-control devices filled with straw, flax, rice, coconut fiber, bark, or composted material. Each roll is wrapped with UV-degradable polypropylene netting for long life or with biodegradable materials like burlap, jute, or outer coconut husks. Fiber rolls have applications similar to compost filter socks previously described. Fiber rolls are commonly placed around stormwater drains to prevent silt from entering the stormwater system.

Gravel Filter Berm

A **gravel filter berm** is a temporary ridge made up of loose gravel, stone, or crushed rock used as an efficient form of sediment control. The gravel filter berm slows and filters runoff and diverts it from an open traffic area. Gravel filter berms are often used where traffic needs to be rerouted because roads are under construction, or in traffic areas within a construction site. Gravel filter berms are constructed using 3/4 to 3 in gavel or crushed rock. Multiple berms can be used depending on the length of the slope. Gravel berms need regular maintenance, because they break down during use.

Sediment Basin and Rock Dam

Sediment basins are used to confine sediment from stormwater runoff in an excavated pool or natural depression. Sediment basins can be designed to retain water permanently, or to only hold water temporarily and release it at a determined rate. The pool depth can be increased by building an earth or rock and gravel perimeter dam. Water can be released from the pool with a perforated outlet riser and drain pipe that allows the water to reach a maximum level, as shown in Figure 3.49. A sediment basin can be a temporary sediment control system for a construction site, or permanently designed into the landscaping to provide long-term sediment control and help slow heavy rain runoff.

Sediment Trap

Sediment traps are temporary excavated pools that allow sediment from construction runoff to collect. The sediment traps are excavated in natural or man-made drainage ways

Figure 3.49   A sediment basin can be a temporary sediment control system for a construction site, or permanently designed into the landscaping to provide long-term sediment control and help slow heavy rain runoff. (*Courtesy of Thomas Carpenter, 5205 Stonebridge Road, West Des Moines, IA.*)

where runoff leaves the construction site. Sediment traps are commonly used at the outlets of stormwater systems where additional sediment and waste can be collected.

Silt Fence

The most obvious stormwater runoff control that you have probably seen on or near construction sites is the **silt fence,** shown in Figure 3.50, is an example of a silt fence used for a check dam. **Silt fence** is a temporary sediment barrier made of porous fabric. Silt fence is used anyplace where it is desired to control sediment runoff and keep sediment from leaving the construction site, from flowing onto roadways, and from entering storm sewers. It primarily functions by ponding water behind it where sedimentation can take place. If you cannot create a ponding effect with the silt fence, it is probably not an appropriate location for the fence. Silt fences keep soil on the construction site until groundwork is completed and permanent soil stabilization measures are in place. Silt fence is supported by 2 × 2 in. (50 mm) wood stakes or steel T-posts are placed at 5 to 8 ft (1.5 to 2.4 m) centers. The stakes are driven into the ground to hold the silt fence in place.

Silt Fence Installation The following is taken in part from the article *Proper Silt Fence Installation* by Tom Carpenter, Certified Professional in Erosion and Sediment Control (CPESC). Properly installed silt fence is the most effective temporary sediment control device available. It can withstand concentrated flows, heavy winds, and potentially retain up

Figure 3.50   An example of a silt fence used for a check dam. (*Courtesy of Thomas Carpenter, 5205 Stonebridge Road, West Des Moines, IA.*)

to 20 tons (18 metric tons) of sediment per run of fence. Silt fence is inexpensive and fairly easy to install. Proper silt fence installation has these six practices to be effective: placement, quantity, installation, compaction, posting, and attachment. Silt fence placement refers to where and in what shape silt fence is placed on site. The key to proper placement is that silt fence must pond and filter water. For example, the silt fence is improperly placed if water can run around the end or ends. Silt fence must be placed where it can store water, with the ends higher than the interior. The location may not be good for silt fence use if the space is too small or too steep for ponding to occur. An arc shape or J-hook shape is often used to create a storage area. Figure 3.51 shows how the J-hook placement creates ponding area for sedimentation. Long runs should be avoided, and broken up into smaller segments. Long runs usually concentrate runoff in one area where runoff can overflow the fence. Restrict run lengths to less than 200 linear feet (61 m). On-site observations and adjustment should be made as needed to make sure the silt fence is properly placed. The first goal in silt fence placement is creating a storage area for runoff and filtration, and the second goal is storing the greatest quantity of water without overflows. It is difficult to place long silt fence runs parallel to contours. A better plan is to place multiple arc shaped runs on a large slope, as shown in Figure 3.52. Silt fences can also be installed in a circle, such as around a catch basin for a storm drain to keep sediment from entering the inlet. Figure 3.53 shows a silt

Figure 3.51 An arc shape or J-hook shape is often used to create a storage area. This photo shows how the J-hook placement creates ponding area for sedimentation. (*Courtesy of Thomas Carpenter, 5205 Stonebridge Road, West Des Moines, IA.*)

Figure 3.52 A good silt fence plan is to place multiple arc-shaped runs on a large slope. (*Courtesy of Thomas Carpenter, 5205 Stonebridge Road, West Des Moines, IA.*)

Figure 3.53 Silt fence around a catch basin for a storm drain to keep sediment from entering the inlet. This is an example of a technically correct installation, but poor placement, because the silt fence opening in the circle faces up hill, allowing water to run directly into the storm drain. (*Courtesy of Thomas Carpenter, 5205 Stonebridge Road, West Des Moines, IA.*)

fence placement around a catch basin. Look carefully at the installation in Figure 3.53. This is an example of a technically correct installation, but poor placement, because the silt fence opening in the circle faces uphill, allowing water to run directly into the storm drain. The opening should have been on the downhill side, the installation should have been closed off, or the best installation would have been an arc-shaped silt fence above the catch basin.

Silt fence quantity refers to the length of silt fence used at the control area. A good rule of thumb for silt fence quantity is a maximum of 100 linear feet (30.5 m) per 10,000 ft² (929 m²) of area. A 1-in./h (25-mm/h) rain event produces 6200 gal (23,470 L) of water on just 10,000 ft², so for effective design, use multiple arc-shaped runs to create multiple storage areas, which also reduces the flow velocities.

The two generally accepted methods of silt fence installation are **trenching** and **static slicing.** Trenching is the digging of a ditch to a desired width and depth with a trenching machine or excavation machine such as a backhoe. When trenching is used, the trench is 6 to 12 in. (147 to 294 mm) wide and 10 to 12 in. (245 to 294 mm) deep. One edge of the silt fence fabric is placed to the bottom of the trench and the trench is then backfilled over the silt fence in the ditch. The trench is compacted by driving over with the tires of the tractor or backhoe. **Compaction** refers to increasing the soil density by applying pressure. Quality compaction has a positive effect on silt fence effectiveness, such as resists wash-outs, minimizes water saturating the soil, and stabilizes the installation. **Static slicing** is done with a tractor using an attached slicing machine that inserts a narrow custom-shaped blade at least 10 in into the ground, and at the same time pulls silt fence fabric into the opening created, as the blade is pulled through the ground. The tip of the blade is designed to slightly disrupt the soil upward, preventing horizontal compaction of the soil and at the same time creating the best soil condition for compaction. Static slicing installation does not excavate soil or leave the spoil spread out, provides consistent and dependable silt fence installation, and reduces many labor-related installation problems over trench installation. The tractor is then used to compact the area by driving over the slice joint with the tires.

Post-setting and driving, followed with tying the silt fence fabric to the posts, final-izes the installation. **Post-setting** refers to driving properly spaced posts to a desired depth in the soil. Posts should be 4 to 5 ft (1.2 to 1.5 m) apart where water concentrates, and 6 to 7 ft (1.8 to 2.1 m) in low pressure areas. Posts should be as deep in the ground as the fabric is above the ground. Steel posts are easier to install to the proper depth and have a better life span than wood posts, but hard wood posts are common and less expensive than steel.

Silt fence **attachment** refers to properly securing the fabric to the posts, providing sup-port for at least 18 in. (441 mm) of water and sediment. Silt fence fabric is attached to steel posts using three plastic ties or wire, installed diagonally within the top 8 in. (196 mm) of each post. Silt fence fabric is attached to wood posts with multiple staples, or a small lathe over the fabric is used to secure fabric to wood posts with screws. Figure 3.54 shows an illustration of proper silt fence installation.

Silt Fence Maintenance.   A silt fence has done its job when it is full of sediment. When this happens, a new silt fence should be installed down slope from the full silt fence, because trying to remove sediment from behind a silt fence can be very difficult or impossible. A silt fence requires maintenance by repairing or replacing if it is falling down, or if there is a washout under the fence.

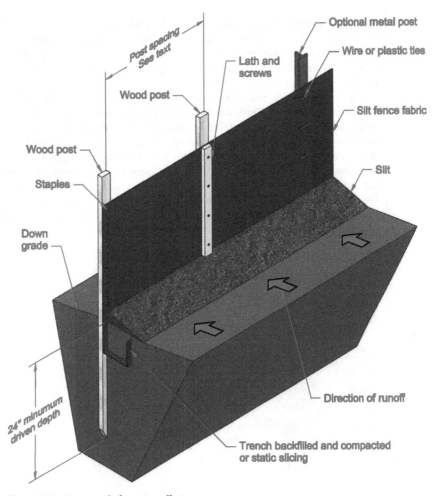

Figure 3.54   Proper silt fence installation.

COMMERCIAL CONSTRUCTION EROSION CONTROL APPLICATION

This chapter demonstrated and illustrated commonly used erosion and sediment control practices. This commercial construction erosion control application, provided by ECBVerdyol, is an example of improved technology being used to provide erosion control in extreme conditions and location.

Arid Lands Project in Mercedes, Texas

The following conditions and issues were identified at this location where immediate quality erosion control intervention was needed:

- Poor soil fertility and quality—less than 0.8 percent organic content
- No supplemental watering scheduled
- High visibility and erosion potential made revegetation critical on the site
- Sandy soil

A significant problem related to this project is a side slope from a roadway that ended at the fence line of private property, making long-term vegetation necessary to prevent possible erosion. Climatic conditions in the area are exceedingly hot and dry, with summer temperatures reaching triple digits, and slight average rainfall. Adequate densities of vegetation are regularly difficult to achieve in this region.

Due to low fertility and poor topsoil quality available on the site, the contractor recommended using **Verdyol Biotic Earth™**, a **Biotic Soil Amendment (BSA)**, to achieve long-term vegetation on the site at acceptable densities and without the need for constant supplemental watering and fertilizer applications. Verdyol Biotic Earth is the original BSA used to achieve sustainable vegetation on less than ideal soils. Important components include sustainably harvested peat moss, flexible flax fibers, and beneficial soil bacteria and mycorrhizae. BSAs are used to improve poor soils by adding organic matter, water holding capacity, and soil life. BSAs are generally hydraulically applied. In late summer of 2016, the contractor hydraulically applied Verdyol Biotic Earth at a rate of 3500 lb/acre, seed, and **tackifier** at a rate of 35 lb/acre over the site. Initial site preparation is shown in **Figure 3.55a**. Straw erosion control

(a)

(b)

Figure 3.55 (*a*) Initial site preparation demonstrates the before erosion control completion. ECBVerdyol. (*b*) The vegetation on the site 2 years since erosion control installation. ECBVerdyol.

blankets were installed in the following days to provide temporary erosion control to the site. Tackifiers are a tacking agent material, of various chemical compositions, used as an additive in hydraulically applied erosion control or soil amendments to increase product to soil adhesion.

The results yielded quality vegetation that was well established on the site 6 weeks after the application, even with the site being seeded in the hottest and most difficult season for installation. Soil loss due to erosion was also minimized. The client has been pleased with the results achieved by this project. Figure 3.55b shows the vegetation on the site 2 years after installation.

Go to Downloads & Resources tab at **www.mhprofessional.com/Commercial Building Construction-Ancillaries** to access the images provided in this chapter and correlate with the content of this chapter. Here you can look at the chapter figures on your computer screen to pan and zoom in and out for better observation, because full drawings reduced to fit on a textbook page are often difficult to read. The image bank also includes the complete set of plans for the Brookings South Main Fire Station used as examples throughout this textbook.

Go to **Downloads & Resources tab at www.mhprofessional.com/Commercial Building Construction-Ancillaries** to access the test and creative thinking problems for this chapter. The chapter test can be used for review or to evaluate content knowledge depending on your course objectives. Creative thinking problems are provided to help expand your knowledge by researching given subjects.

Print reading problems have questions that ask you to find information on the Brookings South Main Fire Station plans. The print reading problems can help you become familiar with the format of construction documents and reinforce your ability to seek specific information. This is an important skill for you to master as you prepare to enter the construction industry. Go to **Downloads & Resources tab at www.mhprofessional.com/Commercial Building Construction-Ancillaries** to access the complete set of plans for the Brookings South Main Fire Station.

CHAPTER 4

Concrete Construction and Foundation Systems

INTRODUCTION

Different types of construction methods relate directly to the materials used, the area of the country where the construction takes place, the type of structure to be built, and even the office practices of the architect or engineer where the designs are created. You should have knowledge of various construction materials and techniques. This chapter introduces the practices and materials used for concrete construction and foundation construction practices. Additional resources should be used for reference, because each construction method discussed has volumes of general and vendor information available. Another valuable way to learn about construction is to visit job sites to talk to contractors and see firsthand how buildings are constructed.

CONCRETE CONSTRUCTION

Concrete is a mixture of Portland cement, sand, gravel (stones, crushed rock), and water. This mixture is poured into **forms** that are built of wood or other materials to contain the concrete mix in the desired shape until it is hard. Concrete is a fundamental material used for building **foundations**. A foundation is the system used to support the building loads and is usually made up of walls, footings, and piers. The term **foundation** is used in many areas to refer to the **footing.** The footing is the lowest member of the foundation system used to spread the loads of the structure across supporting earth or other material. Concrete is also used in commercial applications for wall and floor systems. Commercial buildings typically use concrete foundations with steel-reinforcing, depending on the structural requirements.

Earth is generally excavated before a concrete structure is built. **Excavation** refers to removing earth for construction purposes. Before a concrete structure can be built, earth usually needs to be excavated down to firm, undisturbed supporting soil. Excavation is described in detail in Chapter 3, *Construction Site and Excavation*. Detailed engineering specifications are often placed on the drawing identifying the amount of **bearing pressure** required. The bearing pressure is normally the number of pounds per square foot of pressure the soil is engineered to support. A concrete material symbol is generally used when concrete foundations are provided in section view, as shown in Figure 4.1. **Concrete slabs**

149

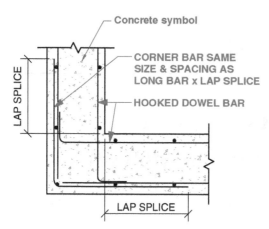

Figure 4.1 A concrete material symbol is used when concrete structures are provided in section view. (*Courtesy of JLG Architects, Alexandria, MN.*)

are also normally found on the foundation drawings. A concrete slab, called a **slab** is a concrete floor system, typically made of poured concrete at ground level. Concrete slabs also normally need to be constructed on firm, undisturbed soil, or other material with engineering specifications provided. The soil compaction specifications are determined by a **geotechnical engineer.** A geotechnical engineer is a civil engineering involved in engineering the performance of construction taking place on the surface or in the ground, using principles of soil and rock mechanics to investigate and monitor site conditions, earthwork and foundation construction. In many cases, a **soil support system** of compacted gravel is specified below the concrete slab. The soil support system can be made up of a **subgrade, subbase,** and **base** gravel layers. Subgrade is the native soil that meets engineering requirements or soil improved by compacting to engineering specifications. **Subbase** is a layer of compacted gravel, such as crushed rock with 10 to 20 percent **fines,** on top of the subgrade. Base is the layer of material on top of the subbase and directly under the slab. Fines describes small miscellaneous gravel, dirt, and debris sizes found in any gravel and crushed rock. All gravel has fines, but the amount of fines allowed depends on the gravel specifications. The under-slab gravel provides for a clean and level or accurate slope surface on which to pour the concrete slab, and it provides for compaction as determined by structural engineering. Not all soil support system layers are always required. Under-slab gravel specifications are determined by engineering and specified on the plans. Concrete and gravel material symbols are used on section view drawings, as shown in Figure 4.2. A drawing note representing gravel under the slab might read:

4″ MIN. 3/4 **MINUS** COMPACTED GRAVEL OVER FIRM UNDISTURBED SOIL

The term **minus** in the previous note refers to gravel containing fines, which is good for compacting. Alternately, **clean gravel** contains no fines or small particles of ground rock, which is normally not used for compacting but is used for good drainage.

Concrete can be poured in place at the job site, formed at the job site and lifted into place, or formed off-site and delivered ready to be erected into place. Concrete alone has excellent **compression strength** qualities but has little tension strength. Compressive strength is the force applied by weight. Steel added to concrete improves the **tension** properties or **tensile strength** of the material. Tension is caused by stretching and tensile strength is the strength the material has against pulling forces. Concrete poured around steel bars placed in the forms is known as **reinforced concrete.** The steel reinforcing bars are called **rebar.** Steel is the best choice for reinforcing concrete because its **coefficient of**

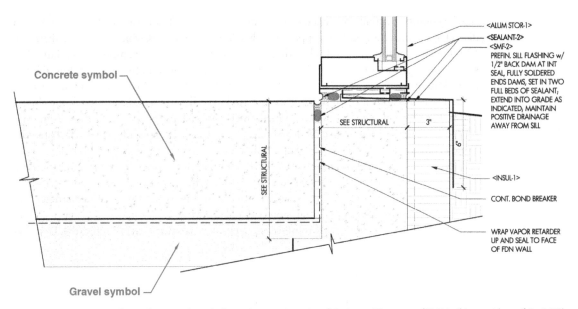

Concrete symbol

<ALUM STOR-1>
<SEALANT-2>
<SMF-2>
PREFIN. SILL FLASHING w/
1/2" BACK DAM AT INT
SEAL, FULLY SOLDERED
ENDS DAMS, SET IN TWO
FULL BEDS OF SEALANT;
EXTEND INTO GRADE AS
INDICATED; MAINTAIN
POSITIVE DRAINAGE
AWAY FROM SILL

SEE STRUCTURAL   3"

SEE STRUCTURAL

SEE STRUCTURAL

<INSUL-1>

CONT. BOND BREAKER

WRAP VAPOR RETARDER
UP AND SEAL TO FACE
OF FDN WALL

Gravel symbol

Figure 4.2   Concrete and gravel material symbols used on section view drawings. (*Courtesy of JLG Architects, Alexandria, MN.*)

thermal expansion is almost the same as cured concrete. The resulting structure has concrete to resist the compressive stress and steel to resist the tensile stress caused by the loads acting on the structure. Coefficient of thermal expansion is the amount of expansion per unit length of a material resulting from 1° change in temperature.

Concrete Reinforcing

Steel reinforcing is available in a number of sizes. Steel reinforcing is generally **deformed** steel bars. Deformed reinforcing bars have raised ridges to hold better in concrete by increasing the bond between the concrete and steel, as shown in Figure 4.3. Steel reinforcing bars are

Figure 4.3   Deformed reinforcing bars have raised ridges to hold better in concrete by increasing the bond between the concrete and steel.

sized by number, usually preceded by the number symbol (#), starting at #3, which is 3/8″ in diameter, and increasing in size at approximately 1/8″ intervals to #18, which is 2-1/4″ in diameter. #4 rebar is a commonly used rebar that is 1/2″ in diameter, for example. Canadian rebar sizes are designated in millimeters in 5 to 10 mm increments and the size is followed by M to indicate metric, for example, 10M, 15M, 20M, 25M, 30M, 35M, 45M, and 55M. Metric rebar size is the same as inch rebar size converted to millimeters and rounded to the closest 5 mm. For example, #4 rebar has a 0.5 in. diameter and 0.5 in. × 25.4 = 12.7 mm. Rounding 12.7 to 13 makes a 13M metric rebar specification, where M represents metric. The formula is inch × 25.4 = mm, where 25.4 is a multiplier used to convert inches to millimeters.

Welded wire reinforcement (WWR) is another steel concrete reinforcing method. WWR is steel wires spaced a specified distance apart in a square or rectangular grid, and the wires are welded together, as shown in Figure 4.4. Figure 4.4a shows the diameter and spacing of welded wire reinforcing and a detail drawing of a concrete slab application. Figure 4.4b shows welded wire reinforcing used in a structural detail. Figure 4.4c shows

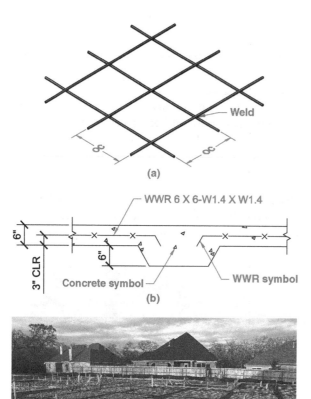

Figure 4.4   Welded wire reinforcement is steel wires spaced a specified distance apart in a square or rectangular grid, and the wires are welded together. (a) The diameter and spacing of welded wire reinforcing. (b) Welded wire reinforcing used in a structural detail. (c) Welded wire reinforcing being used on a construction project.

welded wire reinforcing being used on a construction project. Welded wire reinforcement is specified by giving the spacing of the wire grid, such as 6″ × 6″, which means 6″ on center (OC) each way, followed by the wire type and size of the wires used. The wire type is given as plain or deformed. **Plain wire** is smooth and is designated with a W, and deformed wire is designated with a D. Deformed steel reinforcing was described earlier. Wire sizes are specified by giving the wire diameter area. A few of the available wire sizes are W1.4, D1.4, W2.9, D2.9, W11, D11, W15, D15, W20, D20, W45, and D45. Wires are available in many sizes from W1.4 (3/16″ diameter) to W45 (3/4″ diameter). Most wire sizes are available as either plain or deformed. The wire size numbers relate to the wire area, and determined by dividing the wire designation by 100. For example, W1.4 is 0.014 in.^{2} (1.4 ÷ 100 = 0.014), and W45 is 0.45 in.^{2} The WWR grid can have equally spaced wires and the same wire size, or the spacing and size can be different each way. Sample WWR specifications are written:

$$WWR\ 6 \times 6\text{-}W15 \times W15$$

$$WWR\ 4 \times 18\text{-}D20 \times W8$$

The × symbol in the welded wire reinforcing notes refers to by and is also used as the times symbol when referring to the number of times or when used as the multiplication symbol.

Metric welded wire reinforcement specifications have an M preceding the W or D, and the wire area is given in square millimeters. The metric equivalent for a W2.9-in. wire is MW19, for example. The wire spacing is given in millimeters. A metric welded wire reinforcement specification is

$$WWR\ 102 \times 102\text{-}MW20 \times MW20$$

WWR is purchased in sheets. Sheet sizes can vary, but they are normally 8′ × 15′ and 8′ × 20′, and some can go to 12′ widths and 40′ lengths, depending on shipping and handling equipment to lift and move bundles of sheets. Sheet sizes can also be controlled by state requirements, where restrictions exist for greater than standard loads of 8′ × 40′ and 40,000 lb.

ASTM Standards Related to Steel Reinforcing Bars

The American Society of Testing Materials (ASTM) (www.astm.org) document A615/A615M, *Standard Specification for Deformed and Plain Carbon-Steel Bars for Concrete Reinforcement*, provides standards related to steel reinforcing bars. ASTM A496/A496M, *Standard Specification for Steel Wire, Deformed, for Concrete Reinforcement*, and ASTM A497/A497M, *Standard Specification for Steel Welded Wire Reinforcement, Deformed, for Concrete*, provide standards related to WWR.

CRSI Standards Related to Steel Reinforcing Bars

Consult the *Manual of Standard Practices*, published by the Concrete Reinforcing Steel Institute (CRSI) (www.crsi.org), for complete information and examples related to steel concrete reinforcing.

WRI Standards Related to Steel Reinforcing Bars

Refer to the Wire Reinforcement Institute (WRI) (www.wirereinforcementinsitute.org) documents *TECH FACTS, Excellence Set in Concrete* for detailed information about welded wire reinforcement.

Standard Structural Callouts for Concrete Reinforcing

Commercial construction workers in the office and in the field need to be able to read and interpret structural **callouts** and specifications. The term **callout** is used in the construction industry to identify a given product, process or specification using words, numbers, abbreviations, and symbols to identifying an illustration or part of a drawing feature. The following identifies and describes specific concrete reinforcing products and their related callouts.

Anchor Bolts

An **anchor bolt (AB)** is a bolt embedded in concrete used to hold structural members in place. Anchor bolts are shown and specified on a drawing, by providing the quantity, diameter, type, length, spacing on center (OC), and projection (PROJ) of the thread out of the concrete. The quantity can be displayed followed by a dash or in parentheses. The W is the abbreviation for the word with and can also be abbreviated W/. The following is an example anchor bolt specification:

$$(12)\text{-}3/4'' \ \varnothing \times 12'' \ \text{STD AB } 24'' \ \text{OC W/3'' PROJ}$$

Rebar Specifications

Rebar products and terminology were introduced earlier in this chapter. Rebar is specified on a drawing by giving the quantity (if required), bar size, length (if required), spacing in inches or millimeters on center, horizontal, or vertical spacing in inches or millimeters, and bend information (if required). The (if required) note means that this specific information is given only if needed. These specifications are not needed if general space requirements control the application or if the information is given in general specifications. **Deformed steel rebar** is assumed in commercial construction unless otherwise specified. Deformed steel rebar is rod-shaped steel used for reinforcing concrete, having surface irregularities such as ridges to improve the bond with concrete. **Rebar** is an abbreviated term used in the construction industry when referring to steel reinforcing bar. The following provides typical rebar specifications:

$$\#4 @ 12'' \ \text{OC}$$

$$(25) \ \#8 @ 24'' \ \text{OC HORIZ AND } 16'' \ \text{OC VERT}$$

$$\#6 @ 24'' \ \text{OC EA WAY (OR EW)}$$

$$\#5 @ 16'' \ \text{OC} \times 12'' \ \text{OC}$$

Notice the at symbol (@) is used in the previous notes. This symbol or the word AT can be used to separate the material specification from the designated spacing. The word AT or the @ symbol is also sometimes omitted, which is an acceptable option, depending on company or personal preference.

Dimensions of rebar bends are provided in feet and inches, generally without the (') and (") marks given. Metric rebar bend dimensions are given in millimeters. The minimum bend diameter is also often given in the engineering specifications. The following provides typical rebar bend specifications:

$$\#5 \ \text{AT } 16'' \ \text{OC VERT W/ } 90° \times 12 \ \text{BEND}$$

Rebar bend information can also be provided in a bend diagram, as shown in Figure 4.5.

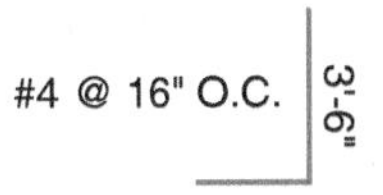

#4 @ 16" O.C. | 3'-6"

1'-6"

Figure 4.5 Rebar bend information can be provided in a bend diagram.

WWR Specifications

WWR products and terminology were introduced earlier in this chapter. WWR is specified by giving the designation WWR, the wire spacing in inches OC, and the wire size. The following provides typical WWR specifications:

$$\text{WWR } 6 \times 6\text{-W1.4} \times \text{W1.4}$$

Reinforcing steel rebar **lap splices** can be specified by giving the length of the lap and a lap location dimension, for example:

#5 REBAR 16″ OC W/ 24″ MIN SPLICE AT FOOTING

Alternately, WWR lap splices can be given in a general note providing the amount of allowable splice overlap, in inches or millimeters, at the crossing wires. A lap splice is the most common method of creating a single structural object from splicing together two rebar or WWR segments. A lap splice is made by overlapping and wiring together two lengths of reinforcement. The most important part of a lap splice is the overlap length between the reinforcement as provided in engineering specifications.

Clear distances are given from the surface of the concrete to the rebar. This dimension is assumed to be to the edge of the rebar or clarified by the abbreviation CLR, as in 3″ CLR. If this dimension is designated to the centerline of the rebar, then OC is specified. The term **clear distance** can have many applications that basically mean the minimum clear spacing between features or objects. For example, the minimum clear space between the edge of concrete to the rebar or WWR, the minimum clear spacing between reinforcement, or the minimum clear spacing between any specific construction members. In many applications, the maximum spacing between members can be specified.

When structural members are embedded in concrete, the size of holes for rebar to pass through is specified with the structural member callout, for example:

#5 AT 16″ OC W/ 13/16 Ø HOLES

Recommended hole sizes for rebar passing through steel or timber with clearance are as follows:

Bar #	Hole Ø	Bar #	Hole Ø
4	11/16	8	1-1/4
5	13/16	9	1-3/8
6	1	10	1-9/16
7	1-1/8		

When rebar must be driven through timber, a tighter hole **tolerance** is recommended. The term **tolerance** refers to the total amount that a specific dimension is permitted to vary. Use of the word tolerance in this application implies a tighter fit between the rebar and the member that the fit used in other applications where more clearance is permitted. The following note is an example of the specification for rebar through a wood timber:

#8 REBAR × 6′-0″ @ 12″ OC W/ 1-1/8″ Ø HOLES AT TIMBER

The following chart provides basic applications for holes used through timber when rebar is driven through timber:

Bar #	Hole Ø	Bar #	Hole Ø
4	9/16	8	1-1/8
5	11/16	9	1-1/4
6	7/8	10	1-7/16
7	1		

Slab thickness, concrete wall, or concrete beam and column cross-sectional dimensions are given in inches, feet and inches, or mm for metric values. Examples of a complete slab-on-grade callouts can read as follows:

4″ CONC SLAB ON FIRM UNDISTURBED SOIL OR 4″ SAND FILL

6″ THICK 3000 PSI CONC SLAB WITH WWF 6 × 6—W2.9 × W2.9 3″ CLR AT 4″ MIN 3/4″ MINUS FILL

Concrete footing thickness are given in inches, feet, and inches, or millimeters for metric values. The footing width is given followed by the thickness all in one note.

Poured-in-Place Concrete

Commercial and residential applications for **poured-in-place concrete,** also called **cast-in-place concrete,** are similar except the size of the **casting** and the amount of reinforcing are generally more extensive in commercial construction. Poured-in-place concrete is the concrete construction method previously described where concrete is poured into forms. The term **casting** refers to the resulting concrete structure when describing poured-in-place concrete. In addition to the foundation and **on-grade** floor systems, concrete is commonly used for walls, columns, and floors aboveground. The term **on-grade** refers to concrete construction formed on the ground, or the elevation of the desired construction surface. The **grade** of a physical feature, landform, or constructed line refers to the tangent of the angle of that surface to the horizontal. The term **grade** is also called slope, incline, gradient, pitch, or rise. The terms **grade** and **on-grade** are also used when referring to the earth on which construction is performed, such as **slab on-grade.** Slab on-grade is concrete construction formed on the ground or other engineered construction surface that is generally 4 to 6 in. (100 to 150 mm) thick depending on loads during final use and based on structural engineering.

Lateral soil pressure acting on concrete structures tends to bend the wall inward, thus placing the soil side of the wall in **compression** and the interior side of the wall in **tension.** The term **compression** refers to pressing together, or forcing something into less space. The term **tension** refers to something being stretched. Lateral soil pressure is the pressure that soil exerts on the structure in the horizontal direction. Steel reinforcing is used to increase the ability of concrete to withstand **tensile stress,** as shown in Figure 4.6. Tensile stress or tensile strength is a measurement of the force required to pull something such as rope, wire, or a structural beam to the point where it breaks. Steel-reinforced walls and columns are constructed by setting steel reinforcing in place and then surrounding it with wood or other material forms to contain the concrete when poured. The word **pour** refers to the process of flowing the concrete into the forms. Once the concrete has been poured and allowed to set,

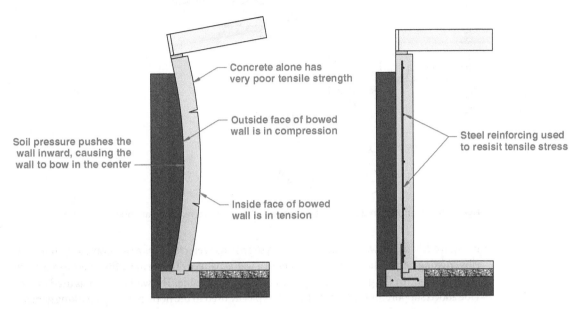

Figure 4.6 Steel reinforcing is used to increase the ability of concrete to withstand tensile stress. (*Previously published in Modern Residential Construction Practices, Routledge, 2017 and reproduced here with permission.*)

the forms are removed. The term **set** refers to the time it takes concrete to cure or harden after it has been poured.

Vertical and horizontal rebar placed in the forms use steel ties to hold the rebar together. **Ties** are wrapped around vertical steel in a column or placed horizontally in a wall or slab to help keep the structure from separating, and they keep the rebar in place while the concrete is being poured into the forms. Figure 4.7 shows two examples of column reinforcing. The drawings required to detail the construction of a rectangular concrete column are shown in Figure 4.8.

Concrete is also used on commercial projects to build above ground floor systems. The floor slab can be supported by a steel deck or be self-supported. The steel deck system is typically used on structures constructed with a steel frame, as shown in Figure 4.9. Two of the most common poured-in-place concrete floor systems are the **waffle** and **ribbed** concrete floor methods also referred to as waffle and ribbed concrete slabs. A waffle concrete floor shown

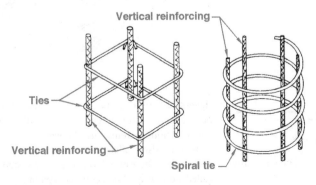

Figure 4.7 Column reinforcing examples.

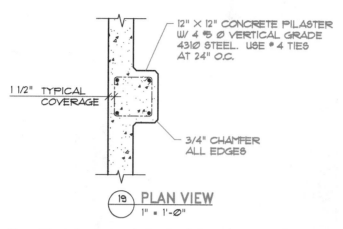

Figure 4.8   A drawing required to detail the construction of a rectangular concrete column.

in Figure 4.10*a*. The waffle concrete floor system is also referred to as a two-way rib system with ribs crossing perpendicular to each other in a waffle pattern. The term **waffle** refers to a pattern of ribs, formed by the grid-like design on each side or one side. The waffle system is used to provide added support for the floor slab and is typically used in the floor systems of parking garages. A one-way ribbed concrete floor system, shown in Figure 4.10*b*, has individual linear ribs spaced an engineered distance apart and supported by columns at their ends. The term **ribbed**

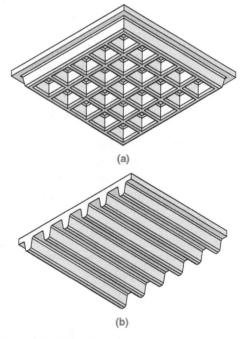

Figure 4.10   Two of the most common poured-in-place concrete floor systems are the waffle and ribbed concrete floor systems. (*a*) The waffle concrete floor system is also called a two-way ribbed concrete system. (*b*) The ribbed concrete floor system is also called a one-way ribbed concrete system.

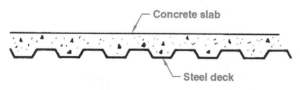

Figure 4.9   The steel deck system is typically used on structures constructed with a steel frame.

refers to a pattern of raised features serving to strengthen or support the structure. The ribbed system is often used in small-scale commercial structures such as office buildings. The ribs serve as floor **joists** to support the slab but are actually part of the slab. Joists are horizontal structural members used in repetitive patterns to support floor and ceiling loads. Spacing of the ribs varies, depending on the span and the amount and size of reinforcing used in the floor system. The span is the horizontal distance between two supporting members.

NOTE

You can find cast-in-place concrete and poured-in-place referred to by a synonym **in situ cast concrete.** The term **in situ** refers to a **concrete** structure made in its final position. In situ is concrete poured in on-site forms where the concrete becomes rigid when hardened. The opposite of in situ concrete is precast concrete that is poured into forms off-site in a manufacturing plant and then delivered to the construction site for installation. Precast concrete is described in the next section.

Precast Concrete

Precast concrete construction consists of forming and pouring the concrete component off-site at a fabrication plant and transporting the component to the construction site. Figure 4.11 shows a precast **beam** being lifted into place. Drawings and specifications for precast components show how precast members are to be constructed and methods of transporting and lifting the member into place. Precast members often have exposed metal flanges so the member can be connected to other parts of the structure. Common details used for wall connections are shown in Figure 4.12. A beam is a horizontal construction member used to support floor systems, and wall or roof loads.

Many concrete structures are precast and prestressed. **Prestressed concrete** is made by placing steel cables, wires, or bars held in tension between the concrete forms while the concrete is poured around them. Once the concrete has hardened and the forms are removed, the cables act like big springs. As the cables attempt to regain their original shape,

Figure 4.11 A precast beam being lifted into place.

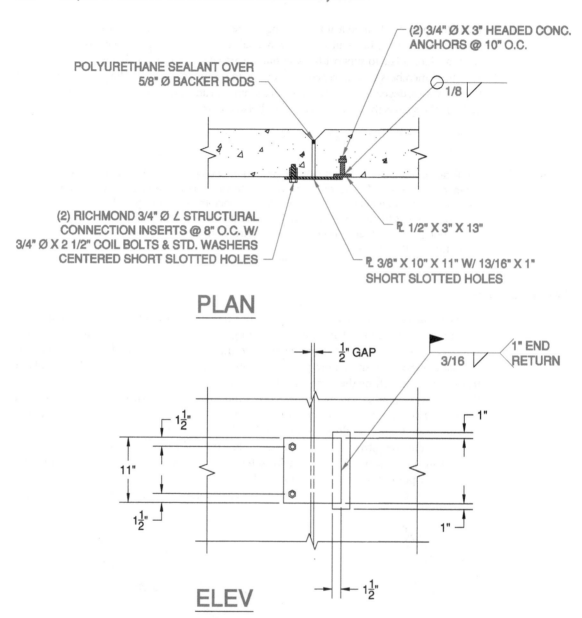

(2) 3/4" Ø X 3" HEADED CONC. ANCHORS @ 10" O.C.

POLYURETHANE SEALANT OVER 5/8" Ø BACKER RODS

1/8

(2) RICHMOND 3/4" Ø ∠ STRUCTURAL CONNECTION INSERTS @ 8" O.C. W/ 3/4" Ø X 2 1/2" COIL BOLTS & STD. WASHERS CENTERED SHORT SLOTTED HOLES

℄ 1/2" X 3" X 13"

℄ 3/8" X 10" X 11" W/ 13/16" X 1" SHORT SLOTTED HOLES

PLAN

½" GAP

3/16   1" END RETURN

1½"

11"

1"

1½"

1"

1½"

ELEV

NOTE: CONNECTIONS TO BE SPACED @ 5'-6" O.C. BEGINNING 12" FROM SLAB.

1	PANEL CONNECTOR
3	SCALE 1" = 1'-0"

Figure 4.12   Precast members often have exposed metal flanges so the member can be connected to other parts of the structure. Common details used for wall connections.

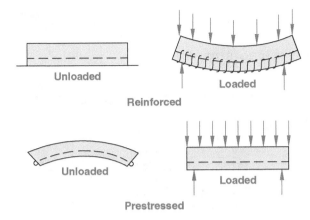

Figure 4.13 The difference between forces being applied before and after installation on reinforced concrete and prestressed concrete.

compression pressure is created within the concrete member. The compressive stresses built into the concrete member help prevent cracking, bending, and deflecting. Prestressed concrete members are generally reduced in size in comparison with the same design features of a standard precast concrete member. Prestressed concrete components are commonly used for the structural beams of structures such as buildings and bridges. Figure 4.13 shows the difference between forces being applied before and after installation on reinforced concrete and prestressed concrete. Common prestressed concrete shapes are shown in Figure 4.14. Figure 4.15 shows a precast concrete panel drawing. A precast concrete beam

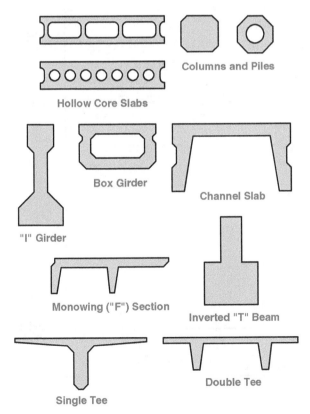

Figure 4.14 Common prestressed concrete shapes.

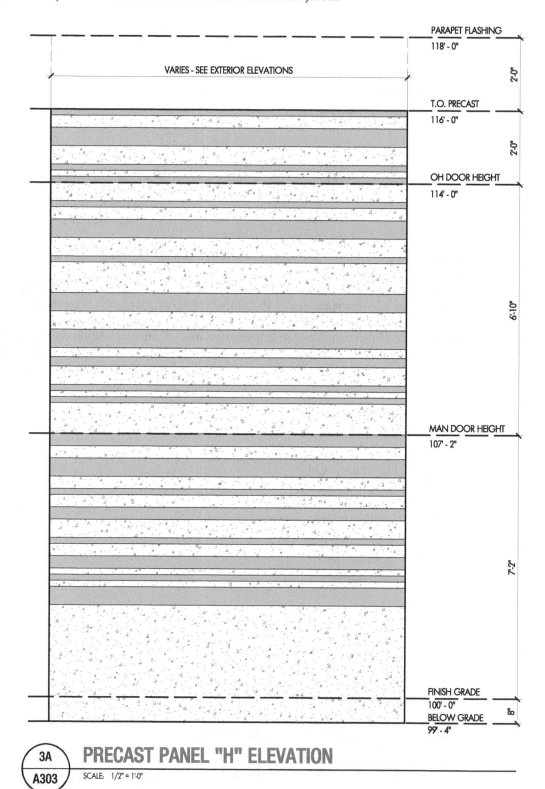

PARAPET FLASHING
118' - 0"

2'-0"

VARIES - SEE EXTERIOR ELEVATIONS

T.O. PRECAST
116' - 0"

2'-0"

OH DOOR HEIGHT
114' - 0"

6'-10"

MAN DOOR HEIGHT
107' - 2"

7'-2"

FINISH GRADE
100' - 0"
BELOW GRADE
99' - 4"

8"

3A
A303

PRECAST PANEL "H" ELEVATION

SCALE: 1/2" = 1'-0"

Figure 4.15   A precast concrete panel elevation. (*Courtesy of JLG Architects, Alexandria, MN.*)

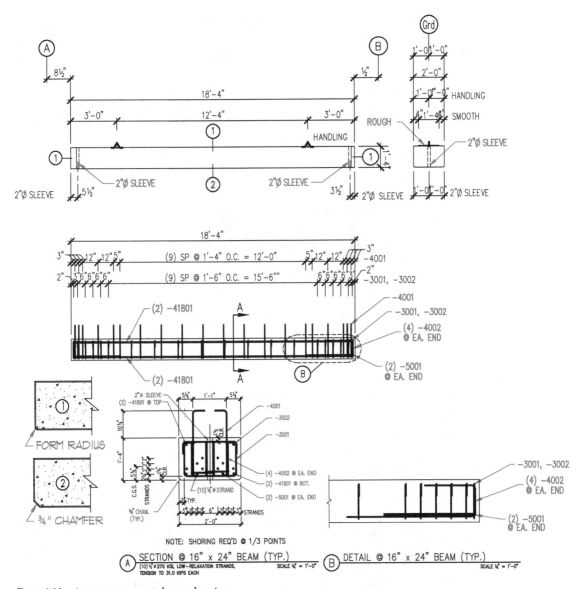

Figure 4.16   A precast concrete beam drawing.

drawing is displayed in Figure 4.16, and a precast concrete slab drawing is shown in Figure 4.17. Construction details are commonly used in prestressed concrete construction. Figure 4.18 shows a precast concrete panel installed on a poured concrete wall and footing.

Tilt-Up Precast Concrete

Tilt-up construction is a precast concrete method using formed wall panels that are lifted or tilted into place. Panels can be formed and poured either at or off the job site. Forms for a wall are constructed in a horizontal position and the required steel placed in the form. Concrete is then poured into the forms and around the steel and allowed to harden.

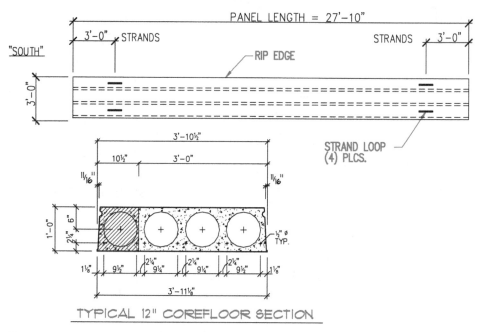

Figure 4.17   A precast concrete slab drawing.

The panel is lifted or tilted into place once it has reached its desired hardening and design strength.

Tilt-Up Panel Plan, Elevations, and Wall Details

The **panel plan** is used in tilt-up construction to show the location of the panels, as shown in Figure 4.19. **Panel elevations** are used when the exterior elevations do not clearly show information about items located on or in the walls. Similar to exterior elevations, the panel elevations show door locations and reinforcing within walls and around openings. Dimensions associated with panel elevations provide both horizontal and vertical dimensions for openings and other features, as shown in Figure 4.20. **Wall details** are used to show the connection points of the concrete panels used in tilt-up construction and connection details at the walls for other types of structures, as shown in Figure 4.21.

A plan view is normally shown on the construction drawings to specify the panel locations. The location and size of steel placement and openings are also displayed and specified in the panel elevation. Figure 4.22 shows tilt-up construction projects. Figure 4.22a shows tilt-up construction braces. Figure 4.22b shows tilt-up construction forms. Figure 4.22c shows tilt-up construction rebar in forms. Figure 4.22d shows tilt-up panel with window cutouts.

Hybrid Concrete Construction

Hybrid concrete construction is a commercial construction system that combines precast concrete and cast-in-place concrete to take advantage the different qualities of each practice. The accuracy, speed, and high-quality finish of precast components are combined with the economy and flexibility of cast-in-place concrete.

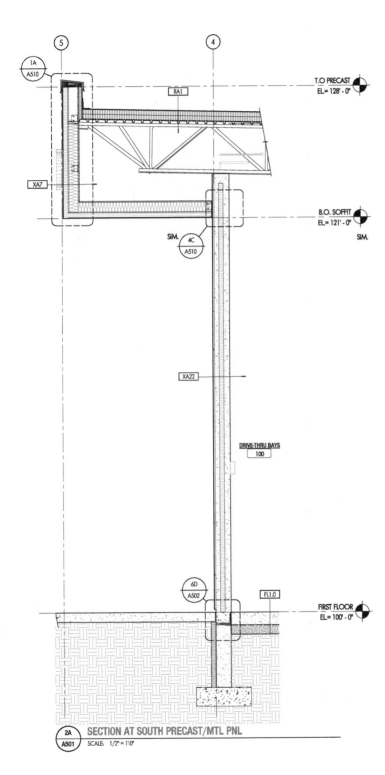

Figure 4.18   A precast concrete section of a precast concrete panel installed on a poured concrete wall and footing. (*Courtesy of JLG Architects, Alexandria, MN.*)

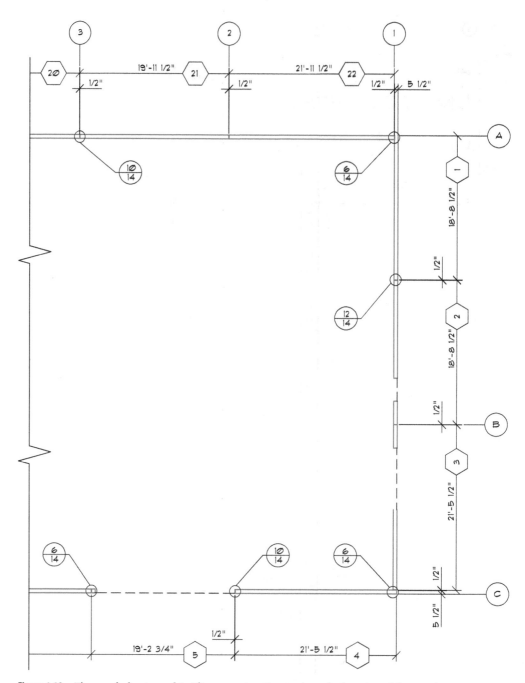

Figure 4.19 The panel plan is used in tilt-up construction to show the location of the panels.

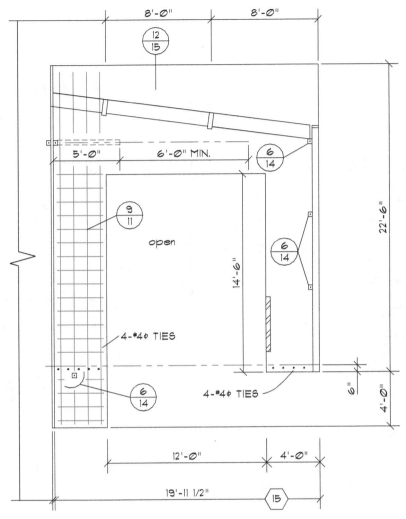

Figure 4.20 Dimensions associated with panel elevations provide both horizontal and vertical dimensions for openings and other features.

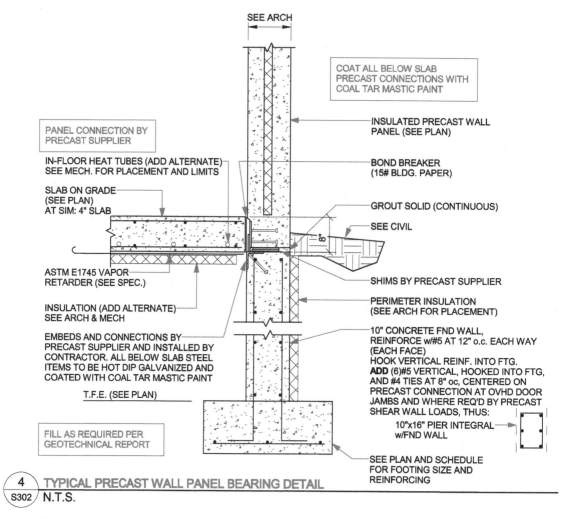

SEE ARCH

COAT ALL BELOW SLAB
PRECAST CONNECTIONS WITH
COAL TAR MASTIC PAINT

INSULATED PRECAST WALL
PANEL (SEE PLAN)

PANEL CONNECTION BY
PRECAST SUPPLIER

IN-FLOOR HEAT TUBES (ADD ALTERNATE)
SEE MECH. FOR PLACEMENT AND LIMITS

BOND BREAKER
(15# BLDG. PAPER)

SLAB ON GRADE
(SEE PLAN)
AT SIM: 4" SLAB

GROUT SOLID (CONTINUOUS)

SEE CIVIL

ASTM E1745 VAPOR
RETARDER (SEE SPEC.)

SHIMS BY PRECAST SUPPLIER

INSULATION (ADD ALTERNATE)
SEE ARCH & MECH

PERIMETER INSULATION
(SEE ARCH FOR PLACEMENT)

EMBEDS AND CONNECTIONS BY
PRECAST SUPPLIER AND INSTALLED BY
CONTRACTOR. ALL BELOW SLAB STEEL
ITEMS TO BE HOT DIP GALVANIZED AND
COATED WITH COAL TAR MASTIC PAINT

10" CONCRETE FND WALL,
REINFORCE w/#5 AT 12" o.c. EACH WAY
(EACH FACE)
HOOK VERTICAL REINF. INTO FTG.
ADD (6)#5 VERTICAL, HOOKED INTO FTG,
AND #4 TIES AT 8" oc, CENTERED ON
PRECAST CONNECTION AT OVHD DOOR
JAMBS AND WHERE REQ'D BY PRECAST
SHEAR WALL LOADS, THUS:

T.F.E. (SEE PLAN)

10"x16" PIER INTEGRAL
w/FND WALL

FILL AS REQUIRED PER
GEOTECHNICAL REPORT

SEE PLAN AND SCHEDULE
FOR FOOTING SIZE AND
REINFORCING

4 TYPICAL PRECAST WALL PANEL BEARING DETAIL
S302 N.T.S.

Figure 4.21   Wall details are used to show the connection points of the concrete panels used in tilt-up construction and connection details at the walls for other types of structures. (*Courtesy of JLG Architects, Alexandria, MN.*)

Figure 4.22   A tilt-up construction project. (*a*) Tilt-up construction braces. (*b*) Tilt-up construction forms. (*c*) Tilt-up construction forms. (*d*) Tilt-up panel with rebar and window cutouts. (*Photos courtesy of constructionphotographs.com.*)

COMMERCIAL CONSTRUCTION PRECAST CONCRETE APPLICATION

Precast concrete enhances the beauty of the Academy Museum of Motion Pictures.
By Mark Crawford

It is one of the most eye-catching structures in the country. The 300,000-ft^2 Academy Museum of Motion Pictures, dedicated to the art and science of the film industry, opening in 2019 in Los Angeles. Designed by Renzo Piano, founder of the Renzo Piano Building Workshop, the museum features 50,000 ft^2 of galleries, exhibits, and theaters.

Perhaps the most impressive part of the project is the Sphere Building, which includes a 1000-ft^2 theater and a spectacular, glass-domed terrace overlooking the famed Hollywood Hills. The bottom half of this globe-shaped building consists of structural concrete with a decorative cladding of precast concrete panels, giving it a stylish, finished exterior.

A Perfect Precast Solution

The original architectural proposal for the Sphere focused on a steel structure clad with metal panels. This design relied on steel trusses spanning across the theater, each supported by curved columns just inside the spherical walls. Because the required truss depths blocked the projection lines inside the theater, the design team decided to change the structure to a shallow concrete dome, which eliminated the problem of the trusses. There was some concern that using cast-in-place concrete would not create the smooth, clean, natural finish that Piano envisioned. Thus, the team decided to use precast concrete panels to provide the fine details and sharp lines they envisioned.

The shape of the museum theater is a sphere cut by intersecting planes. The jointing on the sphere has parallel slices east/west and north/south, similar to an egg slicer. As a result, the shapes become more skewed the farther they get from the sphere midline.

"Using precast concrete allows us to maintain the precision and crispness of the joint pattern and the sphere geometry, and also ensure a high-quality, consistent as-cast finish," said Mark Hildebrand, president and chief engineer for Willis Construction Co. in San Juan Bautista, CA, which provided the precast concrete for the project.

The precast panels serve as the skin of the Sphere and function as the permanent formwork to support the structural concrete during placement operations, as shown in **Figure 4.23**.

"In other words, the architectural finish was installed before the structure was built," said Hildebrand. "This required a steel structure upon which to erect the precast panels."

A system of steel frames was constructed to support the Sphere's 1- to 2-ton precast panels. Workers hung the curved panels on the steel frames, then sprayed **shotcrete** against the back of the panels. "This allowed eliminating the trusses, but required continuous support around the perimeter rather than the discreet columns, which led to the use of continuous shotcrete walls," said Derrick Roorda, principal with BuroHappold Engineering, one of the contractors on the project. "The architects felt strongly that the shotcrete would need precast cladding in order to guarantee a suitable finish. Cost considerations led to the idea of erecting the precast first and using it as formwork for the shotcrete primary structure." Greg Wade, project manager for MATT Construction Corporation and general manager for the project, said they have used shotcrete and precast, but they've never combined the two. Shotcrete is described in detail later in this chapter.

A Complex Design Process

The complex sequencing demanded that many architectural details for the precast finish needed to be worked out and incorporated into the structural design process. All the coordination was completed with the aid of computer models. "The architect, structural engineer and precast manufacturer each developed models for their own scope and we had weekly discussions to resolve problems and talk about upcoming issues," said Hildebrand. "As other trades came on board, their models were developed, discussed and incorporated as well."

Figure 4.23 The precast panels serve as the skin of the Sphere and function as the permanent formwork to support the structural concrete during placement operations. (*Courtesy of the National Precast Concrete Association—Buro Happold Engineering.*)

Once the precast model was complete and approved, molds were fabricated and **shop tickets** drawn directly from the model. Shop tickets, also called **shop drawings** are drawings or sets of drawings produced by the contractor, supplier, manufacturer, subcontractor, or fabricator for use in a fabrication shop to manufacture a construction item.

Unique Precast Panels

The precast panels are double-curved with non-orthogonal returns. The word non-orthogonal refers to no square corners in this application. "Precast gave us the possibility to control the shape to tighter tolerances," said Luigi Priano, an associate, with Renzo Piano Building Workshop. "It took a long period of research to find the optimal mix to achieve the proper crispness, color, and texture. In addition, the craftsmanship of the precast manufacturer greatly helped to control the consistency of the finish."

The Sphere required 727 precast components with 578 unique shapes, many of them one of a kind. The 4-in.-thick panels include welded wire fabric reinforcements cast with 7000-psi concrete.

A huge challenge was giving the mold crew enough information on a 2D shop drawing to build the 3D mold and maintain tight tolerances. The digital model also included all the penetrations in the panels for attaching the glass canopy at the top of the structure, to ensure their exact location on the face of the sphere.

Most of the shapes have no square corners or edges so the typical techniques for layout and alignment in the field did not apply. "Also, during erection, there were no layout lines on the structure because the structure wasn't there yet," said Hildebrand. "Each panel was tagged with four points that corresponded to specific coordinates of the project's global coordinate system. The panels were then set and aligned with positions verified by the surveyor."

The Willis team programmed its computer numerically controlled (CNC) router machine to carve whole molds or mold parts directly from the 3D computer files of the design model. One of the challenges in using this technology in combination with traditional molding techniques is creating all the panel shapes needed from the fewest number of base molds. "We used the CNC to carve 30 base molds from which we could fabricate the 578 different shapes using shut-offs and other mold parts built up off the base," said Hildebrand. "Because the base molds are segments of a sphere, there are no 90-degree angles. The carpenters in the mold shop cannot just 'square it up' in the traditional way." "The shop drawings have lots of dimensions so that components added on to the base mold can be triangulated into position with some degree of certainty." Close coordination between the mold crew and the detailing crew, who are translating the 3D computer shapes into paper drawings, was required to come up with the dimensions that are most useful and effective. "If necessary, we can also model into the computer setting jigs for getting the right angles or skews and then carve those on the CNC machine as construction aids to the carpenters," said Hildebrand.

A Grand Opening

Precast panel work on the Sphere Building is in progress with the panels still needing blind connections to cover the top and bottom of the Sphere. Once the panels are in place, a ring beam around the top of the sphere is installed and covered with concrete deck to tie the structure together to add strength and rigidity.

This one-of-a-kind project integrates both the structural and architectural benefits of precast concrete to create bold shapes, graceful lines, and eye-catching colors and features. Hildebrand eagerly admits he likes the many challenges on this project, such as using precast as permanent formwork. "We even put strain gauges on the connections for the mockup and tracked the forces over time as the different lifts of shotcrete were applied," he said. "This gave us confidence in the ability of our panels and connections to take the strain of the concrete-placing operations in the field, as any failure would have been catastrophic."

Hildebrand also cannot stress enough the importance of communication, collaboration, and teamwork for such a complex project.

FOUNDATION SYSTEMS

The **foundation** is the construction system used to support the structural loads and distribute the loads to the ground. Foundation systems are typically made of **concrete** or a combination of concrete and **concrete block** along with wood or steel framing members. The key features of a foundation are the **footings, foundation walls, stem walls,** and **piers.** Concrete is set in forms along with steel reinforcement to make a solid structure when cured. Concrete blocks are prefabricated construction blocks made from concrete. Concrete block construction is described in Chapter 5, *Masonry Construction.* Footings are the lowest member of the foundation system used to spread the loads of the structure on the supporting ground. Foundation walls are the vertical walls of the foundation system that connect between the footings and the structure above. Stem walls are used at the interior of the foundation system to support structural members at mid **span.** The stem wall structure usually has a continuous concrete footing with a concrete foundation wall and a wood or steel frame wall tied to the footing or foundation wall. The term **span** refers to the horizontal distance between two supporting members. A pier is a cylindrical, square, or rectangular-shaped cube made of concrete or concrete block and used to support individual foundation members. Piers can also be called footings. Figure 4.24 shows the parts of a typical foundation, stem wall, and pier system.

There are several different types of foundation systems used in residential and commercial construction throughout the United States and Canada. The basic foundation systems are the **crawl space, concrete slab, concrete wall,** and **post and beam** or **pole construction.**

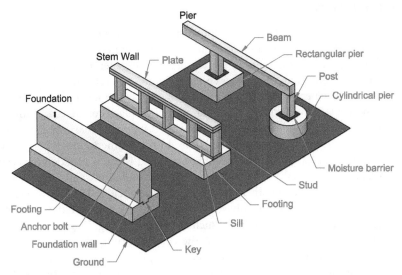

Figure 4.24 The parts of a typical foundation, stem wall, and pier system. (*Previously published in Modern Residential Construction Practices, Routledge, 2017 and reproduced here with permission.*)

CONCRETE CONSTRUCTION PRACTICES

Concrete is **poured** into **forms** that are built of wood, steel, or other materials and used to contain the concrete mix in the desired shape until it is hard. Concrete is delivered to the construction site in a semiliquid state and is poured in the **concrete forms.** Concrete forms are temporary or permanent structures or molds into which concrete is poured. Concrete is delivered in a truck with a large container that keeps the concrete spinning or mixing in the semiliquid state so it can easily flow into the forms. A pour is the process of flowing the concrete into the forms.

Concrete has initial hardening in about 2 to 3 hours depending on air temperature, amount and type of cement in the mix and other environmental conditions. You should stay off new concrete during the first 3 days as much as possible. Concrete has its most rapid strength gain over the first 7 days, so careful construction activities can take place during this time. Concrete reaches full strength in 28 days. Concrete **compressive strength** is commonly specified on the foundation plan and related sections and details. A typical specification might read: CONCRETE TO HAVE A MINIMUM COMPRESSIVE STRENGTH OF 2500 PSI AT 28 DAYS, where PSI is the abbreviation for pounds per square inch and 28 days is when concrete reaches full strength. The metric equivalent is mega pascal (MPa), where 2500 psi converts to 17 MPa. Compressive strength is the force applied by weight. Concrete has excellent compressive strength but poor **tensile strength.** This is why steel reinforcing is added to concrete to increase the tensile strength. Tensile strength is the strength the material has against pulling forces.

Lateral soil pressure acting on concrete structures tends to bend the wall inward, placing the soil side of the wall in compression and the interior side of the wall in tension. Steel reinforcing is used to increase the ability of concrete to withstand this tensile stress as described and demonstrated earlier. Steel-reinforced walls and columns are constructed by setting steel reinforcing in place and then surrounding it with forms to contain the concrete. The forms are removed when the concrete has been poured and allowed to set harden.

Pouring Concrete into Forms

The process of pouring concrete into forms is called **poured-in-place concrete.** Commercial and residential applications for poured-in-place concrete are similar except that the size of the **casting** and the amount of reinforcing are generally more extensive in commercial construction. Poured-in-place concrete is the concrete construction method previously described where concrete is poured into forms. The term **casting** refers to the resulting concrete structure when describing poured-in-place concrete. In addition to the foundation and slab floor systems, concrete is used for walls, columns, and floors above ground. Concrete is a fundamental material used for building foundations. The foundation is the system used to support the building loads and is usually made up of walls, footings, and piers. Commercial buildings use concrete foundations with steel reinforcing. Concrete forms can be made with wood, steel, or other materials as previously described. Wood forms generally made of plywood are commonly used for the foundation construction of small commercial buildings, such as office buildings. Figure 4.25 shows a photograph of wood concrete forms ready for the concrete to be poured. Notice how the forms are supported and braced, because concrete applies a lot of pressure on the forms until the concrete hardens. Figure 4.26 shows a worker pouring concrete into the forms. Concrete can be poured directly from the concrete truck if there is good access around the site. When access is limited, most contractors have the concrete delivered with a concrete pump truck. The concrete pump can be truck or trailer-mounted with a remote-controlled boom hose, using an articulating robotic arm to accurately place concrete. Figure 4.27 shows a worker filling in voids and smoothing the top of the foundation wall after concrete has been poured.

Figure 4.25   Wood concrete forms ready for the concrete to be poured.

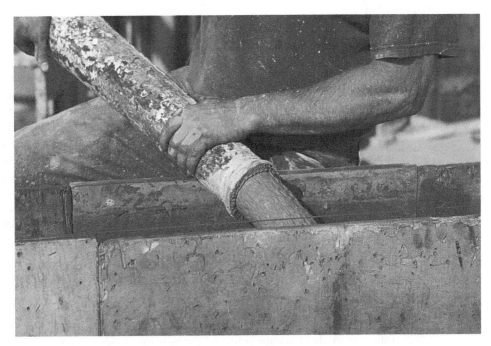

Figure 4.26   A worker pouring concrete into the forms.

Figure 4.27   A worker filling in voids and smoothing the top of the foundation wall after concrete has been poured.

Soil, Excavating, and Grading

This discussion provides a brief review of excavating practices discussed in Chapter 3, *Construction Site and Excavation*. Earth is generally excavated prior to building a concrete structure. **Excavation** refers to removing earth for construction purposes. Before a concrete structure can be built, earth needs to be excavated down to firm, undisturbed supporting soil. Detailed engineering specifications are often placed on the plans identifying the amount of bearing pressure required. The soil **bearing pressure** is the number of pounds per square inch of pressure the soil is engineered to support. Concrete slabs also need firm, undisturbed supporting soil, or compacted fill that meets specific engineer requirements. A **slab** is a concrete floor system, typically made of poured concrete at ground level, but also poured on engineered supporting structures above ground. In many cases, compacted gravel is specified below the concrete slab. The gravel provides for a level or accurate slope surface upon which to pour the concrete slab, and provides for compaction as needed. A drawing note or specification for under-slab gravel might read:

4″ MIN. 3/4 MINUS COMPACTED GRAVEL OVER FIRM UNDISTURBED SOIL

Anchor Bolts and Hold-Down Anchors

You previously learned how anchor bolts are specified. This content provides additional information about anchor bolts and expands into hold-down anchors. Foundation walls and other concrete structures can have **anchor bolts** embedded in the concrete and projecting out far enough to fasten through the **mud sill** or to attach other structural features, as shown in Figure 4.28. Anchor bolts are steel L-shaped bolts that are imbedded in concrete and extend out far enough to fasten the mud sill with a washer and nut at each bolt. The **mud sill** is a continuous **pressure-treated wood** member that provides a barrier between the foundation wall and connection to the framing above. Pressure-treated means that wood has had a liquid preservative forced inside to protect against deterioration due to rot or insect damage. The mud sill is generally 2× pressure-treated material, such as 2 × 4, 2 × 6,

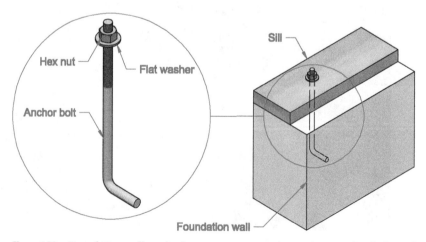

Figure 4.28   Foundation walls and other concrete structures can have anchor bolts embedded in the concrete and projecting out far enough to fasten through the mud sill or to attach other structural features. (*Previously published in Modern Residential Construction Practices, Routledge, 2017 and reproduced here with permission.*)

or 2 × 8 depending on the design requirements. Anchor bolts are also sized and spaced to design requirements, and a typical note on plans or specifications might read: 1/2″ × 10″ ANCHOR BOLTS W/ FLAT WASHER AND HEX NUT 48″ OC AND 12″ MIN. FROM ENDS AND SPLICES, where 1/2″ is the diameter, 10″ is the length, with (W/) a flat washer and hexagon nut, 48″ is the spacing on center, and an anchor bolt a minimum (MIN.) from the ends and splices of the mud sill.

Hold-down anchors, also called **hold-downs** or **holdowns,** are used in a variety of construction applications. This discussion about hold-down anchors refers to fasteners that are embedded in concrete foundation walls and attach to wood or steel structural walls above, or attach wood or steel structural walls between floors, and especially at **shear walls** to control uplift in the wall system. The hold-down is connected to the concrete foundation or structural slab by an embedded or adhesive-installed anchor bolt at the bottom. The hold-down is connected to a wood or steel post with screws, nails, or bolts at the top. See Figure 4.29. In structural engineering, a shear wall, also called a **braced wall line,** is a wall composed of shear panels, also known as braced panels, to counter the effects of **lateral load** acting on a structure. A shear panel is typically part of a wood or steel frame stud wall that is covered with structural sheathing such as plywood, but other materials such as diagonal bracing systems, sheet steel, and steel-backed shear panels can be used. All exterior walls and some interior walls have braced panels depending on structural engineering requirements. Interior shear walls can be bearing or nonbearing. The sheathing on a shear panel is securely nailed, screwed, or bolted to the stud wall framing at the edges and in the field with nails or screws of a specific size and spacing based on structural engineering.

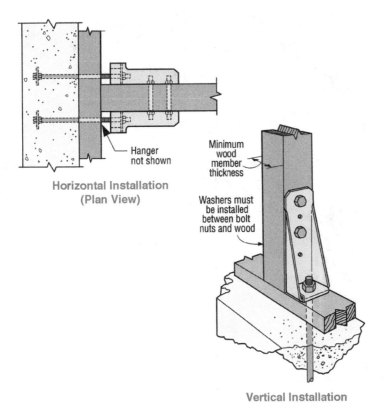

Figure 4.29   The hold-down is connected to a wood or steel post with screws, nails, or bolts at the top. (©*Simpson Strong-Tie Inc. and/or its affiliates. All Rights Reserved.*)

An engineered shear wall has the strength and stiffness to make the structure safe against **lateral loads.** The construction of a structural sheathing-reinforced shear wall is shown in Figure 4.30. A lateral load is a force working on a structure applied parallel to the ground, and diagonally to the structure. Wind and **seismic** loads are the most common lateral loads. Seismic loads are caused by earthquakes.

A shear wall can also be constructed with diagonal wood or metal T-bracing by cutting a diagonal slot in the studs where the bracing can be inserted and screwed or nailed to the studs, as shown in Figure 4.31. A structural engineer determines the system to be used and provides drawings and specifications included on the set of plans.

An interior braced wall can also be constructed as needed for shear panels by using a continuous **stem wall** or **pony wall** at mid-span in the foundation. Stem wall or pony wall is

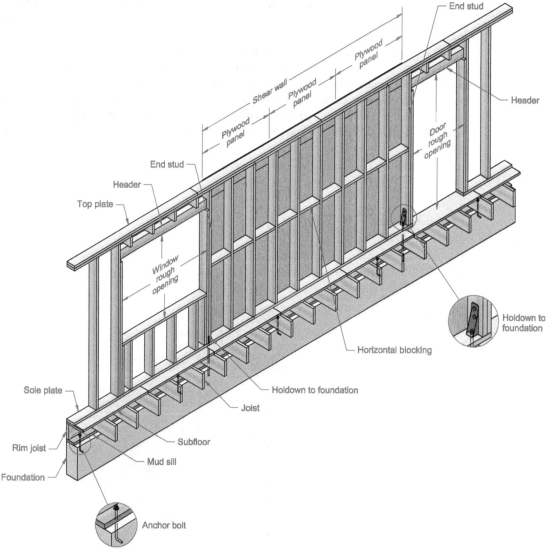

Figure 4.30 The construction of a structural plywood-reinforced shear wall. (*Previously published in Modern Residential Construction Practices, Routledge, 2017 and reproduced here with permission.*)

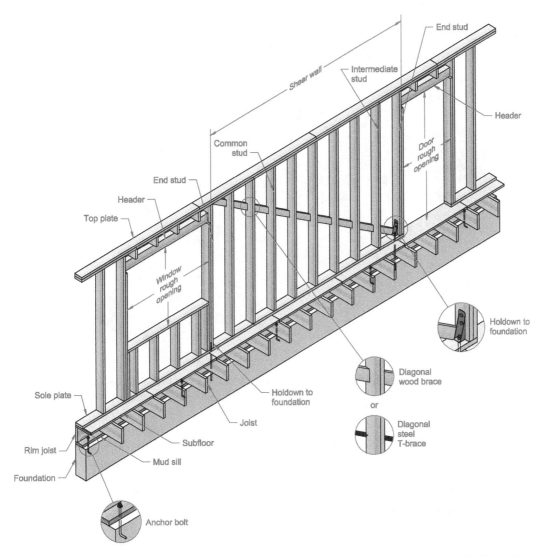

Figure 4.31   A shear wall can also be constructed with diagonal wood or metal T-bracing by cutting a diagonal slot in the studs where the bracing can be inserted and screwed or nailed to the studs. (*Previously published in Modern Residential Construction Practices, Routledge, 2017 and reproduced here with permission.*)

a general term used to describe any short height wall, or partial wall, also called a **knee wall.** This discussion describes a pony wall constructed at the interior of a foundation to support the mid-span of joists above and to act as an interior shear wall, as shown in Figure 4.32. A structural engineer provides the shear wall construction requirements and nailing or screw specifications on the plans.

Other types of structural systems can be used to anchor posts to footings and beams. These are called **post bases** and **post caps** and are used to provide a rigid connection between the footing and post and between the post and beam above, as shown in Figure 4.33. The structural members shown in Figure 4.33 are prefabricated by a manufacturer for standard applications. Many commercial post and base structural systems are engineered using plate steel, weldments, and bolted connections as needed for specific applications. These engineered connections are detailed on the plans and generally fabricated in a shop. The

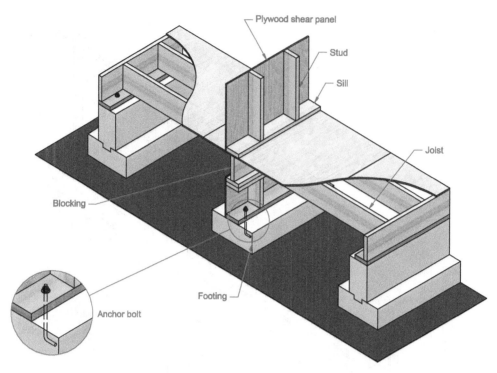

Figure 4.32   A pony wall constructed at the interior of a foundation to support the mid-span of joists above and to act as an interior shear wall. (*Previously published in Modern Residential Construction Practices, Routledge, 2017 and reproduced here with permission.*)

concrete footings, called **piers**, are cylindrical or cube-shaped and placed to provide mid-span support for beams and joists. Piers are cylindrical, square, or rectangular-shaped cube made of concrete or concrete block and used to support individual foundation members.

CRAWL SPACE FOUNDATION

The **crawl space system** has a perimeter footing **keyed** to a foundation wall to support the outside structure, with interior **piers** and **stem walls** to support the mid-span structure above. The footing and foundation wall can be formed and poured at one time, which is referred to as **monolithic.** Monolithic means formed and poured as a single unit. More often the footing is formed and poured first and then the foundation wall is formed over the footing and poured at a different time. The term **keyed** means that the foundation wall is keyed to the footing to keep the wall connected to the footing. The key is a slot in the footing that is created when the footing is poured. Additionally, the foundation wall is normally connected to the footing with steel rebar set in the footing and extending upward into the foundation wall. The term **crawl space** refers to a space that is at least 18 in. (450 mm) between the bottom of a typically wood structure and the ground that provides access to install and service plumbing, and heating and cooling systems where appropriate. Crawl space foundation examples are shown in Figure 4.34*a*, *b*, and *c*. **Crawlspace ventilation** is used to prevent moisture buildup and possible moisture damage to floor construction materials and other structural elements exposed to the crawlspace. Ventilation also helps to prevent the possible build-up of gasses within the crawl space. Crawl space vents are generally placed along the top of the foundation wall based on locations and specifications found

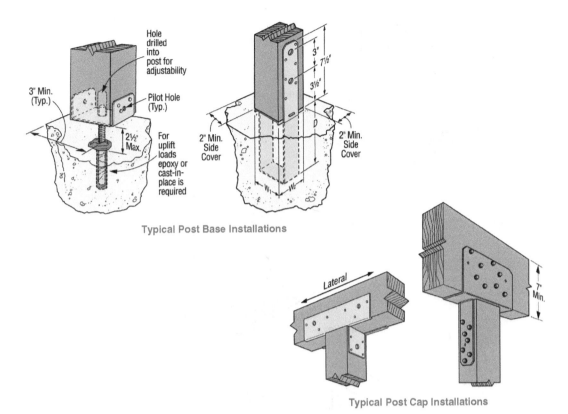

Typical Post Base Installations

Typical Post Cap Installations

Figure 4.33 Post bases and post caps are used to provide a rigid connection between the footing and post and between the post and beam above. (©*Simpson Strong-Tie Inc. and/or its affiliates. All Rights Reserved.*)

on the foundation plans. **Foundation vents** are available in standard sizes that match foundation plan specifications and are made of steel, aluminum, or plastic. Foundation vents are closable with built-in doors, louvers, or insulating foam inserts. This allows the foundation vents to be closed during the winter. A typical foundation vent is shown in Figure 4.34*d*.

There are two basic types of crawl space construction, which are the **post and beam** and **joist construction.**

Post and beam construction uses wood **posts** supported by concrete piers or footings with **beams** above. The beams are generally spaced 24 to 48 in. (600 to 1200 mm) on center and run between the foundation walls. Wood **decking** is fastened above and perpendicular to the beams. The decking is typically 1-1/8 in. tongue and grove (T&G) plywood more commonly used today over 24 to 48 in. on center beams. Posts are vertical wood members that connect between a pier or footing and support the beam above. Posts can be any size depending on structural engineering, but they are commonly 4 × 4 or 4 × 6. Beams are horizontal construction members that are used to support floor systems, wall, or roof loads. The beams in a post and beam system generally rest in a pocket provided at the foundation wall, and the top of the beams are flush with the top of the mud sill anchored to the top of the foundation wall. **Figure 4.35** shows the basic components of the post and beam foundation system, and a partial foundation plan of the same construction.

Joist construction uses standard dimensional lumber or **engineered wood products** as joists that span between foundation walls and can be supported at mid-span by a post and beam system or a stem wall. Joists are dimensional lumber such as 2 × 8, 2 × 10, or 2 × 12,

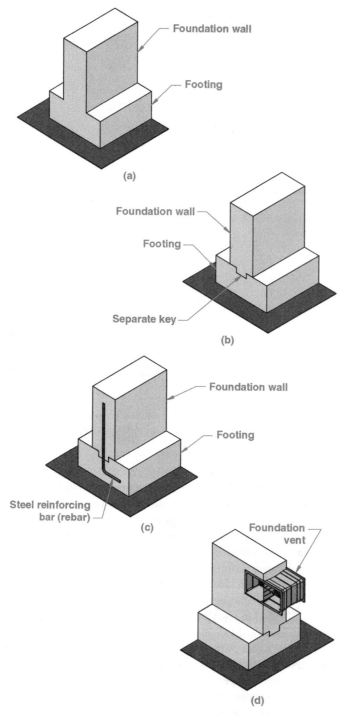

Figure 4.34   The crawl space foundation. (*a*) Monolithic footing and foundation wall. (*b*) Foundation wall keyed to footing. (*c*) Foundation wall keyed to footing with steel rebar. (*d*) A typical foundation vent. (*Previously published in Modern Residential Construction Practices, Routledge, 2017 and reproduced here with permission.*)

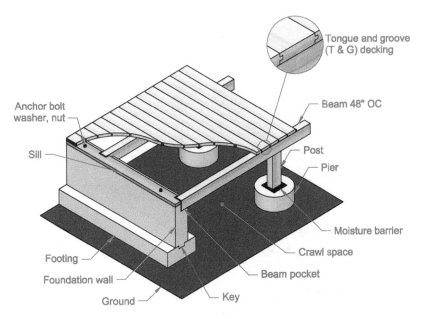

Post and Beam Foundation System

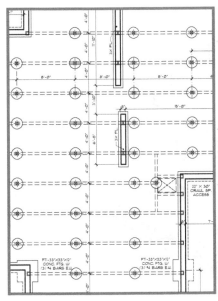

Partial Foundation Plan

Figure 4.35 The basic components of the post and beam foundation system, and a partial foundation plan of the same construction. (*Previously published in Modern Residential Construction Practices, Routledge, 2017 and reproduced here with permission.*)

or engineered wood products that are spaced 12, 16, or 24 in. on center depending on the span and structural engineering requirements. Figure 4.36 shows the difference between joist systems using dimensional lumber and engineered wood products. Engineered wood products are a combination of smaller components used to make structural products that have been engineered for specific applications and fabricated at a manufacturing facility and delivered to the construction site. Engineered wood products used as joists are called

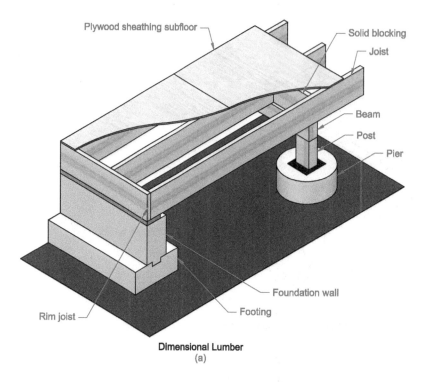

Dimensional Lumber
(a)

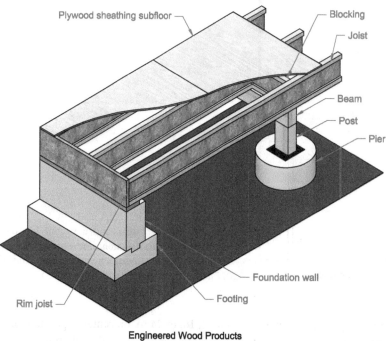

Engineered Wood Products
(b)

Figure 4.36 The difference between joist systems using dimensional lumber and engineered wood products. (*a*) Dimensional lumber joist system. (*b*) Engineered wood products joist system. (*Previously published in Modern Residential Construction Practices, Routledge, 2017 and reproduced here with permission.*)

I-joists, because they are I-shaped, as shown in Figure 4.36b. I-joists offer ease of installation, economy, dimensional stability, strength, and versatility when used for floor joist or other joist construction. I-joists consist of top and bottom flanges of different widths, connected to engineered wood webs in a variety of depths. The flanges resist bending stresses, and the web provides shear performance.

The joists generally rest on top of the mud sill. A **rim joist** is used at the perimeter of the foundation to support the ends of the joists and provide for perimeter support. **Plywood sheathing**, also called subflooring, is fastened to the floor joists to reinforce the structure and provide a backing for finish materials. Plywood is sheets of material generally 4 × 8 ft (1200 × 2400 mm) made of thin layers of wood called **veneer.** The veneer is glued together with the grain of adjoining layers at right angles to each other. Figure 4.37 shows the basic components of the crawl space foundation system using joist construction.

Staking the Crawl Space Foundation for Excavation

For a crawl space foundation, corner stakes are set by a professional surveyor using the site plan to establish the exact building corners. A building perimeter line is established by running string lines between each corner stake and marking the lines with paint on the ground.

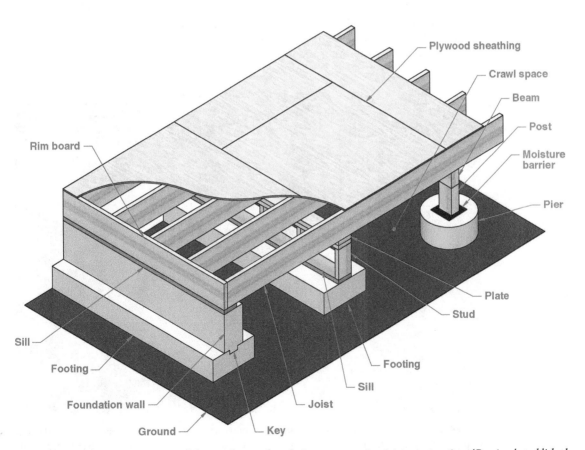

Figure 4.37 The basic components of the crawl space foundation system using joist construction. (*Previously published in Modern Residential Construction Practices, Routledge, 2017 and reproduced here with permission.*)

The surveyor then sets 2 to 3 ft (590 to 880 mm) offset stakes, which provides for the additional space needed to set concrete forms for footings and foundation walls. The excavator removes the earth to the desired depth everywhere inside the offset lines. The excavation for a crawl space is generally 18 to 24 in. (440 to 880 mm) below grade. Now the concrete crew can set forms for footings and foundation walls.

Crawl Space Foundation Forms and Pouring Concrete

After excavation is finished, it is time to set concrete forms for footings and pour concrete footings. A string line is run along the outside footing perimeter at the finished footing height and a laser level is used to make sure all forms are set straight and level. Properly set forms are a critical step in establishing a good foundation. A successful construction project seriously depends on an accurate foundation. Steel reinforcing is placed in the footing and foundation wall based on the construction specifications, as described later in this chapter. Rebar is used in commercial construction based on structural engineering requirements.

Concrete forms for light commercial construction are commonly made from solid wood or plywood, but steel forms are also available, and insulated forms are described in detail later in this chapter. The forms for the footing are set first, as shown in Figure 4.38. The forms in this example are made out of wood panels. Vertical stakes are driven into the ground to hold the forms. Diagonal bracing is used to provide additional support and strengthen the structure. After the footing forms are securely set, the concrete is poured,

Figure 4.38   Concrete forms for light commercial construction are commonly made from solid wood or plywood, but steel forms are also available, and insulated forms are described in detail later in this chapter. The forms for the footing are set first as shown here.

leveled, and smoothed to the top of the forms. The footing **key,** or **keyway,** is set while the concrete is soft. The key is a slot that runs down the center of the footing and along the entire length of the footing. The footing keyway is used to aid the steel reinforcements in resisting the lateral pressure exerted on the base of the foundation walls. The key is made with a form, a board, such as a 2 × 4 on edge or by pressing the end of the 2 × 4 into the concrete at frequent intervals.

The foundation wall forms can be set when the footing concrete has hardened for a day and the footing forms have been **stripped,** which means the forms are removed. Plywood is often used to make foundation wall forms, and aluminum or steel prefabricated form systems are also popular. The first step is to mark the foundation wall location on the top of the footing. This is done by driving a nail in the footing at outside foundation wall corner points and then making a chalk line between the points to show where the wall forms are placed. Outside and inside forms are placed at the same time, generally working from a foundation corner. Spacers are used to keep the form panels evenly spaced the wall thickness. The form panels are leveled, braced, and reinforced together the correct distance apart at the top as installation continues.

Rebar is placed after the footing forms are set, or at the same time. Sometimes it is easier to set the outside form, set the rebar and then set the inside form so the inside form structure is out of the way for placing rebar. Rebar is set in the forms exactly as specified in the plans. For example, the plans might have a note that points to the foundation detail and reads: 3-#4 CONTINUOUS REBAR EQUALLY SPACED. As you look at the foundation detail in Figure 4.39a you can see a dimension that reads: 1-1/2″ CLR. This means that the rebar surface needs to be set a minimum of 1-1/2 in. from the bottom of the footing. The rebar needs to be placed as shown and noted on the drawing, and the rebar needs to be stable so it does not move when pouring the concrete. Rebar can be positioned with prefabricated supports called **chairs.** Place enough of the properly sized supports or blocks under the rebar to keep it from moving and sagging. The rebar is wired together and overlap at splices as described earlier. Wire is cut from a roll and twisted around the rebar with a pliers, or by using prefabricated wires, or snap ties. Wire and supplies can be purchased from a concrete supply company. Vertical rebar is specified on the drawings in Figure 4.39a. The note reads: #4 REBAR 24″ OC, and the rebar is bent having a 24-in. vertical leg and an 8-in. horizontal leg. Enough pieces of rebar need to be bent to this specification in advance and ready to install. The vertical rebar is then wired to the horizontal rebar exactly as shown and dimensioned on the plans. See Figure 4.39b. The reinforced concrete footings are now ready to pour. Use caution when pouring the first concrete around the rebar to make sure the bars stay in place. The concrete should be tamped while being poured to help the semiliquid material flow completely around the rebar and fill all voids in the forms.

The vertical steel reinforcing is already set in the footing as previously described, but horizontal wall rebar needs to be placed as the wall forms are assembled, if required. Horizontal wall rebar can be tied to the vertical rebar or fastened to spacers used to set the forms. Steel spreaders are used throughout the foundation wall forms and at the top of the forms to keep the forms stable and spaced the desired distance apart. Anchor bolt brackets are placed at the top of the forms to act as spacers and proper placement for anchor bolt locations. Figure 4.40a shows the foundation wall forms set on the footing. Figure 4.40b shows the forms set for a crawl space foundation wall. Look closely and you can see the footings have already been poured and the forms stripped. The foundation wall forms are in place. Figure 4.40c shows a concrete pump truck used to easily distribute concrete to the construction site.

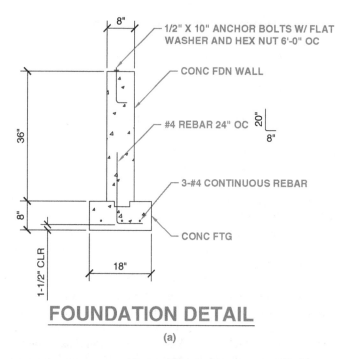

FOUNDATION DETAIL

(a)

(b)

Figure 4.39 (a) Find the dimension that reads: 1-1/2″ CLR. This means that the rebar surface needs to be set a minimum of 1-1/2 in. from the bottom of the footing. The rebar needs to be placed as shown and noted on the drawing, and the rebar needs to be stable so it does not move when pouring the concrete. Rebar can be positioned with prefabricated supports called chairs. (b) Vertical rebar is wired to the horizontal rebar exactly as shown and dimensioned on the plans.

(a)

(b)

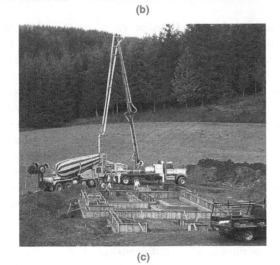

(c)

Figure 4.40   (*a*) Foundation wall forms set on the footing on a small commercial building. (*b*) Forms set for a crawl space foundation wall on a small commercial building. (*Madsen Designs Inc.*) (*c*) A concrete pump truck used to easily distribute concrete to the construction site. (*Courtesy of Madsen Designs Inc.*)

Figure 4.41 Foundation wall prefabricated forms set on the footing. (*Courtesy of Alsina Forms Co.*)

Prefabricated concrete forms help speed up the forming process and produce accurate foundations. Figure 4.41 shows the foundation wall prefabricated forms set on the footing. Additional discussion and examples are provided later in this chapter in the concrete foundation wall construction process.

CONCRETE SLAB FOUNDATION

The **concrete slab foundation system** is used on level sites in warm climates where the ground does not typically freeze in the winter. Freezing can cause the slab to move and crack. There is an expanded concrete perimeter footing and interior footings where required to support structural loads from above and distribute the loads to the ground. The concrete slab and the footings must be placed on firm undisturbed ground or other material that meets specific engineering requirements. The concrete slab foundation system generally uses 3-1/2″ (90 mm) or thicker reinforced concrete that is referred to as a **slab** that is a flat concrete pad poured directly on the ground or on compacted gravel over the ground. The concrete slab is generally thicker for commercial construction projects depending on specific engineering requirements and based on the intended building use. Compacted gravel can be used as needed to even out the grade by filling in low places and provide a uniform elevation for the concrete slab. The gravel is generally 3/4 in. minus. The term **3/4 in. minus** refers to the size of the rock, where 3/4 in. (19 mm) is the largest piece that fits through a 3/4 in. screen, and the approximate amount of **fines** in a product. Minus material can have 60 to 70 percent fines. The term **fines** describes small miscellaneous gravel, dirt, and debris sizes found in any gravel and crushed rock. All gravel has fines, but the amount of fines allowed depends on the gravel specifications. Some specifications require clean gravel be used and compacted to civil engineering requirements. Additional construction-related terms are **subgrade, subbase,** and **base.** Subgrade is the firm undisturbed excavated soil, or compacted soil. Subbase is a layer of compacted gravel, such as crushed rock with 10 to 20 percent fines, on top of the subgrade. Base is the layer of compacted gravel, such as 3/4 minus, on top of the subbase and directly under the concrete slab. A base layer on top of the subbase makes it easier to get a final flat grade, and maintain uniform slab thickness. The base also allows the concrete slab to slide easily as it shrinks and contracts while drying. This helps reduce the possibility of cracks in the slab. The subbase and base material must be compacted and at least 4 in. (100 mm) thick. The subbase and base thickness and compaction specifications need to be designed for the site and structure by a civil or structural engineer. Figure 4.42 shows perimeter footing options and the features of concrete slab construction, and the related partial foundation plan drawing.

Concrete piers are placed below the slab within the perimeter at locations where needed to support structural loads distributed to the ground below. Concrete piers can be cylindrical, cubed, or rectangular cubed to support a concentrated load such as a post, as shown in Figure 4.43. A continuous pier footing is used to support a lineal load such as a bearing wall, as shown in Figure 4.44.

Concrete slabs have **control joints** placed periodically across the slab to control cracking. Cracking occurs in concrete slabs when the concrete solidifies, expands, contracts, and settles after being poured. The control joints help restrict the cracking to the locations where the control joints are placed. Control joints can be a cut made in the slab, or a prefabricated joint that separates the slab at specific locations. Control joints also provide a clean surface to bond **between pours,** to control expansion, and to isolate stress in the concrete. Between pours means that it is often necessary to pour a portion of the slab either until when the concrete truck is empty or at the end of the day, and then continue the slab with another truck at a later time. The joint between pours can be arbitrary or designed in the plans. The slab edge between pours must be clean and can be a straight-flat edge or a detailed construction joint, as shown in Figure 4.45. The joints are also commonly spaced throughout the slab to help control cracking. Concrete slabs can crack due to expansion when the concrete solidifies after being poured, and if settling occurs. Any cracking can be isolated at the joints. Proper groundwork, subbase and base compacting and reinforcing steel help reduce settling and cracking in concrete slabs.

Steel reinforcing is placed in the slab to strengthen, help stabilize the concrete, and minimize cracking. Welded wire reinforcing is the most commonly used reinforcing for concrete slabs, but a square or rectangular rebar pattern can also be used depending on the structural engineering requirements. Figure 4.46 shows the basic components of a concrete slab foundation system.

Staking for Concrete Slab Construction

For concrete slab construction, the footing excavation is done to about 12 in. (300 mm) past the corner stakes to provide room for setting concrete forms. A string line is run at finished floor height and a laser level is used to make sure all form boards are set level and at the correct elevation. Pier locations are also marked on the ground and excavated to the proper size and depth at this time. Insulation, plumbing, and heating systems are installed in the ground before the slab is poured. This is described in more detail later in this chapter.

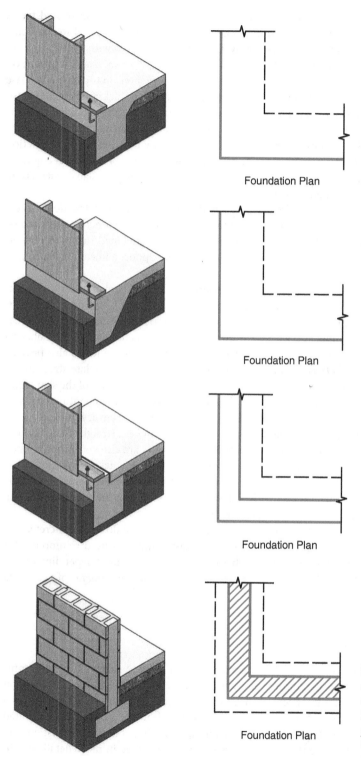

Foundation Plan

Foundation Plan

Foundation Plan

Foundation Plan

Figure 4.42 Perimeter footing options and the features of concrete slab construction, and the related partial foundation plan drawing. (*Previously published in Modern Residential Construction Practices, Routledge, 2017 and reproduced here with permission.*)

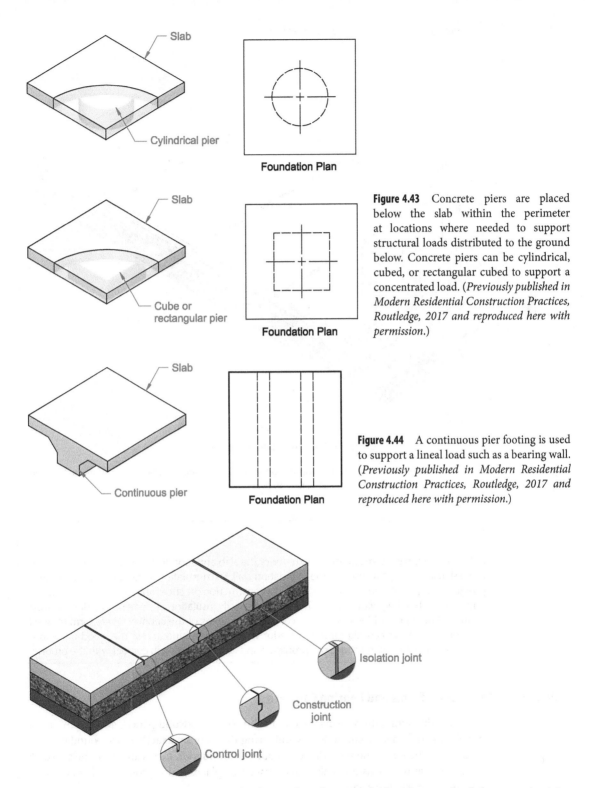

Figure 4.43 Concrete piers are placed below the slab within the perimeter at locations where needed to support structural loads distributed to the ground below. Concrete piers can be cylindrical, cubed, or rectangular cubed to support a concentrated load. (*Previously published in Modern Residential Construction Practices, Routledge, 2017 and reproduced here with permission.*)

Figure 4.44 A continuous pier footing is used to support a lineal load such as a bearing wall. (*Previously published in Modern Residential Construction Practices, Routledge, 2017 and reproduced here with permission.*)

Figure 4.45 The slab edge between pours must be clean and can be a straight-flat edge or a detailed construction joint. (*Previously published in Modern Residential Construction Practices, Routledge, 2017 and reproduced here with permission.*)

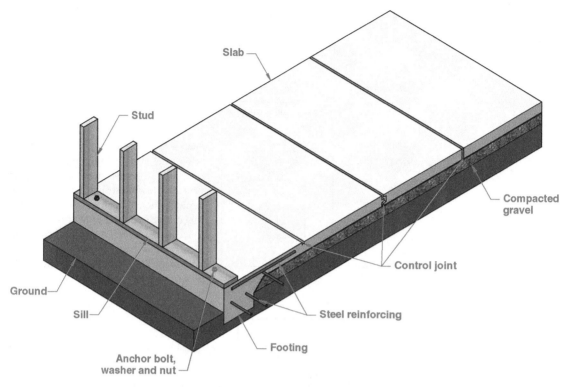

Concrete Slab Foundation System

Figure 4.46   The basic components of a concrete slab foundation system. (*Previously published in Modern Residential Construction Practices, Routledge, 2017 and reproduced here with permission.*)

> **NOTE**
>
> Slab insulation is used against the exterior of the slab and footing before backfilling, or under the slab and along the inside of the foundation wall. Slab insulation techniques vary in different regions, so you should always confirm local construction practices and building code requirements. The building plans accurately show and specify insulation placement. Exterior slab insulation should be avoided or carefully considered in areas of the country where termites are a problem. Termites can dig through exterior slab insulation and access the wood framing in the walls. Local building codes can prohibit using foam insulation in contact with the ground.

Concrete Slab Foundation Forms and Pouring Concrete

Concrete slab foundation forms are often solid wood, plywood panels, aluminum, or steel prefabricated forms. Accurate stakes and string lines are set exactly to the foundation corners and to the top of the slab. The top of outside foundation forms are set to the slab elevation. Foundation forms are staked and braced in place, and a professional surveyor can return to confirm the accuracy.

Most concrete slab foundation systems have steel reinforcing used to help stabilize the concrete and minimize cracking. Welded wire reinforcing is the most commonly used reinforcing for concrete slabs, but rebar spaced in a square or rectangular pattern is also often used. Perimeter footings and pier footings often have rebar to help support the structural loads. Carefully read the plans for drawings and specifications showing and describing required steel reinforcing. The steel reinforcing must be accurately set in the foundation and in the slab so it does not move when the concrete is poured. Rebar and welded wire reinforcing is placed with required ground clearance with prefabricated supports or chairs as previously described. Intersecting rebar is wired together in its exact position and for stability when the concrete is poured. Figure 4.47 shows a concrete slab foundation detail and the same concrete forms and steel reinforcing placed prior to pouring the concrete.

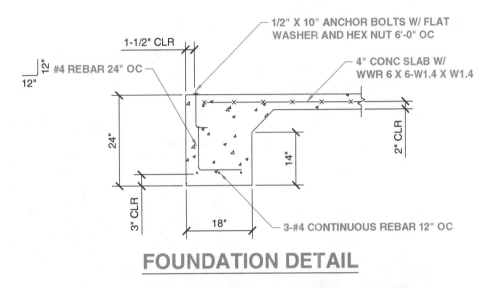

FOUNDATION DETAIL

Figure 4.47   A concrete slab foundation detail and the same concrete forms and steel reinforcing placed prior to pouring the concrete. (*Previously published in Modern Residential Construction Practices, Routledge, 2017 and reproduced here with permission.*)

Figure 4.48   Pouring a concrete slab foundation. (*a*) The concrete truck is in position pouring concrete. (*b*) A worker pours concrete into the slab form from a pump truck. (*c*) A worker pours concrete into the slab area and to the top of the forms. (*d*) Workers leveling and smoothing concrete using a screed. (*e*) A worker is smoothing the concrete with a power trowel.

Figure 4.48 shows several photographs of an actual concrete slab foundation being poured. Figure 4.48*a* shows the concrete truck in position ready to pour concrete. The worker moves the shoot to where the concrete needs to be placed. Notice the welded wire reinforcing on the grade and the forms in the foreground. Color-coded plumbing pipes are also in the slab and project out at the perimeter. Figure 4.48*b* shows a worker pouring concrete into the slab from a pump truck boom. This allows easy concrete pour access to all areas of the slab. The steel reinforcing must be placed and strongly secure, because the workers walk on the rebar while pouring concrete. Figure 4.48*c* shows a worker pouring concrete into the slab area and to the top of the forms. Figure 4.48*d* shows a worker initially smoothing and leveling concrete poured in the slab using a **screed.** A screed is a flat board, or a specially made aluminum tool as shown here, used to level, grade, and smooth a **concrete** slab after it has been poured in the forms. **Screed** is the process of using a screed to level and remove excess wet concrete to the top surface of a slab and to the accurate grade. Figure 4.48*e* shows a worker finishing the concrete slab by smoothing with a **trowel.** A trowel is a flat-bladed tool used for leveling, spreading, shaping, and smoothing concrete and **mortar.** Figure 4.48*f* shows a worker smoothing the concrete slab with a **power trowel.** A power trowel has a base with rotating trowel blades for use when concrete is just hard enough for walking on the surface. A power trowel does the final smoothing of the top of the concrete floor. There are walk-behind power trowels used by most contractors as shown here, and riding power trowels for larger commercial projects. Mortar is a mixture of cement, sand, lime, and water used to bond masonry such as bricks, concrete blocks, or rockwork.

Finishing is the final concrete process. Concrete finishing generally involves smoothing the surface as previously described unless the concrete will be **broomed, stamped,** or have an **exposed aggregate** finish. Broomed concrete has the surface textured with a broom while still wet. A broomed finish improves traction or to create a distinctive texture on the concrete surface. Stamped concrete is a process of using rubber molds pressed into the concrete to create patterns resembling brick, slate, cobblestone, flagstone, or tile. When used with concrete colors, the stamped surface has a decorative appearance without the cost of natural stone. Exposed aggregate concrete is a method of finishing concrete by washing the cement and sand away from the top surface of the concrete, exposing the aggregate immediately under the surface. Exposed aggregate provides an excellent appearance on driveways, patios, and other exterior surfaces. Concrete for exposed aggregate is generally ordered with special gravel that gives a quality appearance when exposed. The exposed aggregate process requires a skilled craftsperson to be done correctly.

In addition to surface smoothing and the other surface preparations described previously, concrete edges and joints are formed with an **edging trowel** to produce a radius along the slab edge that helps resist chipping and damage after the forms are removed. Slab edging is done immediately after surface smoothing. Edging is done especially on patios, curbs, sidewalks, and driveways, where the process creates an attractive edge that is resistant to chipping. There are hand edging trowels that require kneeling next to the slab to perform the edging operation, and edging trowels that attach to a handle, allowing the worker to walk alongside the slab to edge the concrete. Edging trowel are available that produce edge radii from 1/8 to 1 in. (3 to 25 mm). Concrete slabs with or without expansion joints often use narrow grooves made with a special trowel when finishing the concrete. The **joint trowel** creates a groove without extending the opening through the concrete. This groove provides enough space for the expansion of the top of the concrete without cracking. This joint works well with narrow concrete slabs such as sidewalks, but wider floor and driveways slabs should have expansion joints. Figure 4.49 shows a concrete joint trowel being used on a construction project. The joint and edging trowel are used in a similar manner.

Figure 4.49 A concrete joint trowel being used on a construction project. The joint and edging trowel are used in a similar manner.

CONCRETE WALL FOUNDATIONS

The **concrete wall foundation** uses a steel reinforced concrete footing keyed to a steel rein- forced concrete or concrete block foundation wall that extends from the footing to the floor framing above. The foundation wall is normally treated with waterproofing on the outside, and there is drain gravel and a drain tile to keep water from entering the below-grade struc- ture. The foundation wall can also be insulated on the outside with rigid insulation or on the inside as needed to meet code and green building certification requirements. The basement floor is a concrete slab that rests on the footing or on compacted gravel fill at the founda- tion wall. Finish materials are placed on the slab to complete the concrete floor. Figure 4.50 shows the basic components of a concrete wall foundation system.

Staking Concrete Wall Construction

A below grade concrete wall construction dig-out can be 8 to 10 ft (2400 to 3000 mm) deep or more, depending on the design requirements. Corner stakes are set by a professional sur- veyor using the site plan to establish the exact building corners. A building perimeter line is established by running string lines between each corner stake and marking the lines with paint on the ground. The surveyor then sets 4 to 6 ft (590 to 880 mm) offset stakes, which provides for the additional space needed to set concrete forms for footings and foundation walls. The excavator removes the earth to the desired depth everywhere inside the offset lines.

Concrete Wall Forms and Pouring Concrete

After excavation is finished, it is time to set concrete forms for footings and pour concrete footings. A string line is run along the outside footing perimeter at the finished footing

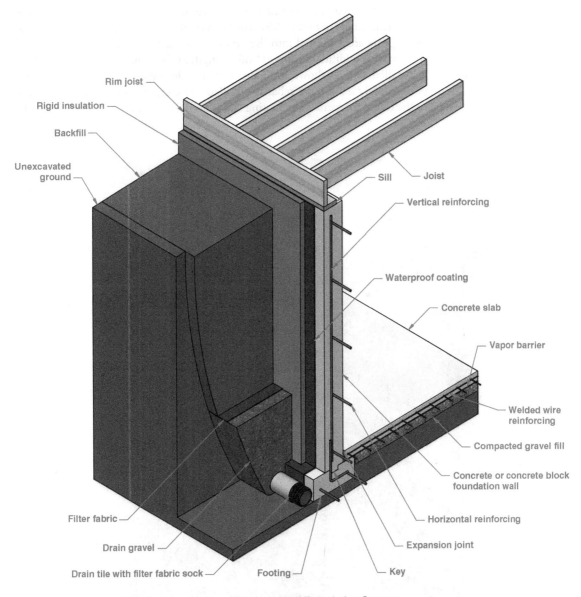

Rim joist

Rigid insulation

Backfill

Unexcavated ground

Sill

Joist

Vertical reinforcing

Waterproof coating

Concrete slab

Vapor barrier

Welded wire reinforcing

Compacted gravel fill

Concrete or concrete block foundation wall

Filter fabric

Drain gravel

Drain tile with filter fabric sock

Footing

Key

Horizontal reinforcing

Expansion joint

Concrete Wall Foundation System

Figure 4.50   The basic components of a concrete wall foundation system. (*Previously published in Modern Residential Construction Practices, Routledge, 2017 and reproduced here with permission.*)

height and a laser level is used to make sure all forms are set straight and level. Properly set forms are a critical step in establishing a good foundation. Steel reinforcing is placed in the footing and foundation wall based on the construction drawings and specifications.

Concrete wall construction is similar to crawl space foundation construction except that the concrete walls are much thicker and higher and additional rebar is commonly used for concrete wall construction. Concrete forms for commercial construction can be made from solid wood or plywood, but aluminum or steel forms are more typically used, and insulated forms can be used as described in detail later in this chapter. The forms for the footing are set first, as shown in Figure 4.51. The forms in this example are made out of solid wood. Form supports are anchored into the ground to hold the forms. Diagonal bracing can be used to provide additional support and strengthen the structure for higher forms. The rebar is installed next, as shown in Figure 4.51, and described in the following. After the footing are securely set and the rebar is installed, the concrete is poured, leveled, and smoothed to the top of the forms. The footing key, if used, is set while the concrete is soft.

Figure 4.51   Rebar is installed in the forms. (*Photo courtesy of constructionphotographs.com.*)

The concrete wall forms can be set when the footing concrete has hardened for a day and the footing forms have been stripped. Plywood is often used to make concrete wall forms, and aluminum or steel prefabricated form systems are also popular for commercial construction. The first step is to mark the concrete wall location on the top of the footing. This is done by driving a nail in the footing at outside wall corner points and then making a chock line between the points to show where the wall forms are placed. Outside and inside forms are placed at the same time, generally working from a foundation corner. Spacers are used to keep the form panels evenly spaced for the wall thickness. The form panels are leveled, braced, and reinforced together the correct distance apart at the top as installation continues.

Rebar is always used in the footings and concrete walls, as determined by structural engineering. Rebar is placed after the footing forms are set, or at the same time. Sometimes it is easier to set the outside form, set the rebar and then set the inside form. This is done so the inside form structure is out of the way for placing rebar. Rebar is set in the forms exactly as specified on the plans. For example, the plans might have a note that points to the foundation detail and reads: #5 CONTINUOUS REBAR SPACED 6″ OC EW (each way). The foundation detail can also have a dimension that reads: 1-1/2″ CLR. This means that the surface of the rebar needs to be set a minimum of 1-1/2 in. from the bottom of the footing providing clearance for concrete to fill the space. The rebar needs to be placed as shown and noted on the drawing, and the rebar needs to be stable so it does not move when pouring the concrete. Rebar can be positioned with prefabricated support chairs. Place enough of the properly sized supports under the rebar to keep it from moving or sagging. The rebar should be wired together and overlap at each splice. The amount of overlap is typically specified on the plans or in specifications. Cut wire from a roll and twist the wire around the rebar with a pliers, or use prefabricated wires or snap ties for rebar connections. Wire and supplies can be purchased from a concrete supply company. Vertical rebar can be specified on the drawings with a note that reads, for example: #5 REBAR 12″ OC, and the rebar is bent having a 24-in. vertical leg and an 8-in. horizontal leg. Enough pieces of rebar needs to be bent to this specification in advance and ready to install. The vertical rebar is then wired to the horizontal rebar exactly as shown and dimensioned on the plans. You are now ready to pour the reinforced concrete footings. Use caution when pouring the first concrete around the rebar to make sure the bars stay in place. The concrete should be tamped while being poured to help the semiliquid material flow completely around the rebar and fill all voids in the forms.

Vertical steel reinforcing is already set in the footing as previously described, but horizontal wall rebar needs to be placed as the wall forms are assembled. Horizontal wall rebar can be tied to the vertical rebar or fastened to spacers used to set the forms. Steel spreaders are used throughout the forms and at the top of the forms to keep the forms stable and spaced the desired distance apart. Anchor bolt and hold down anchor brackets can be placed at the top of the forms to act as spacers and proper placement for anchor bolt and locations, if used. Foundation wall forms are set on the footing as previously described in this chapter. A concrete pump truck is generally used to easily distribute concrete into the forms.

Vibrating Concrete

Vibrating concrete applies to all concrete construction and should be confirmed if recommended or required. Vibrating concrete is also very important to remove air bubbles that occur when pouring concrete and to help make the concrete stronger. Concrete vibration requirements are often given on plans and specifications for commercial and sometime residential concrete construction. Concrete vibration specifications can also be required by building codes. There are different types of concrete vibrators.

The **internal vibrator** is the most common type of concrete vibrator used in construction. The internal vibrator works by pushing the vibrator into the wet concrete, and slowly pulling it out at the rate of about 1 in./s as it vibrates the concrete. Insert the internal vibrator vertically into the concrete and turn it on for about 10 seconds before moving it outward. Continue vibrating until no more bubbles are observed coming out of the concrete. Avoid vibrating too long, because doing so can cause concrete segregation in which the denser aggregates settle to the bottom while the lighter cement separates upwards. Avoid allowing the vibrator to touch the forms as this can damage the forms and can affect the concrete surface appearance.

When using an internal vibrator, it is very important to withdraw the device slowly, or you will not get enough vibration to eliminate the majority of its air bubbles. However, you do not want to leave the vibrator in the concrete too long as it could cause the water to separate from the cement, hurting the look and structural integrity of the concrete.

The solution is to withdraw the vibrator at the rate of about 1 in./s. Start by inserting the head fully, then turning it on and leaving it in place for 10 seconds. Insert it as vertically as possible, so it naturally sinks into the concrete. Then remove 1 in. each second until it is completely withdrawn. If there are still bubbles coming out once you have fully removed the vibrator, repeat the process until no bubbles emerge. The right amount of vibration is important. Under vibration can result in week concrete that can fail, and over vibrating can cause the water and aggregates to separate, which can reduce concrete strength.

A **formwork vibrator** attaches to the outside of concrete form and is used to vibrate air bubbles out of the concrete in the forms. Form vibrators should be spaced about 6 ft apart.

A **surface vibrators,** also called a **jumper vibrator,** is used for concrete slabs 6 in. deep or less, and can provide a smooth surface on the concrete slab. The surface vibrators is placed directly on top of the poured concrete.

Automated Concrete Forms

High-volume concrete contractors for commercial construction often use automated concrete forms designed for quick installation and removal. A selection of automated concrete form manufacturers is available in the industry providing similar applications. Examples of automated concrete forms are aluminum frame, galvanized steel frame with plywood face, and steel frame modules. Automated concrete forms provide long service life and create a clean and smooth concrete surface. Automated concrete form systems provide reusable concrete form modules for fast, simple, and safe installation and removal, as shown in Figure 4.52.

Quick and easy assembly is done with clamps that join, align, and tighten the modular panels. A typical joint and alignment system is made using a clamp or fastener system that allows you to accurately and quickly join and align the panels. Prefabricated bracing systems are also available, as shown in Figure 4.53. Figure 4.54 shows the smooth concrete surface achieved after the concrete is poured and the automated forms are removed.

Concrete Slab

The concrete slab constructed with concrete walls is similar to the slab on grade foundation system previously described. Ground work is necessary before the concrete slab can be poured. The concrete slab must be poured on firm undisturbed soil or compacted gravel fill just like the foundation slab system. The concrete slab can be poured after the concrete walls, or poured later in the construction project, such as after the roof is on, providing protection.

Figure 4.52 Automated concrete form systems provide reusable concrete form modules for fast, simple, and safe installation and removal as shown here. (*Courtesy of Alsina Forms Co.*)

Figure 4.53 A typical joint and alignment system is made using a clamp or fastener system that allows you to accurately and quickly join and align the panels. Prefabricated bracing systems are also available. (*Courtesy of Alsina Forms Co.*)

Figure 4.54   A smooth concrete surface achieved after the concrete is poured and the automated forms are removed. (*Courtesy of Alsina Forms Co.*)

The reason for waiting is to avoid damaging the slab with construction work in progress. The slab should be carefully protected from damage during construction if poured immediately after the concrete walls. After the subgrade and grade pre-slab fill are brought to the correct elevation, a **vapor barrier** or **retarder** is installed that lies between the grade and the slab. Vapor barriers help prevent moisture from transferring from the ground through the slab, can reduce possible moisture-related mold and mildew growth or even block gases that can be a health risk for occupants. Vapor barriers are specified by their **permeability,** which is the ability to let water vapor pass through. Permeability is expressed as perms. The acceptable amount of permeability depends on the application, such as example 0.3 perms is normally recommended for construction. There are several products that can meet vapor barrier specifications and the type and value of vapor barrier specified on the plans.

Depending on the design requirements, there can be an expansion joint used between the concrete slab and the foundation wall to help reduce pressure on the wall from slab expansion. Alternately, perimeter insulation and under-slab insulation can be specified on the plans. This insulation is usually rigid insulation that comes in sheets. The slab can be designed with an expanded perimeter resting on the foundation footing. The expanded slab perimeter supports the slab, reduces settling, and cracking. Concrete slab rebar or welded wire reinforcing is used to reducing cracking. Figure 4.55 shows a concrete slab without an expanded perimeter and with an expanded perimeter.

It is often desirable to **backfill** the foundation as soon as possible. Backfill is the earth or other material used in the process of backfilling, which closes the large space created by

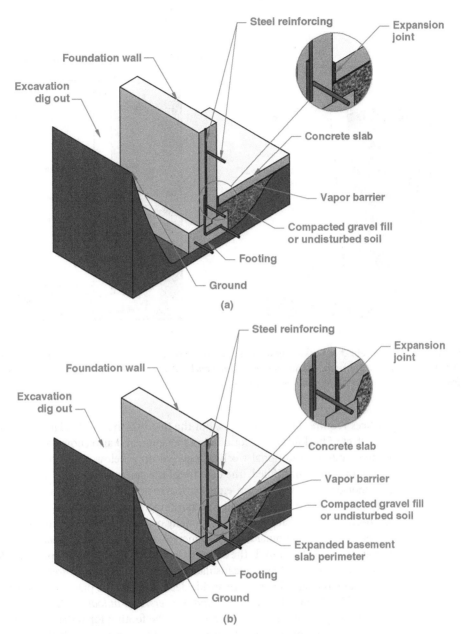

Figure 4.55   (*a*) A concrete slab without an expanded perimeter and (*b*) with an expanded perimeter. (*Previously published in Modern Residential Construction Practices, Routledge, 2017 and reproduced here with permission.*)

the excavation. Backfill makes it easier to access the structure for continuing construction. Additional preparations need to take place before the concrete wall can be backfilled. First, it is very important to tie the top of the wall in place with the floor framing. Failure to do this can cause the wall to bend, tip, or fall. For typical wood frame construction, the framing process begins by bolting down the mud sill followed by constructing the floor joist system. Floor framing is described in detail in Chapter 8, *Wood Construction*. Other construction

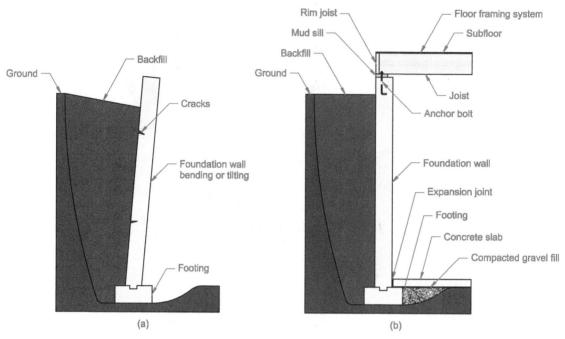

Figure 4.56 The possible effect of backfilling the foundation too soon, and shows components of the wood frame floor framing system to help stabilize the foundation wall. (*Previously published in Modern Residential Construction Practices, Routledge, 2017 and reproduced here with permission.*)

practices require different process, but the intent is the same. Figure 4.56 shows the possible effect of backfilling the foundation too soon, and also shows components of the floor framing system. Additional work is required on a below grade concrete wall before the excavation can be backfilled. The exterior of the concrete wall gets a waterproof coating to keep moisture and water from entering the structure. There are many products available that can be used for concrete wall waterproofing. The plans should specify what to use and subcontractors are available to professionally perform this application. Rigid insulation can also be placed on the outside of the concrete wall to add additional waterproofing and to insulate the wall, if required. The ability of insulation to act as waterproofing depends on the product. The use of below grade insulation depends on the local area. Concrete walls can be insulated on the outside or on the inside depending on the plan requirements. Insulation is described in Chapter 10, *Insulation and Barriers, and Indoor Air Quality and Safety.*

A perforated drain pipe is placed along the footing for water drainage away from the structure. The drain pipe usually has a filter fabric covering called a sock over it to keep fines from clogging the perforations and entering the pipe. Drain gravel is then placed over the pipe and commonly fills two-thirds of the excavation, unless otherwise specified on the plans. Filter fabric is placed over the drain gravel before the remaining backfill is placed in the excavated area. The filter fabric allows water to penetrate while keeping dirt from plugging the drain rock below. The drain pipe and drain gravel acts as drainage system to keep water from entering the structure. Figure 4.57 shows the complete concrete footing, wall, and slab waterproofing system.

Under-Slab Electrical, HVAC, and Plumbing. The concrete slab foundation system was described earlier in this chapter, including excavation, form setting, rebar placement, and

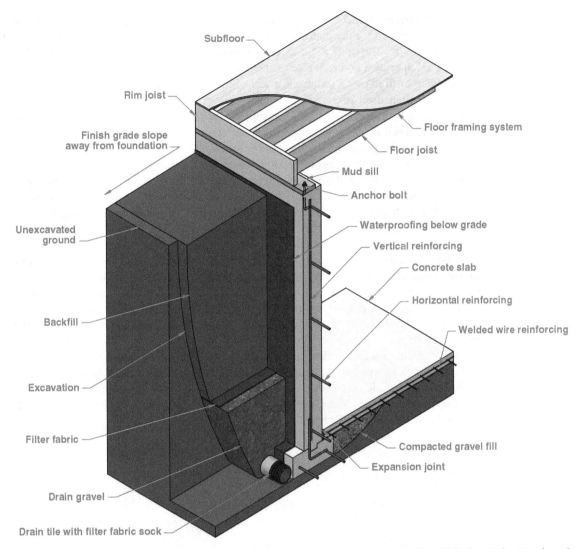

Figure 4.57   The complete concrete footing, wall, and slab waterproofing system. (*Previously published in Modern Residential Construction Practices, Routledge, 2017 and reproduced here with permission.*)

pouring and finishing the concrete slab. There can also be under-slab electrical; under-slab heating, ventilating, and air-conditioning (HVAC); and under-slab plumbing depending on the plans and specification. Each of these installations is performed by the specific electrical, HVAC, or plumbing subcontractor. Careful coordination is necessary, because the work of one subcontractor can interfere with the others and all of this groundwork must be properly completed before slab concrete is poured. These under-slab features need to be roughed-in and properly covered before the rebar or welded wire reinforcing is installed. Locations for these features also need to be excavated below grade by using a trencher or back hoe. The footing and foundation forms are generally in place before this ground work starts. If the forms are in place, a section of the forms need to be removed or left off the foundation for excavation equipment to access inside the foundation area.

The **electrician** trenches and installs the under-slab **electrical conduit** and fittings in the locations shown on the plans. The electrician is the subcontractor who performs the electrical work for the project. Electrical conduit, also called conduit, is electrical piping system used to protect and rout electrical wiring. Electrical conduit can be made of metal, plastic, fiber, or fired clay. Flexible electrical conduit is also available for special purposes. **Bedding** is often used under the conduit to bring the bottom of the trench to grade, and provide a firm, stable, and uniform support for the conduit. Bedding is usually 4 in. (100 mm) or less thick. Bedding is specific material used under pipe and other groundwork for uniform grade, protection, and support. Specific bedding requirements can be specified on the plans and the types of bedding material are described later in this discussion. Figure 4.58 shows electrical conduit in an under-slab trench. The conduit is backfilled with dirt that came from the excavation if the dirt is clean and free of rocks that could damage the conduit, or backfilled with sand to protect the pipe. Forty-five to ninety degree turns, called **sweeps,** can be made in the conduit as long as they have a large radius that makes it easy to pull wire through the conduit during the electrical work.

The **mechanical** subcontractor performs the ground work by installing the **ductwork** for the HVAC under-slab system. The term **mechanical** is used to describe the HVAC system and the subcontractor who performs the installation. Ductwork is round or rectangular metal or plastic pipes that are solid or flexible, and installed for distributing warm or cold air from the furnace or air-conditioning system to locations in the building and back to the furnace or air-conditioning system. The mechanical subcontractor trenches locations for the ductwork, installs the ductwork, and backfills around the ductwork using clean material from the trench excavation. Bedding is often used under the ductwork to bring the bottom of the trench to grade, and provide a firm, stable, and uniform support for the ductwork. Bedding is usually 4 in. (100 mm) or less thick. Figure 4.59 shows ductwork in an under-slab trench that has been partially backfilled.

Figure 4.58   Electrical conduit in an under slab and projecting out above concrete.

Figure 4.59   Ductwork in an under-slab trench that has been partially backfilled.

The **plumber** is a subcontractor who performs the ground work by installing the **plumbing** for the building. Plumbing is the pipes, fixtures, and other equipment of a water, gas, or sewage system in a building. The plumber trenches locations for the plumbing, installs the plumbing, and backfills around the plumbing using clean material from the trench excavation or sand to protect the pipes during and after backfill. Bedding is usually used under the plumbing to bring the bottom of the trench to grade, and provide a firm, stable, and uniform support for the pipes. Bedding is usually 4 in. (100 mm) or less thick. Figure 4.60 shows plumbing in an under-slab trench that has been partially backfilled.

Figure 4.60   Plumbing in an under-slab trench that has been partially backfilled.

Backfilling can take place after bedding material has been placed and the electrical, mechanical, and plumbing systems have been installed. Quality backfilling is important to provide a firm, stable, and uniform support for the conduit, duckwork, and plumbing. The backfill also needs to be properly compacted to support the concrete slab above and minimize the possibility of cracks in the slab. Backfilling should take place in two stages in most cases. Initial backfill places material to a level of 6 to 12 in. (150 to 300 mm) above the top of the installation. The initial backfill is compacted to give side support for the installation with care taken not to damage the installation. The final backfill provides material above the initial backfill to final under-slab grade. The final backfill is compacted for supporting the concrete slab with care taken not to damage the installation. Deep trenches can require more than two levels of backfill. Specific backfill instructions are sometimes provided on the plans or in the specifications.

Bedding and backfill materials can be specified on the plans, in the specifications, or accomplished by following general practices. General practices typically specify that materials must be free of sharp objects, sticks, large clumps, frozen material, organic materials, and rocks. Materials used in bedding and backfilling are specified as Class I through V by the American Society for Testing Materials (ASTM). Class I is angular crushed rock, graded 1/4 to 1-1/2 in. (6 to 38 mm) in size with little or no fines. Class II is clean, coarse-grained material, such as gravel, coarse sands, and gravel sand mixtures with 1-1/2 in. (38 mm) maximum in size. Class III is coarse-grained materials with fines including more than 50 percent silt or clay gravels or sands with 1-1/2 in. (38 mm) maximum in size. Class IV is fine-grained material, such as fine sand and soils containing 50 percent or more clay or silt. Class V material includes organic silts and clays, peat and other organic materials, and are not recommended for use as bedding material.

INSULATING CONCRETE FORMS (ICF)

Insulating concrete forms (ICFs), also called **insulated concrete forms** are created by combining a high-density plastic structural web embedded in two outer layers of foam. This web provides a support structure for each side of the form and for reinforcing materials. The wall is created by sandwiching concrete between each of these layers. ICFs are designed to provide stronger, more comfortable, quiet, and highly energy-efficient structures than conventional concrete walls or wood-framed buildings. ICFs are easy to construct and their lifespan is hundreds of years longer compared to regular construction methods.

ICF foundation walls are constructed using a common size of 4-ft-long and 16-in.-high interlocking blocks. Height and length can vary depending on the manufacturer. A cavity is created with evenly spaced webs that give the blocks their strength. The cavity width can vary to 12 in. Common wall widths are 8 to 12 in. for commercial walls. Most manufacturers offer corner blocks, 45° blocks, brick ledge forms and taper top blocks along with various accessories and fittings. Some blocks are fully reversible, adding to the ease of construction. ICFs allow for unlimited design flexibility. Radius walls, multiple elevations, and other architectural features are easily constructed using ICFs. Figure 4.61 shows installation of

Figure 4.61   Installation of the ICF interlocking blocks for a crawl space foundation wall. (*Courtesy of BuildBlock Building Systems, LLC.*)

the ICF interlocking blocks for a crawl space foundation wall. Notice the cavity created with evenly spaced webs that give the blocks their strength.

Horizontal rebar is installed and tied to vertical rebar as the ICF forms are being placed, as shown in Figure 4.62.

Bracing or alignment systems made specifically for ICFs are installed when a wall is too high to reach from the ground. Most bracing systems have adjusting **turnbuckles** that make the walls perfectly straight as the pour is finished. Concrete is poured in the forms, as shown in Figure 4.63. A turnbuckle is a device for adjusting the tension or length of ropes, cables, or rods and generally made of two threaded fasteners, one with a left-hand thread and the other with a right-hand thread.

ICFs meet or exceed standards set by building codes. The foam protects the concrete and remains in place after the pour, so the building season can be extended in cold climates. The ICF forms provide a moist cure of the concrete, which creates stronger concrete than conventionally poured walls that have forms stripped usually within a day of pouring. ICF walls can withstand hurricanes and tornado force winds. ICF walls offer roughly a 70 per-cent noise reduction compared to conventional construction. ICF walls can achieve up to a 4-hour fire rating. ICF construction can often be accomplished with little or no increase in cost over conventional practices. ICF walls are almost immediately ready for interior and exterior finishes. Figure 4.64 shows typical ICF construction details provided by an ICF manufacturer.

Figure 4.62   Horizontal rebar is installed and tied to vertical rebar as the ICF forms are being placed. (*Courtesy of BuildBlock Building Systems, LLC.*)

Figure 4.63 Most bracing systems have adjusting turnbuckles that make the walls perfectly straightened as the pour is finished. Concrete is poured in the forms. (*Courtesy of BuildBlock Building Systems, LLC.*)

PIER FOUNDATION

A **pier foundation** is used to support **post and beam** or **pole construction**. Post and beam or pole construction is also referred to as **timber-frame construction** or **pole buildings,** generally using large posts and beams or timbers for the horizontal and vertical members. The term **post and beam** was used earlier with the post and beam foundations system. This post and beam construction practice is similar, but the foundation system is much different. Pier foundations are a good option where it is difficult to access the site for a conventional crawl space, slab, or concrete wall foundation, because excavation requires much less earth

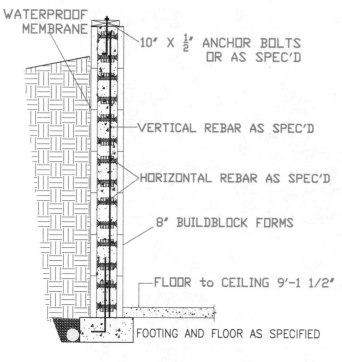

WATERPROOF
MEMBRANE

10' X ½' ANCHOR BOLTS
      OR AS SPEC'D

VERTICAL REBAR AS SPEC'D

HORIZONTAL REBAR AS SPEC'D

8' BUILDBLOCK FORMS

FLOOR to CEILING 9'-1 1/2'

FOOTING AND FLOOR AS SPECIFIED

TYPICAL 8" BUILDBLOCK BASEMENT 9'- 4" HIGH

Figure 4.64   Typical ICF construction details provided by an ICF manufacturer. (*Courtesy of BuildBlock Building Systems, LLC.*)

removal, and less concrete for the foundation. Pole buildings are constructed using a combination of post and beam, timber framing, and conventional stud framing depending on the requirements of the specific building. The vertical poles or posts are used as the building supports to which the horizontal framing beams are fastened. The posts or poles can be treated and embedded into the ground, or tied to steel reinforced concrete footings on the ground. Pole buildings are commonly used for utility buildings and structures such as barns, sheds, shops, warehouses, waterfront piers, and aircraft hangars. Most agriculture and utility pole buildings are uninsulated and can have one or more sides left open for easy access. Pier foundations with post and beam construction is also commonly used to build decks. Figure 4.65 shows two options for post or pole construction at the foundation. Figure 4.65a shows the pole embedded in the ground. Figure 4.65b shows the pole anchored to a reinforced concrete pier or footing.

The pole building foundation is generally a pattern of reinforced concrete footings with post bases used to support the vertical posts. Figure 4.66 shows a partial pole building foundation system and one of the footings detailed.

Setting and Pouring a Pier Foundation

The pole building foundation is generally a pattern of reinforced concrete footings with post bases used to support the vertical posts. The exact location of each footing is critical. A survey crew professionally locates each pier to the centers. Piers can be located to their centers regardless if they are round square or rectangular, because the post base is generally placed in the center of the pier. String lines can be run between centers with the string line stakes out of the way so they are not disturbed during pier excavation. Figure 4.67 shows

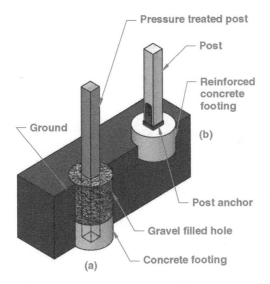

Figure 4.65 Options for post or pole construction at the foundation. (*a*) The pole is embedded in the ground. (*b*) The pole is anchored to a reinforced concrete pier or footing. (*Previously published in Modern Residential Construction Practices, Routledge, 2017 and reproduced here with permission.*)

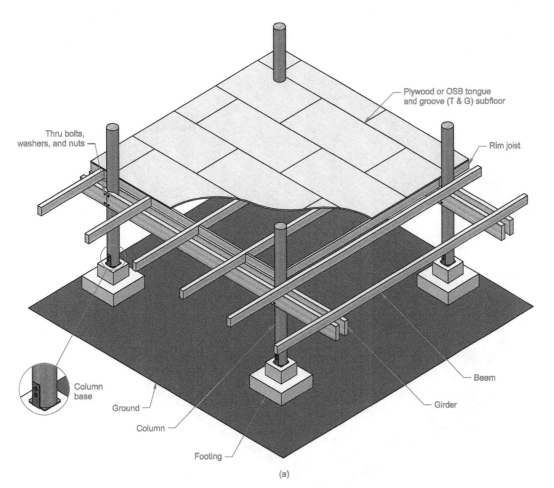

Figure 4.66 (*a*) A partial pole building foundation system and (*b*) one of the footings detailed. (*Previously published in Modern Residential Construction Practices, Routledge, 2017 and reproduced here with permission.*)

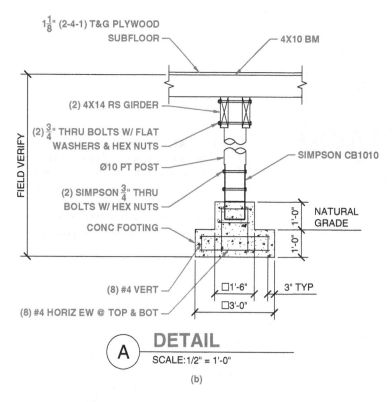

DETAIL

SCALE:1/2" = 1'-0"

(b)

Figure 4.66   (*Continued*)

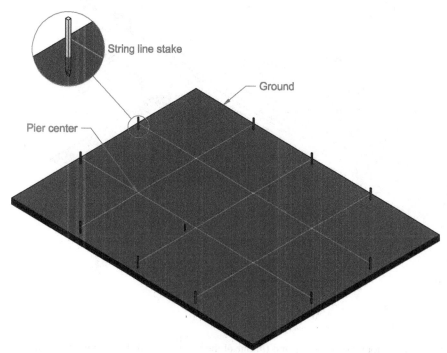

Figure 4.67   String lines set to locate the pier centers. (*Previously published in Modern Residential Construction Practices, Routledge, 2017 and reproduced here with permission.*)

Figure 4.68   Pier footings formed and concrete to be poured.

string lines set to locate the pier centers. Spray paint can be used to provide temporary center marks at each pier location. The string lines can be moved as needed to get excavation equipment on the site. Piers are usually cylindrical or cubic in shape. Excavating to firm undisturbed soil is very important as with any footing excavation. The construction crew often uses a back hoe to dig the holes for large footings, or the entire area is excavated to firm undisturbed soil if required. Each hole is excavated large enough to contain the forms without crowding and to give room to adjust the forms to their exact centers. Forms are constructed out of wood or prefabricated forms are used after the footings are excavated. Prefabricated forms are typically cylindrical in shape, but cube-shaped forms are also available. Using the correct size prefabricated forms saves time, but very large and custom size footings require forms made with wood or other material. Rebar can be placed as required when the forms are set in their exact locations. Now, the concrete can be poured in the forms and post bases placed and leveled. The concrete is smoothed and leveled around the post bases to complete each footing. Figure 4.68 shows pier footings formed and concrete to be poured. The excavated soil around the footings can be used for backfill after the forms are removed.

SHOTCRETE

Shotcrete, also called **gunite** or **sprayed concrete** is concrete or **mortar** transported through a hose and **pneumatically** projected at high velocity onto a surface. Shotcrete can be used to spray concrete on any surface and can be used for structural concrete, while common uses are for mine and tunnel walls, and in-ground swimming pools. The surface where shotcrete is sprayed is typically reinforced with rebar, welded wire reinforcing, steel mesh, or fiberglass. The term **pneumatically** projected refers to the use of a pneumatic mechanism operated by air or gas under pressure. Mortar is a mixture of lime with cement, sand, and water, used in construction to bond bricks, concrete blocks, or rockwork. The force shotcrete is ejected from the pneumatic nozzle places and compacts the material at the same time. Shotcrete can be sprayed onto any type or shape of surface, including vertical or overhead areas.

Two types of shotcrete are dry mix and wet mix. Dry mix shotcrete, commonly called gunite, is used by placing dry material in a hopper and transferring it pneumatically through a hose to the nozzle where water is added at a precise rate. The dry material and water do not completely mix until the material hits the surface. The dry mix process allows the water content to be adjusted for accurate and consistent placement in overhead, vertical, and odd-shaped locations. Wet-mix shotcrete delivers mixed concrete to the pneumatic nozzle. The wet-process typically makes less dust and waste material compared to the dry-mix process. The wet-mix concrete is mixed with water and required additives, allowing larger volumes to be sprayed in less time than with the dry process.

PILING FOUNDATIONS

A **piling foundation** is also called a **deep foundation,** because the foundation transfers building loads to the earth farther down from the surface than conventional spread footing foundations described earlier. The pilings are set to the depth of a subsurface layer based on below grade material and civil and structural engineering requirements. Pile foundations are used when there is a layer of unstable ground material such as sand, fill, or water at and below the surface. The unsound layer in the earth cannot support the weight of the building, so the structural loads must be extended to a layer of stronger soil or rock below the weak layer, or engineered to support the loads in a different manner. Deep foundations are also used when a building has heavy, concentrated loads that require more structural support than what can be provided by a spread footing foundation, such as for high-rise buildings, bridges, and other massive structures.

Piling foundations use columns made of materials such as timber, steel, precast concrete driven, or cast-in-place concrete, composite pilings, and helical piles installed deep into the earth. Piling foundations support the structure and transfer the load at a desired depth, and resist lateral forces with **end bearing** or **skin friction piles,** as shown in Figure 4.69.

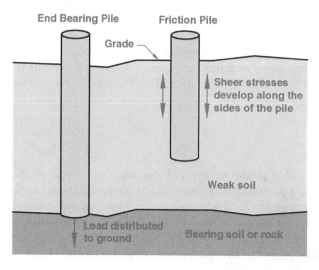

Figure 4.69   End-bearing or skin friction piles.

End-bearing piles are columns driven into the ground and the bottom end of the pile rests on a layer of bearing soil or rock. The load of the building is transferred through the pile onto the ground layer at the bottom. Skin friction piles, as the name implies, develop most of the pile-bearing capacity by **shear stresses** along the sides of the pile, which is basically friction between the surface of the pile and the earth where the pile is driven. Shear stresses are forces caused when two construction features move over each other. Skin friction piles are often used where the ground is soft as in fill or sand, and when a hard layer is too deep to reach economically. The entire surface of a skin friction pile transfers the load to the earth. In addition to their structural advantages and uses, pilings can improve the load-bearing capacity of the soil, eliminate settlement, transfer loads through easily eroded soils to a stable underlying bearing stratum. Pilings also anchor structures subjected to **hydrostatic uplift** or overturning, and to construct a retaining structure using a series of overlapping cast-in-place piles, or interconnecting piles. Hydrostatic uplift is caused by water pressure pushing up on the piling or the structure.

Geological engineers work together with civil and structural engineers to determine soil conditions when piling foundations are designed. Different soils and soil conditions are evaluated to determine the soil bearing conditions and to design the piling foundation so it does not overload the ground beyond its bearing capacity. This is done, in part, to determine the **zone of influence** on the earth around the piling or pilings, as shown in Figure 4.70. The space between pilings is engineered to distribute the loads evenly over the entire zone of influence. The zone of influence is a zone surrounding a piling or group of pilings, in which the stresses caused by the piling construction is at a desired engineered level for supporting the structural loads.

After pilings are driven or constructed into the ground, they are generally cut off at a specific elevation. A **pile cap** is then constructed on individual or a group of pilings. A pile cap is a thick reinforced concrete structure that extends over the piles, and serves as a base on which a column, columns, or walls can be constructed. The load of this structure, over the pile cap, is distributed to all piles in the group. The pile cap can have steel reinforcing projecting from the top for use in connecting the reinforcing for columns or other concrete structures above. The pile cap shape is designed to encompass and overlap the combination of piles in the group, as shown in Figure 4.71.

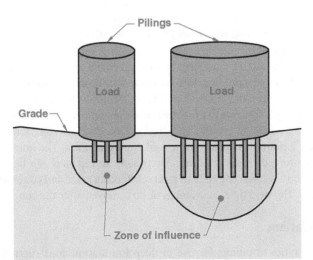

Figure 4.70 The zone of influence is a zone surrounding a piling or group of pilings, in which the stresses caused by the piling construction is at a desired engineered level for supporting the structural loads.

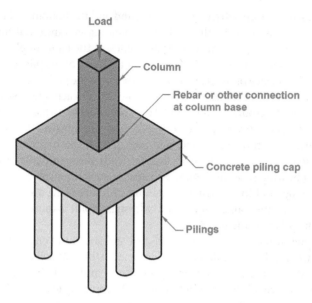

Figure 4.71 The pile cap shape is designed to encompass and overlap the combination of piles in the group.

Installing Pilings

There are two general classifications for piling installation, which are displacement and replacement pilings. **Displacement piles,** also called **driven piles,** are columns installed without removing material from the ground. Special equipment is used to drive, drill, or screw the piles into the ground while displacing soil laterally, and compacting the soil evenly into the immediately surrounding earth. Depending on ground material, conditions, and structural engineering, displacement piles can support high loads. Displacement piling is used when disturbance to the surrounding ground is kept to a minimum but is not used to penetrate obstructions and rocky ground. Displacement piling are delivered to the construction site in the case of timbers, steel, precast concrete, composite, and helical piles. Cast-in-place pilings are cast on the ground and then driven into the ground. A **pile driver** is commonly used to drive piles, which is a machine that holds the pile vertically and hammers the pile into the ground, as shown in Figure 4.72. A pile driver works by repeatedly lifting a heavy weight and dropping it on top of the pile. Piles are generally driven to engineering specifications or to the **refusal point,** which is the location where a pile cannot be driven any further. **Replacement piling,** also called **bore piling,** refers to digging or drilling a hole where a variety of pilings can be constructed. Replacement piling is used for installation in rocky and other difficult ground but does increase disturbance to the surrounding ground. There are a variety of methods for constructing replacement pilings. Holes for replacement pilings can be filled with concrete during the boring process or after boring is complete. The hole can be bored into stable ground and steel reinforcing inserted, followed by pouring concrete to create a poured-in-place reinforced concrete piling. The hole can also be filled with a precast reinforced concrete piling. Alternately, round steel can be inserted into the hole for stabilization and then concrete or steel reinforcing place and concrete can be poured into the steel tube. The steel tube can remain or be removed after the concrete is poured.

Types of Pilings for Deep Foundations

There are several types of pilings for use in deep foundation construction. The types are based on the material used for pilings, the environment such as soil type and conditions,

Figure 4.72   Pile driver. (*Courtesy of Hammer & Steel, Inc.*)

along with civil and structural engineering requirements. This discussion focuses on timber, steel, precast concrete driven, or cast-in-place concrete, composite pilings, and helical piles for deep foundations. See Figure 4.73.

Timber Pilings

Timber pilings are wood. The commonly used wood for pilings in the United States is Douglas-Fir from the Pacific Northwest used in the West and Southern Pine commonly used in the East. Timber pilings can be round or square. Round pilings can be up to 120′ long and square pilings up to 50′ long. Timber pilings can last for many years below grade but can deteriorate when exposed to the environment above grade. Timber pilings can be treated to reduce the effects of decay and to protect against insect infestation.

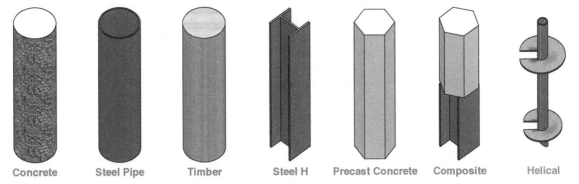

Concrete   Steel Pipe   Timber   Steel H   Precast Concrete   Composite   Helical

Figure 4.73   Types of pilings.

Steel Pilings

Steel pilings are generally made from round steel called steel tubing or pipe piles, or what is referred to as H steel. Steel pipe is available in strengths that are standard, extra strong, and double-extra strong. The wall thickness increases with each type. Pipe pilings can be open end or closed end. Soil enters the bottom of the pipe when the open end pipe is driven into the ground. The soil can remain in the pipe, depending on the structural engineering, or the soil can be removed with a jet of water or by using an auger drilled down into the pipe. Pipe pilings can also have a closed end by welding a steel plate on the bottom or by placing a premanufactured steel shoe on the bottom. Pipe piles can be filled with concrete to improve bending stability and reduce corrosion and maintain strength if the pipe corrodes. Steel pipes can be subject to corrosion, but use of pipe containing steel alloys such as nickel or chromium can improve corrosion resistance.

The construction industry refers to H piles as the type of structural steel used as pilings driven in the ground for deep foundations. Structural steel is available in several different manufactured shapes. The wide flange (W), American standard (S), and miscellaneous (M) shapes all have a I shape cross section. S shape steel has tapered flanges, making them stronger than equivalent size W shape steel. The W shape is commonly used for columns because of having wider flanges with an H shape typically used for pilings. H piles are good for use when driving through hard or rocky layers and to increased depths over timbers. Sections of H piles can be welded together or joined by **mechanical drive-fit splicers** to increase the length. Mechanical drive-fit splicers are premanufactured fasteners used to connect a variety of steel piling shapes together. The splicers generally match the piling material shape with flared ends that slip on easily and fit tightly into the two sides of the H piling or over the pipe pile. A weld can be made to hold the splicer to the pile. To help protect steel pilings from corrosion, coal-tar epoxy or **cathodic protection** can be applied to slow or eliminate the corrosion process. **Cathodic protection (CP)** is a technique used to control the corrosion of a metal surface by connecting the metal to be protected to a more easily corroded metal that acts as the terminal to transfer the corrosion away from the piling. Also common is to over-design the piling size to allow for normal corrosion activity without sacrificing durability. Steel pilings are cut off to a desired elevation after driving and then an engineered piling cap can be constructed.

Concrete Pilings

Two types of concrete piles are poured-in-place and precast pilings briefly introduced earlier in this discussion. When using non-reinforced poured-in-place concrete, holes drilled for pilings can be filled with concrete during the drilling process or after boring is complete. The hole can be bored into stable ground and steel reinforcing inserted, followed by pouring concrete to create a poured-in-place reinforced concrete piling. The hole can also be filled with a precast reinforced concrete piling. Alternately, steel pipe or tubing can be inserted into the hole for stabilization and then non-reinforced concrete or steel reinforcing place and concrete can be poured into the steel tube. The steel tube can remain or be removed after the concrete is poured. Precast concrete pilings are typically made in a manufacturing facility where they are built to specific structural engineering specifications. The precast pilings are then transported to the construction site where they are driven into a drilled hole for a replacement concrete piling, or driven into the earth as a displacement piling.

Composite Pilings

A **composite pile** is a pile made of two or more different types and shapes of material fastened together, end to end, to form a single pile. Typical combinations of materials include a section of timber below a section of H-pile steel. In this practice, the timber might be driven into a ground water layer where timber has a durability advantage over steel. Alternately, the timber piling section can be followed by a precast concrete section for economy of timber and the strength of concrete. Another composite application uses an H-pile steel section for its ability to penetrate hard layers or rock, followed by a section of reinforced cast-in-place concrete or precast concrete.

Fiberglass Pilings. **Fiberglass pilings** are also considered a composite piling, because of the material from which they are made. This usage defines **composite** as a material made from two or more different materials that are combined to create a stronger material than original individual materials. The composite commonly used for fiberglass pilings is fiber-reinforced polymer (FRP) made from a fiber-reinforced polymer matrix that provides durability, flexibility, and strength. Composite fiberglass pilings are often used in marine projects where the material is strong and lightweight for easy transport and handling, totally corrosion-resistant, and durable.

Helical Pilings

Helical piles, also called **screw pilings,** or **earth anchors,** are steel tubes with helical blades attached. Helical piles are drilled into the ground with the helical blades acting like a drill that is pushed into the ground. The installation usually uses earthmoving equipment with a rotary attachment. After the helical pile is driven into the ground, a pile cap is constructed on top of the pile to prepare for continued construction. Helical pilings can be installed faster than other types of pilings and with minimum site disturbance.

> **NOTE**
>
> Building inspections are required throughout the construction process. Be sure to confirm the necessary inspections and when they are required. Failure to have an inspection at the right time or skipping inspections can result in the building official putting a stop work order on your project along with possible fines. The building official could require an engineer to verify that the uninspected portion of the project meets code requirements, which can be expensive. If you accidently miss a required inspection, call the building official as soon as you realize the mistake and explain the situation. The inspector will probably evaluate the construction project in an effort to determine if the missed inspection portion is built to code. The inspector may be satisfied by asking specific questions, or can require a portion of the construction be exposed for observation. Regardless, the inspector needs to be confident that the construction is completed to code. Review Chapter 1 about building inspections and confirm the required inspections with the local building department.

> Go to Downloads & Resources tab at **www.mhprofessional.com/Commercial Building Construction-Ancillaries** to access the images provided in this chapter and correlate with the content of this chapter. Here you can look at the chapter figures on your computer screen to pan and zoom in and out for better observation, because full drawings reduced to fit on a textbook page are often difficult to read. The image bank also includes the complete set of plans for the Brookings South Main Fire Station used as examples throughout this textbook.

Go to **Downloads & Resources tab at www.mhprofessional.com/Commercial Building Construction-Ancillaries** to access the test and creative thinking problems for this chapter. The chapter test can be used for review or to evaluate content knowledge depending on your course objectives. Creative thinking problems are provided to help expand your knowledge by researching given subjects.

Print reading problems have questions that ask you to find information on the Brookings South Main Fire Station plans. The print reading problems can help you become familiar with the format of construction documents and reinforce your ability to seek specific information. This is an important skill for you to master as you prepare to enter the construction industry. Go to **Downloads & Resources tab at www.mhprofessional.com/Commercial Building Construction-Ancillaries** to access the complete set of plans for the Brookings South Main Fire Station.

CHAPTER 5

Masonry Construction

INTRODUCTION

Masonry is one of the most durable, long lasting, and maintenance-free construction materials available. The material is referred to as **masonry units,** which are laid next to each other and bound together by **mortar.** Mortar is a mixture of lime with cement, sand, and water, used in construction to bond bricks, concrete blocks, or rockwork.

Common masonry materials are brick, stone, marble, granite, travertine, limestone, cast stone, concrete block, glass block, and tile. Buildings constructed using masonry are attractive, durable, easy to maintain, quiet, and fire resistant. Masonry buildings generally display a quality that is unmatched in any other exterior appearance found in architecture. Commercial projects can be built with **structural masonry walls,** or more commonly with **masonry veneer.** Structural masonry walls are typically constructed using concrete masonry units (CMUs). Masonry veneer walls are made of a single nonstructural exterior layer of masonry that takes the place of traditional siding over wood-frame, steel-frame, or other construction. Masonry work is done by a specialty subcontractor known as a **mason,** who is skilled in masonry construction. Masonry units are **laid** by the mason. The term **laid** means placing one masonry unit at a time to build the masonry structure. Masonry units are laid with mortar surrounding each unit and continuing to desired heights and lengths. The masonry structure gains strength as the mortar cures or hardens, along with any structural steel or ties described later in this chapter.

Masonry is a good **thermal mass** that can be heated and cooled by exposure to the sun and the exterior temperatures. A good thermal mass is a dense material that can effectively absorb and store heat and release the heat as the structure cools at night. This makes masonry construction a good choice for solar architectural design. Masonry is cool when the outside temperature is cool, making it necessary to insulate inside the masonry cavity or to provide an insulated interior wall.

Masonry also provides excellent fire proof construction, and is typically used when designing and building a **firewall.** A firewall is a wall or partition designed to deter or prevent the spread of fire in the building.

In addition to other characteristics, masonry walls provide excellent sound insulation when used in commercial construction. The acoustical performance of masonry is improved by filling the masonry unit cavities with insulation or by insulating between **wythes.** Wythe is a continuous vertical section of masonry one unit in thickness.

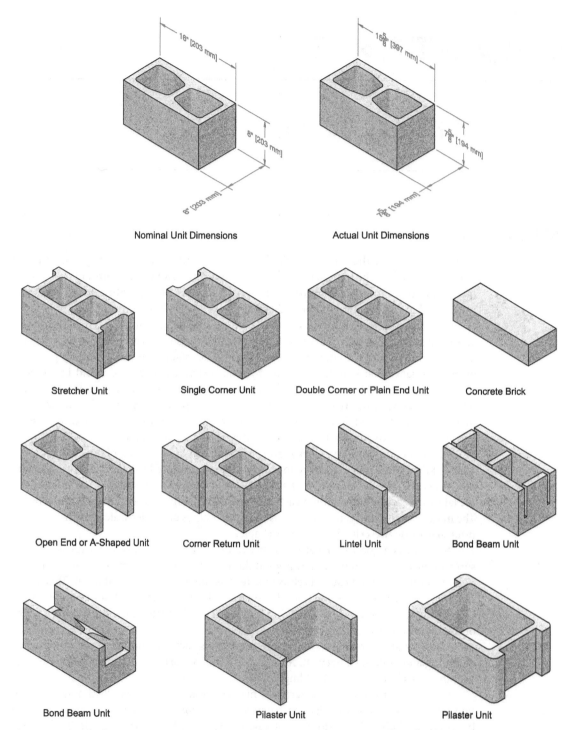

Figure 5.1   Standard concrete block sizes and specifications. (*Previously published in Modern Residential Construction Practices, Routledge, 2017 and reproduced here with permission.*)

COMMON MASONRY CONSTRUCTION MATERIALS

The most common types of masonry materials are concrete and clay masonry units. **Concrete masonry units,** also called **concrete blocks,** can be used to build foundations and exterior walls of buildings above and below grade. Standard concrete block sizes and specifications are shown in Figure 5.1. Concrete blocks and **clay masonry units** can be solid or hollow, and can have a variety of surface finishes for different appearances.

Clay masonry units are what you normally think of as **bricks.** Bricks are commonly made from clay, but can be made from other natural minerals. Bricks are made by extruding soft clay into the typical rectangular cube shape or other shapes. The clay shapes are heated to 1200°F (650°C) in a **kiln** for hardening and to add optional coloring and glazing. A kiln is a furnace or oven used to heat products to a desired temperature. Additional surface preparation and cutting can be done to create desired products. Brick sizes can vary between manufacturers, but standard sizes are established by the American Society for Testing and Materials (ASTM). Brick dimensions and terminology are shown in Figure 5.2. There are a variety of different brick sizes classified by **actual size, nominal size,** and by brick type. Actual size, also referred to as specified size, is the physical dimension of each brick. Nominal size is the actual size plus the width of the mortar joint. Typical mortar joints are ⅜ or ½ in.

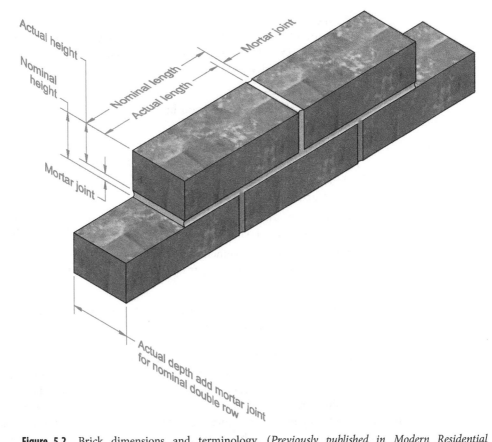

Figure 5.2 Brick dimensions and terminology. (*Previously published in Modern Residential Construction Practices, Routledge, 2017 and reproduced here with permission.*)

Most bricks are manufactured so the nominal sizes fit into a 4-in. grid, which is the same as the modules of other building materials, construction spacing, and building modules. The following is a list of typical bricks and their actual sizes:

Brick Type	Actual Sizes in Inches Depth × Height × Length
Standard	3-5/8 × 2-1/4 × 8
Modular	3-5/8 × 2-1/4 × 7-5/8
Norman	3-5/8 × 2-1/4 × 11-5/8
Roman	3-5/8 × 1-5/8 × 5/8
Jumbo	3-5/8 × 2-3/4 × 8
Economy	3-5/8 × 3-5/8 × 7-5/8
Engineer	3-5/8 × 2-13/16 × 7-5/8
King	2-3/4 × 2-5/8 × 9-5/8
Queen	2-3/4 × 2-3/4 × 7-5/8
Utility	3-5/8 × 3-5/8 × 11-5/8

Visiting the showroom of a brick manufacturer or construction materials supplier, you can discover a large variety of brick colors, finishes, and shapes. Bricks can also have a different appearances and architectural applications based on the orientation used when they are laid. Common brick orientation patterns are shown in Figure 5.3 and described as follows:

- **Stretchers** are the most common brick pattern with units oriented horizontally with the full face exposed.
- **Headers** are used to break up the typical stretchers by providing units oriented perpendicular to the face of the wall with the end of each brick exposed.
- **Soldiers** add accent style to the brick wall by placing units oriented vertically with the full face exposed.
- **Rowlocks** are oriented perpendicular to the face of the wall similar to headers, except they are used with the end and face exposed at sills and at the top of walls.
- **Herringbone** brick placement provides an interesting pattern by placing bricks in angular rows of parallel groups with any two adjacent rows angled in opposite directions.

The terms **course** and **wythe** refer to the masonry structure, as shown in Figure 5.4. A course is a row of bricks, and courses are multiple rows of bricks laid on top of each other to build up a structure such as a wall. Bricks are laid with staggered vertical joints for appearance and strength. A wythe is a continuous vertical section of masonry one unit in thickness. A wythe can be separate from or linked with adjacent wythes. A single wythe used for appearance and is not structural is called veneer. Masonry walls can be single or multiple wythe. A multiple-wythe wall can be constructed with a one type of masonry unit laid to increase wall thickness and structural strength, or different masonry units can be used for economy and appearance, such as a concrete block wythe for a structural application and an architectural brick wythe for appearance.

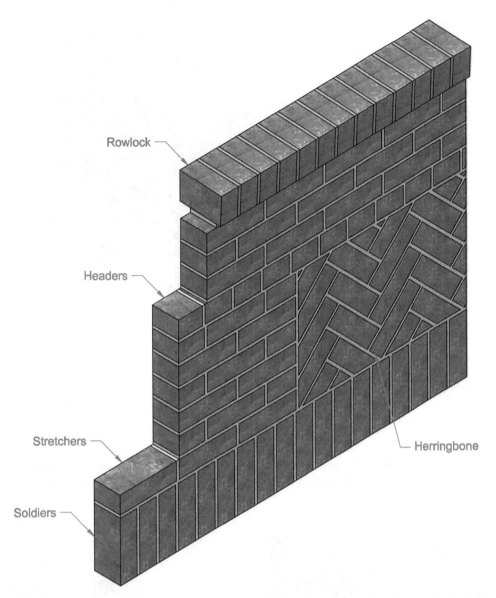

Figure 5.3 Common brick orientation patterns. (*Previously published in Modern Residential Construction Practices, Routledge, 2017 and reproduced here with permission.*)

When building a single-wythe wall, such as brick veneer, headers and rowlocks must be cut for the length to match the width of the wythe, as shown in Figure 5.5a. A header or rowlock used over a double wythe is laid full length, as shown in Figure 5.5b. Multiple wythes can be built without being linked together and with the space between wythes filled with mortar, as shown in Figure 5.6a, or connected using metal ties or masonry units, as shown in Figure 5.6b.

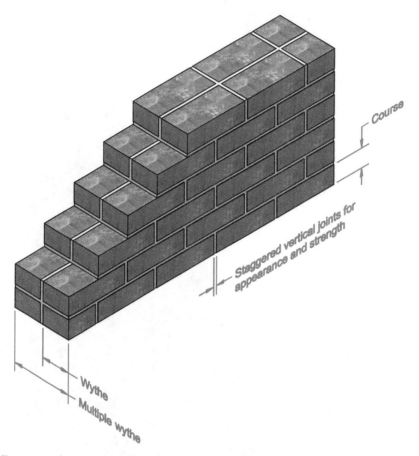

Figure 5.4 The terms course and wythe refer to the masonry structure. (*Previously published in Modern Residential Construction Practices, Routledge, 2017 and reproduced here with permission.*)

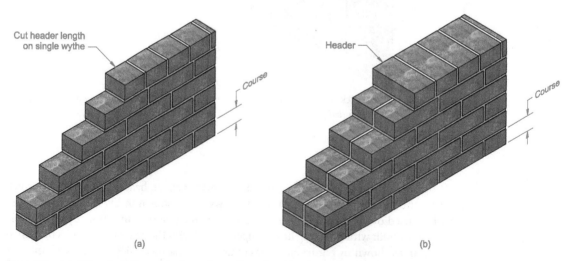

Figure 5.5 (*a*) When building a single-wythe wall, such as brick veneer, headers and rowlocks must be cut for the length to match the width of the wythe. (*b*) A header or rowlock used over a double wythe is laid full length. (*Previously published in Modern Residential Construction Practices, Routledge, 2017 and reproduced here with permission.*)

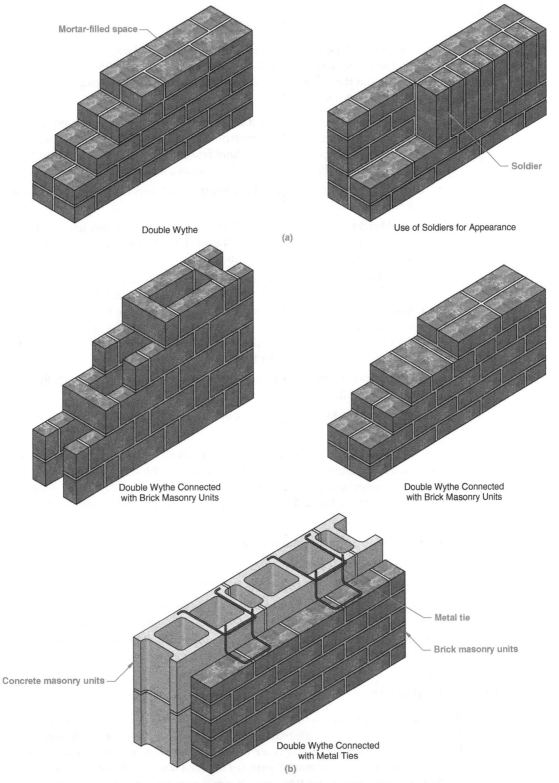

Mortar-filled space

Double Wythe

(a)

Soldier

Use of Soldiers for Appearance

Double Wythe Connected
with Brick Masonry Units

Double Wythe Connected
with Brick Masonry Units

Metal tie

Brick masonry units

Concrete masonry units

Double Wythe Connected
with Metal Ties

(b)

Figure 5.6 (*a*) Multiple wythes can be built without being linked together and with the space between wythes filled with mortar. (*b*) Multiple wythes can be built without being linked together and with the space between wythes connected using metal ties or masonry units. (*Previously published in Modern Residential Construction Practices, Routledge, 2017 and reproduced here with permission.*)

Bricks are selected based on the desired application, and standards that establish strength, durability, and aesthetic requirements. Brick specifications are established by the ASTM in the United States and the Canadian Standards Association (CSA) in Canada. The following describes the basic brick specifications:

- **Building bricks** are typically used as **backing material** in nonstructural and structural applications where appearance is not a requirement. Backing material is a comparatively low-quality brick used behind **face brick** or other masonry. Face brick is used for visual quality on the exposed surface of a building wall or other structure.

- **Face bricks** are used in nonstructural and structural applications where appearance is a requirement.

- **Glazed bricks** have a **glaze** surface. Glaze is a layer or coating that has been fused to a ceramic object through firing to high temperatures in a kiln. A glaze can provide color, beautify, strength, and waterproofing.

- **Hollow bricks** are used in the manufacture of building bricks and facing bricks. Hollow brick cavities are used to place wall anchoring and for steel reinforced masonry where the cavities have steel reinforcing and are filled with grout. The voids in hollow bricks are often cylindrical holes that can be called cores or cells. Brick with hollow spaces are easier to extrude than solid bricks. Hollow bricks are also lighter, making them easier to carry and handle.

- **Paving bricks** are used as the wearing surface of patios, walks, and roads for people and vehicle traffic.

- **Thin veneer bricks** have normal face dimensions with reduced thickness for application to surfaces with adhesive.

Clay masonry units are classified by characteristics such as appearance and grade. Factors related to masonry appearance include amount of chips, **dimensional tolerances,** distortion, and squareness. Dimensional tolerance refers to how tightly the manufactured brick is held to its nominal size. The following appearance letter designations can be applied to brick manufacturing:

- **X** specifies extreme or extra control.
- **S** refers to standard production.
- **A** designates architectural or aesthetic specifications.

Masonry grade relates to durability, where Grade SW stands for severe weathering and is recommended in most areas of the United States and Canada. Grade MW refers to moderate weathering for use in areas where freezing cycles are not expected. Grade NW are negligible weathering units for use in air-conditioned and moisture-free locations.

Mortar for Masonry Construction

Mortar, also called **grout,** is a mixture of Portland cement, sand, lime, and water. Mortar becomes a paste-like material after mixing the components with water, and hardens to create a strong solid structure when it cures. Mortar is used in construction to bond together any type of masonry units. A **mortar joint** is a mortar-filled space between masonry units. The mortar component proportions can change depending on the desired construction properties. There are different mortar types depending on the required strength. Mortars for new construction are classified as Types M, N, and S. Type O or polymer mortars can be required to duplicate the original mortar for repairs to older buildings. The following briefly describes new construction mortars and their use:

- Type M mortar is used for below grade applications. Type M mortar yields high compressive strength and has the highest proportion of Portland cement, with 3 parts Portland cement, 1 part lime, and 12 parts sand.

- Type N mortar is the most common mortar used for nonstructural above-grade masonry veneer applications. Type N mortar has good bond qualities and resistance to water penetration. Type N mortar uses 1 part Portland cement, 1 part lime, and 6 parts sand.

- Type S mortar is used in structural masonry applications. Type S mortar mixes with 2 parts Portland cement, 1 part lime, and 9 parts sand.

The mortar components are mixed with water at the construction site to reach a paste-like mixture. Additional water can be added as needed to maintain workability. Mortar that is unused after 2 hours from the initial mix should be discarded, because the ability to bond is reduced. A portable concrete mixer, shown in Figure 5.7, has a revolving drum to contain

Figure 5.7 A portable concrete mixer has a revolving drum to contain and mix mortar and is commonly used to mix mortar components at the construction site, and continue turning to maintain the workability of the mortar. (*Previously published in Modern Residential Construction Practices, Routledge, 2017 and reproduced here with permission.*)

and mix mortar and is commonly used to mix mortar components at the construction site, and continue turning to maintain the workability of the mortar. The concrete mixer gives the mason time to use the mortar before it cures.

CONCRETE BLOCK CONSTRUCTION

Poured-in-place concrete construction is commonly used for foundation walls, but concrete block construction can also be used for foundation walls. Concrete block construction can also be used for building exterior walls. Concrete block construction is one of the applications of masonry construction that includes other products such as brick, rock, stone, and glass blocks. A **concrete block** is also called a **concrete masonry unit (CMU),** cement block or foundation block. A concrete block is a rectangular concrete form used in construction. Concrete block construction can be used for light commercial building foundations but is also used in aboveground construction. In commercial applications, concrete blocks are used to form the wall systems for many types of buildings. Concrete blocks provide a durable construction material and are relatively inexpensive to install and maintain. In light commercial applications, foam-filled blocks provide excellent insulating characteristics and are often used in desert climates.

Concrete blocks are commonly manufactured in nominal size modules of $8 \times 8 \times 16$ in., $4 \times 8 \times 16$ in., or $6 \times 8 \times 16$ in. There are also 2-, 4-, 6-, 8-, 10-, 12-, and 16-in. modular units. The metric conversion of a nominal $8 \times 8 \times 16$-in. concrete masonry unit is $200 \times 200 \times 400$ mm. The actual size of the block is smaller than the nominal size so mortar joints can be included in the final size. A **mortar joint** is also called a **grout joint.** A mortar joint is a mortar-filled space between concrete blocks, bricks, and other masonry materials. An architect or engineer determines the size of the structure based on modular principles of concrete block construction. Wall lengths, opening locations, and wall and opening heights must be based on the modular size of the block being used. Failure to maintain the modular layout can result in a big increase in labor costs to cut and lay the blocks.

Concrete blocks are often reinforced with a wire mesh at every other course of blocks. Where the risk of seismic activity must be considered. Concrete block structures are often required to have reinforcing steel placed within the wall to help tie the blocks together. The steel is placed in a block that has a channel or cell running through. This cell is then filled with grout or concrete to form a **bond beam** within the wall. The bond beam solidifies and ties the concrete block structure together. A typical bond beam concrete block structure is detailed in Figure 5.8. Figure 5.9 shows typical concrete block reinforcing methods. Openings for windows and doors require steel reinforcing for both poured-in-place and concrete block construction. See Figure 5.10.

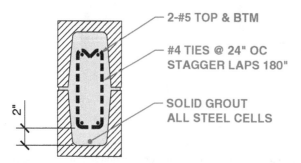

Figure 5.8 A detail of a typical bond beam concrete block structure. (*Previously published in Modern Residential Construction Practices, Routledge, 2017 and reproduced here with permission.*)

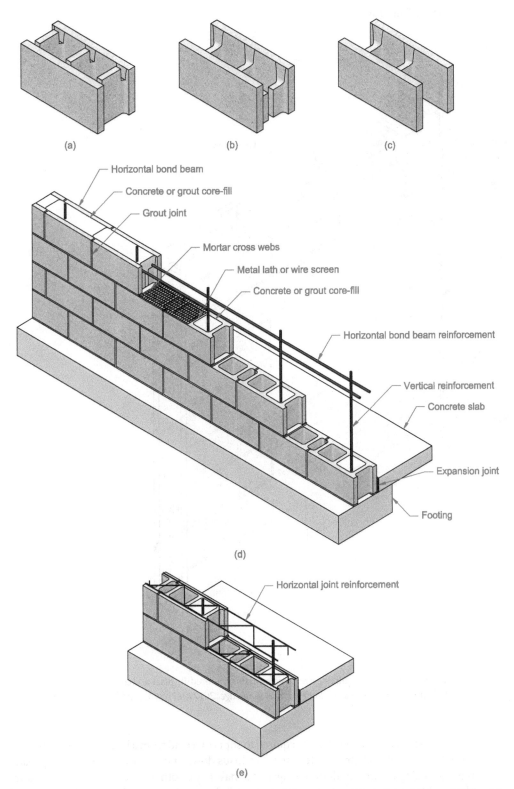

(a)

(b)

(c)

Horizontal bond beam

Concrete or grout core-fill

Grout joint

Mortar cross webs

Metal lath or wire screen

Concrete or grout core-fill

Horizontal bond beam reinforcement

Vertical reinforcement

Concrete slab

Expansion joint

Footing

(d)

Horizontal joint reinforcement

(e)

Figure 5.9 Typical concrete block reinforcing methods. (*Previously published in Modern Residential Construction Practices, Routledge, 2017 and reproduced here with permission.*)

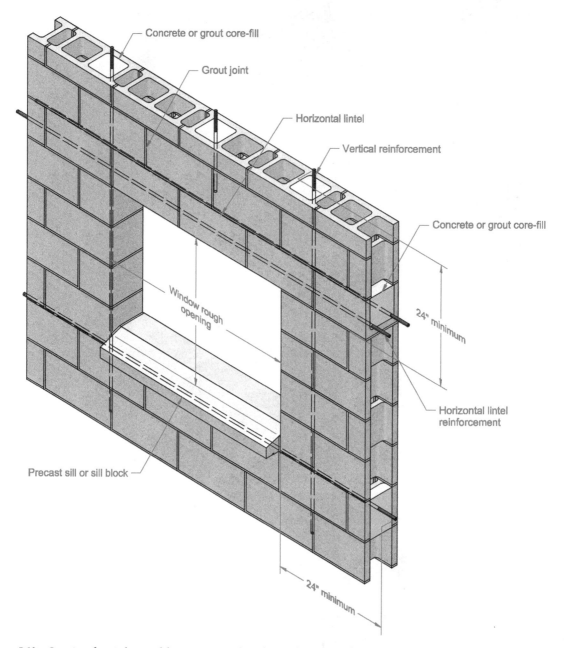

Figure 5.10 Openings for windows and doors require steel reinforcing for both poured-in-place and concrete block construction. (*Previously published in Modern Residential Construction Practices, Routledge, 2017 and reproduced here with permission.*)

When the concrete blocks are required to support a load from a beam, a **pilaster** is often placed in the wall to help transfer the beam loads down the wall to the footing. A pilaster is a reinforcing column built into or against a masonry or other wall structure. Pilasters are also used to provide vertical support to long walls by providing reinforcing columns placed periodically throughout the length of the wall. Concrete block pilasters examples are shown in Figure 5.11. A concrete block pilasters structural detail is shown in Figure 5.12.

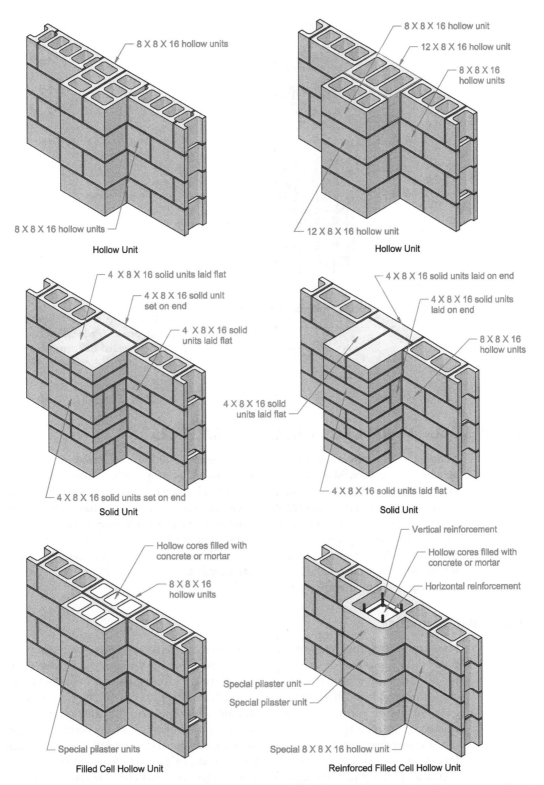

Figure 5.11 Concrete block pilasters examples. (*Previously published in Modern Residential Construction Practices, Routledge, 2017 and reproduced here with permission.*)

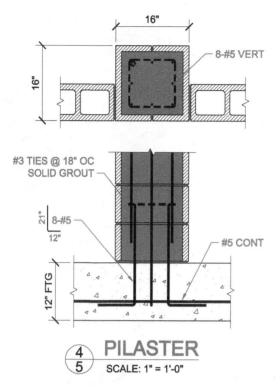

Figure 5.12 A concrete block pilasters structural detail. (*Previously published in Modern Residential Construction Practices, Routledge, 2017 and reproduced here with permission.*)

COMMON MASONRY WALL CONSTRUCTION PRACTICES

Masonry construction applications for commercial buildings are **masonry veneer** and **structural masonry.** Masonry veneer walls are made of a single nonstructural exterior layer of masonry that takes the place of traditional siding over wood-frame or steel-frame, or structural masonry construction. Masonry veneer over wood-frame construction, shown in Figure 5.13a, is a commonly used masonry construction application. Structural masonry walls are typically constructed using reinforced concrete masonry units, as shown in Figure 5.13b. Masonry veneer and structural masonry walls must be laid on a concrete footing when bearing on grade, or on a structural steel beam or reinforced concrete beam when supported by the structure.

Masonry Veneer Construction

Masonry veneer provides an attractive and durable exterior covering that replaces traditional siding on wood- or steel-frame construction. Masonry veneer can be used to cover the entire exterior of a building, to provide an attractive front elevation, or for architectural accent on specific locations of the building. Masonry veneer can also be used to cover or create an interior partition. The most common masonry veneer construction is over wood framing and steel framing, but masonry veneer can be used over any construction practice. The masonry veneer wall rests on an expanded concrete foundation wall or concrete footing, as shown in Figure 5.14. Figure 5.14a shows an expanded concrete foundation with a ledge to support the masonry veneer. The expanded concrete foundation wall is below grade for structural support and the masonry veneer is above grade for appearance.

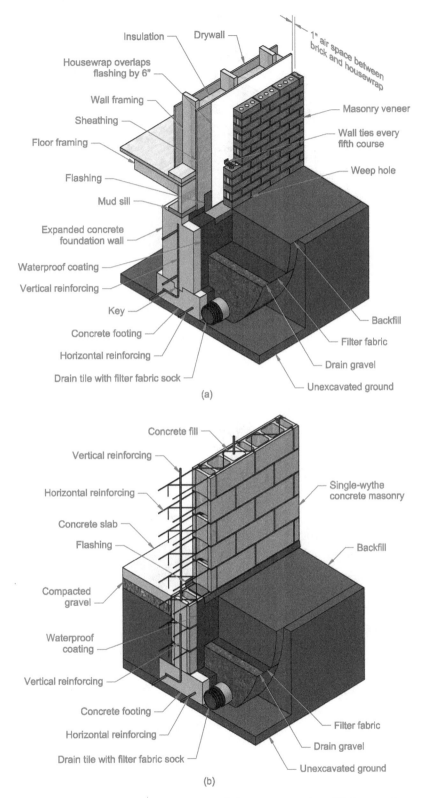

Figure 5.13 (*a*) Masonry veneer over wood-frame construction. (*b*) Structural masonry walls are typically constructed using reinforced concrete masonry units. (*Previously published in Modern Residential Construction Practices, Routledge, 2017 and reproduced here with permission.*)

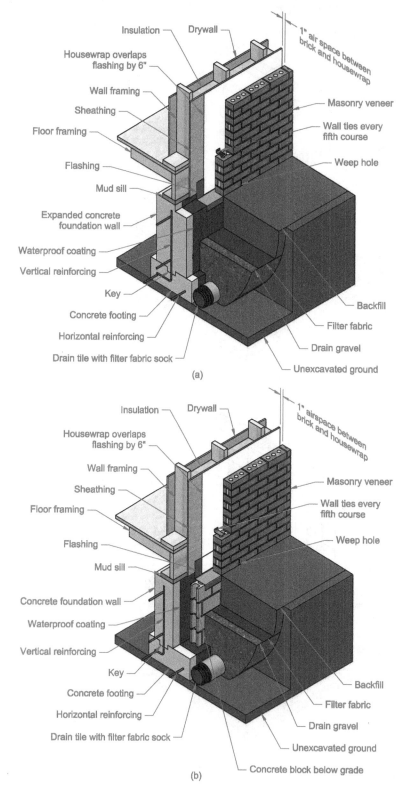

Figure 5.14 The masonry veneer wall rests on an expanded concrete foundation wall or concrete footing. (*a*) An expanded concrete foundation with a ledge to support the masonry veneer. (*b*) A standard width concrete foundation wall with an expanded concrete footing and an economical concrete block ledge below grade to support the masonry veneer on the concrete footing. (*Previously published in Modern Residential Construction Practices, Routledge, 2017 and reproduced here with permission.*)

Figure 5.14*b* shows a standard width concrete foundation wall with an expanded concrete footing and an economical concrete block ledge below grade to support the masonry veneer on the concrete footing. The following describes each masonry veneer construction component above the concrete or concrete block ledge, as shown in Figure 5.14.

Flashing between the concrete or concrete block ledge and the masonry veneer acts to collect and divert moisture from the wall. The flashing projects past the foundation wall to keep moisture from entering the wall. Moisture from condensation and rain can be a serious problem in masonry wall deterioration if quality flashing material and installation are not used. Copper flashing is widely used for corrosion resistance, and its flexibility in forming leak-free corners and ends. Flashing must be installed watertight, with special care taken to seal seams, corners, and at the end of masonry. **End dams,** also called **side dams,** are made by bending the end of flashing up to stop water from flowing across the flashing and into the adjacent construction. Special door and window flashing assemblies are made for use where masonry is used around **rough openings.** Rough opening is any unfinished opening that is framed to specific measurements to accommodate the finish product. Other flashing products include stainless steel, lead-coated copper, and synthetic material available from your masonry supplier. Carefully check manufacturer warranties to confirm reliability of flashing applications and materials. **Housewrap** is used over the exterior sheathing. Housewrap or house wrap is the term that describes a variety of synthetic products that have replaced tar paper for use as a vapor barrier.

It is important for the housewrap to overlap the flashing by at least 6 in. (150 mm) to be an effective part of the moisture control system.

Masonry veneer can be any of the masonry materials and patterns described in this chapter and used for architectural appearance, such as brick and stone. The first masonry course rests directly on the concrete or concrete block ledge over the flashing. **Wall ties** are used to transfer lateral forces, such as wind loads, on the masonry veneer wall back to the wood- or steel-frame structure. Corrugated wall ties are normally used for masonry veneer over wood-frame construction and are not recommended for brick veneer over steel stud construction. Corrugated wall ties are strong and easy to attach and are available in 16, 18, and 22 gauge galvanized steel for durability and corrosion resistance. Corrugated wall ties are generally spaced at each stud location horizontally and 16 in. (400 mm) on center vertically. Wall ties for brick veneer over steel stud construction are specially designed for the application where deflection in steel construction can be transferred to masonry veneer and cause cracks in the masonry. Carefully follow the engineering found on plans and specifications for proper construction applications. Figure 5.15 shows brick veneer construction used on an expanded concrete footing with a concrete slab floor. The connecting wall system uses steel stud framing and special flexible wall ties allows the steel structure to flex as needed.

Weep holes are openings in the first course of masonry that allow water to drain out through the bottom of the wall, and help dry the structure by providing air circulation behind the masonry veneer. Weep holes are usually made by leaving the mortar out between every third or fourth masonry unit along the first course.

An **air space** between the masonry veneer and the housewrap on the exterior sheathing allows for ventilation, allows moisture from condensation and rain to drain down the weep holes, and aids in keeping the structure dry. The recommended air space between masonry veneer and wood-frame construction is 1 in. (25 mm), with up to 2 in. (50 mm) of air space used for steel-frame and other masonry structures.

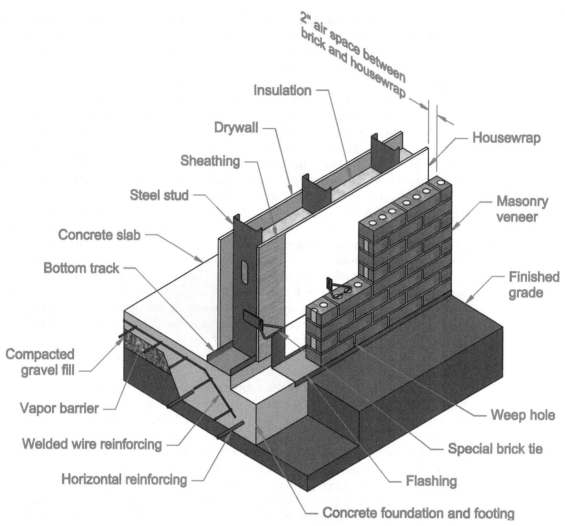

Figure 5.15 Brick veneer construction used on an expanded concrete footing with a concrete slab floor. The connecting wall system uses steel stud framing and special flexible wall ties allows the steel structure to flex as needed. (*Previously published in Modern Residential Construction Practices, Routledge, 2017 and reproduced here with permission.*)

Controlling Vertical Movement for Masonry Veneer

Masonry movement can be caused by expansion and shrinkage from changes in temperature, moisture, settlement, and forces of nature. **Movement joints** are used to separate masonry construction into segments in an effort to prevent wall damage such as buckling and cracking. Movement joints are applied to masonry walls to control vertical and horizontal movement. Masonry veneer walls are supported vertically at the base of the wall by the concrete foundation and footing, and the wall ties transfer lateral forces on the masonry veneer wall back to the structure, as described earlier. The masonry veneer wall is otherwise self-supporting up to heights of about 12 ft (3600 mm) or as determined by structural engineering. Masonry veneer used on one-story buildings generally require only the foundation support and the wall ties, but multistory structures can require additional vertical support

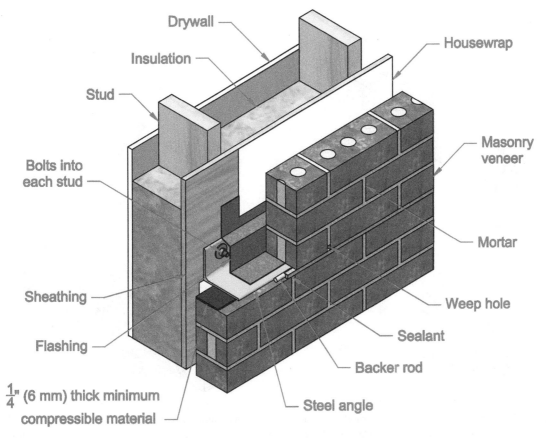

Figure 5.16   Multistory structures can require additional vertical support and expansion control for the masonry veneer typically provided at each floor line. This support, called a shelf angle, is a horizontal steel angle that provides a break in the veneer for movement when the masonry expands and the connected structure shrinks, or other movement occurs in the wall. The shelf angle is bolted to the frame structure at the desired height. (*Previously published in Modern Residential Construction Practices, Routledge, 2017 and reproduced here with permission.*)

and expansion control for the masonry veneer typically provided at each floor line. This support, called a **shelf angle,** is a horizontal steel angle that provides a break in the veneer for movement when the masonry expands and the connected structure shrinks, or other movement occurs in the wall. This is done with a shelf angle bolted to the frame structure at the desired height, as shown in Figure 5.16.

Controlling Horizontal Movement for Masonry Veneer

Horizontal movement in a masonry veneer wall can be controlled using **expansion joints.** Expansion joints are used to control horizontal movement by separating masonry into sections to prevent cracking, and stop water penetration and air infiltration in the masonry wall. Expansion joints are made using elastic materials placed in a continuous, unobstructed vertical joint through the masonry wythe, as shown in Figure 5.17. Figure 5.17a shows a waterproof sealant and **backer rod** used as a control joint. A backer rod is a round open- or closed-cell polyethylene or polyurethane foam product used to fill joints between building materials. Figure 5.17b shows a waterproof sealant, backer rod, and foam pad

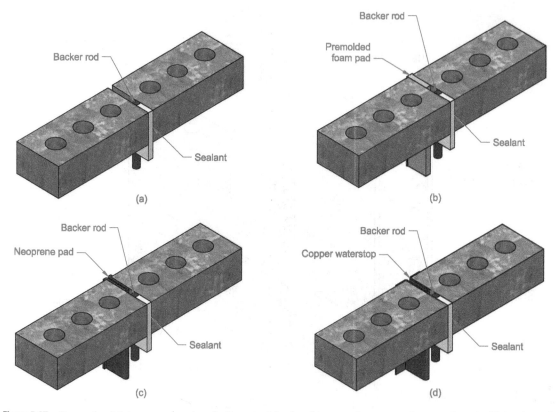

Figure 5.17   Expansion joints are made using elastic materials placed in a continuous, unobstructed vertical joint through the masonry wythe. (*a*) A waterproof sealant and backer rod used as a control joint. A backer rod is a round open- or closed-cell polyethylene or polyurethane foam product used to fill joints between building materials. (*b*) A waterproof sealant, backer rod, and foam pad used as a control joint. (*c*) A waterproof sealant, backer rod, and neoprene pad used as a control joint. (*d*) A waterproof sealant, backer rod, and copper waterstop used as a control joint. (*Previously published in Modern Residential Construction Practices, Routledge, 2017 and reproduced here with permission.*)

used as a control joint. Figure 5.17*c* shows a waterproof sealant, backer rod, and **neoprene** pad used as a control joint. Neoprene is synthetic material similar to rubber that is resistant to heat and weathering. Figure 5.17*d* shows a waterproof sealant, backer rod, and **copper waterstop** used as a control joint. A copper waterstop is a premanufactured copper product that fits into the joint. Copper is a soft metal that easily compresses. The foam pad, neoprene pad, and copper waterstop are also used to keep mortar or debris from entering the joint. A general definition for waterstop is a component of a concrete or masonry structure, intended to prevent the passages of water running continuously through the joints.

Masonry Veneer Construction Supported by Structural Features

Masonry veneer walls must be laid on a structural bearing support directly on grade, or on **structural steel** or **structural concrete** when supported by the structure. Structural steel is manufactured in a variety of shapes for use as load-bearing structural members

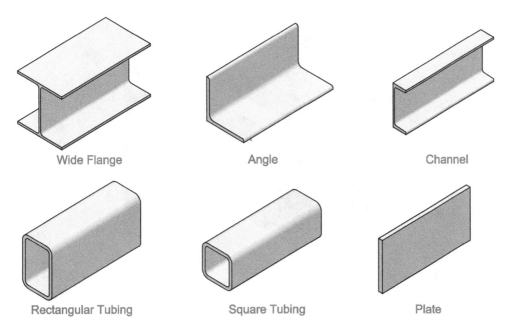

Figure 5.18 Common structural steel shapes. (*Previously published in Modern Residential Construction Practices, Routledge, 2017 and reproduced here with permission.*)

in building construction or other structures. Common structural steel shapes are shown in Figure 5.18. The wide flange beam and angle are commonly used in commercial construction to support masonry veneer over openings and directly on wood framing. **Lentil** is the term used in masonry construction to describe the horizontal support over a door or window opening. Figure 5.19*a* shows a steel angle used as a lintel to support masonry veneer over a typical door or window opening, and Figure 5.19*b* shows a steel wide flange beam used to support masonry veneer over a large span opening.

Prestressed concrete was discussed in Chapter 4, *Concrete Construction and Foundation Systems*. A prestressed concrete beam has steel reinforcing stretched from both ends in the concrete form or mold. The steel reinforcing is released after the concrete cures, allowing the finished concrete beam to have a slight arch, called **camber** that makes it better for supporting heavy loads than precast concrete. The camber is always on top when the beam is placed in position. Prestressed concrete beams are delivered to the construction site and lifted into place by a crane where needed in the building. Figure 5.20 shows a prestressed concrete beam used as a header to support masonry veneer over an opening.

Masonry Arch Construction

A masonry **arch** is a self-supporting curved masonry structure over an opening that provides support for the structure above. A typical masonry arch is constructed with tapered bricks called **voussoirs.** Voussoirs can be cut at the job site or ordered from the masonry manufacturer for use in typical arch applications. The side angles of the *v*oussoir masonry units are determined by the arch radius. A **pillar** or wall is needed to support the arch

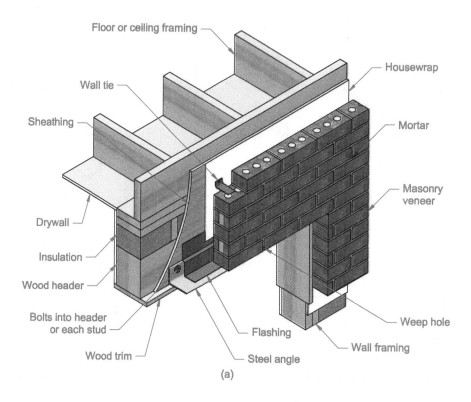

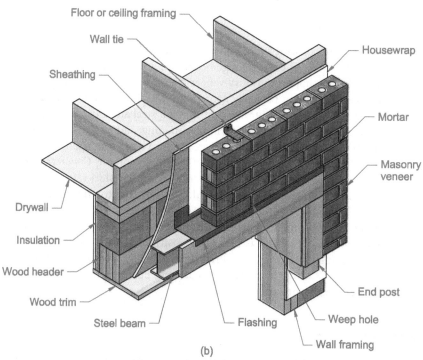

Figure 5.19 Masonry support over openings. (*a*) A steel angle used as a lintel to support masonry veneer over a typical door or window opening. (*b*) A steel wide flange beam used to support masonry veneer over a large span opening. (*Previously published in Modern Residential Construction Practices, Routledge, 2017 and reproduced here with permission.*)

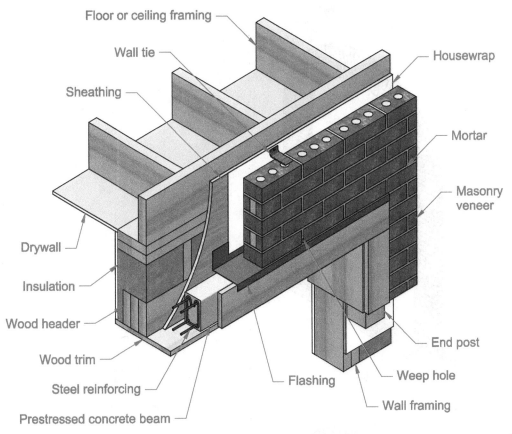

Floor or ceiling framing

Wall tie

Sheathing

Housewrap

Mortar

Masonry veneer

Drywall

Insulation

Wood header

Wood trim

Steel reinforcing

Prestressed concrete beam

Flashing

End post

Weep hole

Wall framing

Figure 5.20 A prestressed concrete beam used as a header to support masonry veneer over an opening. (*Previously published in Modern Residential Construction Practices, Routledge, 2017 and reproduced here with permission.*)

structure at the base where the forces of the arch push outward, as shown in Figure 5.21. A pillar is a vertical structure of stone, wood, or metal, used as a support or decoration. The masonry arch can be built over a prefabricated steel angle template formed in the desired arch. Traditionally, masons build an arch template out of wood and plywood. The template has two matching sides placed apart equal to the desired masonry width. A surface is created with consecutive short pieces of wood, such as 2 × 4s placed along the top edge of the arc, or by bending a plywood strip along the top edge of the arc, or by manufacturing a steel template to use for frequent jobs. The template structure needs to be solid and stable to hold the weight of the masonry units during construction. Masonry units are laid over the template starting with a central **key brick** in the top center and an equal number of masonry units laid on each side down to the supporting structure.

Structural Masonry Construction

Structural masonry is a structural system for building or commercial building construction that can be combined with wood, steel, and reinforced concrete applications. Structural masonry is commonly used in commercial construction. Structural masonry walls are typically constructed using reinforced concrete masonry units that must be laid on a concrete

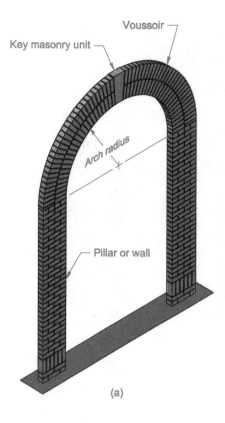

(a)

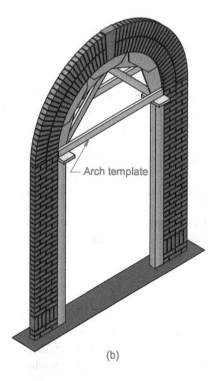

(b)

Figure 5.21   A pillar or wall is needed to support the arch structure at the base where the forces of the arch push outward. (*a*) Characteristics of a masonry arch. (*b*) Masonry arch construction. (*Previously published in Modern Residential Construction Practices, Routledge, 2017 and reproduced here with permission.*)

footing bearing on grade. Concrete masonry walls are normally reinforced horizontally and vertically. Horizontal reinforcement is normally prefabricated welded wire products that are manufactured for specific CMU sizes. The horizontal reinforcing is mortared into the joints, as shown in Figure 5.22. The horizontal reinforcing improves strength of the masonry and controls cracking from shrinkage. Vertical reinforcement is placed in CMU cells and the cells are filled with concrete or mortar to strengthen the structure, as shown in Figure 5.22.

Single-wythe concrete masonry construction shown in Figure 5.22 can be used for light commercial construction where insulation and interior finish is not required. Wood or steel framing can be added to the inside of the structure to provide insulation and support

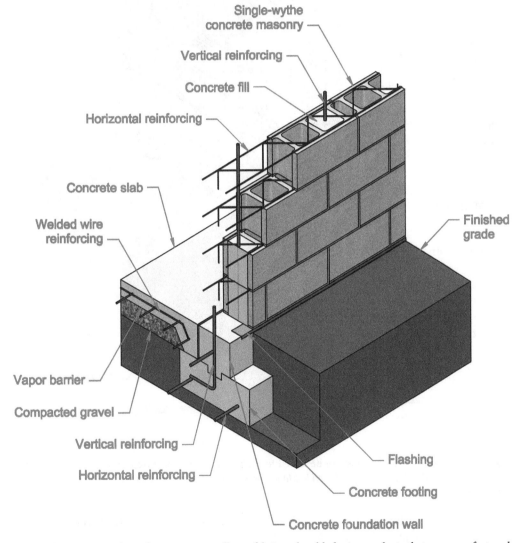

Figure 5.22 Horizontal reinforcing is normally prefabricated welded wire products that are manufactured for specific CMU sizes and mortared into the joints. Vertical reinforcement is placed in CMU cells and the cells are filled with concrete or mortar to strengthen the structure. (*Previously published in Modern Residential Construction Practices, Routledge, 2017 and reproduced here with permission.*)

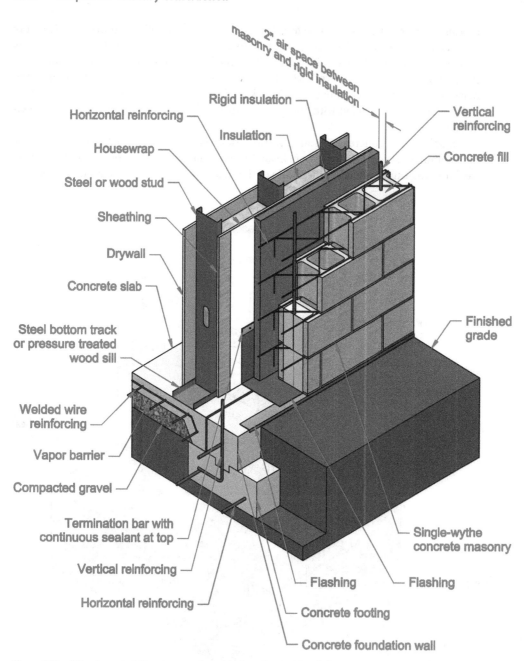

Figure 5.23 Wood or steel framing can be added to the inside of the structure to provide insulation and support for interior wall material. (*Previously published in Modern Residential Construction Practices, Routledge, 2017 and reproduced here with permission.*)

for interior wall material, as shown in Figure 5.23. The following describes the features found in Figure 5.23:

- Steel or wood framing is used on the interior with insulation placed between studs. A steel bottom track is used with steel studs and a pressure treated sill is used with wood studs.
- Structural plywood or OSB sheathing is used over wood framing.

- A vapor barrier is applied to the exterior surface of the concrete block wall to help keep moisture from entering the wall.
- A stainless steel, aluminum, or plastic **termination bar** is used to secure the top edge of the flashing to the sheathing or steel stud framing. The termination bar is placed under the vapor barrier and the top edge is sealed.
- Rigid insulation can be used to provide desired energy efficiency to the structure.
- A 1-in. (25-mm) minimum air space is required and a 2-in. (50-mm) air space is recommended.
- A copper, aluminum, stainless steel, or galvanized steel drip edge is placed at the edge of the concrete foundation wall, and the flashing is sealed over the drip edge.

A double-wythe wall is more common for most applications where concrete block construction is combined with masonry veneer and rigid insulation included, as shown in Figure 5.24. The following describes the features found in Figure 5.24:

- Reinforced concrete block wall has horizontal and vertical reinforcing as previously described.
- A vapor barrier is applied to the exterior surface of the concrete block wall to help keep moisture from entering the wall.
- A stainless steel, aluminum, or plastic termination bar is used to secure the top edge of the flashing to the concrete block wythe. The termination bar is placed under the vapor barrier and the top edge is sealed.
- Rigid insulation is used to provide desired insulation value to the structure.
- A 1-in. (25-mm) minimum air space is required and a 2-in. (50-mm) air space is recommended.
- A copper, aluminum, stainless steel, or galvanized steel drip edge is placed at the edge of the concrete foundation wall, and the flashing is sealed over the drip edge.
- The masonry veneer wythe is constructed, as described earlier.

Using Control Joints in Concrete Block Construction

Concrete block walls are constructed with **control joints,** which are vertical joints at specific locations in the wall. Control joints allow the concrete masonry wall structures to shrink independently between joints, and to transfer lateral wind loads from one side of the joint to the other. Figure 5.25 shows a typical control joint in a concrete block wall. **Sash blocks** and **shear lugs** are used at control joints to transfer lateral wind loads across the joint. A sash block is a concrete block unit manufactured with a vertical groove where the shear lug is placed. Shear lugs are made with hard rubber or plastic and placed in the sash block groove running the entire height of the concrete block wall to transmit externally applied loads to the structure. Mortar is left out of the control joint and caulking is used to seal along the exterior of the joint.

Using Isolation Joints in Masonry Construction

Isolation joints are used in concrete block and masonry veneer construction to create a joint between the masonry and other material that allows both structures to contract and expand independently. Isolation joints are used where masonry joins other building materials or where new masonry joins existing masonry. Figure 5.26a shows an isolation joint used at two adjoining concrete block walls with a steel anchor embedded in mortar-filled cores, transferring loads, and allowing the walls to move independently. Figure 5.26b shows a steel isolation joint anchor used between two adjoining masonry walls.

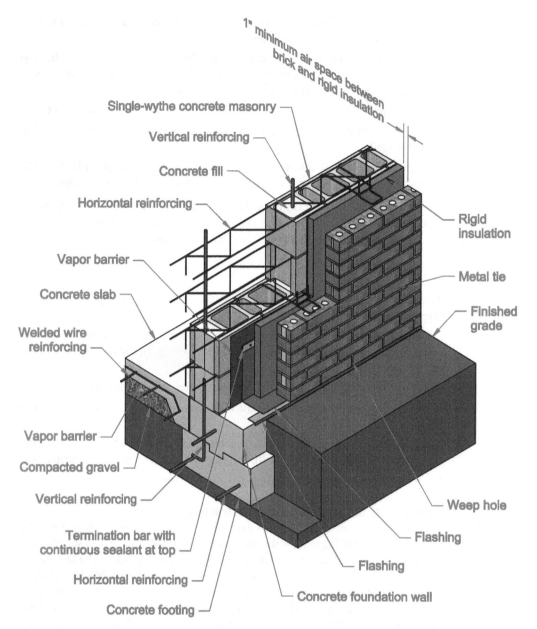

1" minimum air space between brick and rigid insulation

Single-wythe concrete masonry

Vertical reinforcing

Concrete fill

Horizontal reinforcing

Vapor barrier

Concrete slab

Welded wire reinforcing

Vapor barrier

Compacted gravel

Vertical reinforcing

Termination bar with continuous sealant at top

Horizontal reinforcing

Concrete footing

Rigid insulation

Metal tie

Finished grade

Weep hole

Flashing

Flashing

Concrete foundation wall

Figure 5.24   A double-wythe wall is more common for most applications where concrete block construction is combined with masonry veneer and rigid insulation included. (*Previously published in Modern Residential Construction Practices, Routledge, 2017 and reproduced here with permission.*)

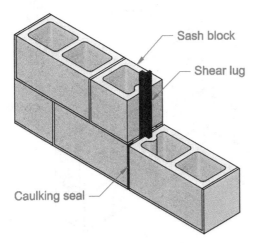

Figure 5.25   A typical control joint in a concrete block wall. (*Previously published in Modern Residential Construction Practices, Routledge, 2017 and reproduced here with permission.*)

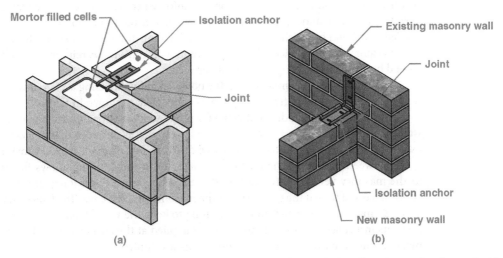

Figure 5.26   (*a*) An isolation joint used at two adjoining concrete block walls with a steel anchor embedded in mortar-filled cores, transferring loads and allowing the walls to move independently. (*b*) A steel isolation joint anchor used between two adjoining masonry walls. (*Previously published in Modern Residential Construction Practices, Routledge, 2017 and reproduced here with permission.*)

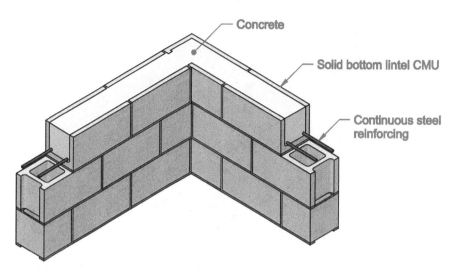

Figure 5.27   Solid bottom lintel concrete masonry units are filled with concrete and with continuous reinforcing used to create a bond beam around the top of the wall perimeter. (*Previously published in Modern Residential Construction Practices, Routledge, 2017 and reproduced here with permission.*)

Top of Wall Applications

A **bond beam** can be used to complete the top of a masonry wall for attaching the roof construction. A bond beam is a horizontal reinforced concrete or concrete masonry beam used to strengthen and tie a masonry wall together at the top or in other locations where needed. The bond beam shown in Figure 5.27 demonstrates how solid bottom lintel concrete masonry units are filled with concrete and with continuous reinforcing used to create a bond beam around the top of the wall perimeter.

Anchor bolts can be embedded in the bond beam for attaching a 2 × 4 or 2 × 6 pressure treated sill where the roof framing system is connected, as shown in Figure 5.28.

Figure 5.29 shows additional detail for proper **eave** construction at a masonry veneer wall. Eave is the term used to describe the lowest part of the roof that projects from the exterior wall, also referred to as a **cornice** or **overhang.** The 3/4 in. (20 mm) minimum space between the top of the masonry veneer and wood framing allows for movement in the masonry veneer. A **frieze board** is used between the masonry and the eave for appearance and to seal the gap between the masonry and the eave. The frieze board can be 1× or 2× material that is placed over flashing to keep the wood from moisture decay and prevents insect access. A caulking sealant is applied at the corner where the frieze board meets the masonry to provide additional moisture control.

Masonry Construction at Openings

Special construction applications are important at masonry openings to stop water from entering the structure. Figure 5.30 shows a construction detail for proper masonry veneer drainage wall system construction at a window opening. The same methods apply over door openings and at other openings in the masonry veneer. Window frames, door frames, and other openings must be attached to the framing, and not attached to the masonry veneer. As you look at Figure 5.30 notice that the housewrap is carefully folded over the wood framing. This is done around the entire opening to help keep moisture from entering the wood. The **through-wall flashing,** also called a **sill pan** flashing at the bottom, is a

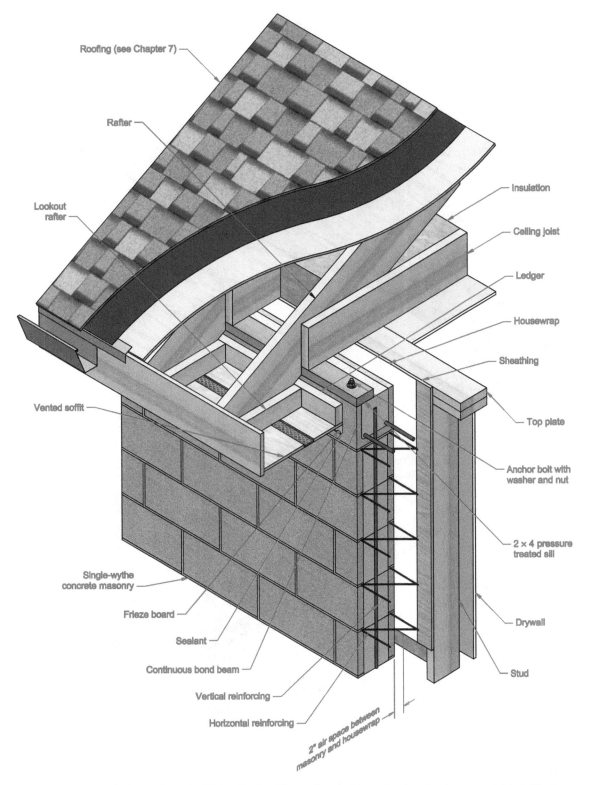

Figure 5.28 Anchor bolts can be embedded in the bond beam for attaching a 2 × 4 or 2 × 6 pressure treated sill where the roof framing system is connected. (*Previously published in Modern Residential Construction Practices, Routledge, 2017 and reproduced here with permission.*)

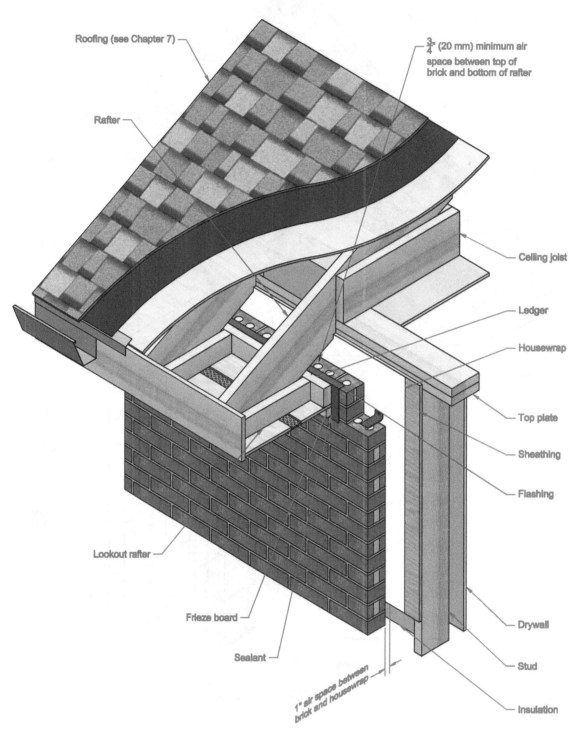

Roofing (see Chapter 7)

Rafter

¾" (20 mm) minimum air space between top of brick and bottom of rafter

Ceiling joist

Ledger

Housewrap

Top plate

Sheathing

Flashing

Lookout rafter

Frieze board

Sealant

Drywall

Stud

1" air space between brick and housewrap

Insulation

Figure 5.29 Additional detail for proper eave construction at a masonry veneer wall. (*Previously published in Modern Residential Construction Practices, Routledge, 2017 and reproduced here with permission.*)

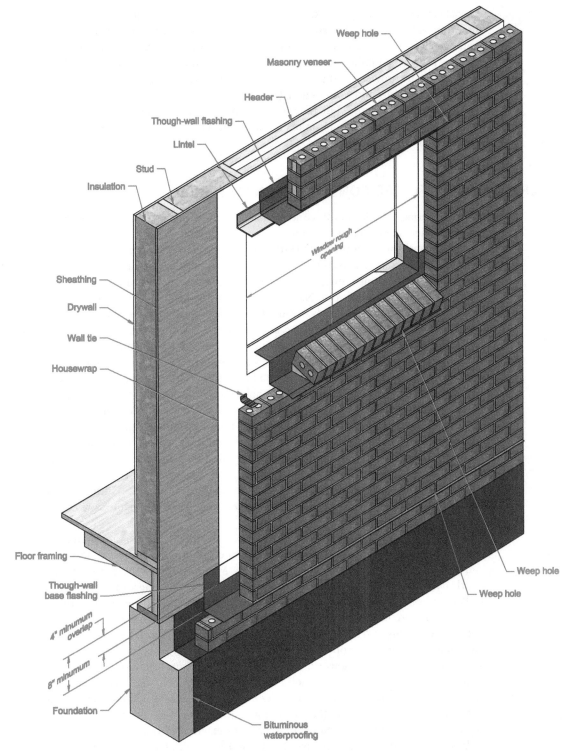

Figure 5.30 A construction detail for proper masonry veneer drainage wall system construction at a window opening. (*Previously published in Modern Residential Construction Practices, Routledge, 2017 and reproduced here with permission.*)

special fabricated flashing with sealed seams to keep water from entering the structure and allow rainwater to flow out. A space is provided between the masonry veneer and the wood frame to allow for movement in the masonry. The masonry rowlock sill at the base of the window is laid at a minimum angle of 15° from horizontal, or 1/4 in./ft (25 mm/100 mm). Weep hopes are provided in the rowlock sill to allow water to flow out over the flashing. It is important to seal around window and door frames with a recessed foam backer rod covered by a sealant. A minimum 1/4 in. (6 mm) wide sealant joint is recommended around window and door openings, and minimum of 1/2 in. (13 mm) is recommended between the sill and the bottom of the window frame.

OTHER MASONRY-RELATED CONSTRUCTION MATERIALS

Other materials, in addition to concrete blocks and bricks, used in construction include natural stone and manufactured stone.

The term **natural stone** refers to products that are quarried or mined from the earth and used as building materials and other applications. Natural stone products commonly include materials such as granite, limestone, marble, quartzite, travertine, slate, sandstone, brownstone, bluestone, and coral rock. Natural stone materials are generally more expensive than manufactured masonry materials because they are extracted from the earth and processed, which is labor and machinery intensive, and where materials are often difficult to access or in short supply. The Natural Stone Council established the trademarked **Genuine Stone**® as a defining name for natural stone. This name was created because manufactured stone manufacturers often marketed artificial stone as natural stone. Only Genuine Stone is acceptable in some construction applications. For example, granite is widely used for counter tops, and marble is often preferred for its rich appearance on floors and architectural features.

Manufactured stone, also called **artificial stone, cast stone, engineered stone, reconstructed stone,** or **simulated stone** is a concrete masonry product that looks like natural stone for use in architectural applications in the building construction industry. Manufactured stone is a widely used replacement for natural stone and is less expensive and more uniform than natural stone. Manufactured stone can be made to so closely duplicate natural stone for use on new architecture that is built next to existing old architecture constructed with natural stone.

Manufactured stone can be made from white and grey cement or a combination, also using manufactured or natural sand, crushed stone, or natural gravel. Manufactured stone can be colored with natural mineral pigments. Manufactured stone can be made to be nearly nonporous and can resist the corrosion. Manufactured stone is made from a mixture of Portland cement with high-quality fine and coarse aggregate, depending on the desired finish. A mixture including a high percentage of fine aggregate creates a smooth, consistent texture for closely resembling natural stone. The mixture can also include siliceous and aluminous additives to improve the glue that holds the concrete together. Manufactured stone can be made with a low water-to-cement ratio mixture referred to as a dry mixture uniformity. The dry mix is made solid by placing into a mold and using pneumatics, or electric tamping device and vibration under pressure. Manufactured stone can also be made with traditional wet concrete that is the same as construction concrete. This concrete is typically mixed and compacted using internal or external vibration applied to the mold, or by using self-compacting additives for the desired results. Other additives are also used when making dry mix manufactured stone. Additives are also used to improve plasticity and flexibility and to increase compaction in the mold. Some plasticizers have chemical properties that react with the cement to increase product strength. Waterproofing formulas added to manufactured stone can improve strength up to 20 percent and reduce absorption by 40 percent. The increased strength and reduced absorption results in improved freeze and thaw characteristics.

Manufactured stone can be made with a smooth and glossy surface using a mix of marble powder, resin, and pigment cast using vacuum oscillation to form slabs that are finished by cutting, grinding, and polishing. High-strength polyester resin can be added to improve hardness, strength, and gloss, and water-resistent characteristics. Commonly used engineered stone is engineered marble and engineered quartz. Engineered marbles are commonly used as flooring, and engineered quartz is typically used for counter tops, window sills, and floor and wall coverings.

COMMERCIAL CONSTRUCTION MASONRY APPLICATION

This commercial construction application is provided, in part, by Acme Brick and developed by Ray Don Tilley. This masonry project demonstrates the beauty, creativity, durability, and historic significance of using brick masonry construction. The project is the Central Academic Building, Texas A&M University in San Antonio, shown in Figure 5.31. The architect is Muñoz & Company, San Antonio. The general contractor is Bartlett Cocke, San Antonio, and the masonry contractor is Shadrock & Williams, Helotes, Texas. Acme Brick created Texas A&M San Antonio custom modular brick.

This project opens a new chapter for the revered A&M system, and an entire new product line for Acme Brick. The project designed to interpret regional tradition and architectural history in striking ways. Acme Brick worked closely with the design team to create a **blade cut brick** for a smoother alternative to **wire cut brick.** Blade cut brick is brick cut with masonry saw blades designed to cut through brick, stone, ceramic, tile, and other types of masonry materials. Wire cut brick is brick cut from clay shaped by extrusion before burning, where the long bar of extruded clay is cut into bricks by a set of wires. Acme Brick now provides select blade cut blends available as new options for creative brick design.

Figure 5.31　This commercial construction masonry application project is the Central Academic Building, Texas A&M University, San Antonio. (*Courtesy of Acme Brick and Ray Don Tilley.*)

Bringing History to New Tradition

True to its name, the Central Academic Building (CAB) is the **frontispiece** of Texas A&M San Antonio. Like missions that served as for settlement of San Antonio, the CAB stands as an organizing headquarters and the design standard to which future buildings can follow. In architecture, a frontispiece is the combination of elements that frame and decorate the main, or front, door to a building. New structures can continue the distinctive historical and cultural identity established using blade cut Acme Brick. Acme Brick engineers helped architects make best use of standard brick and create structurally sound special shapes where needed. According to Geoff Edwards, AIA, Principal, Muñoz & Company, "We designed Texas A&M San Antonio to reflect historic traditions and sources of architecture in South Texas, drawing on the examples of Mission San Jose, Spanish colonial buildings, and even Granada. With brick specifically, we looked at how many variations, patterns, and shadow effects we could achieve with a single modular brick. Acme was willing to reconsider the manufacturing process to create just the right consistent look across the growing campus with a less textured blade cut brick, and then help us shave costs by creating a thin brick for monumental archways. Acme's local sourcing was important in ongoing efforts to build responsibly now, and over the full development of the 694 acre site."

Brick for Lasting Architecture

Across the centuries, architects stand firm on masonry for appealing designs that last. Brick is particularly pleasing for capturing a connection with history through shapes and textures that beautifully matches admired forms. Functional arcades create shelter from weather, and arched openings and pilasters elevate otherwise ordinary building elements. Remarkably, virtually all of the intricate and varied patterns and treatments achieved were created by talented masons manipulating simple modular bricks. Architects honored the rhythms and scale of Mission San José by creating grand multistory entrances and ordered wings divided by dentils and soldier courses that serve as bases for arcades, and as caps and cornices for pilasters and roof lines. At a distance blade cut brick presents a smooth, soft-appearing facade that heightens the contrast with diagonally cut and turned saw tooth units in grand spandrel patterns, and with alternating recessed soldier units lining numerous architectural details throughout the project.

Go to Downloads & Resources tab at **www.mhprofessional.com/Commercial Building Construction-Ancillaries** to access the images provided in this chapter and correlate with the content of this chapter. Here you can look at the chapter figures on your computer screen to pan and zoom in and out for better observation, because full drawings reduced to fit on a textbook page are often difficult to read. The image bank also includes the complete set of plans for the Brookings South Main Fire Station used as examples throughout this textbook.

Go to **Downloads & Resources tab at www.mhprofessional.com/Commercial Building Construction-Ancillaries** to access the test and creative thinking problems for this chapter. The chapter test can be used for review or to evaluate content knowledge depending on your course objectives. Creative thinking problems are provided to help expand your knowledge by researching given subjects.

Print reading problems have questions that ask you to find information on the Brookings South Main Fire Station plans. The print reading problems can help you become familiar with the format of construction documents and reinforce your ability to seek specific information. This is an important skill for you to master as you prepare to enter the construction industry. Go to **Downloads & Resources tab at www.mhprofessional.com/Commercial Building Construction-Ancillaries** to access the complete set of plans for the Brookings South Main Fire Station.

CHAPTER 6

Steel Construction

INTRODUCTION

Steel construction refers to buildings built using steel and can be divided into three categories:

1. Prefabricated steel structures
2. Steel-framed structures
3. Cold-formed steel (CFS)

This chapter describes and illustrates commercial steel construction materials and practices, including fabrication methods and related drawing examples.

PREFABRICATED STEEL STRUCTURES

Prefabricated buildings, premanufactured buildings, and **metal buildings,** as they are often called, have become a common type of construction for commercial and agricultural structures in many parts of the country. Standardized premanufactured steel buildings are often available in modular units with given spans, wall heights, and lengths in 12 or 20 ft (3600 or 6000 mm) increments. Most manufacturers provide a wide variety of design options for custom applications that are required by the client. One advantage of these structures is faster erection time as compared with other commercial construction methods.

Prefabricate Steel Structural System

The prefabricated steel structural system has the **frame** that supports the walls and roof. There are several different types of prefabricated steel structural systems commonly used, as shown in Figure 6.1. The wall system is constructed using horizontal **girts** attached to the vertical structure and metal wall sheets attached to the girts. A girt is a horizontal structural member in a framed wall that provides lateral support to the wall panel to resist wind loads. Girts are attached horizontally to the vertical wall structure and used to attach the metal siding in steel construction, as shown in Figure 6.2. The roof system is

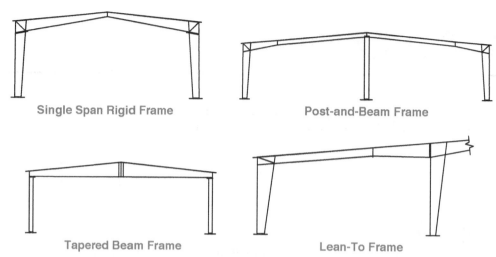

Figure 6.1   Several different types of common prefabricated steel structural systems.

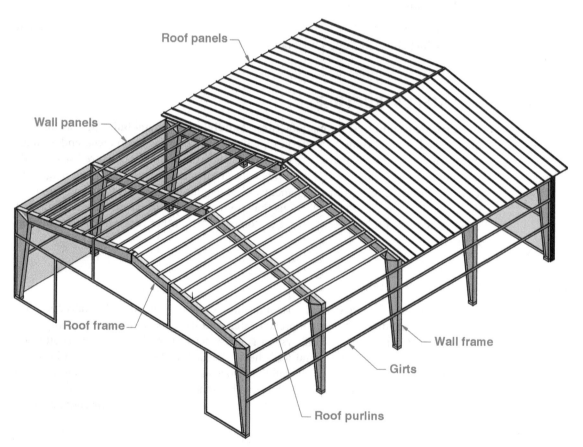

Figure 6.2   The roof system is constructed using horizontal purlins attached to the structure and metal sheets attached to the purlins.

constructed using horizontal **purlins** attached to the structure and metal sheets attached to the purlins, as shown in Figure 6.2. Purlins are attached horizontally to the roof framing system and are used to attach the metal sheets of roofing material in steel construction. Steel wall and roof sheets are available from many vendors in a variety of patterns and can be purchased plain, galvanized, or pre-painted. A sample of steel roofing pattern designs is shown in Figure 6.3. A widely used steel **cladding** material is called **standing seam.** Standing seam metal roofing and wall material is a concealed fastener metal panel system with vertical interlocking legs and a broad, flat area between the two legs. The word cladding is a wide range of materials as a general term used to describe the application of one material over another to provide usually an exterior layer over construction.

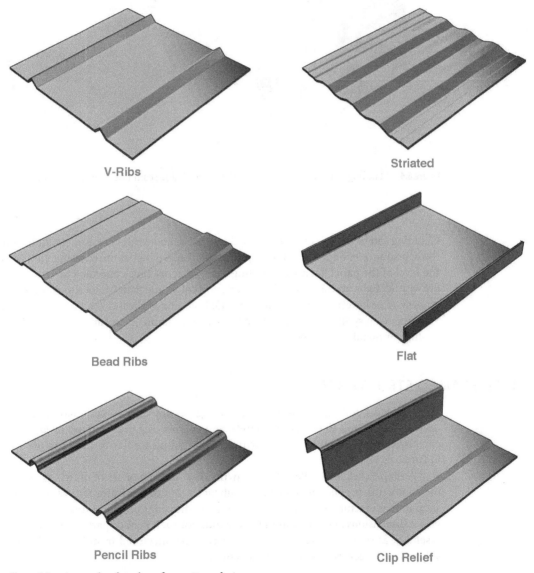

Figure 6.3   A sample of steel roofing pattern designs.

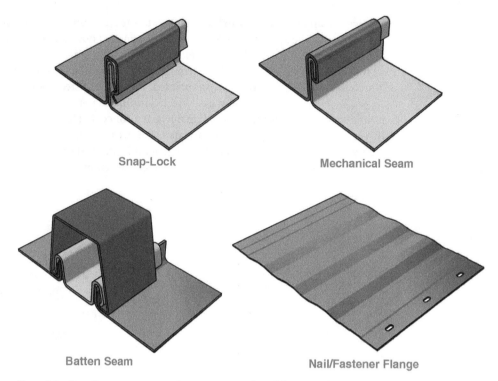

Snap-Lock   Mechanical Seam

Batten Seam   Nail/Fastener Flange

Figure 6.4   Standing seam systems have a variety of available seam fastening applications.

Cladding can be used to provide thermal insulation and weather resistance, and to improve the appearance of buildings. Standing seam vertical legs act as raised seams that rise above the level of the panel flat area. Standing seam systems have concealed fastening systems, are weather tight, and are available in a variety of colors and patterns for appearance. Standing seam systems have a variety of available seam fastening applications, as shown in Figure 6.4. Standing seam systems are generally made with aluminum or steel and can be used for metal roofing, or metal wall siding.

STEEL-FRAMED STRUCTURES

Steel-framed buildings require structural engineering and a combination of structural drawings and **shop ticket.** Architectural and engineering drawings are used to show the construction of the building or portion of a building similar to the example shown in Figure 6.5.

A shop ticket, also called **shop drawing,** is a drawing or set of drawings produced by the contractor, supplier, manufacturer, subcontractor, or fabricator for use in a fabrication shop to manufacture a construction item. Shop drawings provide a drawing with detailed views, dimensions and fabrication information for each item or part in the structural steel assembly, as shown in Figure 6.6. Each part is manufactured from its shop drawing and shipped to the construction side for assembly.

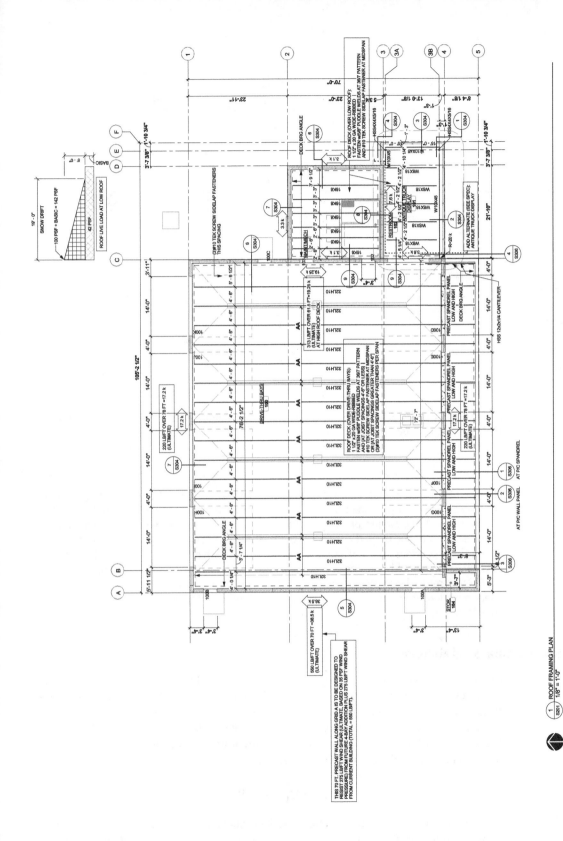

Figure 6.5 Architectural and engineering drawings are used to show the construction of the building or portion of a building similar to the example shown here. (*Courtesy of JLG Architects, Alexandria, MN.*)

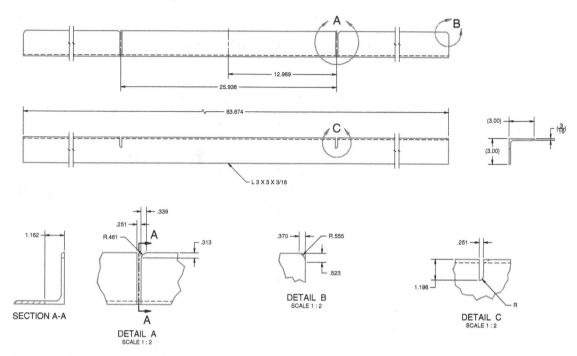

Figure 6.6 Shop drawings provide detailed views, dimensions and fabrication information for each item or part in the structural steel assembly. (*Courtesy Madsen Designs, Inc.*)

STEEL STANDARDS

Detailed information related to the steel construction industry can be found in the *Manual of Steel Construction* published by the American Institute of Steel Construction, Inc. (AISC) (www.aisc.org). The *Manual of Steel Construction* is a primary reference that helps you determine dimensions and properties of common steel shapes and steel construction practices. The manual also provides local and national building codes. Each office and commercial construction classroom should have one or more copies of the *Manual of Steel Construction*.

Another manual that provides information on dimensions for detailing and properties for design work related to steel structural materials is *Structural Steel Shapes*, published by the U.S. Steel Corporation (USS) (**www.uss.com**).

Common Structural Steel Materials

Structural steel is commonly identified as plates, bars, or shape configurations. **Plates** are flat pieces of steel of various thickness used at the intersection of different members and for the fabrication of custom connectors. **Figure 6.7** shows an example of a steel connector that uses top, side, and bottom plates. Plates are typically specified on a drawing by giving the thickness, width, and length in that order and with or without inch marks. The symbol ℙ is often used to specify plate material. For example,

$$PL\frac{1}{4} \times 6 \times 10$$

Bars are the smallest of structural steel products and are manufactured in round, square, rectangular, flat, or hexagonal cross sections. Bars are often used as supports or

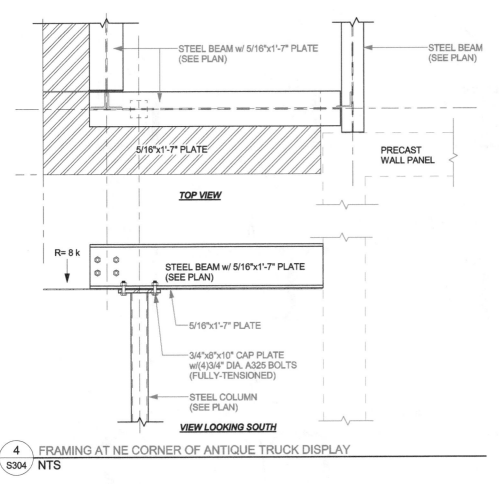

STEEL BEAM w/ 5/16"x1'-7" PLATE
(SEE PLAN)

STEEL BEAM
(SEE PLAN)

5/16"x1'-7" PLATE

PRECAST
WALL PANEL

TOP VIEW

R= 8 k

STEEL BEAM w/ 5/16"x1'-7" PLATE
(SEE PLAN)

5/16"x1'-7" PLATE

3/4"x8"x10" CAP PLATE
w/(4)3/4" DIA. A325 BOLTS
(FULLY-TENSIONED)

STEEL COLUMN
(SEE PLAN)

VIEW LOOKING SOUTH

4 | FRAMING AT NE CORNER OF ANTIQUE TRUCK DISPLAY
S304 | NTS

Figure 6.7   A steel connector that uses top, side, and bottom plates. (*Courtesy of JLG Architects, Alexandria, MN.*)

braces for other steel parts or connectors. Flat bars are usually specified on a drawing by giving the width, thickness, and length, in that order. For example,

$$\text{BAR } 3 \times \frac{1}{2} \times 1'\text{-}6''$$

Structural steel is also available in several different manufactured shapes called Wide flange, S shape, M shape, HP shape, angle, channel, T shape, tubing, and pipe, as shown in **Figure 6.8**. When a steel shape is shown and specified on a drawing, the shape identification letter is followed by the member depth, then by symbol ($\times$), and the weight in number of pounds per linear foot. For example,

$$\text{W12} \times 22 \quad \text{or} \quad \text{C6} \times 10.5$$

The dimensional values and specifications for the sample wide flange and channel examples are shown in **Figure 6.9**. In the AISC *Manual of Steel Construction* specific information regarding dimensions for detailing and dimensioning is clearly provided along with typical connection details. The representative *Manual of Steel Construction* pages for the W 12 × 22 wide flange and the C 6 × 10.5 channel from Figure 6.9 are shown in **Figure 6.10**.

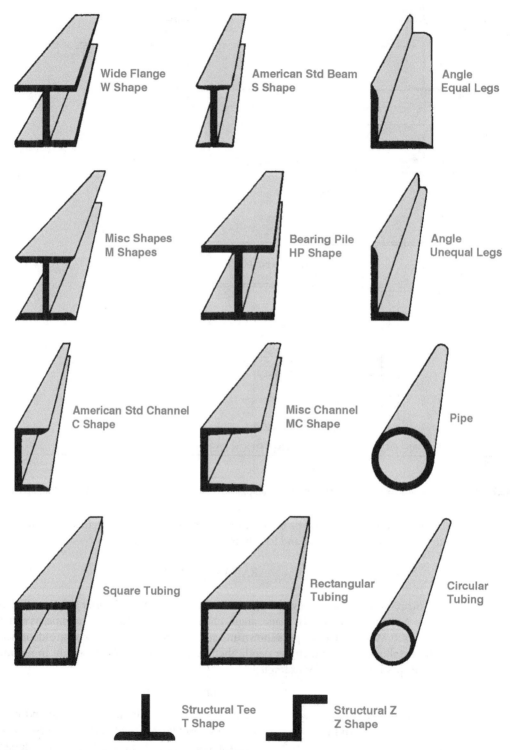

Figure 6.8 Structural steel is available in several different manufactured shapes.

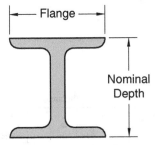

The designation for a wide flange looks like this:

W 12X22

Where: W is the shape
12 is the nominal depth
22 is the number of pounds
per lineal foot

Wide Flange Shapes

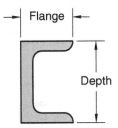

The designation for a channel shape looks like this:

C 6X10.5

Where: C is the shape
6 is the depth
10.5 is the number of pounds
per lineal foot

Channels

Figure 6.9   The dimensional values and specifications for the W12 × 22 or C6 × 10.5 wide flange and channel examples.

The W, S, and M shapes all have an I-shaped cross section and are often referred to as **I beams.** The three shapes differ in the width of their **flanges.** The flanges are the horizontal steel portions on the top and the bottom of the actual full member. In addition to varied flange widths, the S-shape flanges are tapered, making them stronger than equivalently sized **W beams** and suitable for train rail or monorail beams. The W shape is commonly used for columns. All can be used for horizontal or vertical members. The W beam is a **wide flange** W-shape structural steel shape with an I or H form.

Angles are structural steel components that have an L shape. The legs of the angle can be either equal or unequal in length but are usually equal in thickness. Steel angles are used for a wide variety of steel construction applications, including brackets, braces, trim, reinforcing, and larger angles for beams when engineered.

Channels have a squared C cross-sectional area and are designated with the letters C or MC. Channels have many applications that include use as girts, purlins, studs, braces, joists, and rafters.

Z shapes have a web between a leg and flange, forming a Z shape as the name implies. Z shapes are commonly used for wall girts and roof purlins to support siding and roof panels.

Structural tees are produced from W, S, and M steel shapes. Common designations include WT, ST, and MT. Structural tees are made by cutting a W, S, M, or HP shape down the middle, through the web, leaving a flange and stem in the shape of a T. Structural tees can be used to support steel decking and concrete slabs, as concrete reinforcing, and commonly as track for hoists, curtain walls, and amusement rides.

Structural tubing is manufactured in square, rectangular, and round cross-sectional configurations. These members are used as columns to support loads from other members. Tubes are also commonly used for beams and truss members. Tubes are specified by the size of the outer wall followed by the thickness of the wall.

Steel pipe is commonly used for columns, bracing, and pilings, or any use similar to steel tubing. Available steel pipe strengths are standard, extra strong, and double-extra strong. The wall thickness increases with each type.

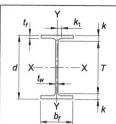

Table 1-1 (continued)
W-Shapes
Dimensions

Shape	Area, A (in.²)	Depth, d (in.)		Web Thickness, t_w (in.)	$\frac{t_w}{2}$ (in.)	Flange Width, b_f (in.)		Flange Thickness, t_f (in.)		Distance k — k_{des} (in.)	k_{det} (in.)	k_1 (in.)	T (in.)	Workable Gage (in.)
W14×38^c	11.2	14.1	14 1/8	0.310	5/16	3/16	6.77	6 3/4	0.515	1/2	0.915	1 1/4	13/16	11 5/8	3 1/2^g
×34^c	10.0	14.0	14	0.285	5/16	3/16	6.75	6 3/4	0.455	7/16	0.855	1 3/16	3/4	↓	3 1/2
×30^c	8.85	13.8	13 7/8	0.270	1/4	1/8	6.73	6 3/4	0.385	3/8	0.785	1 1/8	3/4		3 1/2
W14×26^c	7.69	13.9	13 7/8	0.255	1/4	1/8	5.03	5	0.420	7/16	0.820	1 1/8	3/4	11 5/8	2 3/4^g
×22^c	6.49	13.7	13 3/4	0.230	1/4	1/8	5.00	5	0.335	5/16	0.735	1 1/16	3/4	11 5/8	2 3/4^g
W12×336^h	98.9	16.8	16 7/8	1.78	1 3/4	7/8	13.4	13 3/8	2.96	2 15/16	3.55	3 7/8	1 11/16	9 1/8	5 1/2
×305^h	89.5	16.3	16 3/8	1.63	1 5/8	13/16	13.2	13 1/4	2.71	2 11/16	3.30	3 5/8	1 5/8		
×279^h	81.9	15.9	15 7/8	1.53	1 1/2	3/4	13.1	13 1/8	2.47	2 1/2	3.07	3 3/8	1 5/8		
×252^h	74.1	15.4	15 3/8	1.40	1 3/8	11/16	13.0	13	2.25	2 1/4	2.85	3 1/8	1 1/2		
×230^h	67.7	15.1	15	1.29	15/16	11/16	12.9	12 7/8	2.07	2 1/16	2.67	2 15/16	1 1/2		
×210	61.8	14.7	14 3/4	1.18	1 3/16	5/8	12.8	12 3/4	1.90	1 7/8	2.50	2 13/16	1 7/16		
×190	56.0	14.4	14 3/8	1.06	1 1/16	9/16	12.7	12 5/8	1.74	1 3/4	2.33	2 5/8	1 3/8		
×170	50.0	14.0	14	0.960	15/16	1/2	12.6	12 5/8	1.56	1 9/16	2.16	2 7/16	1 5/16		
×152	44.7	13.7	13 3/4	0.870	7/8	7/16	12.5	12 1/2	1.40	1 3/8	2.00	2 5/16	1 1/4		
×136	39.9	13.4	13 3/8	0.790	13/16	7/16	12.4	12 3/8	1.25	1 1/4	1.85	2 1/8	1 1/4		
×120	35.2	13.1	13 1/8	0.710	11/16	3/8	12.3	12 3/8	1.11	1 1/8	1.70	2	1 3/16		
×106	31.2	12.9	12 7/8	0.610	5/8	5/16	12.2	12 1/4	0.990	1	1.59	1 7/8	1 1/8		
×96	28.2	12.7	12 3/4	0.550	9/16	5/16	12.2	12 1/8	0.900	7/8	1.50	1 13/16	1 1/8		
×87	25.6	12.5	12 1/2	0.515	1/2	1/4	12.1	12 1/8	0.810	13/16	1.41	1 11/16	1 1/16		
×79	23.2	12.4	12 3/8	0.470	1/2	1/4	12.1	12 1/8	0.735	3/4	1.33	1 5/8	1 1/16		
×72	21.1	12.3	12 1/4	0.430	7/16	1/4	12.0	12	0.670	11/16	1.27	1 9/16	1 1/16	↓	↓
×65^f	19.1	12.1	12 1/8	0.390	3/8	3/16	12.0	12	0.605	5/8	1.20	1 1/2	1	↓	↓
W12×58	17.0	12.2	12 1/4	0.360	3/8	3/16	10.0	10	0.640	5/8	1.24	1 1/2	15/16	9 1/4	5 1/2
×53	15.6	12.1	12	0.345	3/8	3/16	10.0	10	0.575	9/16	1.18	1 3/8	15/16	9 1/4	5 1/2
W12×50	14.6	12.2	12 1/4	0.370	3/8	3/16	8.08	8 1/8	0.640	5/8	1.14	1 1/2	15/16	9 1/4	5 1/2
×45	13.1	12.1	12	0.335	5/16	3/16	8.05	8	0.575	9/16	1.08	1 3/8	15/16	↓	↓
×40	11.7	11.9	12	0.295	5/16	3/16	8.01	8	0.515	1/2	1.02	1 3/8	7/8	↓	↓
W12×35^c	10.3	12.5	12 1/2	0.300	5/16	3/16	6.56	6 1/2	0.520	1/2	0.820	1 3/16	3/4	10 1/8	3 1/2
×30^c	8.79	12.3	12 3/8	0.260	1/4	1/8	6.52	6 1/2	0.440	7/16	0.740	1 1/8	3/4		
×26^c	7.65	12.2	12 1/4	0.230	1/4	1/8	6.49	6 1/2	0.380	3/8	0.680	1 1/16	3/4	↓	↓
W12×22^c	6.48	12.3	12 1/4	0.260	1/4	1/8	4.03	4	0.425	7/16	0.725	15/16	5/8	10 3/8	2 1/4^g
×19^c	5.57	12.2	12 1/8	0.235	1/4	1/8	4.01	4	0.350	3/8	0.650	7/8	9/16		
×16^c	4.71	12.0	12	0.220	1/4	1/8	3.99	4	0.265	1/4	0.565	13/16	9/16		
×14^{c,v}	4.16	11.9	11 7/8	0.200	3/16	1/8	3.97	4	0.225	1/4	0.525	3/4	9/16	↓	↓

^c Shape is slender for compression with $F_y = 50$ ksi.
^f Shape exceeds compact limit for flexure with $F_y = 50$ ksi.
^g The actual size, combination, and orientation of fastener components should be compared with the geometry of the cross section to ensure compatibility.
^h Flange thickness greater than 2 in. Special requirements may apply per AISC *Specification* Section A3.1c.
^v Shape does not meet the h/t_w limit for shear in AISC *Specification* Section G2.1(a) with $F_y = 50$ ksi.

Figure 6.10 The representative pages for the W12 × 22 wide flange and the C6 × 10.5 channel from *Manual of Steel Construction*. (©*American Institute of Steel Construction. Reprinted with permission. All rights reserved.*)

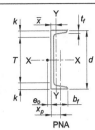

Table 1-5
C-Shapes
Dimensions

Shape	Area, A	Depth, d	Web Thickness, t_w		$\dfrac{t_w}{2}$	Flange Width, b_f		Flange Average Thickness, t_f		Distance k	Distance T	Distance Workable Gage	r_{ts}	h_o
	in.²	in.	in.		in.	in.		in.		in.	in.	in.	in.	in.
C15×50	14.7	15.0	15	0.716	11/16	3/8	3.72	3 3/4	0.650	5/8	1 7/16	12 1/8	2 1/4	1.17	14.4
×40	11.8	15.0	15	0.520	1/2	1/4	3.52	3 1/2	0.650	5/8	1 7/16	↓	2	1.15	14.4
×33.9	10.0	15.0	15	0.400	3/8	3/16	3.40	3 3/8	0.650	5/8	1 7/16	↓	2	1.13	14.4
C12×30	8.81	12.0	12	0.510	1/2	1/4	3.17	3 1/8	0.501	1/2	1 1/8	9 3/4	1 3/4 ᵍ	1.01	11.5
×25	7.34	12.0	12	0.387	3/8	3/16	3.05	3	0.501	1/2	1 1/8	↓	↓	1.00	11.5
×20.7	6.08	12.0	12	0.282	5/16	3/16	2.94	3	0.501	1/2	1 1/8	↓	↓	0.983	11.5
C10×30	8.81	10.0	10	0.673	11/16	3/8	3.03	3	0.436	7/16	1 1/16	8	1 3/4 ᵍ	0.924	9.56
×25	7.35	10.0	10	0.526	1/2	1/4	2.89	2 7/8	0.436	7/16	1 1/16		1 3/4 ᵍ	0.911	9.56
×20	5.87	10.0	10	0.379	3/8	3/16	2.74	2 3/4	0.436	7/16	1 1/16		1 1/2 ᵍ	0.894	9.56
×15.3	4.48	10.0	10	0.240	1/4	1/8	2.60	2 5/8	0.436	7/16	1 1/16	↓	1 1/2 ᵍ	0.868	9.56
C9×20	5.87	9.00	9	0.448	7/16	1/4	2.65	2 5/8	0.413	7/16	1	7	1 1/2 ᵍ	0.850	8.59
×15	4.40	9.00	9	0.285	5/16	3/16	2.49	2 1/2	0.413	7/16	1	↓	1 3/8 ᵍ	0.825	8.59
×13.4	3.94	9.00	9	0.233	1/4	1/8	2.43	2 3/8	0.413	7/16	1		1 3/8 ᵍ	0.814	8.59
C8×18.75	5.51	8.00	8	0.487	1/2	1/4	2.53	2 1/2	0.390	3/8	15/16	6 1/8	1 1/2 ᵍ	0.800	7.61
×13.75	4.03	8.00	8	0.303	5/16	3/16	2.34	2 3/8	0.390	3/8	15/16		1 3/8 ᵍ	0.774	7.61
×11.5	3.37	8.00	8	0.220	1/4	1/8	2.26	2 1/4	0.390	3/8	15/16	↓	1 3/8 ᵍ	0.756	7.61
C7×14.75	4.33	7.00	7	0.419	7/16	1/4	2.30	2 1/4	0.366	3/8	7/8	5 1/4	1 1/4 ᵍ	0.738	6.63
×12.25	3.59	7.00	7	0.314	5/16	3/16	2.19	2 1/4	0.366	3/8	7/8	↓	↓	0.722	6.63
×9.8	2.87	7.00	7	0.210	3/16	1/8	2.09	2 1/8	0.366	3/8	7/8			0.698	6.63
C6×13	3.82	6.00	6	0.437	7/16	1/4	2.16	2 1/8	0.343	5/16	13/16	4 3/8	1 3/8 ᵍ	0.689	5.66
×10.5	3.07	6.00	6	0.314	5/16	3/16	2.03	2	0.343	5/16	13/16		1 1/8 ᵍ	0.669	5.66
×8.2	2.39	6.00	6	0.200	3/16	1/8	1.92	1 7/8	0.343	5/16	13/16	↓	1 1/8 ᵍ	0.643	5.66
C5×9	2.64	5.00	5	0.325	5/16	3/16	1.89	1 7/8	0.320	5/16	3/4	3 1/2	1 1/8 ᵍ	0.616	4.68
×6.7	1.97	5.00	5	0.190	3/16	1/8	1.75	1 3/4	0.320	5/16	3/4	3 1/2	–	0.584	4.68
C4×7.25	2.13	4.00	4	0.321	5/16	3/16	1.72	1 3/4	0.296	5/16	3/4	2 1/2	1 ᵍ	0.563	3.70
×6.25	1.84	4.00	4	0.247	1/4	1/8	1.65	1 5/8	0.296	5/16	3/4		–	0.549	3.70
×5.4	1.58	4.00	4	0.184	3/16	1/8	1.58	1 5/8	0.296	5/16	3/4		–	0.528	3.70
×4.5	1.34	4.00	4	0.125	1/8	1/16	1.52	1 1/2	0.296	5/16	3/4	↓	–	0.506	3.70
C3×6	1.76	3.00	3	0.356	3/8	3/16	1.60	1 5/8	0.273	1/4	11/16	1 5/8	–	0.519	2.73
×5	1.47	3.00	3	0.258	1/4	1/8	1.50	1 1/2	0.273	1/4	11/16		–	0.496	2.73
×4.1	1.20	3.00	3	0.170	3/16	1/8	1.41	1 3/8	0.273	1/4	11/16		–	0.469	2.73
×3.5	1.09	3.00	3	0.132	1/8	1/16	1.37	1 3/8	0.273	1/4	11/16	↓	–	0.456	2.73

ᵍ The actual size, combination, and orientation of fastener components should be compared with the geometry of the cross section to ensure compatibility.
– Indicates flange is too narrow to establish a workable gage.

Figure 6.10   (*Continued*)

Structural Steel Callouts

It is important for you to recognize, identify, and properly read structural steel **callouts** found on plans and shop tickets. A callout is used in the construction industry to identify a given product, process or specification using words, numbers, abbreviations, and symbols to identifying an illustration or part of a drawing feature.

Many structural steel materials are specified by shape designation, flange width, and weight in pounds per linear foot as previously described. For example,

$$W24 \times 120$$

The following structural steel shapes fall into this group:

W—wide flange shapes

S—American standard beams

M—miscellaneous beam and column shapes

C—American standard channels

MC—miscellaneous channel shapes

WT—structural tees cut from W shapes

ST—structural tees cut from S shapes

MT—structural tees cut from M shapes

T—structural tees

Z—zee shapes

HP—steel H piling

The specifications for square and rectangular structural steel tubing, and sample structural steel shape designations, are provided in *Manual of Steel Construction.*

Structural materials that are specified by shape designation, type, and diameter or outside dimension, and wall thickness are in the following table:

Designation	Shape	Example	Meaning
PL	Plate	$PL \frac{1}{2} \times 6 \times 8$	THK × WIDTH × LENGTH
BAR	Square bar	$BAR\ 1\frac{1}{4}^*$	$1\frac{1}{4}''$ WIDTH × THICKNESS
	Round bar	$BAR\ 1\frac{1}{4}\varnothing$	$1\frac{1}{4}''$ DIAMETER
	Flat bar	$BAR\ 2 \times \frac{3}{8}$	WIDTH × THICKNESS
PIPE	Pipe	PIPE 3ØSTD [STANDARD]	OD [OUTSIDE DIA] SPEC
TS	Structural tubing Square*	TS 4 × 4 × .250	OUTWIDTH × OUTHEIGHT × WALL THICKNESS
	Rectangular	TS 6 × 3 × 0.375	SAME
	Round	TS 4 OD × 0.188	OD (outside diameter) × WALL THICKNESS

Designation	Shape	Example	Meaning
L	Angle Unequal leg	$L3 \times 2 \times \dfrac{1}{4}$	LEG × LEG × THICKNESS
	Equal leg	$L3 \times 3 \times \dfrac{1}{2}$	LEG × LEG × THICKNESS

*Some companies use o for the square symbol to designate a square feature dimension.

NOTE

If length dimensions are required, they are given at the end of the callout in feet and inches except for plates, which should be given only in inches.

When plate material is bent, the minimum bend radius is given with the plate callout, and the length of the bend legs are dimensioned on the drawing. For example,

$$\frac{3}{8} \times 10 \text{ W/MIN BEND R} \frac{5}{8}''$$

Location dimensions for structural steel components are provided, as shown in Figure 6.11.

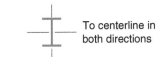

To centerline in both directions

Wide Flanges and Other I Shapes

Centerline along X–X axis and back face of web along Y–Y axis

Channels

Note: An exception for wide flanges and channels is when specifying top of beam.

Outside face of legs

Angles

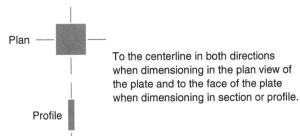

Plan

To the centerline in both directions when dimensioning in the plan view of the plate and to the face of the plate when dimensioning in section or profile.

Profile

Plates

Figure 6.11 Places to look for location dimensions.

Shop Drawings

Shop drawings were briefly introduced earlier in this chapter. When working in the commercial construction industry, you can be associated with or work in a steel fabrication facility where steel building components are manufactured. **Shop drawings,** also called **shop tickets,** and **fabrication drawings,** are used to extract each individual component of a structural engineering drawing down into individual fabrication parts. Many large companies do both structural and shop drawings. Depending on the complexity of the structure, the structural and shop drawings can be combined in the same set of plans. Figure 6.12 shows an example of a shop drawing.

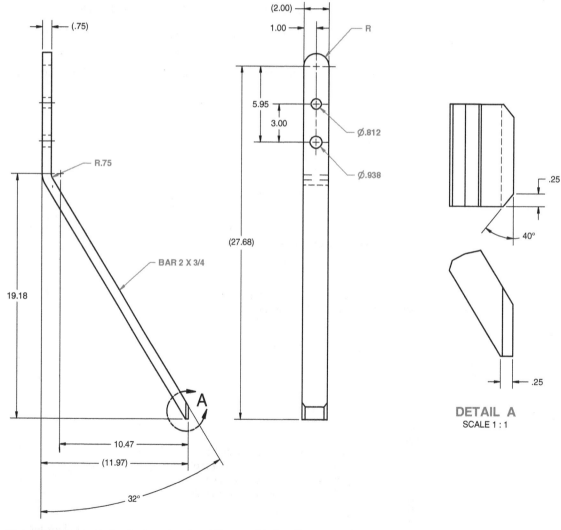

Figure 6.12   An example of a shop drawing. (*Courtesy Madsen Designs, Inc.*)

COMMON CONNECTION METHODS

There are a variety of common connection methods for steel construction, including bolted, riveted, and welded connections. The following discussions describe each of these connection practices. Often, a combination of connection methods are used to take advantage of the best characteristics of each for the specific application. Structural engineers determine the material and connection process used for the specific design.

Bolted Connections

Bolted connections are one of the most common connection practices used in steel construction. Bolted connections have advantages over other methods, including lower installation cost, assembly with basic tools, quick connections, minimum noise, and ease of installation over rivets and welds. Bolted connections can also carry their design load immediately after connection. **Bolts** are used for many connections in lumber and steel construction. A bolt is a threaded fastener with a square or hexagonal (hex) head on one end connected to a body called the shank. Other head types are available, but the square and hex head are most commonly used in construction. The end opposite of the head is threaded to fit a nut for fastening into a threaded feature. The bolt is generally used in construction to fasten two or more pieces together in combination with a washer under the head and a washer under the nut, as shown in Figure 6.13. The nut is also square or hex and

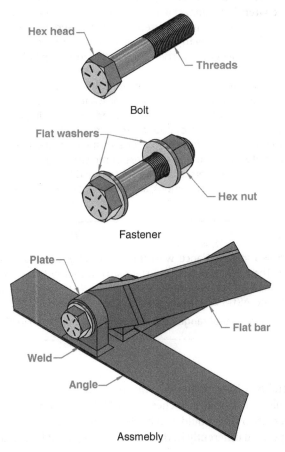

Figure 6.13 A bolt is a threaded fastener with a head on one end connected to a body called the shank. The end opposite of the head is threaded to fit a nut or the bolt can be fastened into a threaded feature. The bolt is generally used in construction to fasten two or more pieces together in combination with a washer under the head and a washer under the nut.

has the same thread specification as the bolt so they can fit together. A bolt specification includes the diameter, length, and strength of the bolt. **Washers** or plates are specified so the bolt head has a flat bearing surface and does not pull through the hole made for the bolt or compromise the bolt head or nut when tightened. The nut has the same thread and strength specification as the mating bolt. As previously mentioned a washer is usually used with a bolt and nut. There are many different washer types, but a flat washer is most commonly used in construction. A washer is a disc-shaped object with a center hole to allow a fastener to pass through. A washer is used to create a cushion for a bearing surface, to relieve friction, and as an optional fastener locking device. A specific washer is used that properly fits with the specified bolt and nut. The drawing specifications provide the information you need to determine the correct bolt, washer, and nut for the application.

NOTE

Tightening notes or specifications can be provided on the drawings for bolted connections. Tightening specifications are determined by structural engineering and specified in accordance with the American Institute of Steel Construction (AISC) standards and engineering calculations. Steel construction applications influence the fastener characteristics along with the type of connections, materials, structural loads, and types of structures. The following are basic tightening conditions:

Snug tight means that no specific level of fastening tension is required. This condition is typically applied after a few impacts of an **impact wrench** or the full effort of an ironworker with an ordinary **spud wrench**. An impact wrench is an electric or pneumatic power wrench with interchangeable toolhead attachments, used for installing and removing nuts, bolts, and screws. A spud wrench is a wrench with an adjustable or standard box wrench on one end and a tapered spike on the other end. The spike can be used to line up bolt holes in structural assemblies.

Fully tensioned is when the bolt or nut is turned tighter than snug tight to a positive setting on a **torque wrench** to engineering specifications. A torque wrench is used for setting and adjusting the tightness of nuts and bolts to a desired value reading on the wrench dial that registers as tightening applied in foot pounds (ft-lb) of torque.

Slip-critical connection are transferred from one element to another through friction forces that can be created by the extreme tightness of the structural bolts holding the connection together. These bolts, washers, and nuts are fastened to the structure with specific structural engineering applications, conditions, and treatment specifications.

BOLT STANDARDS

Bolt strength is classified in accordance with the American Society for Testing Materials (ASTM) specifications. The ASTM International standards for heavy hex structural bolts are titled *Standard Specification for Structural Bolts, Steel, Heat Treated, 120/105 ksi Minimum Tensile Strength*, ASTM A325, and ASTM A325M for metric fasteners. These ASTM standard defines mechanical properties for bolts that range from $\frac{1}{2}$ to $1\frac{1}{2}$ in. (13 to 38 mm) in diameter.

Standard Structural Bolt Callouts

You should understand bolt information on drawings and in specifications. Look for the quantity, diameter, bolt type, length in inches, and ASTM specification. If special washer requirements are given, then this should also be determined. Hex head bolts and hex nuts are assumed unless specified differently in the callout. Examples of bolt callouts are

$2\frac{3}{4}''\varnothing$ BOLTS ASTM A503

$4\frac{1}{2}''\varnothing \times 10''$ BOLTS

$2\frac{5}{8}''\varnothing \times 6''$ CARRIAGE BOLTS

$6\frac{3}{4}''\varnothing$ BOLTS W/MALLEABLE IRON WASHERS AND HEAVY HEX NUTS

$4\frac{1}{2}''\varnothing$ GALVANIZED BOLTS

The hole diameter for bolts is also specified on the drawings. This is important to determine, because a specific bolt used in specific material requires a specific hole diameter for proper assembly. In general, holes are $\frac{1}{16}$ in. (1.5 mm) larger in diameter than the specified bolt for standard steel-to-steel, wood-to-wood, or wood-to-steel construction for hole sizes up to 1 in. (25 mm) in diameter, and $\frac{1}{8}$ in. (3 mm) larger for hole sizes over 1 in. in diameter. Holes are typically $\frac{1}{8}$ in. (3 mm) larger than the specified bolt for standard steel-to-concrete or wood-to-concrete applications unless otherwise specified by the engineer. For example,

$$2\frac{3}{4}''\varnothing \text{ BOLTS FIELD DRILL } \frac{13}{16}''\varnothing \text{ HOLES}$$

The following are recommended standard hole diameters in inches for given bolt sizes:

Bolt Ø	Standard Hole Ø	Concrete Hole Ø	Oversize Hole Ø
$\frac{1}{2}$	$\frac{9}{16}$	$\frac{5}{8}$	$\frac{11}{16}$
$\frac{5}{8}$	$\frac{11}{16}$	$\frac{3}{4}$	$\frac{13}{16}$
$\frac{3}{4}$	$\frac{13}{16}$	$\frac{7}{8}$	$\frac{15}{16}$
$\frac{7}{8}$	$\frac{15}{16}$	1	$1\frac{1}{16}$
1	$1\frac{1}{16}$	$1\frac{1}{8}$	$1\frac{1}{4}$
$1\frac{1}{8}$	$1\frac{1}{4}$	$1\frac{1}{4}$	$1\frac{7}{16}$
$1\frac{1}{4}$	$1\frac{3}{8}$	$1\frac{3}{8}$	$1\frac{9}{16}$
$1\frac{3}{8}$	$1\frac{1}{2}$	$1\frac{1}{2}$	$1\frac{11}{16}$
$1\frac{1}{2}$	$1\frac{5}{8}$	$1\frac{5}{8}$	$1\frac{13}{16}$

Lag bolt specifications typically give the lead or tap hole diameter with the bolt specification. A lag bolt, also called a lag screw, is a heavy wood screw with a tapered point and a square or hex head. For example,

$$2\tfrac{3}{4}''\varnothing \times 8 \text{ LAG BOLTS W/7/16}''\varnothing \text{ LEAD HOLES}$$

A **pilot hole** is required otherwise you are unable to enter a fastener in the desired location. A pilot hole, also called a lead hole or tap hole, is a hole drilled into material to a specified depth or through the material prior to a bolt, screw or other fastener being inserted. The following are pilot hole diameters in inches for lag bolts used in Douglas fir, larch, or Southern pine:

Bolt $\varnothing$	Lead Hole $\varnothing$
$\tfrac{3}{8}$	$\tfrac{1}{4}$
$\tfrac{1}{2}$	$\tfrac{5}{16}$
$\tfrac{5}{8}$	$\tfrac{7}{16}$
$\tfrac{3}{4}$	$\tfrac{1}{2}$
$\tfrac{7}{8}$	$\tfrac{5}{8}$
1	$\tfrac{3}{4}$

Bolts are located to their centerlines on a drawing and correlated to the same location on the construction project. A **counterbore** can be required for some fastener applications. A counterbore is a flat-bottomed cylindrical enlargement of the mouth of a hole with enough depth to hide the bolt head or washer and nut below the surface. When a counterbore note follows the hole note when a counterbore is required. For example,

$$\varnothing 2\tfrac{3}{4}'' \text{ DRILL } \varnothing 3\tfrac{1}{4}'' \text{ CBORE} \times \tfrac{7}{8}'' \text{ DEEP}$$

This note uses terms, such as DRILL and DEEP, and an abbreviation CBORE for counterbore. The same note can also be given using symbols. The counterbore symbol is ⌴, and the depth symbol is ▽. The word DRILL refers to using a **drill** to make the hole. A drill is a rotating cutting tool used for making holes. The word DRILL can be omitted from the process note, leaving the process for making the hole up to the worker. A hole is assumed to go through the material unless a specific depth is given.

Riveted Connections

A **rivet** is a metal pin with a head used to fasten two or more materials together. The rivet is placed through holes in mating parts and the end without a head extends through the parts to be **headed-over**. The term **headed-over** means to form into a head by hammering,

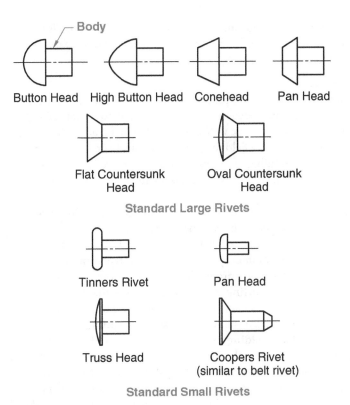

Figure 6.14 Rivets are classified by body diameter, length, head type as shown here.

pressing, or forging. The end with the head is held in place with a solid steel bar called a **dolly** while a head is formed on the other end. Rivets are classified by body diameter, length, head type, as shown in Figure 6.14.

Riveted steel connections are not as common as bolted connections. Bolt connections have a number of advantages described earlier. Construction with riveted connections have additional disadvantages, including the need for preheating the rivet to form the headed-over end, the riveting process creates noise when hammering the end during fabrication, and rivets are difficult to replace for maintenance. Riveted connections do have advantages over welded connections, described next. Rivets and the riveting process can be less expensive than welding, and the riveting process does not require large amounts of electricity needed for making welded connections.

Welded Connections

Welds are created using **welding** processes and are classified according to the type of joint on which they are used. **Welding** is a process of creating a weld by joining two or more pieces of like metals by heating the material to a temperature high enough to cause softening or melting. A **weldment** is an assembly of parts welded together. The parts that are welded become one, and a properly welded joint is as strong as or stronger than the original material. Most welding operations are performed by filling a heated joint between pieces with molten metal. Advantages and disadvantages of welding over riveting were listed in the previous section. Welding advantages over other fastening methods include better strength, better weight distribution and reduction, and a potential time saving with many weldments created in a fabrication shop.

Welding Processes

There are a number of welding processes available for use in industry. The most common welding processes include **oxygen gas welding, shielded metal arc welding, gas tungsten arc welding,** and **gas metal arc welding.** Shielded metal arc welding is the most common welding process used in the field for commercial steel construction and use in the fabrication shop for many welding applications. Gas metal arc welding process is often used to weld parts in the fabrication shop, because of the speed and quality available with these processes.

Oxygen gas welding, commonly known as **oxyfuel welding** or **oxyacetylene welding,** can also be performed with fuels such as natural gas, propane, and propylene. Oxyfuel welding is most typically used to fabricate thin materials, such as sheet metal and thin-wall pipe or tubing, but not commonly used for making structural steel connections. Most cuts made on structural steel parts are made in a fabrication shop, but a process that uses an oxyfuel mixture is **flame cutting** is often performed in the field. The flame cutting process uses a high-temperature gas flame to preheat the metal to a kindling temperature, at which time a stream of pure oxygen is injected to cause the cutting action.

Shielded metal arc welding (SMAW), also called **stick electrode welding,** is the most traditionally used welding method and most commonly used in steel construction fabrication. High-quality welds on a variety of metals and thicknesses can be made rapidly with excellent uniformity. This welding method uses a flux-covered metal electrode to carry an electrical current forming an arc that melts the work and the electrode. The molten metal from the electrode mixes with the melting base material, forming the weld. Shielded metal arc welding is popular because of low-cost equipment and supplies, flexibility, portability, and versatility.

The **gas tungsten arc welding** process is sometimes referred to as **tungsten inert gas welding (TIG),** or as **Heliarc˚,** which is a trademark of the Union Carbide Corporation. Gas tungsten arc welding can be performed on a wider variety of materials than shielded metal arc welding, and it produces clean, high-quality welds. This welding process is useful for certain materials and applications. Gas tungsten arc welding is generally limited to thin materials, high-integrity joints, or small parts, because of its slow welding speed and high cost of equipment and materials.

Another welding process that is extremely fast, economical, and produces a very clean weld is **gas metal arc welding.** This process can be used to weld thin material or heavy plate. It was originally used for welding aluminum using a metal inert gas shield, a process referred to as **metal inert gas (MIG)** welding or **metal active gas (MAG)** welding. The application employs a current-carrying wire that is fed into a joint between pieces to form the weld. This welding process is used in industry with automatic or robotic welding machines to produce rapidly made, high-quality welds in any welding position.

The standard related to welding practices covered in this chapter is AWS A2.4, *Standard Symbols for Welding, Brazing, and Nondestructive Examination.* This American Welding Society (AWS) standard is developed in accordance with the rules of the American National Standards Institute (ANSI) and published by the American Welding Society.

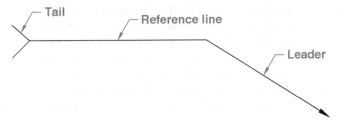

Figure 6.15 Basic components of a welding symbol, beginning with the reference line, tail, and leader.

Welding Symbols

Structural drawings containing weldments use welding symbols to identify the welds. **Welding symbols** identify the location of the weld, the welding process, the size and length of the weld, and other weld information. The **weld symbol** indicates the type of weld and is part of the welding symbol. There are a few basic components of a welding symbol, beginning with the reference line, tail, and leader, as shown in Figure 6.15. The reference line is the line upon which the welding symbol elements are located. The symbol tail is placed at one end of the reference line and is used to identify processes, specifications, and other references. The tail can be omitted if not used for these purposes. The reference line continues with a leader that is capped with an arrowhead that directs the welding symbol to the weld location. Additional information is placed on the reference line to continue the weld specification. Figure 6.16 shows the standard location of welding symbol elements as related to the reference line, tail, and leader.

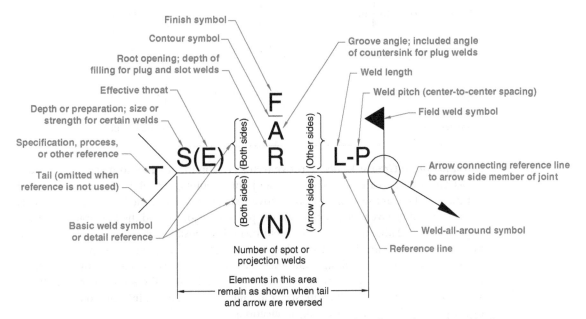

Figure 6.16 Standard location of welding symbol elements as related to the reference line, tail, and leader.

Weld Symbol Leader Arrow Related to Weld Location. Welding symbols are applied to the joint as the basic reference for the welding process used. All joints have an **arrow side** and **other side.** When fillet and groove welds are used, the welding symbol leader arrows connect the symbol reference line to one side of the joint known as the arrow side. The side opposite the location of the arrow is called the other side. If the weld is to be deposited on the arrow side of the joint, then the proper weld symbol is placed below the reference line, as shown in Figure 6.17*a*. If the weld is to be deposited on the side of the joint opposite the arrow, then the weld symbol is placed above the reference line, as shown in Figure 6.17*b*. The same weld symbol is shown above and below the reference line when welds are to be deposited on both sides of the joint, as shown in Figures 6.17*c* and *d*.

Types of Welds

Common welds used in steel construction are the fillet, groove, and plug or slot welds.

Fillet Weld. A **fillet weld** is formed at the internal corner of the angle formed by two pieces of metal. A fillet weld joins two edges at right angles. The cross section of a fillet weld is an approximate right triangle. The sides of the right triangle are the **legs.** The size of the fillet weld is shown on the same side of the reference line as the weld symbol and to the left of the symbol. When both legs of the fillet weld are the same, the size is given once, as shown in Figure 6.18. When the leg lengths are different in size, the vertical dimension is followed by the horizontal dimension, as shown in Figure 6.19. Recommended minimum sizes for **prequalified** fillet welds are shown in the following table: Prequalified welds are most of the common welded joints used in steel construction and are exempt from tests and qualification.

Minimum Prequalified Fillet Weld Size

Base Material Thickness of the Thinner Part Joined (T)	Minimum Size of Fillet Weld
T less than or equal to $\frac{1}{4}$ in. (6 mm)	$\frac{1}{8}$ in. (3 mm)
T greater than $\frac{1}{4}$ in. and less than or equal to $\frac{1}{2}$ in. (12 mm)	$\frac{3}{16}$ in. (5 mm)
T greater than $\frac{1}{2}$ in. and less than or equal to $\frac{3}{4}$ in. (19 mm)	$\frac{1}{4}$ in. (6 mm)
T greater than $\frac{3}{4}$ in.	$\frac{5}{16}$ in. (8 mm)

Introduction to Groove Welds. **Groove welds** are commonly used to make edge-to-edge joints. Groove welds are also used in corner joints, T joints, and joints between curved and flat pieces. There are several types of groove welds described in the following. The differences between groove welds depend on the parts to be joined and the preparation of their edges. When creating a groove weld, weld metal is deposited within the groove and weld penetration fuses with the base metal to form the joint. The selection of specific welds is based on the weld groove design, and the material thickness is an engineering application that requires stress analysis and calculations that are beyond the scope of this textbook. When working in a fabrication industry, the engineer provides the information and specifications needed for accurate welding applications.

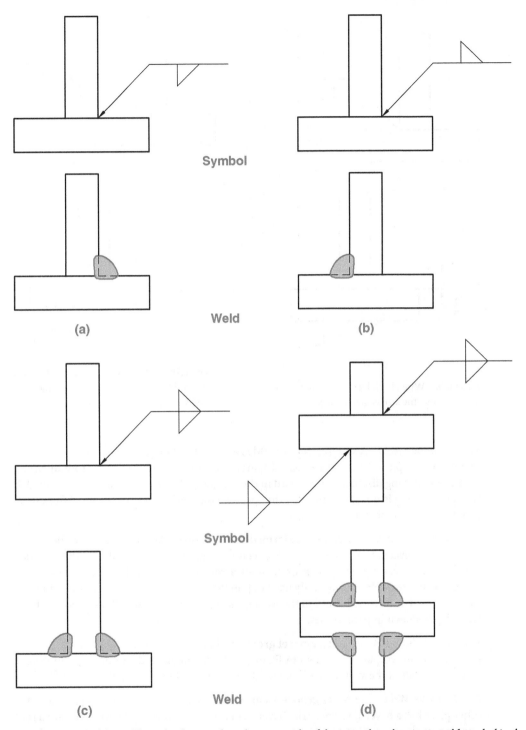

Figure 6.17   (a) If the weld is to be deposited on the arrow side of the joint, then the proper weld symbol is placed below the reference line. (b) If the weld is to be deposited on the side of the joint opposite the arrow, then the weld symbol is placed above the reference line. (c) The same weld symbol is shown above and below the reference line when welds are to be deposited on both sides of the joint. (d) Both sides for two joint welds.

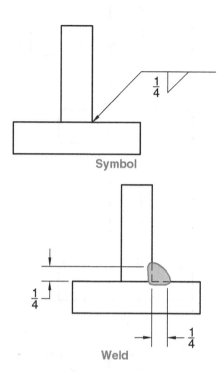

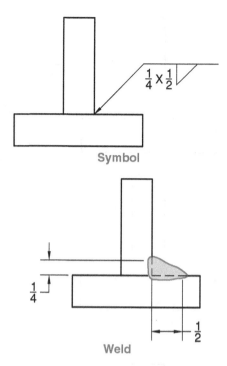

Figure 6.18 When both legs of the fillet weld are the same, the size is given once.

Figure 6.19 When the leg lengths of a fillet weld are different in size, the vertical dimension is followed by the horizontal dimension.

Square Groove Weld. A **square groove weld** is applied to a butt joint between two pieces of metal. The two pieces of metal are spaced apart a given distance known as the **root opening**. If the root opening distance is a standard in the company, then this dimension is assumed. If the root opening is not standard, then the specified dimension is given inside of the square groove symbol, as shown in Figure 6.20.

V Groove Weld. A **V groove weld** is formed between two adjacent parts when the side of each part is beveled to form a groove between the parts in the shape of a V. The included angle of the V can be given with or without a root opening, as shown in Figure 6.21. A **flare-V groove weld** is commonly used to join two rounded or curved parts, as shown in Figure 6.22. The intended depth of the weld itself is given to the left of the symbol, with the weld depth shown in parentheses.

Flare-Bevel Groove Weld. A **flare-bevel groove weld** is commonly used to join flat parts to rounded or curved parts, as shown in Figure 6.22. The intended depth of the weld itself is given to the left of the symbol, with the weld depth shown in parentheses.

Bevel Groove Weld. The **bevel groove weld** is created when one piece is square and the other piece has a beveled surface. The bevel weld can be given with a bevel angle and a root opening, as shown in Figure 6.23.

U Groove Weld. A **U groove weld** is created when the groove between two parts is in the form of a U. The angle formed by the sides of the U shape, the root, and the weld size are generally given, as shown in Figure 6.24.

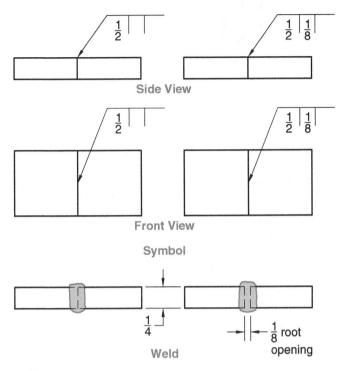

Figure 6.20   The root opening on a square groove weld.

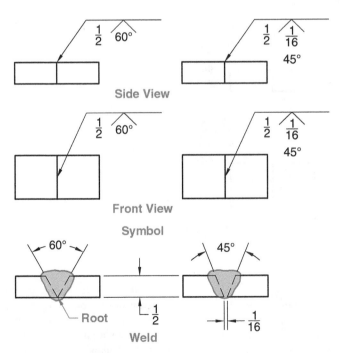

Figure 6.21   V groove weld is formed between two adjacent parts when the side of each part is beveled to form a groove between the parts in the shape of a V.

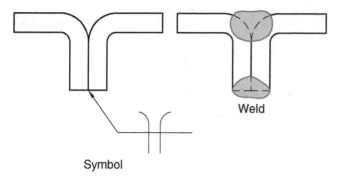

Symbol

Flare-V-Groove

Weld

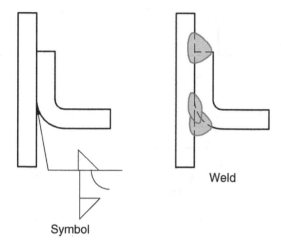

Symbol

Flare-Bevel-Groove

Weld

Figure 6.22 A flare-V groove weld is commonly used to join two rounded or curved parts. A flare-bevel groove weld is commonly used to join flat parts to rounded or curved parts.

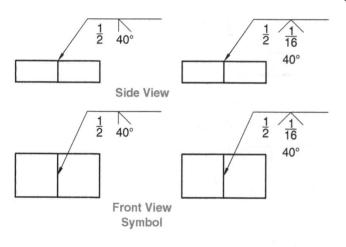

Side View

Front View
Symbol

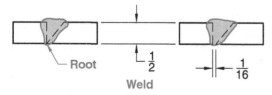

Root

Weld

Figure 6.23 The bevel groove weld is created when one piece is square and the other piece has a beveled surface. The bevel weld can be given with a bevel angle and a root opening.

288

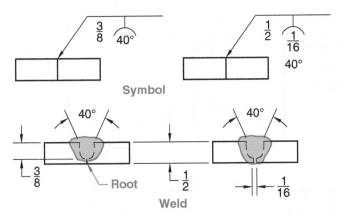

Figure 6.24 A U groove weld is created when the groove between two parts is in the form of a U.

J Groove Weld. The **J groove weld** is necessary when one piece is a square cut and the other piece is in a J-shaped groove. The included angle, the root opening, and the weld size are given, as shown in Figure 6.25.

Plug Weld. A **plug weld** is made in a hole in one piece of metal that is lapped over another piece of metal. These welds are specified by giving the weld size, angle, depth, and pitch, as shown in Figure 6.26a. The same type of weld can be applied to a slot. This is referred to as a **slot weld,** as shown in Figure 6.26b.

Flush Contour Weld. Generally, the surface contour of a weld is raised above the surface face. If this is undesirable, then a flush contour surface symbol is applied to the weld symbol. When the **flush contour weld** symbol is applied without any further consideration, the welder performs this flush effect without any finishing. The other option is to specify a flush finish using another process. The letter designating the other process is placed above the flush contour symbol for an other side application or below the flush contour symbol for an arrow side application. The options include C = chipping, G = grinding, M = machining, R = rolling, and H = hammering. See Figure 6.27.

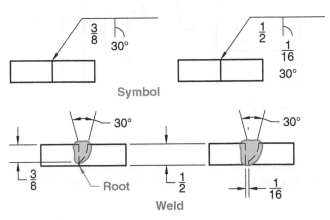

Figure 6.25 The J groove weld is necessary when one piece is a square cut and the other piece is in a J-shaped groove.

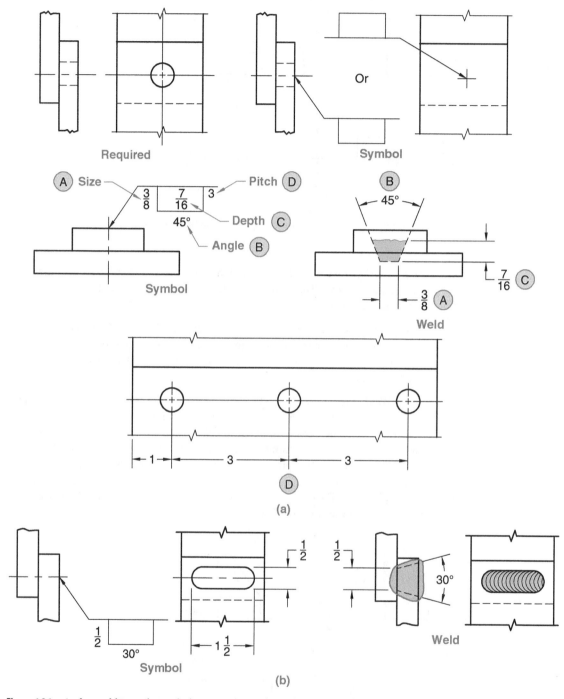

Figure 6.26 A plug weld is made in a hole in one piece of metal that is lapped over another piece of metal. (*a*) Plug welds are specified by giving the weld size, angle, depth, and pitch. (*b*) A plug weld can be applied to a slot, and is referred to as a slot weld.

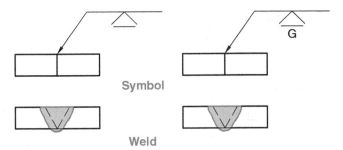

Symbol

Weld

Figure 6.27 A flush surface symbol is applied to the weld symbol when the weld must be flush with the surface.

Weld Joints

Types of weld joints are often closely associated with the types of weld grooves. The weld grooves can be applied to any of the typical joint types. The weld joints used in most weldments are the butt, lap, tee, outside corner, and edge joints shown in Figure 6.28.

Additional Weld Characteristics

Additional weld characteristics can be added to the welding symbol, including field weld, weld penetration, melt through, weld all around, and weld length.

Field Weld

A **field weld** is a weld that is performed on the job site, which is referred to as the field, rather than in a fabrication shop. The reason for this application can be that the individual components are easier to transport disassembled or that the mounting procedure requires job-site installation. The field weld symbol is a flag attached to the reference line at the leader intersection, as shown in Figure 6.29. The field weld symbol can be above or below the reference line, and the flag always points in the direction of the reference line.

Weld Penetration

Unless otherwise specified, a weld penetrates through the thickness of the parts at the joint. The size of the groove weld remains to the left of the weld symbol. Figure 6.30*a* shows the size of grooved welds with partial penetration. Notice in Figure 6.30*b* that a weld with partial penetration can specify the depth of the groove followed by the depth of weld penetration in parentheses, with both items placed to the left of the weld symbol.

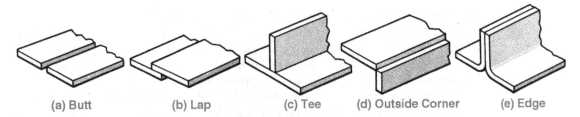

(a) Butt   (b) Lap   (c) Tee   (d) Outside Corner   (e) Edge

Figure 6.28 The weld joints used in most weldments are the butt, lap, tee, outside corner, and edge joints.

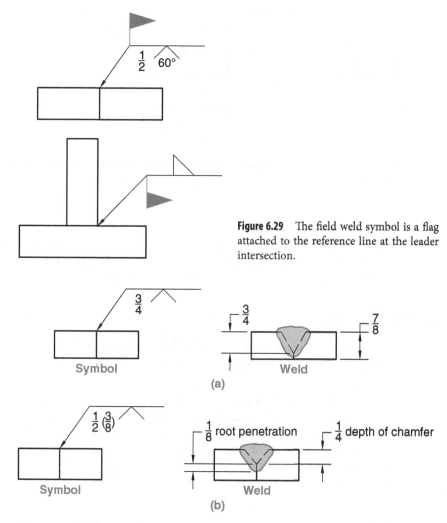

Figure 6.29 The field weld symbol is a flag attached to the reference line at the leader intersection.

Figure 6.30 (*a*) The size of grooved welds with partial penetration. (*b*) A weld with partial penetration can specify the depth of the groove followed by the depth of weld penetration in parentheses, with both items placed to the left of the weld symbol.

The weld size can be omitted when single-groove and symmetrical double-groove welds penetrate completely through the parts being joined, as shown in Figure 6.31. The depth of penetration of flare-formed groove welds is assumed to extend to the tangent points of the members, as shown in Figure 6.32.

Melt Through. **Melt through** is a term that refers to the weld melting through the bottom of the weld or the opposite side of where the weld is being applied. When improperly done or when not specified, melt through can be unacceptable. However, melt through is desired when the melt-through symbol, as shown in Figure 6.33, is used. The melt-through symbol is placed on the side of the reference line opposite the weld symbol. Melt through is used only when complete joint penetration and visible root reinforcement is required in welds made from one side. The desired melt-through distance can be placed to the left of the melt-through symbol on the reference line, as shown in Figure 6.33.

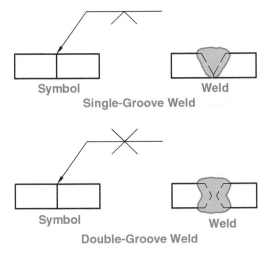

Symbol Weld

Single-Groove Weld

Symbol Weld

Double-Groove Weld

Figure 6.31 The weld size can be omitted when single-groove and symmetrical double-groove welds penetrate completely through the parts being joined.

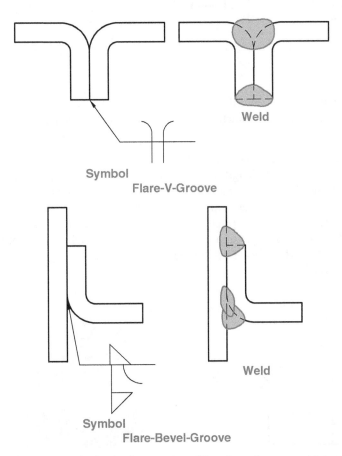

Weld

Symbol

Flare-V-Groove

Weld

Symbol

Flare-Bevel-Groove

Figure 6.32 The depth of penetration of flare-formed groove welds is assumed to extend to the tangent points of the members.

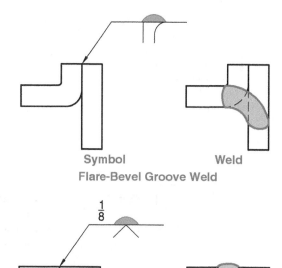

Symbol Weld

Flare-Bevel Groove Weld

$\frac{1}{8}$

Symbol $\frac{1}{8}$ Weld

V Groove Weld

Figure 6.33 The melt-through symbol is placed on the side of the reference line opposite the weld symbol. Melt through is used only when complete joint penetration and visible root reinforcement is required in welds made from one side.

Weld All Around. When a welded connection must be performed all around a feature, the **weld-all-around** symbol is attached to the reference line at the junction of the leader, as shown in Figure 6.34.

Weld Length and Pitch. **Weld length** is given when a weld is not continuous along the length of a part. The weld length is the length of a weld that is not continuous. In some situations, the weld along the length of a feature is given in lengths spaced a given distance apart. The distance from one point on a weld length to the same corresponding point on the next weld is called the **pitch.** The weld pitch is generally from center to center of weld lengths. The weld length and pitch are shown to the right of the weld symbol, as in Figure 6.35. A weld is always continuous unless the welding symbol specifies a weld length and pitch.

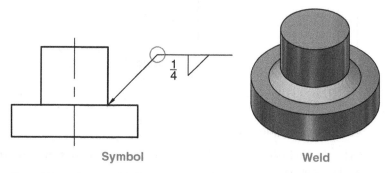

Symbol Weld

Figure 6.34 When a welded connection must be performed all around a feature, the weld-all-around symbol is attached to the reference line at the junction of the leader.

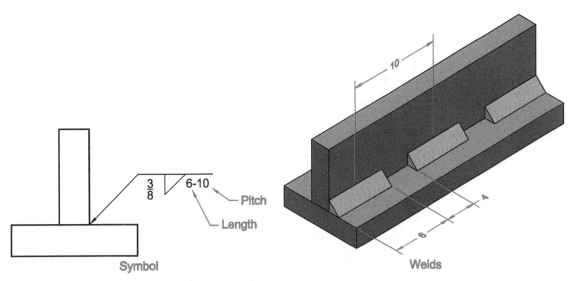

Figure 6.35   The weld length is given when a weld is not continuous along the length of a part.

Welding Process Designation. A **tail** is added to the welding symbol when it is necessary to designate the welding specification, procedures, or other supplementary information needed to fabricate the weld, as shown in Figure 6.36.

Welding Specifications

A **welding specification** is a detailed statement of the legal requirements for a specific classification or type of product. Products manufactured to code or specification requirements commonly must be inspected and tested to ensure compliance.

A number of agencies and organizations publish welding codes and specifications. The application of a particular code or specification to a weldment can be the result of one or more of the following requirements:

- Local, state, or federal government regulations
- Bonding or insurance company requirements
- Customer requirements
- Standard industrial practice

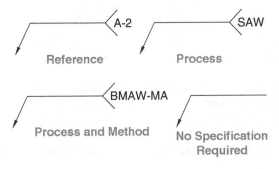

Figure 6.36 The tail is added to the welding symbol when it is necessary to designate the welding specification, procedures, or other supplementary information needed to fabricate the weld.

Commonly used codes include

- No. 1104, American Petroleum Institute (API). Used for pipeline specifications.
- Section IX, American Society of Mechanical Engineers (ASME). Used to specify welds for pressure vessels.
- D1.1, American Welding Society (AWS). Welding specifications for bridges and buildings.
- AASHT, American Association of State Highway and Transportation Officials.
- AIAA, Aerospace Industries Association of America.
- AISC, American Institute of Steel Construction.
- ANSI, American National Standards Institute.
- AREA, American Railway Engineering Association.
- AWWA, American Water Works Association.
- AAR, Association of American Railroads.
- MILSTD, Military Standards, Department of Defense.
- SAE, Society of Automotive Engineers.

> The American Institute of Steel Construction (AISC) and the Structural Welding Code of the American Welding Society (AWS) exempt from tests and qualification most of the common welded joints used in steel construction. These specific common welded joints are referred to as **prequalified.** Work on prequalified welded joints must be in accordance with the Structural Welding Code. Generally, fillet welds are considered prequalified if they conform to the requirements of the AISC and the AWS code.

FABRICATION METHODS

There are an unlimited number of fabrication methods that can be engineered for use on steel construction projects. Many techniques are typical, whereas others require special design. A combination of connection methods can be used depending on the function and location of the connection. For example, it is typical to have bolted and welded processes used on a single connection. The welded connection can be used in applications where a bracket, plate, or other feature can be made in the fabrication shop and then sent to the field, where bolted connections are made at the same location. The holes required for the bolted connections are also fabricated in the shop before the parts are delivered to the field. Both of these processes make quick field assembly.

The following examples provide typical connection applications for different structural applications. Figure 6.37 shows typical basic column base connections, where you can see a combination of bolted and welded connections. Figure 6.38 shows typical beam-to-column and beam-to-beam connections, where you can see a combination of bolted and welded connections.

Figure 6.39 shows typical beam-to-column and end plate-to-beam connections, where you can see a combination of bolted and welded connections. Figure 6.40 shows common horizontal beam-to-horizontal beam connections. Columns and other steel construction members often need to be spliced together to make them longer or for other applications, as shown in Figure 6.41. Many steel structures require bracing for engineering conditions or

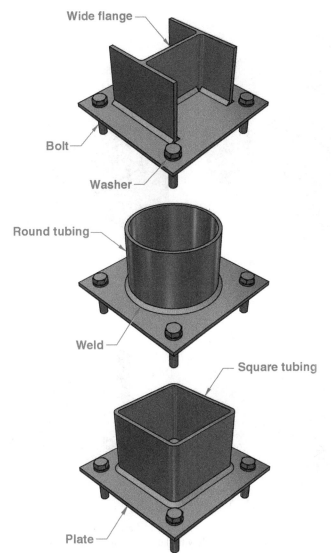

Wide flange

Bolt

Washer

Round tubing

Weld

Square tubing

Plate

Figure 6.37 Typical basic column base connections.

for the type of structure such as bridges and towers. Typical steel structural bracing examples are shown in Figure 6.42.

> The AISC *Manual of Steel Construction* provides a number of common connection details.

LIGHTWEIGHT STEEL-FRAME CONSTRUCTION

Steel framing is also called **cold-formed steel (CFS)** framing. CFS is primarily used for framing light commercial buildings such as office and medical buildings. CFS is the common term used for cold processes of rolling, pressing, and stamping thin **gauge** sheet steel into products that are similar in sizes to traditional dimensional lumber, such as 2 × 4, 2 × 6,

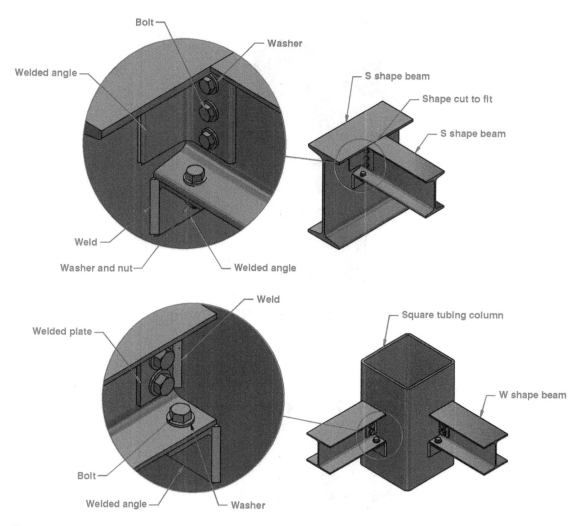

Figure 6.38 Typical beam-to-column and beam-to-beam connections, where you can see a combination of bolted and welded connections. Bolted connections generally use a bolt with flat washer at the bolt head and a flat washer and nut at the threaded end.

2×8, 2×10, and 2×12. The CFS process is done without heat, making the products stronger than the original sheet steel. Gauge is a range of numbers from 10 to 26 specifying thickness for sheet metal and wire. The sheet metal gets thinner as the gauge number gets larger. For example, 10 gauge is 0.1345 in. (3.416 mm), and 26 gauge is 0.0179 in. (0.4547 mm) thick. Sheet metal gauge values vary with the manufacturer. Sheet steel thickness is measured in gauge and **mils,** where 1 mil equals 1/1000 of an inch. Sheet steel thickness is the thickness of the **base metal,** which excludes protective coatings. CFS products used for construction generally have protective coatings such as **galvanize** to help prevent corrosion. Galvanize is a zinc coating. Lightweight prefabricated steel construction materials are used in many types of commercial structures and commonly used to build small commercial buildings. Prefabricated steel construction materials offer lightweight, noncombustible, corrosion-resistant framing for interior partitions and load-bearing exterior walls as high as four stories. Steel members are available for use as studs, joists, rafters, and other construction members.

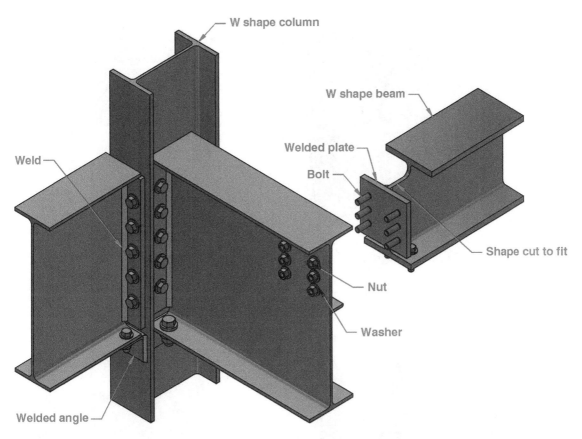

Figure 6.39 Typical beam-to-column and end plate-to-beam connections, where you can see a combination of bolted and welded connections. Bolted connections generally use a bolt with flat washer at the bolt head and a flat washer and nut at the threaded end.

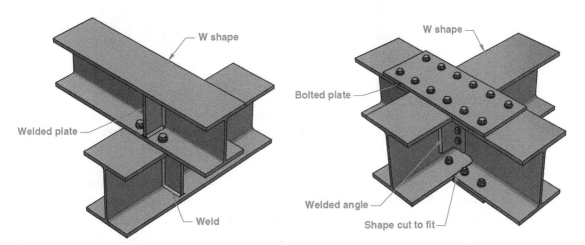

Figure 6.40 Common horizontal beam-to-horizontal beam connections. Bolted connections generally use a bolt with flat washer at the bolt head and a flat washer and nut at the threaded end.

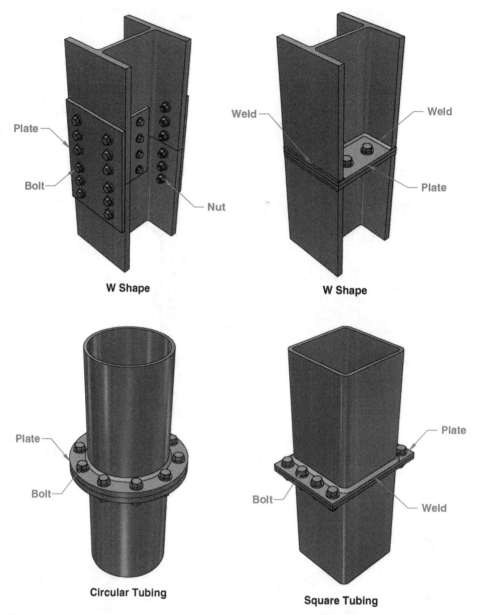

Figure 6.41   Columns and other steel construction members often need to be spliced together to make them longer as shown here. Bolted connections generally use a bolt with flat washer at the bolt head and a flat washer and nut at the threaded end.

Members are designed for rapid assembly and are predrilled for electrical and plumbing conduits. Standard 24 in. (600 mm) spacing of steel members reduces the number of studs required by about one-third when compared with 16 in. (400 mm) OC wood-frame construction spacing. Steel member widths range from $3\frac{5}{8}''$ to $10''$ (90 to 250 mm), but can be manufactured in any width. The material used to make members ranges from 12- to 20-gauge thick steel, depending on the design loads to be supported.

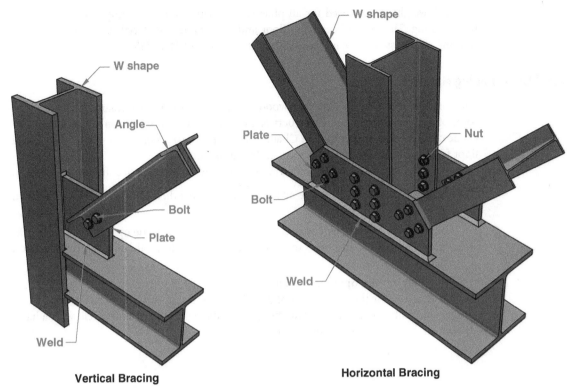

Figure 6.42 Typical steel structural bracing examples. Bolted connections generally use a bolt with flat washer at the bolt head and a flat washer and nut at the threaded end.

Wood framing materials and practices, described in Chapter 8, *Wood Construction*, can be used in light commercial construction, but steel construction has advantages over wood construction. Wood construction has the advantage of familiarity and popularity with builders, because of its historic use in construction. Steel construction requires different practices, skills, and tools than those used in wood construction. Builders who gain experience with steel construction can take advantage the following benefits of steel framing:

- Wood construction material prices change dramatically with the economy and timber markets, while steel prices remain fairly constant.
- Steel products have excellent dimensional tolerances, straightness, and stability without shrinkage, while wood can shrink depending on moisture content and wood quality is not as consistent as steel.
- Steel construction members are lightweight and precut to desired lengths, making easy assembly.
- Steel construction members have preformed holes for easy installation of plumbing, electrical, and mechanical lines.
- Steel framing is noncombustible.
- Steel framing is completely resistant to rot, and wood-destroying insects.

Lightweight steel is used for all phases, of construction, including floors, walls, and roof framing. This section describes floor and wall framing materials and practices. Roof framing is discussed in Chapter 7, *Roof Construction and Materials*.

Steel Floor Framing Materials

The construction members used in wood and steel framing are similar in function, but some of the terminology is different between wood and steel construction members. Wood-frame construction is discussed in Chapter 8, *Wood Construction*. Figure 6.43 shows steel floor framing members used over a concrete foundation wall.

Figure 6.44 demonstrates the following terminology used in light steel-frame construction:

Flange: The part of a C shape or track that is perpendicular to the web, as shown in Figure 6.44.

Lip: The part of a C shape that extends from the flange at the open end, as shown in Figure 6.44. The lip increases the strength characteristics of the member and acts as a stiffener to the flange.

Punchout or **web opening:** A hole or opening in the web of a steel-framing member allowing for the installation of plumbing, electrical, and other utility installation. A punchout can be made during the manufacturing processor in the field with a hand punch, hole saw, or other suitable tool.

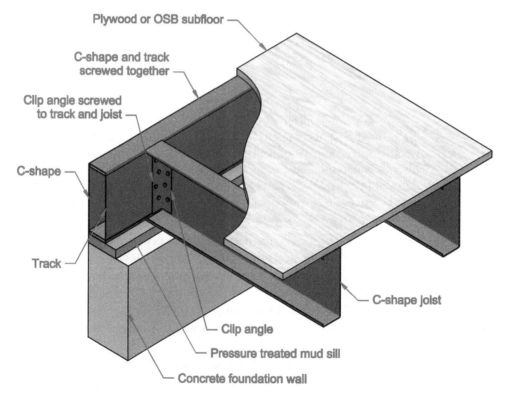

Figure 6.43   Steel floor framing members used over a concrete foundation wall. (*Previously published in Modern Residential Construction Practices, Routledge, 2017 and reproduced here with permission.*)

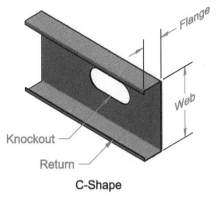

C-Shape

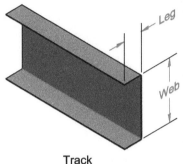

Track

Figure 6.44 The following terminology is used in light steel-frame construction. (*Previously published in Modern Residential Construction Practices, Routledge, 2017 and reproduced here with permission.*)

Track: Used for applications, such as rim joists, for flooring systems. A track has a web and two flanges, but no lips. Track web depth measurements are taken to the inside of the flanges. Rim joist, also called a rim board, band joist, or header, is attached perpendicular to the joists and provides lateral support for the ends of the joists while capping off the end of the floor or deck system.

Web: The part of a C shape or track that connects the two flanges.

Web Stiffener: Additional material attached to the web to strengthen the members.

Steel Wall Framing Materials and Methods

Steel framing can reduce energy performance efficiency because of heat loss through the metal framing. Even though wood has poor insulation value, steel has even less. For this reason, practices should be used to provide insulation between the steel framing and the exterior when using steel construction.

Steel Wall Framing Materials

Steel wall framing materials include **studs, cripples, jambs, bottom track, top track, headers,** and **sheet metal screws** for fastening members together. The horizontal bottom track and top track fit tightly over the vertical studs and are fastened together using sheet metal screws, as shown in Figure 6.45. Sheet metal screws have deep spiral threads along the entire body length and have a pointed end for easy start when threading, as shown in Figure 6.46. Sheet metal screws can have a variety of head types, but the most common are the hex slotted head and the truss head for use when the screw head needs to have a low profile.

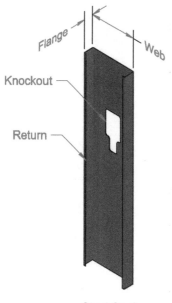

Steel Stud

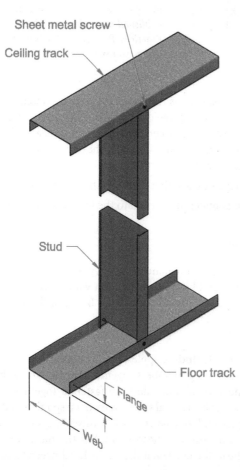

Figure 6.45 The horizontal bottom track and top track fit tightly over the vertical studs and are fastened together using sheet metal screws. (*Previously published in Modern Residential Construction Practices, Routledge, 2017 and reproduced here with permission.*)

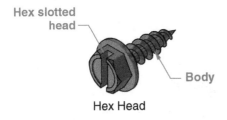

Hex slotted head

Body

Hex Head

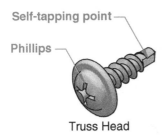

Self-tapping point

Phillips

Truss Head

Figure 6.46 Sheet metal screws have deep spiral threads along the entire body length, and have a pointed end for easy start when threading. (*Previously published in Modern Residential Construction Practices, Routledge, 2017 and reproduced here with permission.*)

Sheet metal screws can be installed in a **pilot hole** or directly into the metal when they have a **self-taping** end. A pilot hole is a small hole drilled through the metal with a diameter designed to accommodate a specific screw size. Self-tapping is the ability of a screw to creating its own thread without the need of a pilot hole. Studs are vertical framing members used to construct walls and partitions, and are usually spaced 16 or 24 in. on center. A cripple is a short stud framed above a door rough opening, or above and below a window rough opening between the header and a top track, or between the sill and bottom track. A jamb is a vertical framing member that forms the sides of a door opening, window opening, or other opening. The jam used in steel wall framing replaces the king stud used in wood wall framing. The steel wall framing jamb is normally made by placing two steel studs with the returns touching each other. Header is a horizontal structural member that supports the load over an opening such as a door or window, or around an opening.

Steel Wall Framing Methods

Steel wall framing layout is similar to wood framing, except some of the terminology is different between steel and wood framing. The tools, techniques, and skills used in steel work is also different from those used in wood framing. Figure 6.47 shows the typical layout used to frame a door and window opening in a steel-frame wall.

Steel Wall Frame Bridging. **Continuous steel wall frame bridging** is used in bearing walls to provide lateral bracing called **bridging**. Continuous bridging between studs in a load bearing wall is important to protect the structure against **buckling** and **rotation**. Buckling is bulging, bending, bowing, or kinking of the steel studs as a result of compression stress on the structure. Rotation is an action of studs rotating around their axis caused by **tension** applied to the wall system by wind or seismic activity. Tension is caused by stretching. The bridging must be continuous between anchorage points.

A continuous steel wall frame bridging application, called a **tension system,** is designed to resist the stud buckling from tension forces. This framing system uses a tightly installed flat strap attached to both flanges of the studs with blocking at intervals within the wall to provide resistance to rotation of the studs. The flat strap is typically attached to each stud flange within 12 in. (300 mm) below the top track. Blocking is placed to resist rotation and

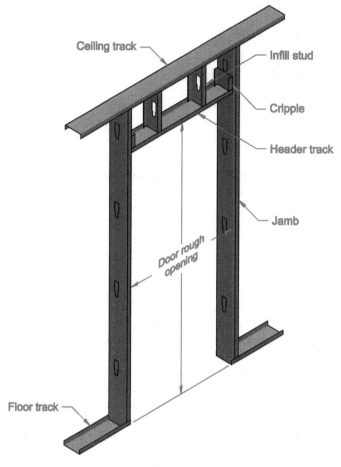

Door Rough Opening

Figure 6.47 Typical layout used to frame a door and window opening in a steel frame wall. (*Previously published in Modern Residential Construction Practices, Routledge, 2017 and reproduced here with permission.*)

is attached to the flat strap on each side of the wall, as shown in Figure 6.48. The blocking is located at intervals determined by structural engineering.

A bridging system designed to resist buckling of the stud in tension and compression is the **cold rolled channel system** that uses a continuous cold rolled steel channel as bridging that runs continuously through aligned **knockouts** in the studs. The cold rolled steel channel is attached to the studs using premanufactured clips that attach the bridging to the stud, as shown in Figure 6.49. Knockouts are prepunched holes at regular intervals to allow rapid installation of electrical conduit, mechanical, piping, and structural applications. The term **cold rolled** refers to a steel forming process when the cold metal is rolled into sheets or other shapes such as steel framing members.

> **NOTE**
>
> Specifications for bridging member sizes, spacing locations, and fastener quantity and sizes are determined by structural engineering and provided on the plans or in specifications.

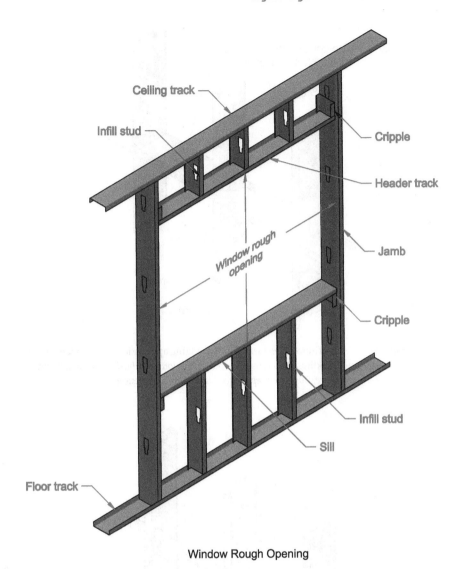

Window Rough Opening

Figure 6.47 (*Continued*)

Steel Wall Frame Deflection Track. Vertical movement can occur in a wall as a result of seismic activity, foundation settling, and pressure caused from movement in the structure above. This vertical movement is referred to as **deflection.** Deflection is the movement of a structure as a result of stress applied to the assembly. Possible deflection is considered when framing steel non-load-bearing partitions. This is done by allowing the top track to move freely as deflection occurs in the structure. A variety of deflection products are designed and manufactured to allow for vertical movement while maintaining wall quality and fire protection. The top track is called a **deflection track** when used in this application. A deflection track is designed and installed to allow vertical movement of the structure without damage to the wall system and wall finish material. One application uses a deep flange track that has 2 in. (50 mm) minimum high flanges. The steel studs are cut $\frac{1}{2}$ to 1 in. (25 to 50 mm) shorter than full length. The top track is placed tightly over the studs and

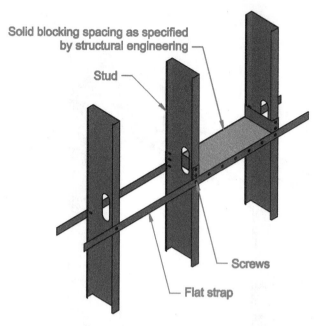

Figure 6.48   A continuous steel wall frame bridging application, called a tension system, is designed to resist the stud buckling from tension forces. (*Previously published in Modern Residential Construction Practices, Routledge, 2017 and reproduced here with permission.*)

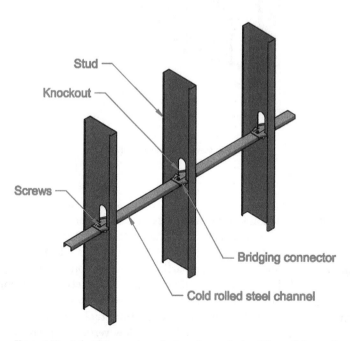

Figure 6.49   A bridging system designed to resist buckling of the stud in tension and compression is the cold rolled channel system that uses a continuous cold rolled steel channel as bridging that runs continuously through aligned knockouts in the studs. (*Previously published in Modern Residential Construction Practices, Routledge, 2017 and reproduced here with permission.*)

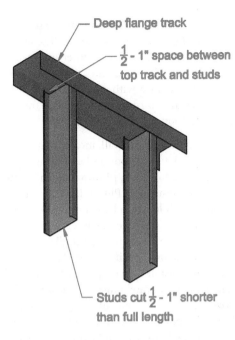

Figure 6.50 A deflection track is designed to allow vertical movement of the structure without damage to the wall system and wall finish material. One application uses a deep flange track. The steel studs are cut shorter than full length. The top track is placed tightly over the studs and held in place by friction. (*Previously published in Modern Residential Construction Practices, Routledge, 2017 and reproduced here with permission.*)

held in place by friction, as shown in Figure 6.50. When using any of the methods described here, the **drywall** is not fastened to the top track, allowing the track to move freely with vertical movement without damaging the studs or drywall. Drywall, also called gypsum board, gyprock, plasterboard, sheetrock, or wallboard, is a plaster panel made of **gypsum** pressed between two thick sheets of paper. Gypsum is a soft white or gray mineral consisting of hydrated calcium sulfate and is used to make plaster, which is mixed with water and allowed to harden for various applications such as making drywall. Another option, shown in Figure 6.51, is a slotted deflection track that provides positive attachment between the top track and stud flanges to prevent the transfer of forces into the drywall. The steel studs are cut $\frac{1}{2}$ to 1 in. (25 to 50 mm) shorter than full length. In this application, screws are fastened through a slot in the top track flanges and loosened after tightening, which allows the

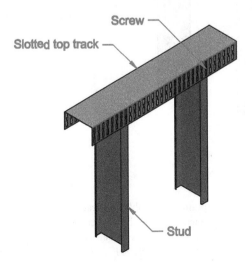

Figure 6.51 This deflection track option uses a slotted deflection track that provides positive attachment between the top track and stud flanges to prevent the transfer of forces into the drywall. (*Previously published in Modern Residential Construction Practices, Routledge, 2017 and reproduced here with permission.*)

top track to move vertically as needed. Also available are bushings placed in the top track flange slots allowing the screws to be tightened into the studs. The bushings allow freedom of movement between the top track and the studs. Another option is a double top track system where one top track is attached to the studs and another top track fits over the first track and is attached to the ceiling structure with a gap between the two tracks, which allows for vertical movement, as shown in Figure 6.52. The separate tracks are used to connect to the top of the wall and to the next floor and then held together by friction.

Steel Wall Framing. When framing a steel-frame wall, measure and cut the bottom and top tracks to length. Screw the bottom track to the floor and screw the top track to the ceiling, using a laser level to perfectly align the top track **plumb** over the bottom track. Mark all stud and jamb locations along the bottom track. Plumb is the term used to describe true vertical. Plumb and install a stud at each location by clamping the stud and tracks tightly with C-clamps or locking pliers. Drill a pilot hole for each screw and then install each screw or use self-taping screws. Use at least one screw in the middle of each assembly location unless otherwise specified by the plans and specifications. Measure and cut headers and cripples for rough door openings and install the components with brackets and screws as needed. Make sure all components are accurately installed, **level,** and plumb. Level is the term used to describe true horizontal. Measure and cut headers, sills, and cripples for rough window openings and install the components with brackets and screws as needed. Install any required bridging and deflection track using manufacturer instructions and engineering specifications.

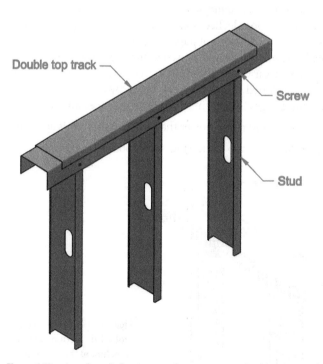

Figure 6.52   Another deflection track option is a double-top track system where one-top track is attached to the studs and another top track fits over the first track and is attached to the ceiling structure with a gap between the two tracks, which allows for vertical movement. (*Previously published in Modern Residential Construction Practices, Routledge, 2017 and reproduced here with permission.*)

COMMERCIAL CONSTRUCTION STEEL CONSTRUCTION APPLICATION

This commercial construction application is provided by the Steel Framing Industry Association. The project is SOVA ON GRANT, a 12-story, 211-unit luxury apartment home building in Denver, Colorado. The following firms participated in the project:

Developer: McWhinney, Denver, CO
Project Architect: Craine Architecture, Denver, CO
General Contractor: Weitz Co., Denver, CO
Exterior Cold-Formed Steel Panelizer: South Valley Prefab, Littleton, CO
Interior Cold-Formed Steel Panelizer: Infinity Structures, Alpharetta, GA

The material used in the project is 43 to 97 mil cold-formed steel framing that was described earlier in this chapter.

Cold-Formed Steel-Framed, Load-Bearing Exterior Finish Panels Saves Developer Time

SOVA is a 12-story luxury apartment building at 19th Ave. and Grant Street in Denver, shown in Figure 6.53. The structure features cold-formed steel-framed, load-bearing exterior finish panels and cold-formed steel-framed, load-bearing interior **panels.** Use of the word panels refers to **panelization,** or assembling the components of the building into panels, including walls, floors, and roofs. The panels are manufactured in a fabrication facility with a controlled environment to precise specifications for easy assembly at the project site. Steel framing is especially suitable for panelization because it is able to be manufactured to accurately meet specifications, and its light weight allows for easier handling of assembled components. Panels can be shipped without sheathed and the sheathing fastened to the panels at the job site, or the panels can be completely assembled, including sheathing and finish at the manufacturing facility as demonstrated in this discussion.

Figure 6.53   SOVA is a 12-story luxury apartment building at 19th Ave. and Grant Street in Denver, CO. (©2019 *Steel Framing Industry Association*.)

Excavation and slab installation began in April 2017. Construction of the 211 apartments was completed in February 2019. By using CFS-framed load-bearing exterior finish panels, the general contractor began the interior work at the same time the exterior panels were being erected. The developer, McWhinney of Denver, opened the property months ahead of schedule.

Seeing Opportunity

Saving money on construction of this project would not be easy at SOVA. The site had a **zero-lot line.** Zero-lot-line refers to real estate property where the structure is allowed, by zoning, to be on or very near to the **property line.** Property line is the legal boundary line of a piece of land. Each trade, on the SOVA project, had almost no **lay-down space** during work because of the zero-lot line situation. In addition, the trades scrambled to find skilled workers with craft labor supplies limited in Denver. A **laydown space,** also called **laydown area,** is an area used for the receipt, temporary storage, construction equipment use, and for assembly of building components. The laydown area can be on or near the construction site and can the surface covered with gravel or temporary pavement to help keep mud and erosion under control.

The design and construction team used these challenges as an opportunity to put prefabrication to the test. South Valley Prefab, Littleton, CO, prefabricated 483 exterior CFS-framed load-bearing panels for the project. The panel's specialty finishes were applied in the factory.

"You don't often see finishes included on cold-formed steel load-bearing panels," says Travis Vap, president of South Valley Prefab. "We are one of five companies that I know of in the United States that can consistently do this type of work."

Finishing prefab panels in the shop gives building owner and general contractor enormous advantages.

Savings Passed On to Client

The CFS-framed panels featured two exterior finishes—a thin brick finish and a synthetic finish—custom-framed window openings and properly placed vents.

South Valley Prefab's in-house engineer and design team worked closely with SOVA's design team to work out the panel designs, tested the panels and settled on a final engineered product.

"One challenge was how to include the finishes for the edge of slab between floors," Vap says. "In the past, panelizers would leave this off and install it later in a traditional manner. Our panels were 100 percent complete as we erected, so we did not have to come back for any exterior work."

Vap's company incorporated multiple trades into one exterior panel system. "We passed on the schedule savings to our client." he says.

Months Saved

"Most subcontractors don't apply exterior finishes to prefabricated CFS-framed load-bearing panels—at least not prior to their installation—because it's hard to do. It's a specialized way of working," Vap says. "The advantages are it saves both costs and time."

"You eliminate the exterior scaffolding—the time to erect it, the manpower to erect it and the scaffold safety hazards," Vap says. "You also free up the other trades to work."

In traditional **podium construction,** work below the podium lags behind the work above it by about 4 months with scaffolding in the way. By adding the exterior finishes during the panel fabrication process, Vap says SOVA's, "developer saved months in the construction process." This was evident in downtown Denver. When SOVA was under construction, three other major projects were also underway in the same area of town. Even though the other projects broke ground before SOVA, Vap says, "SOVA was completed three to six months ahead of the others." **Podium construction,** also called a **podium building,** is the construction of multiple stories of **wood-frame construction** over one or more levels of concrete structure. Podium

Figure 6.54   Fabricating and installing finished CFS-framed panels. (*©2019 Steel Framing Industry Association.*)

construction uses two types of material connected together, generally, a concrete structural base is often designed with parking garages below or above grade and retail space above grade followed by multiple stories of wood-frame construction for use as apartments or condominiums. Wood-frame construction is the use of wood to build all exterior and interior walls, floor, and roof constructions. Wood construction is described in Chapter 8, *Wood Construction*.

"If you look at the time value of money for a developer and getting early access to his building and early rents, we provided a tremendous dollar value in savings for them," Vap says. "You can't get that through normal construction, or even normal prefab."

Prefabrication adds speed to a construction process. And bringing the exterior finish application to the factory saves even more speed, reduces additional underlying costs, and significantly improves overall quality. **Figure 6.54** shows the building under construction.

"This is very impressive, and it's exciting for construction," Vap says. "We love building, working with great people and delivering best in class quality products. Using cold-formed steel, load-bearing exterior finish panels on multi-family and mid-rise construction is going to make a lot of people happy and become a consistent construction method. We provided a tremendous dollar value in savings for the developer."

Go to Downloads & Resources tab at **www.mhprofessional.com/Commercial Building Construction-Ancillaries** to access the images provided in this chapter and correlate with the content of this chapter. Here you can look at the chapter figures on your computer screen to pan and zoom in and out for better observation, because full drawings reduced to fit on a textbook page are often difficult to read. The image bank also includes the complete set of plans for the Brookings South Main Fire Station used as examples throughout this textbook.

Go to **Downloads & Resources tab at www.mhprofessional.com/Commercial Building Construction-Ancillaries** to access the test and creative thinking problems for this chapter. The chapter test can be used for review or to evaluate content knowledge depending on your course objectives. Creative thinking problems are provided to help expand your knowledge by researching given subjects.

Print reading problems have questions that ask you to find information on the Brookings South Main Fire Station plans. The print reading problems can help you become familiar with the format of construction documents and reinforce your ability to seek specific information. This is an important skill for you to master as you prepare to enter the construction industry. Go to **Downloads & Resources tab at www.mhprofessional.com/Commercial Building Construction-Ancillaries** to access the complete set of plans for the Brookings South Main Fire Station.

CHAPTER 7

Roof Construction and Materials

INTRODUCTION

This chapter introduces you to basic commercial roof frame materials and practices and roofing materials used in commercial construction. There is also a review of typical roof-related drawings such as roof plan, roof framing plan, and details. The roof drawings are given to introduce you to roof terminology and definitions.

ROOF PLANS AND DETAILS

A **roof plan** is a scaled drawing of proposed **roof** construction containing dimensions of the entire roof structure, including shape, size, design and placement of all materials, ventilation, drainage, slopes, valleys, and other features. The term **roof** refers to the supporting structure and exterior surface on top of a building. The roof plan for a small commercial building is shown in Figure 7.1.

Look at the following features found on the roof plan in Figure 7.1 as you study how they are identified on the plan and described in the following:

SLOPED STRUCTURE, SEE STRUCTURAL DRAWINGS. This refers to the roof structural members that are shown in a section view provided in Figure 7.2. The arrow and 1/4":12" next to the slope structure note identifies the roof slope as 1/4" per 12". The term **slope**, also called **pitch**, refers the amount of rise a roof has compared to a horizontal measurement called the run. This roof slope has 1/4" rise for each 12" of run.

CRICKET. There are several crickets on this roof plan. A **cricket** is generally a small sloped surface or roof built to divert water over an area where water would otherwise collect, such as behind a chimney or a place where a slope roof meets a vertical wall.

CURB FLASHING DETAIL. A **curb** can have one of several different purposes, but is basically a raised member used to support roof penetrations above the level of the roof surface, or a raised roof perimeter or projection relatively low in height. One of the curb details found in this set of plans is shown in Figure 7.3. A primary purpose for this detail is to identify and show the materials used to build the curb, including

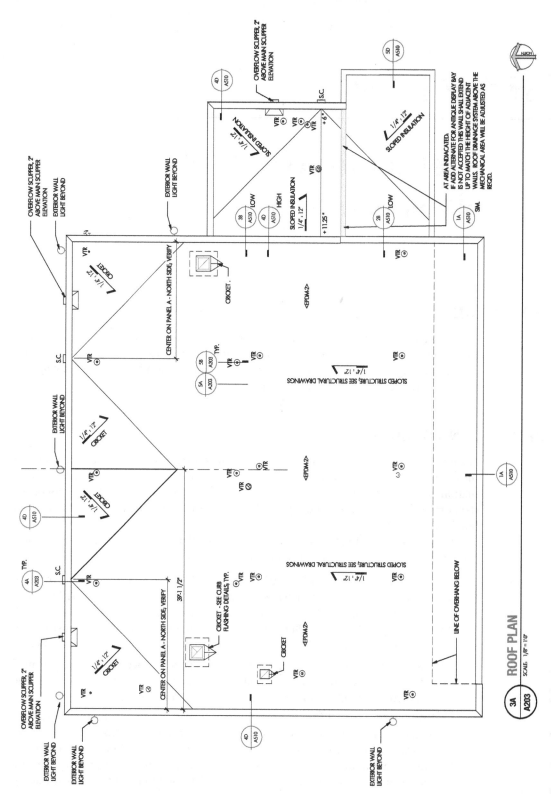

Figure 7.1 The roof plan for a small commercial building. (*Courtesy of JLG Architects, Alexandria, MN.*)

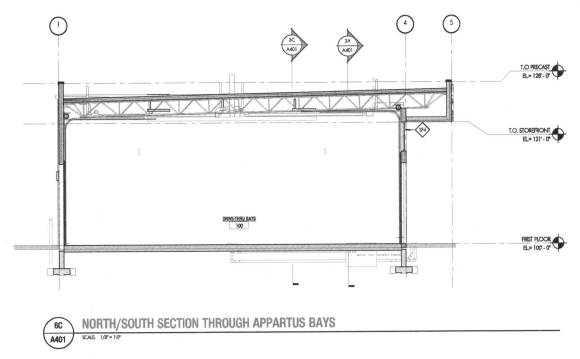

6C **A401** NORTH/SOUTH SECTION THROUGH APPARTUS BAYS

SCALE 1/8" = 1'-0"

Figure 7.2　Roof structural members are shown in this section view. (*Courtesy of JLG Architects, Alexandria, MN.*)

flashing and roof materials. **Flashing** between the concrete or concrete block ledge and the masonry veneer acts to collect and divert moisture from the wall. Flashing is components used to weatherproof or seal the roof system edges at perimeters, penetrations, walls, expansion joints, valleys, drains, and other places where the roof covering is interrupted or terminated. Types of flashing include **base flashing** that is metal or composition flashing at the joint between a roofing surface and a vertical surface, such as a wall or parapet. **Cap flashing** is used to cover the top of various building components. **Counter flashing** is a strip of sheet metal in the form of an inverted L built into a vertical wall of masonry and bent down over the flashing to make it watertight.

VTR. VTR is the abbreviation for vent through roof. A **vent** is an opening designed to convey air, heat, water vapor, or other gas from inside a building or a building component to the atmosphere. Special construction applications are used to seal the vents through the roof to make sure there are no weather-related leaks. Most commercial plans provide specific construction details, as shown in the Figure 7.4 example.

SCUPPER. A **scupper** is an outlet in the side of a building or from a roof for draining water. A scupper can be connected to a **gutter** and **downspout**. Gutter is a channel at the edge of the eave for moving rainwater from the roof to the downspout. Downspout is a vertical pipe that is connected to the gutters for the purpose of moving rainwater from the gutter to the ground or to a rain drain pipe in the ground. The roof of this building has a low slope and scuppers are placed in specific locations to drain water off the roof through the **parapet**. Look at the roof plan in Figure 7.1 and find specific notes related to scupper locations and find the SC abbreviation that also locates scuppers.

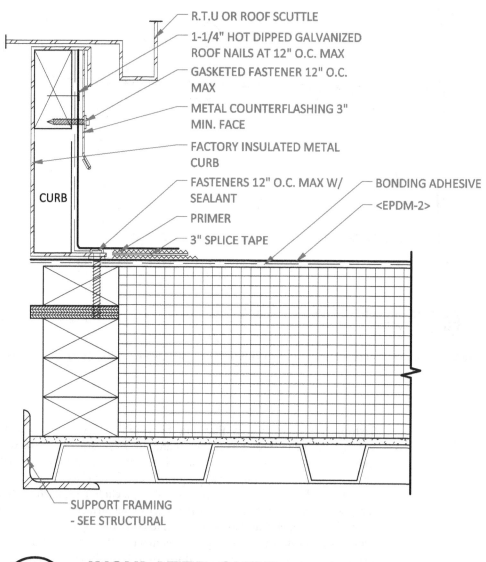

R.T.U OR ROOF SCUTTLE

1-1/4" HOT DIPPED GALVANIZED
ROOF NAILS AT 12" O.C. MAX

GASKETED FASTENER 12" O.C.
MAX

METAL COUNTERFLASHING 3"
MIN. FACE

FACTORY INSULATED METAL
CURB

FASTENERS 12" O.C. MAX W/
SEALANT

PRIMER

3" SPLICE TAPE

BONDING ADHESIVE

<EPDM-2>

CURB

SUPPORT FRAMING
- SEE STRUCTURAL

4B

A203

INSULATED CURB

SCALE: 3" = 1'-0"

Figure 7.3   A curb details found in this set of plans is shown here. (*Courtesy of JLG Architects, Alexandria, MN.*)

A scupper detail is shown in Figure 7.5. A parapet is a protective wall along the edge of a roof, or other structure. The parapet for the roof in Figure 7.1 is show with two parallel lines around the perimeter of the roof.

SLOPED INSULATION. Sloped insulation is applied to this roof in specific corners and other locations to drain and insulate the low-slope roof. Sloped roof insulation is made using lightweight molded **polystyrene** insulation with high compressive

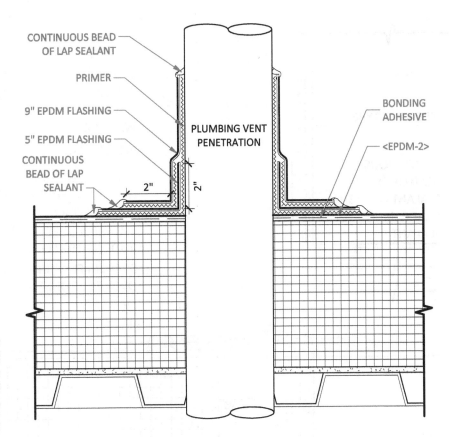

CONTINUOUS BEAD
OF LAP SEALANT

PRIMER

9" EPDM FLASHING

5" EPDM FLASHING

CONTINUOUS
BEAD OF LAP
SEALANT

2"

2"

PLUMBING VENT
PENETRATION

BONDING
ADHESIVE

<EPDM-2>

5B

A203

PLUMBING VENT W/ FIELD FABRICATED FLASHING

SCALE: 3" = 1'-0"

Figure 7.4   A VTR construction detail. (*Courtesy of JLG Architects, Alexandria, MN.*)

strength, good **R-value,** and moisture resistance. Polystyrene is a synthetic resin which is a polymer of styrene, used chiefly as lightweight rigid foams and films. R-value of a material is a measure of thermal resistance to heat flow.

EPDM. **Ethylene propylene diene terpolymer (EPDM)** is an extremely durable synthetic rubber roofing membrane widely used in low-slope buildings.

LINE OF OVERHANG BELOW. Roof plans often show construction features that are below the roof using **hidden lines.** Hidden lines are short equally spaced dashed lines that represent a feature that is not visible in the current view. Look for the hidden lines on the roof plan in Figure 7.1.

EXTERIOR WALL LIGHT BEYOND. The roof plane will often show features installed on or next to the roof, such as the exterior wall light identified with specific notes in Figure 7.1.

The **roof framing plan** for the Brookings South Main Fire Station is shown in Figure 7.6 using structural steel **bar joist** framing. A bar joist example is shown in Figure 7.7. The purpose

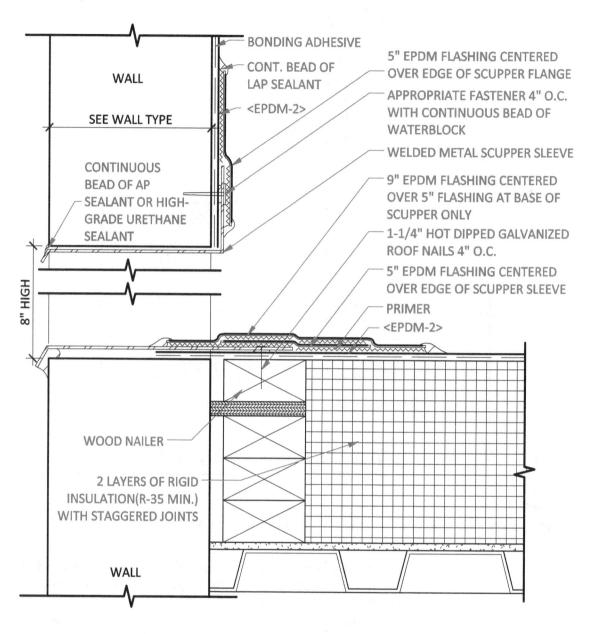

WALL

SEE WALL TYPE

CONTINUOUS BEAD OF AP SEALANT OR HIGH-GRADE URETHANE SEALANT

8" HIGH

WOOD NAILER

2 LAYERS OF RIGID INSULATION(R-35 MIN.) WITH STAGGERED JOINTS

WALL

BONDING ADHESIVE

CONT. BEAD OF LAP SEALANT

<EPDM-2>

5" EPDM FLASHING CENTERED OVER EDGE OF SCUPPER FLANGE

APPROPRIATE FASTENER 4" O.C. WITH CONTINUOUS BEAD OF WATERBLOCK

WELDED METAL SCUPPER SLEEVE

9" EPDM FLASHING CENTERED OVER 5" FLASHING AT BASE OF SCUPPER ONLY

1-1/4" HOT DIPPED GALVANIZED ROOF NAILS 4" O.C.

5" EPDM FLASHING CENTERED OVER EDGE OF SCUPPER SLEEVE

PRIMER

<EPDM-2>

4A
A203

THRU-WALL SCUPPER ROOF DECK

SCALE: 3" = 1'-0"

Figure 7.5 A scupper detail. (*Courtesy of JLG Architects, Alexandria, MN.*)

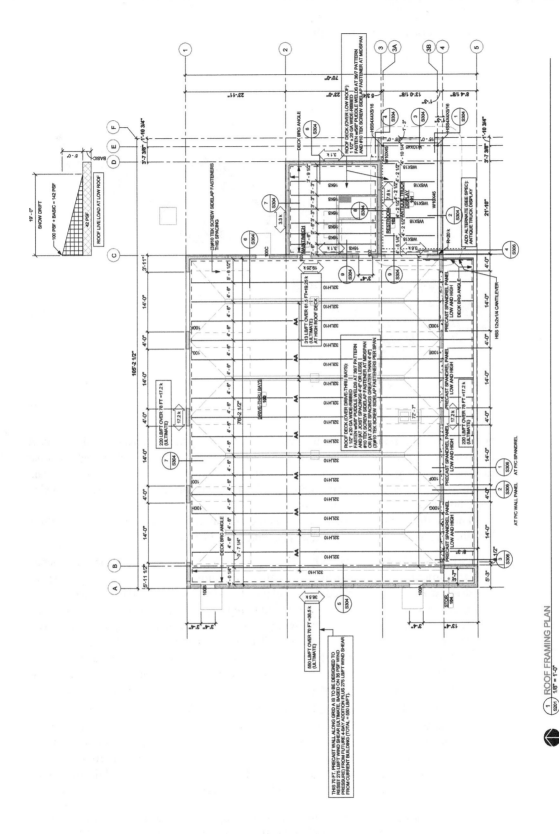

Figure 7.6 The roof framing plan for the Brookings South Main Fire Station. *(Courtesy of JLG Architects, Alexandria, MN.)*

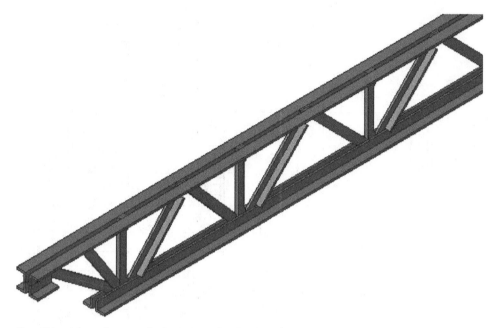

Figure 7.7   A bar joist example. (*Courtesy of Redbird Engineering.*)

of the roof framing plan is to show the major structural components in plan view that occur at the roof level. A bar joist is a welded steel joist with an open web made with a single bent bar running in a zigzag pattern between horizontal upper and lower **chords.** A chord is a main member of a bar joist extending parallel across the span, and with a bottom chord connected to a top chord by a web of structural members.

Look at the following features found on the roof faming plan in Figure 7.6 as you study how they are identified on the framing plan and described in here:

PRECAST SPANDREL PANEL. Precast refers to precast concrete that is concrete structure made by casting concrete with steel reinforcing in a form or mold. A **spandrel panel** is a preassembled structural panel used as a wall or roof panel that can have a variety of textures and colors, and used to replace the appearance of a masonry wall. The precast concrete panels on Figure 7.6 roof framing plan are represented by two parallel line around the perimeter and filled with the concrete symbol. Openings in the precast panels are shown with a perpendicular line segment at each end and two close parallel lines in between. The lines of dimensions around the outside perimeter of the roof framing plan provide the sizes of precast panels and opening locations and sizes.

32LH10. 32LH10 is next to the thick vertical lines that run across the main roof. The thick lines represent the bar joists and the 32LH10 is the bar joist specification. Bar joists are identified by a three-part code, such as <u>32</u> <u>LH</u> <u>10</u> used on this roof. In this example, the 32 is joist depth in inches, the LH represents the joist series, and the 10 specifies the top and bottom chord section number. The low roof on the right side of the fire station has bar joists specified as <u>16</u> <u>K</u> <u>6</u>. The following describes the different bar joist series:

- K series, used on the low roof, is a common open web joists that span from 8′ to 60′ and with depths of 8″ to 30″, and with chord section numbers are 1 to 12.

- CS series, where CS refers to constant shear joists used for nonuniform and concentrated loads. CS series have joist depths of from 10″ to 30″ and chord numbers from 1 to 5.
- LH series, used on the main roof, are long-span joists with spans from 25′ to 96′ and depths from 18″ to 48″, and chord sections from 2 to 17.
- DLH series are deep long-span joists that can span from 89′ to 144′ with depths from 52″ to 72″, and chord numbers from 10 to 19.
- SLH series are super long span with spans from 111′ to 240′ and depths from 80″ to 120″, and chord numbers 15 to 25.

W8X18 and W10X45. These are wide flange roof beams used on the roof structure for the antique truck display area. Wide flange steel members were described in Chapter 6, *Steel Construction*.

ROOF DECK (OVER DRIVE-THRU BAYS). The following describes the roof deck and roof deck fastening over the drive-thru bays:

1-1/2″ × 20 GA WIDE-RIBBED. There are a wide range of steel deck materials available from many manufacturers. You can search the Internet with terms like steel roof deck to find available patterns and dimensions. The roof deck used on the example building is 1-1/2″ high and 20 gauge thick wide-ribbed style similar to the example in Figure 7.8.

FASTEN w/ 5/8″ PUDDLE WELD AT 36/7 PATTERN AND AT JOIST SPACES 4′-6″ OR LESS. The deck is fastened with (w/) 5/8″ **puddle welds** on a 36/7 pattern. A puddle weld is a type of plug weld for joining two sheets of light-gauge material by burning a small hole in the upper sheet and then filling the hole with a puddle of weld metal to fuse the upper sheet to the lower sheet. A plug weld is made in a hole in one piece of metal that is lapped over another piece of metal. The **36/7 pattern** is a common deck material attachment pattern used to place deck fasteners every 36″ on center length and 7″ on center apart in a grid pattern that results in the fasteners being placed in the deck flat channel. Also confirm that welds are on the joist spaces that are 4′-6″ apart. Look for the 4′-6″ dimensions between joists in Figure 7.6.

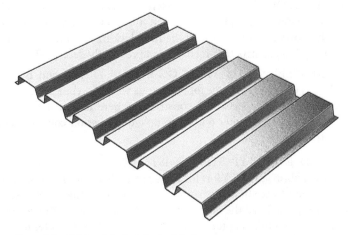

Figure 7.8   Wide-ribbed style steel deck.

#10 TEK SCREW SIDE LAP FASTENERS AT MID SPAN OR AT JOIST SPACES GREATER THAN 4'-6" and (3) #10 TEK SCREW SIDE LAP FASTENERS PER SPAN. #10 is the screw size and **Tek screw** is a self-tapping sheet metal screw that eliminates the need to predrill a pilot hole. The Tek screws are placed at deck mid span and joist spaces over 4'-6", plus 3 Tek screw side lap fasteners per span.

DECK BRG ANGLE. There is an angle around the perimeter at the roof to support the roof deck at that location. The abbreviation BRG stands for bearing support for the deck material. You can see the angle specifications on a roof detail that is described and shown in the next content.

AA. The AA refers to the dimension between the Floor Trench Drains to keep them consistent, which is a general contractor note.

HSS4X4X5/16. This is square structural tubing column installed in the lower right corner of the antique truck display bay.

Roof framing details are required to show the construction methods used at various member intersections in the building. Details can include the following intersections: wall to beam, beam to column, beam splices and connections, truss details, bottom chord bracing plan and details, purlin clips, cantilever locations, and roof drains. Roof-related details were described and shown in the previous content where related to the parts of the roof being discussed. The following provides additional roof details for the Brookings South Main Fire Station. Look at the roof framing plan in Figure 7.6 and find the detail tags that correlate with each of the following construction details. Read the information found on the roof framing plan and the same information on each detail.

Detail 1/S304. This is detail 1 on page S304 titled FRAMING AT SE CORNER OF ANTIQUE TRUCK DISPLAY, shown in Figure 7.9. Here you can see the wide flange beams described earlier and the connections used. Refer back to what you learned about steel construction connections in Chapter 6, *Steel Construction*. Find the HSS4×4×5/16 steel column described earlier.

Detail 2/S304. This is detail 2 on page S304 titled FRAMING AT SOUTH PERIMETER OF ANTIQUE TRUCK DISPLAY, as shown in Figure 7.10. This detail shows wide flange beam connections. The detail also shows the L3-1/2 × 3-1/2 × 1/4 CONT that is the abbreviation for continuous. This is the angle that was described earlier in the roof framing plan as DECK BRG ANGLE. Here, you can also see some welding symbols. Refer back to Chapter 6 and see if you can interpret the welding symbols. This detail also clearly shows a section through the roof deck material that was described earlier. The following is a construction member not previously described:

600S162-54 STEEL STUDS AT 16" OC w/ TOP TRACK and BOTTOM TRACK. 600S162-54 is the product name for steel studs with 6" web, 1-5/8" flange, 1/2" lip, and 16 gauge material. The steel stud construction here uses matching bottom and top tracks. Refer back to Chapter 6 for a complete discussion covering light weight steel stud construction.

Detail 7/S304 is the TYP STEEL JOIST BRG AT PRECAST PANEL, shown in Figure 7.11. The abbreviation TYP is for the word typical, and BRG is for bearing. This is a good detail for you to see how the bar joists connect with the precast panels. TYP is the abbreviation for typical. Read the notes found on this detail and correlate with information you learned throughout this chapter and in Chapter 6. Notice the notes features that are TO BE DESIGNED DETAILED AND SUPPLIED BY PRECAST SUPPLIER. This means that

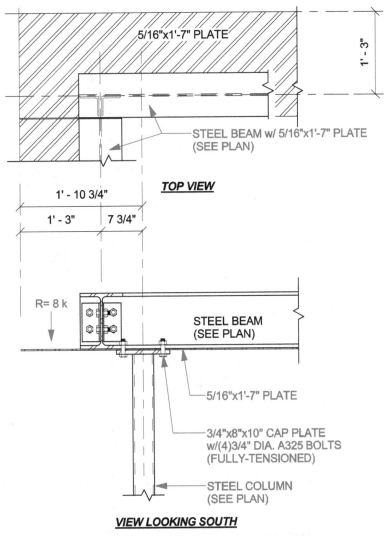

5/16"x1'-7" PLATE

1' - 3"

STEEL BEAM w/ 5/16"x1'-7" PLATE
(SEE PLAN)

TOP VIEW

1' - 10 3/4"

1' - 3"   7 3/4"

R= 8 k

STEEL BEAM
(SEE PLAN)

5/16"x1'-7" PLATE

3/4"x8"x10" CAP PLATE
w/(4)3/4" DIA. A325 BOLTS
(FULLY-TENSIONED)

STEEL COLUMN
(SEE PLAN)

VIEW LOOKING SOUTH

1 FRAMING AT SE CORNER OF ANTIQUE TRUCK DISPLAY
S304 NTS

Figure 7.9   Detail 1/S304. FRAMING AT SE CORNER OF ANTIQUE TRUCK DISPLAY. (*Courtesy of JLG Architects, Alexandria, MN.*)

the identified features are prepared and fabricated at the precast panel manufacturer and delivered to the construction site installed with the precast panels. This requires shop drawing to be prepared in cooperation with the architect for exact dimensions and specifications to meet the requirements.

Detail 2/S306 shows the CANTILEVERED JOIST BRG ON PRECAST WALL PANEL. Here, you can see how the bar joists connect with the precast wall panels and then **cantilever** past to create the designed roof overhang. Cantilever is a structure that is supported at one end and is self-supporting on the other end where it projects into space.

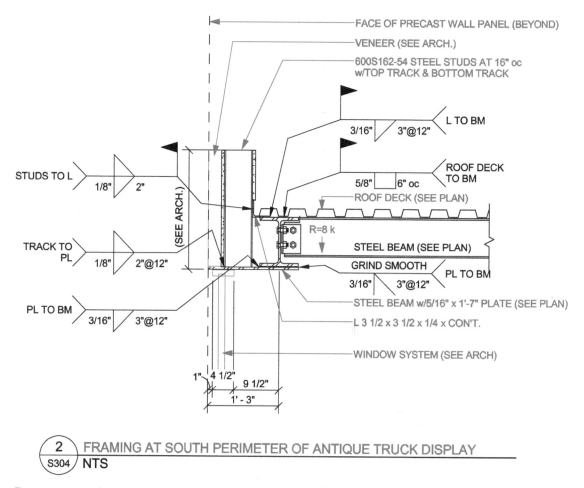

FACE OF PRECAST WALL PANEL (BEYOND)

VENEER (SEE ARCH.)

600S162-54 STEEL STUDS AT 16" oc
w/TOP TRACK & BOTTOM TRACK

L TO BM
3/16"   3"@12"

STUDS TO L
1/8"   2"

ROOF DECK
TO BM
5/8"   6" oc

ROOF DECK (SEE PLAN)

(SEE ARCH.)

R=8 k

TRACK TO
PL
1/8"   2"@12"

STEEL BEAM (SEE PLAN)

GRIND SMOOTH
PL TO BM
3/16"   3"@12"

STEEL BEAM w/5/16" x 1'-7" PLATE (SEE PLAN)

PL TO BM
3/16"   3"@12"

L 3 1/2 x 3 1/2 x 1/4 x CON'T.

WINDOW SYSTEM (SEE ARCH)

1"   4 1/2"
9 1/2"
1' - 3"

2 FRAMING AT SOUTH PERIMETER OF ANTIQUE TRUCK DISPLAY
S304 **NTS**

Figure 7.10 Detail 2/S304. FRAMING AT SOUTH PERIMETER OF ANTIQUE TRUCK DISPLAY. (*Courtesy of JLG Architects, Alexandria, MN.*)

Section C3/A401 is the EAST/WEST SECTION THROUGH APPARATUS BAYS & SUPPLY SPACE, shown in **Figure 7.12**. The most important information found on these typical cross sections is an overview of the full construction through the building. This full section, also called a building section, cuts through the bar joists showing their spacing. Compare the full section in Figure 7.12 with the full section in Figure 7.2. The full section in Figure 7.2 is cut through the building so you can see the bar joist construction. Limited specific information is provided, but there are section cutting planes and detail tags that direct you to other drawings where related information is found in more detail. The A, B, C, D, E, and F are in **balloons** above the section that correlate with the same balloons provided on plan views. These balloons are given to help related locations between drawings. A balloon is a circle placed on a drawing with an identification number or letter inside the circle, and the circle is connected to the view with a leader. The leader is a vertical line connected to a specific location in the section, in this example. The balloon number or letter correlates with the same balloon identifying the same location on other drawings. Look at the roof plan in Figure 7.1 and the roof framing plan in Figure 7.6 and find the balloons that correlate with the section in Figure 7.12.

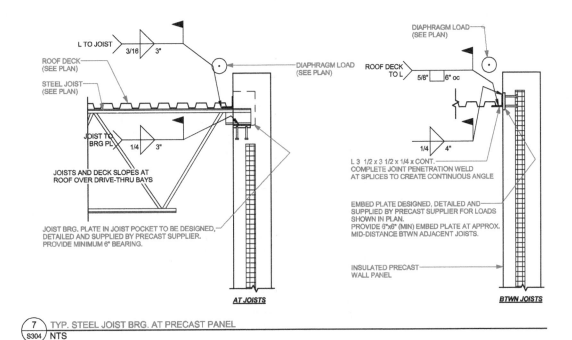

Figure 7.11 Detail 7/S304 is the TYP STEEL JOIST BRG AT PRECAST PANEL. TYP is abbreviation for typical and BRG is abbreviation for bearing. (*Courtesy of JLG Architects, Alexandria, MN.*)

Notice the **elevation symbol** shown on the section in Figure 7.12. The elevation symbol is commonly used on structural drawings to give the elevation of locations from a known zero elevation. The zero elevation might be at the first floor or other good reference point such as the top of a foundation wall. Elevation symbols are used together with elevation values at each elevation symbol location. Notice that the FIRST FLOOR elevation, abbreviated EL = 100′-00″ in Figure 7.12 and the elevation at the DECK BRG is EL = 126′-0″. You can calculate the dimension between elevations by subtracting the lower elevation from the upper elevation, such as 126′-0″ − 100′-0″ = 26′-0″. Elevation values can be given in feet and inches as shown here, in feet and tenths of a foot, or in millimeters.

Roof Drainage Plan

The **roof drainage plan** can be part of a set of structural drawings for some buildings, although it can be considered part of the plumbing or piping drawings, depending on the particular company use and interpretation. The purpose of the roof drainage plan is to show the elevations of the roof and provide for adequate water drainage required on low slope and flat roofs. There is no specific roof drainage plan for the Brookings South Main Fire Station, but roof drainage features and terminology were described earlier as related to the roof plan in Figure 7.2. Some of the terminology associated with roof drainage plans includes the following:

Roof drain (RD)—a screened opening to allow for drainage

Overflow drain (OD)—a backup in case the roof drains fail

Down spout (DS)—usually a vertical pipe used to transport water from the roof

Gutter or **scupper**—a water collector usually on the outside of a wall at the roof level to funnel water from the roof drains to the downspouts

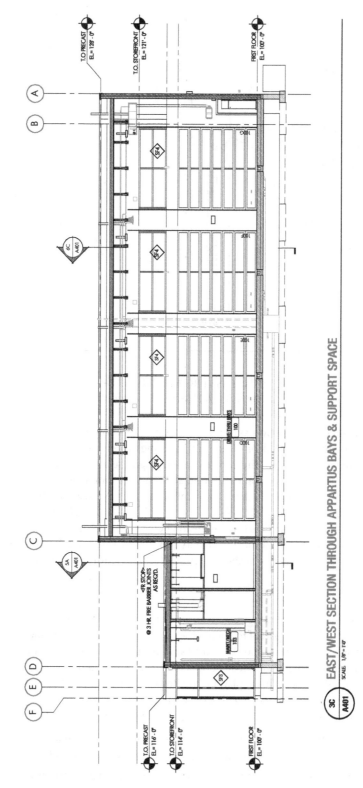

Figure 7.12 EAST/WEST SECTION THROUGH APPARATUS BAYS & SUPPLY SPACE. (*Courtesy of JLG Architects, Alexandria, MN.*)

Exterior Elevations

Exterior elevations, introduced in Chapter 1, *Plans, Specifications, and Construction Management*, are drawings that show the external appearance of the building. Elevations are briefly described here because of the roof-related information they can provide, the correlated balloons, and the related building height elevation dimensions. An elevation is drawn at each side of the building to show the relationship of the building to the final grade, the location of openings, wall heights, roof materials, roof slopes, exterior building materials, and other exterior features. Elevations can be labeled as FRONT, REAR, RIGHT, and LEFT ELEVATION or labeled by compass orientation, such as NORTH, SOUTH, EAST, and WEST ELEVATION.

> **NOTE**
>
> The content provided so far in this chapter is related to information found on the Brookings South Main Fire Station drawings. The examples and terminology provide typical information related to commercial building roof construction, but there are many variations found in the real world. This discussion has given you an understanding about the information found on construction drawings, how to read the information, and how to locate and define similar content for other buildings. This is a starting point for you to use the Internet and other resources to research different types of construction materials and practices.
>
> **General notes** relate to the entire plan and provide important information related to the construction materials and practices used in the design and building process. The general notes found in a set of commercial drawings are often found on the first sheet in each discipline area, but general notes can be found on any sheet related to the sheet content. For example, there is a group of general notes provided for the architectural, civil, structural, mechanical, and electrical disciplines. When working in the commercial construction industry on a specific project, you must take the time to read every general note and become familiar with all the plan views, sections, and details along with the materials, and practices related to these views. You should spend as much time as possible looking at the complete set of Brookings South Main Fire Station drawings found on the correlated website and find the plans for other construction projects to review. Go to Downloads & Resources tab at **www.mhprofessional.com/CommercialBuilding-Construction-Ancillaries** to access the images provided in this chapter and correlate with the content of this chapter along with the complete set of drawings for the fire station.

ROOF TRUSSES

A **truss** is a prefabricated or job-built construction member formed of triangular shapes used to support roof or floor loads over long spans. There are many different truss styles with a few of the most common shown in **Figure 7.13**. Steel parallel **chord** trusses are often used on commercial flat roof construction. One of the other styles with a sloped top chord, shown in Figure 7.13, is typically used on small commercial buildings that resemble residential architecture. A **chord** is a main member of a truss extending across the span. There is a bottom chord and a top chord. The top chord can be parallel to the bottom chord, or with a bottom chord connected to a top chord forming the roof slope. The top and bottom chords are connected by a web of members. Manufactured roof trusses are structurally engineered for each specific building design and fabricated using quality materials, providing uniform pitch and size. A complete building roof structure system is delivered to the project on a crane truck and each roof truss is placed directly on the framed walls, where they are quickly assembled to create the roof structure. Trusses are commonly made using

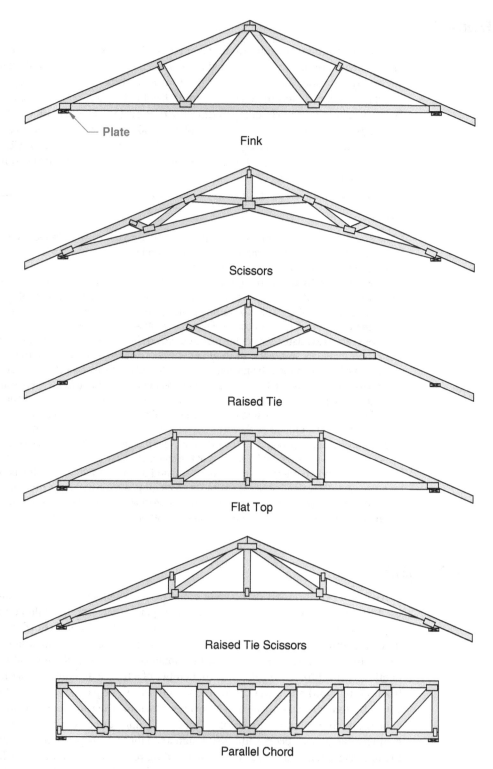

Fink

Scissors

Raised Tie

Flat Top

Raised Tie Scissors

Parallel Chord

Figure 7.13   Common roof truss styles. (*Previously published in Modern Residential Construction Practices, Routledge, 2017 and reproduced here with permission.*)

2 × 4 dimensional lumber or lightweight aluminum or steel members. Aluminum or steel truss plated act as gussets to connect the members at each joint. The use of manufactured trusses can save about 25 percent over other conventional framing methods. When using roof trusses, walls are framed first and ready for the roof system before trusses are delivered to the construction site. The trusses are delivered on a truck with a crane that lifts the trusses individually or in groups onto the framed walls, as shown in Figure 7.14. The construction

(a)

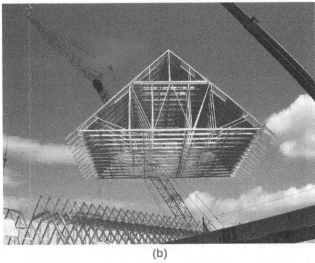

(b)

Figure 7.14 (a) Steel trusses being installed on a project. (*Courtesy of TrusSteel.*) (b) Steel truss installation using a rafting technique, which means by which the trusses are erected, braced, and sheathed on the ground then an entire section of truss framing is hoisted in place. (*Courtesy of TrusSteel | Westminster School Academic Center, Simsbury, Connecticut | Reliable Truss, New Bedford, MA.*)

worker then places each truss in its specific location, fastens the truss ends to the wall structure, and provides lateral bracing to keep the trusses in their desired locations before roof sheathing is applied for final stabilization.

Light steel roof trusses are available for light commercial construction and can be designed and manufactured for any roof style or complexity. Steel trusses are light weight and easy to handle, as shown in Figure 8.52. Figure 8.53 shows the installation of cold-formed metal trusses installed using a **rafting** technique. Rafting is a system where trusses are erected, braced, and sheathed on the ground and then an entire section of truss framing is hoisted in place. The main benefit of this type of installation is safety, where a majority of the work is done on the ground rather than on the upper level of the building.

> **NOTE**
>
> Panelized wood roofing systems are described in this chapter as related to roof frame materials and practices. Additional content is provided in Chapter 8, *Wood Construction*.

PANELIZED COMMERCIAL ROOF CONSTRUCTION

According to *Building Design+Construction,* **panelized wood roofs** have been used in construction for more than 40 years west of the Mississippi, where wood construction resources have always been plentiful. However, the panelized roofing system is new to regions east and in Canada, where steel construction has been dominant.

Panelized wood roofs consist of structural wood panels attached to 2 × 4 or 2 × 6 dimensional lumber **subpurlins** or **stiffeners,** typically spaced 24 in. on center. Subpurlins or stiffeners are light construction framing member resting on or between **purlins** and usually running at right angles to purlins. Purlins are attached horizontally to the roof framing system and are used to attach the metal sheets of roofing material in steel construction or sheathing in penalized wood construction. The panelized wood roof panel construction is done on ground level, and then the entire pre-framed panelized unit is lifted into position on the roof, as shown in Figure 7.15. Once on the roof, only one or two workers are usually needed to complete the assembly sequence.

Panelized Wood Roof Systems

Panelized wood roof systems consist of two common types: an all-wood system and a hybrid system. Both types are most commonly used with tilt-up concrete structures and, in some cases, reinforced masonry walls. Panelized wood roof systems are suited for buildings with low-slope roofs, such as warehouses and big retail stores.

The **all-wood panelized roof system** is anchored by long-span glued laminated timber framing that uses purlins attached to the primary glulam beams using pre-engineered metal hangers, as shown in Figure 7.16. **Glued-laminated timber** or **beam (glulam)** is a structural member made up of layers of lumber that are glued together. The free edge of the wood decking for each panelized unit is nailed to the framing edge of the previously placed unit. Pre-framed panel ends attach to the main glulam beams to complete the assembly. The pre-framed roof sections speed the erection process and add strength, dimensional stability, and high **diaphragm** capacity to the roof. A diaphragm is a structural element that

Figure 7.15 The panelized wood roof panel construction is done on ground level, and then the entire pre-framed panelized unit is lifted into position on the roof. (*Courtesy of the APA—The Engineered Wood Association.*)

FM CLASS 1-135 WITH ARMA ROOF COVERING[b]

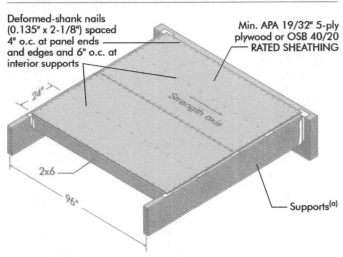

Deformed-shank nails (0.135" x 2-1/8") spaced 4" o.c. at panel ends and edges and 6" o.c. at interior supports

Min. APA 19/32" 5-ply plywood or OSB 40/20 RATED SHEATHING

Strength axis

24"

2x6

96"

Supports[a]

(a) Design in accordance with local building code requirements for roof loads and anchorage. All framing must be minimum net thickness of 1-1/2 inches No. 2 Douglas-fir or southern pine or equivalent. For wood I-joists, follow manufacturer's recommendations for minimum nail spacing.

(b) To install panels with strength axis parallel to supports spaced 24" o.c., as illustrated, see minimum panel requirements listed in Table 2 of APA publication, *Wind-Rated Roofs*, Form G310.

Figure 7.16 The all-wood roof system is anchored by long-span glued laminated timber framing that uses purlins attached to the primary glulam beams using pre-engineered metal hangers. (*Courtesy of the APA—The Engineered Wood Association.*)

transmits lateral loads to the vertical resisting elements of a structure, such as **shear walls.** The most common lateral loads to be resisted are those resulting from wind and earthquake action. Shear refers to forces caused when two construction pieces move over each other. A shear wall, also called a braced wall line, is a braced wall made of a shear panel to oppose the effects of **lateral load** acting on a structure. Lateral load is a force working on a structure applied parallel to the ground, and diagonally to the structure. Wind and seismic loads are the most common lateral loads.

The **hybrid panelized roof system** combines panelized wood components connected to open web steel joists. The entire panelized unit is assembled on the ground and then lifted into position at the roof level, where the steel joists are welded or bolted to primary steel trusses, as shown in Figure 7.17. The free edge of the wood decking for each panelized unit is nailed to the framing edge of the previously placed unit. The pre-framed panel ends attach to the main steel trusses to complete the assembly.

A purlin alternative is a **glulam truss** made with glulam chords and webs interconnected with heavy metal gusset nail plates, as shown in Figure 7.18. Glulam trusses spaced eight feet on center, allow large column grid spacing between the main glulam beams. Pre-framed panelized sections of up to 72 ft long are possible using glulam truss purlins, which allow contractors to erect roof units larger than 500 ft^2 in one lifting sequence. **Structural panels** are the structural wood panels attached to 2 × 4 or 2 × 6 dimensional lumber sub-purlins or stiffeners. **APA-rated** sheathing or APA-rated structural 1 sheathing is recommended for structural decking. The sheathing is typically plywood or oriented strand board (OSB), which comes in varying thicknesses and span ratings. OSB panels are manufactured in sizes up to 8 × 24 ft; larger sizes are sometimes possible. These large dimension panels

Figure 7.17   The hybrid panelized roof system combines panelized wood components connected to open web steel joists. The entire panelized unit is assembled on the ground and then lifted into position at the roof level, where the steel joists are welded or bolted to primary steel trusses. (*Courtesy of the APA—The Engineered Wood Association.*)

Figure 7.18 Glulam truss made with glulam chords and webs interconnected with heavy metal gusset nail plates. (*Courtesy of the APA—The Engineered Wood Association.*)

reduce perimeter fastening and speed construction. APA refers to APA—The Engineered Wood Association. APA rated refers to sheathing that is rated by the APA for use as sub-floor, wall, roof, diaphragm and shear wall sheathing, and construction applications where strength and stiffness are required. Structural wood panels are installed with their long dimension parallel to stiffeners that support panel edges. The wood sheathing acts as a diaphragm element that resists and transfers lateral loads from wind or earthquakes. Panel thickness and fastening schedules are determined by **gravity loads,** required shear load capacity, and the wind uplift resistance needed. **Gravity loads** include **dead load,** which is the weight of the structure, including its walls, floors, finishes, and mechanical systems, and **live load,** which is the weight of contents, occupants, and weight of snow.

Glulam beams, which are the main structural beams, are often designed as **cantilevered beam** systems to optimize structural performance. Single- and double-cantilevered beams can be used depending on building design. **Cantilever beam** means a rigid beam that is fixed to a support, such as a vertical column or wall at one end and again near the other end, and the other end of the beam is detached. The detached end is designed to carry vertical loads. Cantilevered beams are specified with a designation, such as a 24F-V8 Douglas-fir or a 24F-V5 southern pine. The 24F designation indicates a beam with an allowable bending stress of 2400 pounds per square inch (psi) and the V8 part of the designation refers to the lumber quality used to make the beam. Glulam beams can also be specified as

balanced or **unbalanced**. **Balanced beams** are used in applications such as cantilevers or continuous spans, where either the top or bottom of the member can be stressed in tension due to loads. In **unbalanced beams,** the quality of lumber used on the tension side of the beam is higher than the lumber used on the corresponding compression side, allowing a more efficient use of the timber resource.

Simple-span glulam **girders,** spanning column-to-column, in 30- to 60-ft lengths, is another popular design. Unbalanced combinations, such as 24F-V4 Douglas-fir or 24F-V3 southern pine, can be specified. Simple-span girders are recommended for their ease of installation and their ability to be used with seismic connectors. Simple span refers to a construction member such as a beam, girder, or truss that spans from one support to another. A girder is a horizontal structural member made of wood, laminated wood, engineered wood, or steel that spans between two or more supports at the foundation level, or above any floor level.

Purlins in all-wood roofing system are placed 8 ft on center unless engineered otherwise. Purlins can have many forms depending on the spacing between the main glulam beams and design preference, but the most common is **narrow resawn glulam** beams that are 2-1/2 in wide with depths of 18 to 27 in or greater. The term **narrow resawn glulam** refers to cutting large glulams into smaller sizes.

Prefabricated lightweight wood **I-joists** are an economical purlin alternative for glulam beam spacing of approximately 40 ft or less or for relatively light roof design loads. I-joists can have depths up to 30 in., depending on the span and loading conditions. I-joists also make it easier to cut holes in the webs for mechanical ductwork, wiring, and sprinkler lines. An I-joist is an engineered wood product with a web between a top and bottom flange, creating the I shape, shown in Figure 7.19. I-joists have great strength in relation to size and weight when compared with dimensional lumber.

Steel joists used in the hybrid panelized system are typically placed 8 ft on center. Steel joists can span long distances, allowing panelized units to be prefabricated in lengths up to 72 ft, or longer, to allow contractors to erect roof units larger than 500 ft² in one lifting sequence.

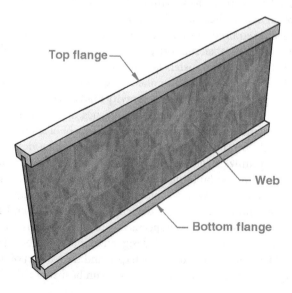

Top flange

Web

Bottom flange

Figure 7.19 An I-joist is an engineered wood product with a web between a top and bottom flange, creating the I shape.

The panelized roof system is connected to walls using wood or steel ledgers available as standard purchase items. Connectors range from 2 × 4 or 2 × 6 stiffener hangers to the more complex cantilever connectors for the main glulam beams. Additional tie plates and straps can be required depending on lateral load design requirements.

Panelized Wood Roof Nailer Attachment and Insulation

Diaphragm shear capacity and wind uplift requirements are important considerations when specifying the attachment of the wood nailer to the top chord of the steel joists for hybrid panelized roof systems. The nailer, which is typically attached to the joist with bolts or screws, makes it possible to connect the wood stiffeners and wood structural panels to joists with standard nailing techniques. The nailer attachment laterally supports the top chord of the steel joist and transfers horizontal shear and wind uplift forces. A diaphragm is a horizontal feature that distributes lateral loads from wind or seismic activity to the shear walls or foundation of a structure.

Insulated low-slope roofs typically involve little or no ventilation, making it important to understand the fundamental techniques for controlling moisture, especially when insulation is used below the roof. Primary moisture sources in wood construction include wetting during construction, condensation, and accumulated moisture from occupant use or natural infiltration.

COMMERCIAL ROOFING MATERIALS

The following content is taken, in part, from West Roofing Systems, Inc. describing the five most common types of commercial roofing along with their average costs and advantages. Reported costs are approximate and can vary in different locations.

Shingle Roofing

Shingle roofing is commonly used for residential roofing, but shingles are also often used for steep slope commercial roofing applications, especially on light commercial buildings such as offices, apartments, and condominiums. Shingles can be made out of a few different materials, including wood, slate, metal, plastic, ceramic, and composite material such as asphalt. Two main types of shingles are asphalt and architectural.

Asphalt Shingles

Asphalt shingles are made with **asphalt** for weatherproofing and are the most common, cost-effective shingle on the market used for architectural applications. These roof systems often come with 15- to 30-year warranties on average. Asphalt is a mixture of dark **bituminous pitch** and sand or gravel, used for purposes such as surfacing roads, flooring, and roofing. Bituminous refers to a natural substance, and pitch is a polymer that can be natural or manufactured, derived from petroleum, coal tar, or plants.

Architectural Shingles

Architectural shingles are multilayered asphalt shingles and heavier than regular asphalt shingles to add architectural appeal, durability, and weatherproofing to the building, as shown in Figure 7.20. Architectural shingles often have a lifetime warranty.

Figure 7.20   Architectural shingles are multilayered asphalt shingles and heavier than regular asphalt shingles to add architectural appeal, durability, and weatherproofing to the building. (*Courtesy of West Roofing Systems, Inc.*)

Shingle Roofing Advantages.   Benefits of commercial shingle roofing include

- For an average commercial roof, asphalt shingles usually cost between $2.50 and $5.00 per square foot in materials and labor to install a typical shingle roofing system. This price range can change depending on the location, architectural complexity, type, and style of the shingles used.
- A wide variety of colors and design options.
- Some types of shingles are **class 4 hail rated.** A class 4 hail-rated product is designed to withstand high winds and hail damage.
- It is easy to spot and repair small areas of damage to asphalt shingles.

METAL ROOFING

Metal roofing is one of the oldest commercial roofing systems on the market. Figure 7.21 shows a worker installing a metal roof on a steep-slope roof. The most commonly used metal roofing is called **standing-seam roofing.** This metal roofing material is manufactured in long sheets that are installed vertically on the roof all the way from the ridge to the eave when possible. The standing seam terminology comes from the

Figure 7.21 Workers installing a metal roof on a steep-slope roof. (*Courtesy of the U.S. Air Force.*)

Figure 7.22 Standing seam roofing terminology comes from the raised seam between each parallel sheet where one sheet is joined to the next, creating an interlocking, and watertight seam.

raised seam between each parallel sheet where one sheet is joined to the next, creating an interlocking, and watertight seam, as shown in Figure 7.22. Metal roofing systems often use **corrugated galvanized steel,** although other materials such as aluminum, tin, and copper are used. Corrugated galvanized steel is a building material made in sheets of hot-dip **galvanized** mild steel, cold-rolled to produce a linear corrugated pattern. Galvanize is a zinc coating used to resist corrosion. Metal roofing material is lightweight, making it an option to be installed over existing roof. After a metal roofing system is installed, a coating can be added for waterproofing, rust protection, and ultraviolet (UV) protection.

A typical metal commercial roofing system can cost between $5.00 and $10.00 per square foot in materials and labor to install. This price range can change depending on the location, architectural complexity, type, and style of the metal panels used.

Metal Roofing Advantages. Metal roofing system advantages include

- A properly installed and maintained commercial metal roof can last 40 years.
- A variety of material and color choices are available for design alternatives.
- Metal roofing systems have the stability to withstand high winds and snow loads.
- Metal roofing can be made from recycled material for sustainable construction practices.
- Metal panels are fire resistant.

SPRAY POLYURETHANE FOAM ROOFING

Spray polyurethane foam (SPF) is an **eco-friendly** roofing option for commercial, industrial, and manufacturing facilities, as shown in Figure 7.23. The term eco-friendly refers to an application or product that is not harmful to the environment. Spray **polyurethane** foam, commonly referred to as SPF, is a material that is sprayed as a liquid so it can expand into a foam, creating a solid layer across the roof sheathing or over an existing roof. Polyurethane is a synthetic resin used mainly as ingredients of paints, varnishes, adhesives, and foams. An SPF roofing system can be used in any climate, and can last over 50 years when correctly installed and maintained.

An SPF roof can cost between $4.00 and $7.00 per square foot in materials and labor to install for an average 20,000 ft² commercial roof system. This price range is for an SPF roof of average thickness and building structure.

Spray Polyurethane Roofing Advantages. SPF roofing system advantages include

- SPF delivers thermal, air, and moisture barriers to provide the highest R-value per inch for energy efficiency.
- The foam material is durable and can expand and contract with the building, decreasing the likelihood of cracks and splitting.
- The continuous solid surface is seamless and waterproof, removing seams that are the most vulnerable area for leaking.

Figure 7.23 Spray polyurethane foam (SPF) is an eco-friendly roofing option for commercial, industrial, and manufacturing facilities. (*Courtesy of West Roofing Systems, Inc.*)

- For reroofing projects, there is minimal stripping of the original roof, eliminating the need for costly roof tear-offs and waste for maximum environmental conservation. The materials that are used for SPF roofing systems are also environmentally friendly with zero ozone depletion potential (ODP), low in the emission of volatile organic compounds (VOCs), free from chlorofluorocarbons (CFCs), and ultra-low global warming potential (GWP).

Single-Ply Membrane Roofing

Single-ply membrane is a roofing option for commercial, industrial, and manufacturing facilities. Single-ply membranes are sheets of rubber and other synthetics that can be laid down by weight, mechanically fastened or chemically adhered to insulation creating a layer of protection on a commercial facility, as shown in Figure 7.24. Single-ply membrane roofing provide a variety of materials to choose from. Their main characteristic is that they are designed to be installed in one layer. The material is usually black or white finish and glued, fastened, or install directly over roof insulation material.

Single-ply membrane roofing usually costs between $3.50 and $7.50 per square foot for **EPDM** and $3.50 to $6.50 per square foot for **TPO,** including materials, labor, and

Figure 7.24   Single-ply membrane is a roofing option for commercial, industrial, and manufacturing facilities. Single-ply membranes are sheets of rubber and other synthetics that can be ballasted, mechanically fastened, or chemically adhered to insulation creating a layer of protection on a commercial facility. (*Courtesy of West Roofing Systems, Inc.*)

warranty for an average 20,000 ft^2 commercial roof. EPDM is the abbreviation and commercial name for ethylene propylene diene terpolymer, which is an extremely durable synthetic rubber roofing membrane widely used on low-slope roofs. The two primary EPDM ingredients are ethylene and propylene that are derived from oil and natural gas. TPO is the abbreviation and commercial name for thermoplastic polyolefin, which is a single-ply reflective roofing membrane made from polyprophylene and ethylene propylene rubber polymerized together. TPO is typically installed in a fully adhered or mechanically attached system, allowing the white membrane to remain exposed throughout the life of the roof.

Single-Ply Membrane Advantages. Single-ply membrane roofing system advantages include

- A commercial single-ply membrane roof can last 30 years.
- EPDM roofing has been used for commercial flat roofing for many years. This amount of time on the market has allowed various laboratory and field studies to be performed and tracked.
- *Customer choice of insulation:* Since single-ply membrane roofing does not include the insulation factor, allowing architects and contractors the option of using a variety of roof insulation products.
- TPO membranes and EPDM can achieve Underwriters Laboratories (UL) **class A fire resistance** listings by adding fire-retardant chemicals during the manufacturing process. Class A fire resistance is the most stringent rating available for building materials by the National Fire Protection Association (NFPA).
- TPO is generally white and highly reflective and the opposite with EPDM that is often described as black roofs due to the natural dark color of the membrane.

BUILT-UP ROOFING

Built-up roofing systems, commonly referred to as tar and gravel roofs, have been in use in the United States for over 100 years. Built-up roofing systems are installed by alternating layers of asphalt or tar and supporting fabrics directly onto the roof. The number of layers, called **plies,** is determined by the architect or contractor. The final layer of a built-up roofing system consists of stone or gravel, as shown in Figure 7.25.

A typical commercial built-up roofing system usually costs between $5.50 and $8.50 per square foot in materials and labor to install. This price range can change depending on the number of piles and materials used.

Built-Up Roofing Advantages. Built-up roofing systems have the following advantages:

- The continuous solid surface does not require joints or seams, making the roof system waterproof.
- Properly installed and maintained commercial built-up roof can last 40 years.
- Built-up roofing systems provide ultraviolet protection.
- Requires minimal maintenance after installation.

Figure 7.25   Built-up roofing systems are installed by alternating layers of asphalt or tar and supporting fabrics directly onto the roof. The number of layers, called plies, are determined by the architect or contractor. The final layer of a built-up roofing system consists of stone or gravel. (*Courtesy of West Roofing Systems, Inc.*)

Go to Downloads & Resources tab at **www.mhprofessional.com/Commercial Building Construction-Ancillaries** to access the images provided in this chapter and correlate with the content of this chapter. Here you can look at the chapter figures on your computer screen to pan and zoom in and out for better observation, because full drawings reduced to fit on a textbook page are often difficult to read. The image bank also includes the complete set of plans for the Brookings South Main Fire Station used as examples throughout this textbook.

Go to **Downloads & Resources tab at www.mhprofessional.com/Commercial Building Construction-Ancillaries** to access the test and creative thinking problems for this chapter. The chapter test can be used for review or to evaluate content knowledge depending on your course objectives. Creative thinking problems are provided to help expand your knowledge by researching given subjects.

Print reading problems have questions that ask you to find information on the Brookings South Main Fire Station plans. The print reading problems can help you become familiar with the format of construction documents and reinforce your ability to seek specific information. This is an important skill for you to master as you prepare to enter the construction industry. Go to **Downloads & Resources tab at www.mhprofessional.com/Commercial Building Construction-Ancillaries** to access the complete set of plans for the Brookings South Main Fire Station.

CHAPTER 8

Wood Construction

INTRODUCTION

Chapter 4, *Concrete Construction and Foundation Systems* covered common foundation systems used in commercial construction, including related wood **framing** practices. The term **framing** refers to the construction of the structural parts of a building. Recapping, the foundation is the construction system used to support the structural loads and distribute the loads to the ground. There are several different types of foundation systems used in construction throughout the United States and Canada. The basic foundation systems, where wood-frame construction can be used, are the crawl space, concrete slab, concrete wall, and post and beam or pole construction. Different types of floor framing systems were also described during the foundation systems discussion, because it was important for you to know how the foundation system and floor system tie together. The foundation systems were described in detail, while the floor framing systems were introduced. This chapter continues **light wood-frame** materials and construction practices related to additional floor framing practices, walls, and roofs. Commercial roof construction was described in Chapter 7 where light wood roof framing was omitted, because light wood framing correlates better with the content found in this chapter.

This chapter also discusses timber-frame construction and introduces cross-laminated timber (CLT), dowel-laminated timber (DLT), and nail-laminated timber (NLT) practices that are large-scale, prefabricated, solid engineered wood panels used in building construction.

LIGHT WOOD FRAMING MATERIALS

Light wood-frame construction is construction that uses small and closely spaced, such as 16 and 24 in. (400 and 600 mm) wood members that are generally assembled by nailing. Wood framing materials are generally **dimensional lumber** and **engineered wood products.**

Dimensional Lumber

Dimensional lumber is lumber that is cut and **planed** to standard width and depth specified in inches, and lengths specified in inches or feet and inches. For example, one piece of lumber might be specified as $2 \times 8 \times 12$, which is 2 in. $\times$ 8 in. $\times$ 12 ft. The term **planed** means to plane lumber by using a planer, which is a machine with cutters that remove material from the surface of the lumber to desired smoothness and finished dimensions. Lumber dimensions are specified by the American Softwood Lumber Standards based on **nominal** and **actual** sizes. Nominal size lumber is the size before it is planed, which is also referred to as **rough sawn lumber.** The nominal size is also the size that is commonly used when referring to lumber in the construction industry. For example, a 2×8 piece of lumber is 2 in. by 8 in. and is called a 2×8. The actual size of lumber is the dimensional lumber after it is planned. So, the 2×8 nominal size lumber is actually 1-1/2 in. by 5-1/2 in., or 1-1/2 $\times$ 5-1/2. Actual size lumber can also be referred to as **S4S,** which means surfaced on four sides. One inch and 2 in. thick lumber is commonly **kiln dried.** Kiln dried is the reduction of moisture content in wood by controlling the heat, air circulation, and humidity. Lumber over 2 in. thick is generally air dried, which is referred to as **green lumber.** Nominal and actual sizes of lumber are shown in Figure 8.1. You can also use the following general rules to determine actual lumber sizes:

- Lumber under 2 in., subtract 1/4 in. for actual dimension.
- Lumber 2 to 8 in., subtract 1/2 in. for actual dimension.
- Lumber over 8 in., subtract 3/4 in. for actual dimension.

Lumber length is specified in feet. Most dimensional lumber comes in length increments of 2 ft. Standard lumber lengths are 6, 8, 10, 12, 14, 16, 18, 20, 22, and 24 ft in the United States and Canada. Industry standard lengths help architects, contractors, and

Nominal Size (inches)	Actual Size (inches)
1	3/4
2	1-1/2
3	2-1/2
4	3-1/2
5	4-1/2
6	5-1/2
7	6-1/2
8	7-1/4
9	8-1/4
10	9-1/4
11	10-1/4
12	11-1/4
13	12-1/4
14	13-1/4
15	14-1/4
16	15-1/4

Figure 8.1 Nominal and actual sizes of lumber.

suppliers make planning, estimating, and construction easier, and it reduces waste during construction.

Most construction materials used in Canada are manufactured in metric **modules** using millimeters (mm) as the unit of measurement. Modules are a selected unit of measure used as a basis for the planning and standardization of building materials. The base metric module is 100 mm. Canadian plywood, for example, is manufactured in sheets that are 1200 × 2400 mm. Plywood in the United States is manufactured in 4 × 8 ft sheets, or larger. Converting 4 × 8 ft sheets to 1200 × 2400 mm is referred to as **soft metric conversion,** where the actual metric conversion of 4 × 8 ft = 1219 × 2438 mm is rounded to the nearest 100 mm modules, resulting in 1200 × 2400 mm. While this modular system is typical with most construction materials, it does not apply to lumber sizes, because Canada supplies a large amount of lumber to the United States. When the same materials, such as lumber, are used in the United States and Canada, the metric equivalent is referred to as a **hard conversion** to metric units. Hard conversion means that the typical inch units are converted directly to metric. For example, a 2 × 4 that is milled to 1-1/2 × 3-1/2 in. converts directly to 38 × 89 mm using the hard conversion method. To convert from inches to millimeters, use the formula 25.4 × inches = millimeters.

Engineered Wood Products

Engineered wood products are a combination of smaller components used to make structural products that have been engineered for specific applications, fabricated in a manufacturing facility, and delivered to the construction site. Floor framing applications for engineered wood products were introduced in Chapter 4, *Concrete Construction and Foundation Systems* and there is additions discussion later in this chapter.

Subflooring Materials

The **subfloor** is a layer of structural material fastened above the floor joists to tie the floor joists together and to provide structural support and a base for the finished floor. Most subfloor construction uses plywood or **oriented strand board (OSB)** subflooring that is glued and nailed or screwed to the floor joists. The plywood or OSB is **tongue and grove (T&G)** material, and the thickness depends on the joist spacing and structural requirements. Thickness varies from 3/4 to 1-1/8 in. (20 to 30 mm). OSB is structurally engineered board, manufactured from cross-oriented layers of thin, rectangular wooden strips compressed and bonded together with wax and resin adhesives. OSB is used for subflooring, exterior wall sheathing, and roof sheathing. In most applications, OSB is dimensionally stronger and stiffer than comparable dimensional plywood. Designers and builders commonly use OSB for increased strength and for cost, which is less than plywood. Figure 8.2 shows plywood or OSB subfloor used over floor joists and engineered wood joists. Most structural plywood and OSB has T&G edges, where one joining edge has a tongue that fits into the groove of the other joining edge, as shown in Figure 8.3. The T&G provides a strong connection that prevents one sheet or board from moving up or down at the joint.

Plywood and OSB is manufactured in 4 × 8 ft sheets, or larger, for use in the United States, and 1200 × 2400 mm sheets for use in Canada. The 4 × 8 ft sheets are used over floor joists spaced 12, 16, 24, or 36 in. on center, depending on the structural requirements. The 1200 × 2400 mm sheets are used over floor joists spaced 100, 200, 300, 400, or 600 mm on center, depending on the structural requirements.

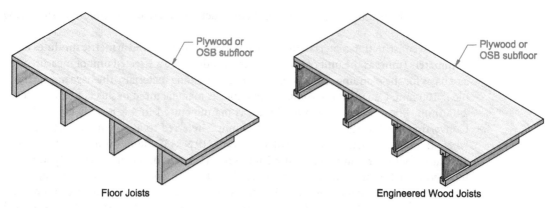

Figure 8.2 Plywood or OSB subfloor used over floor joists and engineered wood joists. (*Previously published in Modern Residential Construction Practices, Routledge, 2017 and reproduced here with permission.*)

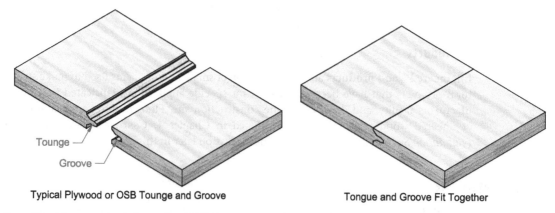

Figure 8.3 Most structural plywood and OSB has tongue and groove (T&G) edges, where one joining edge has a tongue that fits into the groove of the other joining edge. (*Previously published in Modern Residential Construction Practices, Routledge, 2017 and reproduced here with permission.*)

NAILED CONNECTIONS

Nails used for the fabrication of wood-to-wood members are sized by the term **penny** and denoted by the letter d. Penny is a weight classification for nails based on the number of pounds per 1000 nails. For example, one thousand 16d nails weigh 16 lb. Nails are also sized by diameter when over 60d. Nails are specified by the penny weight plus the quantity and spacing if required. The specification for special nails is given when required. Nailing callout examples include

4–16d NAILS

5–20d GALV NAILS EA SIDE, where GALV is the abbreviation for galvanized.

10d NAILS 4″ OC AT SEAMS AND 12″ OC IN FIELD, where OC is the abbreviation for on center.

30d NAILS ALTERNATELY STAGGERED 12″ OC TOP AND BOTTOM BOTH SIDES

8d RING SHANK NAILS 6″ OC

If pilot holes are required for nailing, the diameter of the pilot hole is given after the nail callout. For example, 4–20d NAILS W/5/32″ Ø HOLES. Verify pilot hole diameters with manufacturers' recommendations for types of woods and applications.

PLATFORM AND BALLOON FRAMING

The framing method used in light wood-frame construction in the United States and Canada is called **platform framing.** Platform framing is like building a box, where the floor joists and rim joist form the sides of the box and the subfloor is attached as the top of the box. This floor box rests on the mud sill for the first floor framing, as shown in Figure 8.4a.

The floor box and joists establish the platform upon which the wall structure is built, as shown in Figure 8.4b. The platform construction is repeated at each floor. For example, the second floor platform is framed over the top of the wall framing, as shown in Figure 8.4c.

Balloon framing is a construction system that was used in the United States and Canada in the 1800s through around 1955. Balloon framing uses continuous studs that rest on the mud sill at the foundation and run through each floor and to the top plate at the roof. The floor joists rest on the sill at the first floor, or on the top plate for second and third floors, and attach to the studs, as shown in Figure 8.5. Balloon framing is stronger than platform framing, because the studs are an integral part of the structure from sill to roof, and the joists are fastened directly to the studs. Platform framing took the place of balloon framing, because platform framing makes it much easier to construct one individual floor at a time. It is also difficult to find the long studs needed for balloon framing. Figure 8.5 compares platform and balloon framing. While it is not important for you to know about balloon framing for new construction, you might find balloon framing in an old building when you work on a **remodeling** project. Remodeling is changing an existing structure, either internally or externally. Remodeling can be building an addition, which adds a new portion of building to an existing structure. Remodeling can also be renovation, which is improving by renewing and restoring.

FLOOR FRAMING SYSTEMS

Light wood floor framing was introduced with crawl space and concrete wall construction in Chapter 4 as related to foundation systems, including post and beam and joist construction practices. The following describes additional floor framing practices.

Floor construction uses joists that are standard dimensional lumber or engineered wood products that span between foundation walls, exterior walls, **bearing partitions, beams,** and **headers.** Bearing partitions carry structural weight from above and distribute the weight to the ground. Beams are horizontal construction members that are used to support floor systems, wall, or roof loads. A header is a horizontal structural member that supports the load over an opening such as a door or window, or around an opening. Figure 8.6 shows the floor framing over a concrete foundation wall and continuing to the second floor.

Using Hold-Down Anchors

Attaching the mud sill to the foundation wall and the use of hold-down anchors was introduced in Chapter 4. Hold-down anchors, also called hold-downs or holdowns, are used in a variety of construction applications. This discussion about hold-down anchors refers to fasteners that attach wood structural walls between floors, and especially at shear walls to

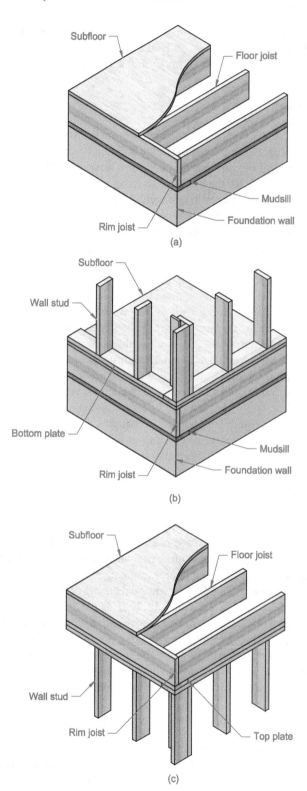

Figure 8.4   (*a*) Platform framing rests on the mud sill for the first floor. (*b*) The floor box and joists establish the platform upon which the wall structure is built. (*c*) The platform construction is repeated at each floor. This example shows an upper floor platform framed over the top of the wall framing below. (*Previously published in Modern Residential Construction Practices, Routledge, 2017 and reproduced here with permission.*)

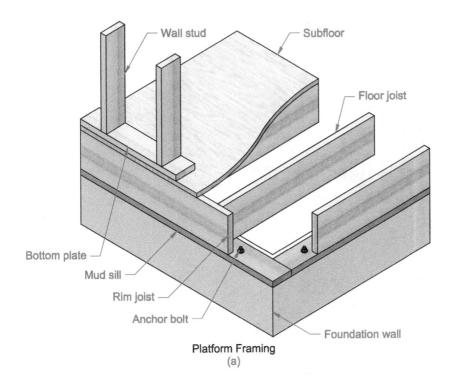

Platform Framing
(a)

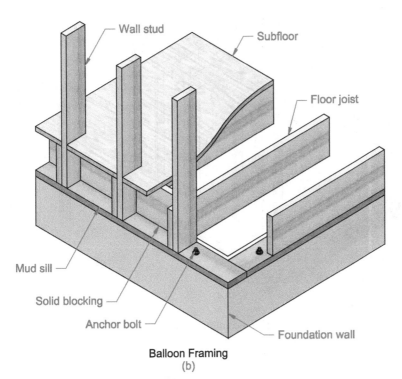

Balloon Framing
(b)

Figure 8.5 A comparison between platform and balloon framing. (*Previously published in Modern Residential Construction Practices, Routledge, 2017 and reproduced here with permission.*)

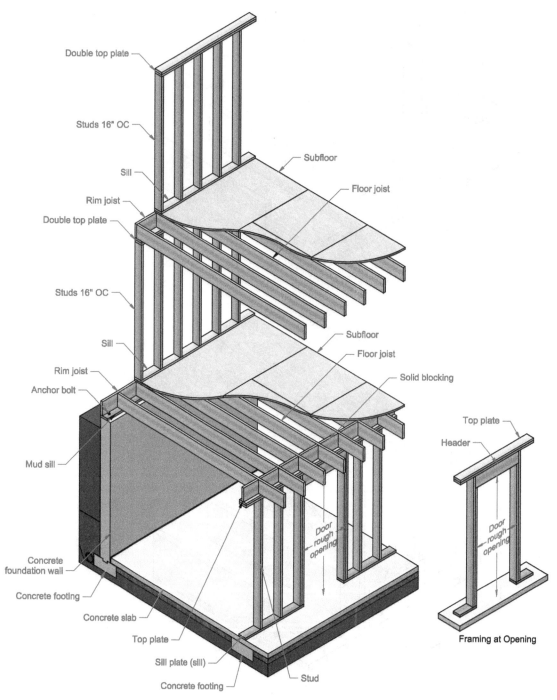

Figure 8.6 Floor framing over a concrete foundation wall and continuing to the second floor. (*Previously published in Modern Residential Construction Practices, Routledge, 2017 and reproduced here with permission.*)

control uplift in the wall system. The hold-down is connected to a wood post or studs with screws, nails, or bolts at the top. When used between floors, there is a steel bracket attached to a stud or post below and one above, with a threaded rod connected to each bracket and extending between floors. A nut on the threaded rod located at each bracket is used to tighten the rod and secure the system, as shown in Figure 8.7.

Floor Framing at Partitions

The specific framing practice always depends on structural engineering and as represented in the set of plans and specifications. There are basic methods used when floor joists intersect perpendicular or run parallel to bearing and non-bearing partitions. Non-bearing partitions do not carry a structural load. Figure 8.8a shows how double **solid blocking** is used when floor joists cross over a bearing partition below and under a bearing partition above. Solid blocking is generally the same size lumber or engineered wood product as the joists where the blocking is placed. Solid blocking is placed between floor joists to help resist lateral loads, to prevent joist rotation, and to provide nailing surface for the bottom plate of a wall or partition framed above. The blocking provides a strong structural support to stabilize the joists and strengthen between the bearing partitions below and above. The blocking also provides a nailing surface for the edge of the finish material when placed at each edge of the top plate, as shown in Figure 8.8b. Always make sure there is nailing surfaces provided

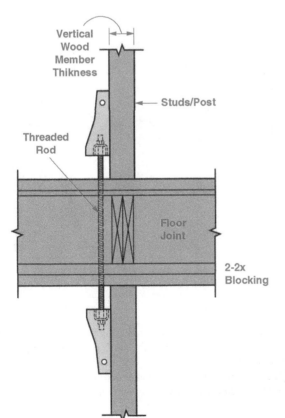

Figure 8.7 Hold-down anchors used between floors. (*©Simpson Strong-Tie Inc. and/or its affiliates. All Rights Reserved.*)

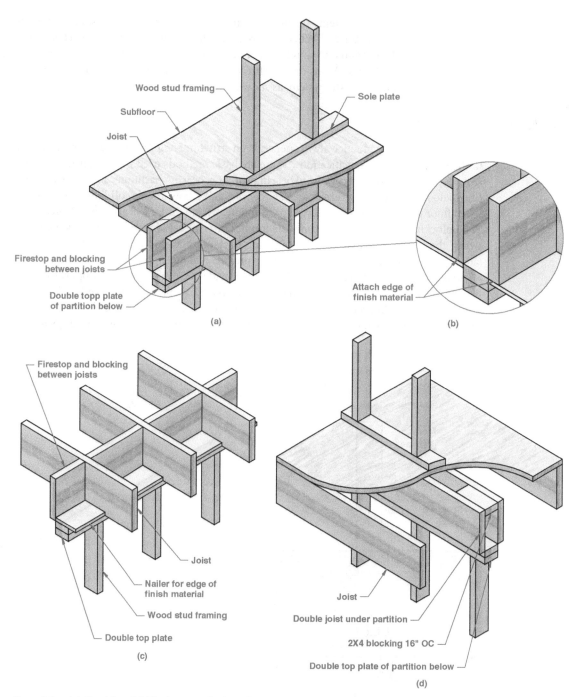

Figure 8.8   (*a*) Double solid blocking used when floor joists cross over a bearing partition below and under a bearing partition above. (*b*) Blocking provides a nailing surface for the edge of the finish material when placed at each edge of the top plate. (*c*) Framing used when floor joists cross over a bearing partition below and there is no partition above. (*d*) A double floor joist is normally provided between bearing partitions below and above when the bearing partitions run parallel to the joists. (*Previously published in Modern Residential Construction Practices, Routledge, 2017 and reproduced here with permission.*)

to secure the edges of finish material where joists intersect partitions. Figure 8.8c shows the framing used when floor joists cross over a bearing partition below and there is no partition above. A double floor joist is normally provided between bearing partitions below and above when the bearing partitions run parallel to the joists, as shown in Figure 8.8d.

Typical framing is less complex when a non-bearing partition is placed perpendicular above floor joists, as shown in Figure 8.9a. Minimum 2 × 4 blocking is used between floor joists that run parallel to a non-bearing partition below, as shown in Figure 8.9b. Also notice the **nailer** that provides a nailing surface for the edge of the finish material below. A nailer is a wood member fastened to the structure and used for attaching other wood members or finish materials. Figure 8.9c shows three different methods for framing when non-bearing partitions run parallel to the floor joists below.

Framing Floor Openings

Openings are framed into the floor joists for anything that needs to go through the floor, such as a **stairwell** or **chimney**. The stairwell is the opening in the floor where the stairs are located. A chimney is a structure containing a passage through which smoke and gases escape from a fireplace, furnace, vent, or other application. Framing at a floor opening requires additional structure to support the empty space. The extra framing is provided by double joists parallel to the opening and **double headers** perpendicular to the opening, as shown in Figure 8.10a. The double header is two structural members that are the same size material as the joists, and used to support the ends of the joists at the opening. A single header can be used at small openings. The two parallel joists are nailed together to fasten and strengthen the structure. The first header is nailed to the joist ends and the second header is nailed to the first header to fasten the structure. Specific nailing requirements can be found on the plans or in specifications. Depending on structural requirements, joists hangers can be specified where the headers meet the double joists and where the joists meet the headers, as shown in Figure 8.10b.

Framing a Cantilever

A **cantilever** is a structure that is supported at one end and is self-supporting on the other end where it projects into space, as shown in Figure 8.11. A cantilever design can be used as a balcony, deck, or any desired self-supporting structure. Figure 8.11a shows the cantilever structure framed perpendicular to the floor joists. The joists of the cantilever structure generally extend at least two joist spaces to where they are attached to a double joist that acts as a header. The sizes and specific length of the cantilever joists and the distance of the cantilever are designed by structural engineering. Solid blocking is provided between the cantilever joists above the exterior bearing wall or partition. Notice the joist hangers in Figure 8.11a that are commonly used to attach joists to headers. A single or double header is nailed to the ends of the cantilever joists to act as a rim joist. Figure 8.11b shows the cantilever structure framed parallel to the floor joists. Solid blocking is provided between the cantilever joists above the exterior bearing wall or partition. A single or double header is nailed to the ends of the cantilever joists to act as a rim joist.

Bracing Used in Floor Framing

Bracing can be used in locations throughout the floor framing to provide stiffness to a floor system, to resist lateral loads, to prevent joist rotation, to provide a nailing surface

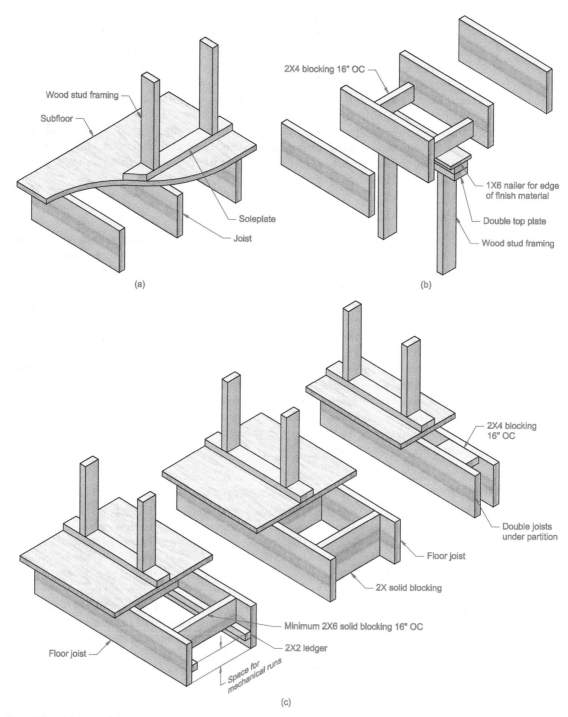

Figure 8.9   (*a*) Typical framing is less complex when a non-bearing partition is placed perpendicular above floor joists. (*b*) Minimum 2 × 4 blocking is used between floor joists that run parallel to a non-bearing partition below. (*c*) Three different methods for framing when non-bearing partitions run parallel to the floor joists below. (*Previously published in Modern Residential Construction Practices, Routledge, 2017 and reproduced here with permission.*)

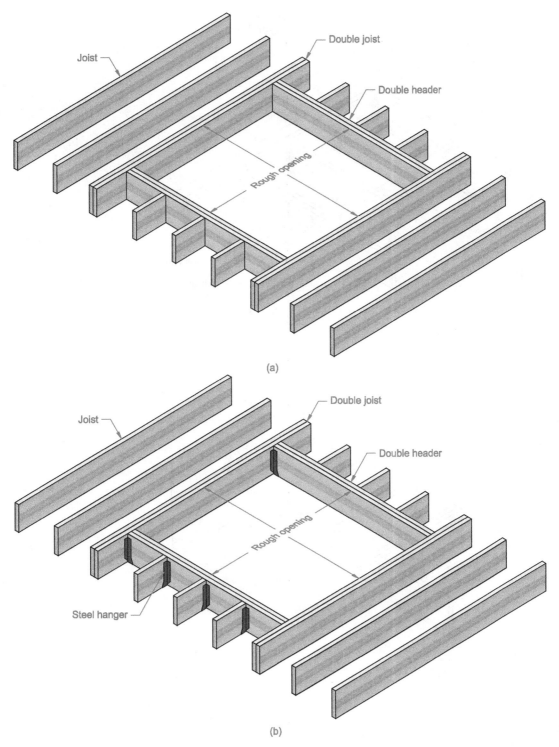

Figure 8.10 (*a*) Framing an opening through floor joists. (*b*) Steel hangers used to support the joists at the header. (*Previously published in Modern Residential Construction Practices, Routledge, 2017 and reproduced here with permission.*)

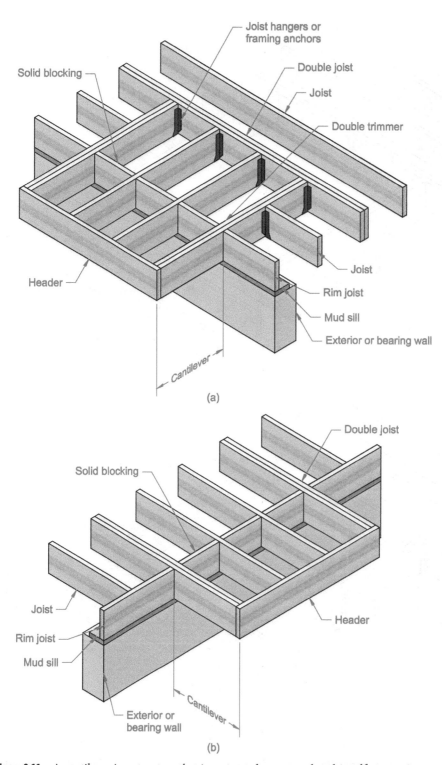

Figure 8.11 A cantilever is a structure that is supported at one end and is self-supporting on the other end where it projects into space. (*Previously published in Modern Residential Construction Practices, Routledge, 2017 and reproduced here with permission.*)

for the bottom plate of a partition framed above, or used to support plumbing and heating equipment. The two types of floor joist bracing commonly used in construction are solid blocking and cross bracing.

The use of solid blocking was introduced in the previous discussion about partitions and joist framing, and cantilever framing. Solid blocking is generally the same size lumber or engineered wood product as the joists where the blocking is placed, but can be smaller dimensional lumber depending on the application. Solid blocking is used when floor joists cross over a bearing partition below and under a bearing partition above. Solid blocking is placed between floor joists to help resist lateral loads, to prevent joist rotation, and to provide nailing surface for the bottom plate of a wall or partition framed above. The blocking provides a strong structural support to stabilize the joists and strengthen between the bearing partitions below and above. The blocking also provides a nailing surface for the edge of the finish material when placed at each edge of the top plate. Refer to the examples shown in Figures 8.8 and 8.9. In addition to these applications and examples, solid blocking should be used 8 ft apart (2400 mm) between 2 × 10 or larger floor joists to prevent joist rotation and to strengthen the structure.

Solid blocking provides more support and strictness than cross bracing, but cross bracing prevents joist rotation and stiffens the structure while allowing plumbing pipes and electrical wires to be run between floor joists. Cross bracing can be easier to install than solid blocking. Two types of cross bracing are 1 × 4 lumber and premanufactured metal ties nailed in a crossing system between floor joists. Figure 8.12 shows solid blocking and cross bracing applications.

Framing at Bearing Partitions and Beams

The floor joist span refers to the horizontal distance between two supporting members. Floor joist size and spacing determines the distance the joists can span. The joist size, spacing, and span are all characteristics determined by structural engineering and provided as information on the plans and notes. Floor joists can span between foundation walls when the foundation walls are close enough together, as shown in Figure 8.13*a*. When the foundation walls are too far apart for the floor joists to span, then a **bearing partition** or supporting **beam** is needed somewhere between the foundation walls to reduce the span, as shown in Figure 8.13*b*. A bearing partition carries structural weight from above and distributes the weight to the ground. The bearing partition is a **stud** wall structure. Studs are vertical framing members used to construct walls and partitions and are usually 2 × 4 or 2 × 6 and spaced 16 or 24 in. on center.

A supporting **beam** is used in place of the bearing partition when an open area is desired below. A beam is a horizontal construction member used to support floor systems, wall, or roof loads. Beams are generally solid lumber or **glued-laminated beams.** Solid lumber beams have sizes such as 4 × 12 or 6 × 14, but the beam size depends on structural engineering and is determined by the span and loads on the beam. A typical drawing note for a beam might read like this on the plans: 6 × 14 BEAM. Glue-laminated beams, referred to as glu-lams or glulams, are popular in framing. A glued-laminated beam (glu-lam) is an engineered wood product structural member made up of layers of small-size lumber that are glued together. The individual pieces of lumber are end-joined to create long lengths referred to as laminations. Glue-laminated beams have many advantages over solid lumber. For example, glued-laminated beams are environmentally friendly, because quality large size and length solid lumber beams are very difficult and expensive to obtain, because they are cut from old-growth logs that are not commonly available today. An equivalent size

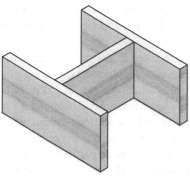

Solid Blocking

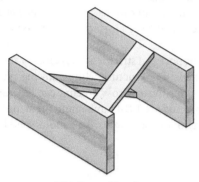

1X4 Cross Bracing

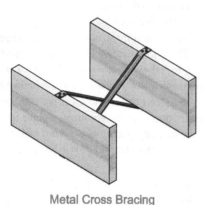

Metal Cross Bracing

Figure 8.12 Solid blocking and cross bracing applications. (*Previously published in Modern Residential Construction Practices, Routledge, 2017 and reproduced here with permission.*)

glued-laminated beam is stronger than a solid beam, is generally better quality, and is more dimensionally stable. Glue-laminated beams can span long distances using lighter-weight members with fewer supports than solid beams. Glued-laminated beams can be manufactured curved, making them more versatile than solid wood beams. A typical drawing note for a glued-laminated beam might read like this on the plans: M4—5-1/8″ × 12″ GLB. The 5-1/8″ × 12″ is the width and depth of the beam, and GLB is the abbreviation for glue-laminated beam. M4 is a **keynote** name for the beam, which refers to the sizing calculation sheet. A keynote is a note found on the drawings. Each keynote has a letter or number, or combination of letters and numbers or symbols next to or pointing at a specific feature on

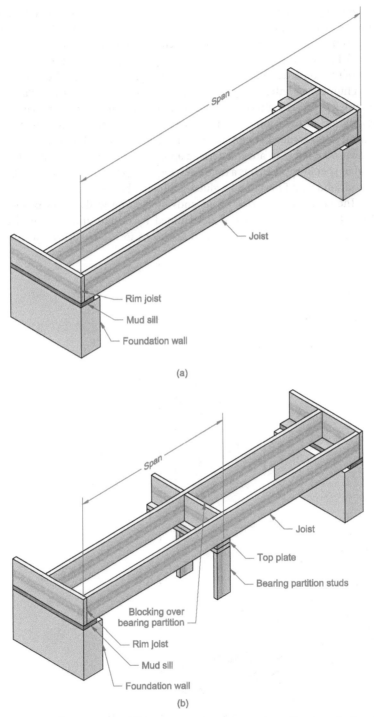

(a)

(b)

Figure 8.13 (*a*) Floor joists can span between foundation walls when the foundation walls are close enough together. (*b*) When the foundation walls are too far apart for the floor joists to span, then a bearing partition or supporting beam is needed somewhere between the foundation walls to reduce the span. (*Previously published in Modern Residential Construction Practices, Routledge, 2017 and reproduced here with permission.*)

the drawing that correspond to the description of the keynote in a **legend** or in a **general note.** For example, the M4 in the glue-laminated beam example refers to a keynote somewhere on the drawing that identifies the engineering calculations used to establish the beam size. A keynote can provide specific instructions on materials or practices to be used in a part of the construction. A keynote can be a symbol such as a hexagon or triangle containing a number or letter. A legend is generally found in the corner of a drawing, such as the lower left or lower right corner, and contains a list of symbols or characteristics and a brief description of each found on the drawing. A general note is a note grouped with other general notes somewhere on the drawing, such as in the lower left or lower right corner. A general note refers to a characteristic applied to the entire drawing. This compares with a **local** or **specific** note that is a note referring to a specific feature on the drawing. Figure 8.14 show a solid wood beam and a glued-laminated beam.

The beam is supported on its ends by bearing walls or **posts** and can be supported along its length with posts as determined by design and structural engineering. Floor joists can be framed over the beam, as shown in Figure 8.15a, or intersecting the beam

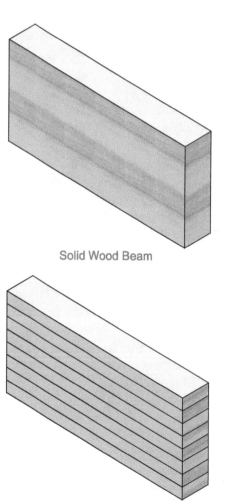

Solid Wood Beam

Glued-Laminated Beam

Figure 8.14 A solid wood beam and a glued-laminated beam. (*Previously published in Modern Residential Construction Practices, Routledge, 2017 and reproduced here with permission.*)

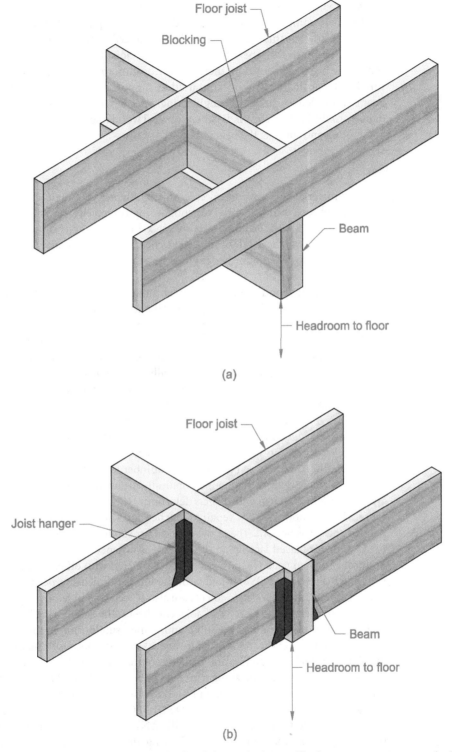

(a)

(b)

Figure 8.15   (*a*) Floor joists can be framed over the beam. (*b*) Floor joists can intersect the beam and secured to the beam with joist hangers. (*Previously published in Modern Residential Construction Practices, Routledge, 2017 and reproduced here with permission.*)

and secured to the beam with joist hangers, as shown in Figure 8.15b. The easiest framing method is running the floor joist over the beam and placing blocking between the joists over the beam, but this technique results in reduced headroom below the beam. Framing the floor joist flush with the top of the beam with **joist hangers** intersecting the beam helps reduce the beam exposure and increases the headroom below. A joist hanger is a manufactured metal angle, bracket, or strap used to support and attach the ends of floor joists to beams as needed for structural engineering. Special hardened nails are used to fasten steel hangers to the joist and beam. Posts are vertical wood members that connect between a pier and footing and support the beam above.

Installing Subflooring

Subfloor material was introduced early in this chapter as a main component of platform framing. Plywood and OSB used for subflooring is typically manufactured in 4 × 8 sheets (1200 × 2400 mm) and the floor joists are spaced 12, 16, or 24 in. on center (200 to 400 mm). This spacing allows the subfloor material to be supported and fasten to evenly spaced floor joists below. Also, the subfloor edges meet exactly on the center of a floor joist where they are securely fastened.

Softwood plywood subflooring and sheathing is manufactured using an A-D grading system. A grade stamp is found on plywood and OSB that identifies the grade, thickness, span design, and other information, as shown in Figure 8.16. Softwood plywood is structural material made of layers of softwood **veneer** glued together, under heat and pressure, with the grains of adjoining layers placed at right angles. Hardwood plywood is generally not used for construction, because it is usually more expensive and manufactured for other than structural applications, such as cabinets, furniture, and finish surfaces. Hardwood plywood is manufactured using the same veneer system as softwood plywood, except hardwood is used as the facing surface veneer for a quality wood appearance. Veneer is thin sheets of wood glued together to form plywood or glued to a wood base material. The A-D plywood letter grading system is used to indicate the quality of the exterior veneer panel. The main difference between grades is the number of knot holes and voids in the layers of veneer, which are defects that happen naturally in softwood veneer. The number of defects and the work needed to fix the defects determines the plywood grade. A-B grade is the highest quality plywood with few repaired patches and voids, and is used when an exposed or painted surface is needed. The plywood used for subflooring is typically C-D grade, where the C veneer is placed face up and the D side is fastened to the joists. C-D grades are economical and often used when the plywood is covered with other materials. Subfloor material, either plywood or OSB, is laid with the longest dimension perpendicular to the floor joists, as shown in Figure 8.17. Sheets are also placed over at least two joist spaces and end joints are staggered at the center of adjacent sheets. The subflooring is generally glued and nailed or screwed to the joists. Gluing and nailing subflooring to the floor joists provides stiffer floor construction and minimizes squeaks in the floor resulting from loose subflooring. Subfloor thickness, gluing, and nailing requirements are shown and noted on the plans. It is common for the nailing along the edges to be space closer than the nailing in the **field.** The field is the internal area of the plywood between the edges. A note that can be found on the plans reads: 3/4″ C-D PLYWOOD SUBFLOOR GLUED AND NAILED TO FLOOR JOISTS WITH **8d** NAILS 6″ OC AT EDGES AND 12″ OC IN THE FIELD. This note means that C-D grade plywood is used for subflooring, the subfloor is glued and nailed to the joists with 8d nails spaced 6 in. apart at the edges and 12 in. apart in the field. The d in the 8d nail

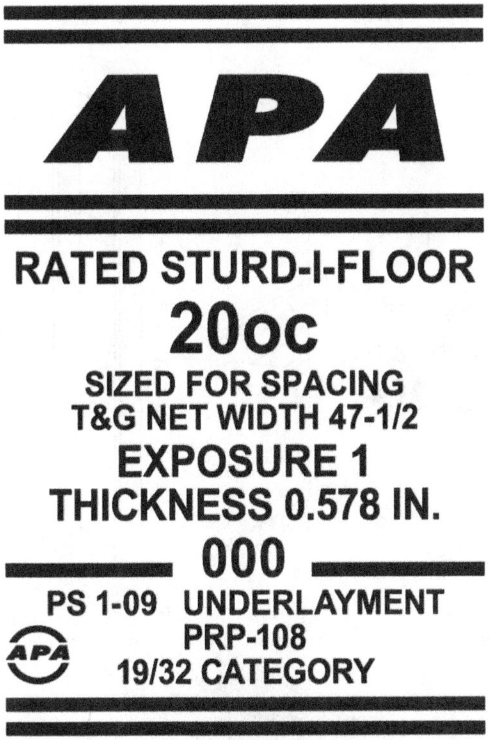

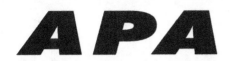

APA
RATED STURD-I-FLOOR
20oc
SIZED FOR SPACING
T&G NET WIDTH 47-1/2
EXPOSURE 1
THICKNESS 0.578 IN.
000
PS 1-09 UNDERLAYMENT
PRP-108
19/32 CATEGORY
APA

(a)

APA
RATED SHEATHING
24/16
SIZED FOR SPACING
EXPOSURE 1
THICKNESS 0.418 IN.
000
PS 2-10 SHEATHING
PRP-108 HUD-UM-40
7/16 CATEGORY
APA

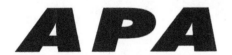

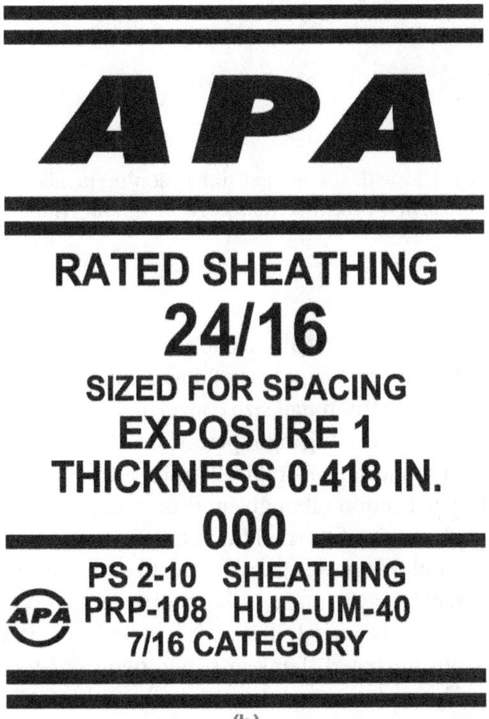

(b)

Figure 8.16 A grade stamp is found on plywood and OSB that identifies the grade, thickness, span design and other information. (*a*) Plywood grade stamp. (*b*) OSB grade stamp. (*Courtesy of the APA—The Engineered Wood Products Association.*)

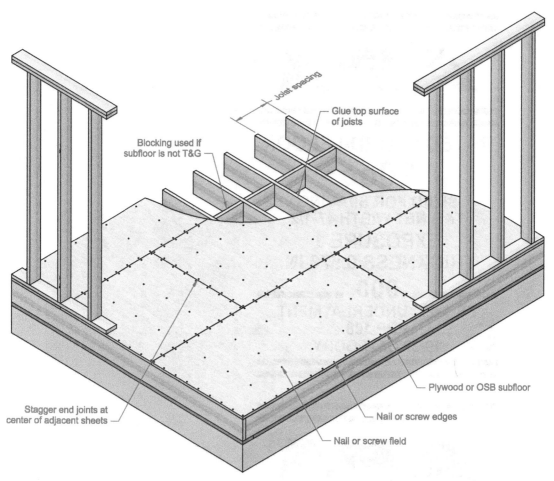

Figure 8.17 Subfloor material, either plywood or OSB, is laid with the longest dimension perpendicular to the floor joists. (*Previously published in Modern Residential Construction Practices, Routledge, 2017 and reproduced here with permission.*)

specification refers to penny, which is an old term still used to specify nail sizes today. The nail size gets larger as the number gets larger. For example, an 8d nail is 2-1/2 in. long and a 16d nail is 3-1/2 in. long. Figure 8.18 gives common nail size conversions. Historically, the term **penny** is from one hundred 2-1/2 in. nails costing eight pennies, resulting in the 8d designation. This is your first introduction to nails. Nails typically used for installing subflooring are 6d and 8d, with a **ring shank** specification often given. Ring shank nails have ridges or grooves along the **shank,** which is the body of the nail that penetrates the wood. Similar to a screw, ring shank nails have better holding power than nails with a smooth shank, because the rings act as wedges to keep the nail firmly in place. Other nail terminology includes **box, common, finish,** and **sinker.** A box nail is a small diameter nail typically used in making boxes. A common nail has a smooth uncoated shank less than one-third the diameter of the **nail head** and is most commonly used framing. A finish nail is a thin nail with a small head designed for setting below the surface of the wood. A sinker nail is used for framing and is thinner than a common nail with a funnel-shaped head, a grid stamped on the top

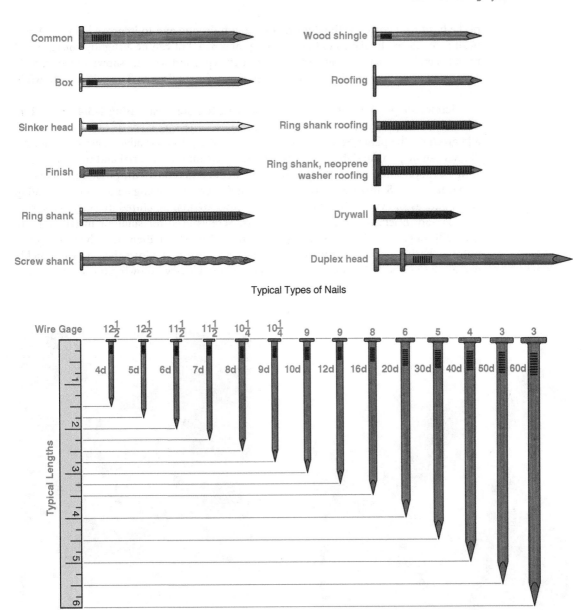

Typical Types of Nails

Typical Sizes of Common and Box Nails

Figure 8.18   Nail terminology, types and sizes. (*Previously published in Modern Residential Construction Practices, Routledge, 2017 and reproduced here with permission.*)

of the head, and coated with adhesive for smooth driving and to improve holding. A nail head is the normally flat round enlarged top of a nail, but can be a slightly enlarged and rounded end as on a finish nail. Nail terminology, types and sizes are shown in Figure 8.18. A nail gun or a screw gun is typically used to fasten the subfloor to floor joists, as shown in Figure 8.19.

Screws are popular for installing subflooring to floor joists, using 1-3/4 to 2-1/2 in. deck screws depending on the subfloor thickness. The screw specifications are generally given on the plans. Screwing the subfloor to the joists has advantages over nailing. Screws can help make tighter contact between the subfloor material and the floor joists. A tighter application allows the glue to secure better and reduces the possibility of having a squeaky floor. Screw installation is quick and easy when using an automatic standup screw gun, as shown in Figure 8.20. Wood screws are manufactured in gauge sizes from 0 to 24 and in lengths of 1/2 in increments. Screws used for subfloor installation are generally #8 gauge, which has an approximate 5/32 in. shank diameter. The wood screw point is sharp for easy starting and has a helical thread that allows the screw to quickly penetrate the wood. A flat head is common on wood screws allowing the screw head to

Figure 8.19 A nail gun is typically used to fasten the subfloor to the floor joists.

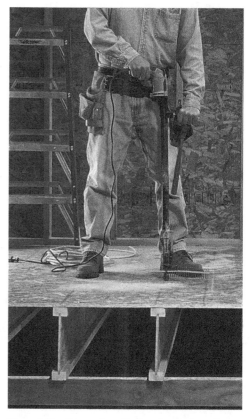

Figure 8.20 Screw installation is quick and easy when using an automatic standup screw gun. (*Courtesy of Louisiana Pacific Corporation.*)

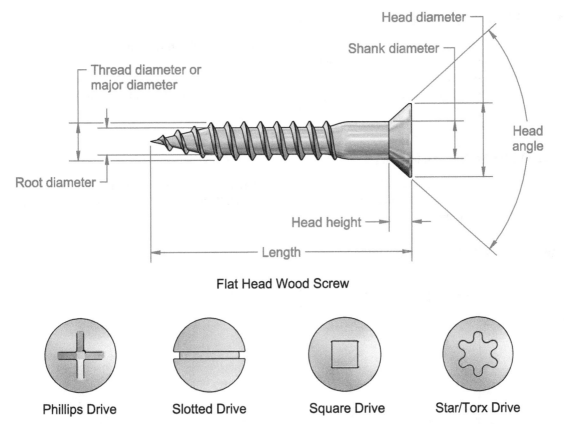

Flat Head Wood Screw

Phillips Drive   Slotted Drive   Square Drive   Star/Torx Drive

Figure 8.21   Parts of a typical wood screw and the different drive types. (*Previously published in Modern Residential Construction Practices, Routledge, 2017 and reproduced here with permission.*)

recess flush with or below the wood surface. Several different drive configurations are available. Figure 8.21 shows the parts of a typical wood screw and the different **drive types.** The drive types refer to the type of driver tool used to drive or turn the screw into the material.

INTRODUCTION TO LIGHT WOOD WALL FRAMING

Walls are typically vertical structures made from wood, steel, concrete, or masonry used to enclose, or divide the floor area based on the design provided in the plans. There are exterior and interior walls. Exterior walls establish outside surfaces of the building, and are covered with siding material to protect the interior of the building from the outside elements and weather. Exterior walls have windows and exterior doors. Interior walls are called partitions. There are bearing and non-bearing partitions. Walls are normally framed flat on the floor and then tilted into place where they are nailed and fastened to the floor and structure with hold-down anchors where specified by structural engineering requirements.

WALL FRAMING SYSTEMS

The two basic wall framing systems, **platform** and **balloon framing,** used in the United States and Canada were introduced earlier in this chapter. It is important for you to know these practices as you continue to learn about wall framing materials and construction in this chapter. The following provides a brief review of platform and balloon framing practices.

WOOD FRAMING MATERIALS

The most commonly used wall framing material is dimensional lumber, but walls are also framed with engineered wood products and steel. Walls can also be constructed using concrete and masonry.

Wall Framing with Dimensional Lumber

Typical building wall framing materials, methods, and terminology are shown in Figure 8.22. The principal components making up the wall structure include

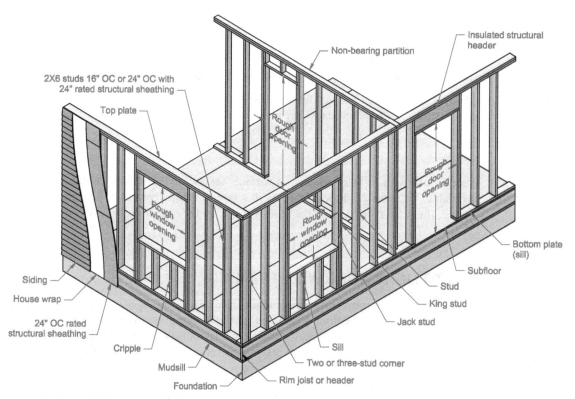

Figure 8.22   Typical building wall framing materials, methods, and terminology. (*Previously published in Modern Residential Construction Practices, Routledge, 2017 and reproduced here with permission.*)

Studs are vertical framing members used to construct walls and partitions, and are usually 2 × 4 or 2 × 6 and spaced 16 or 24 in. (400 or 600 mm) on center. The wall constructed using studs is often referred to as a **stud wall.** The 16 or 24 in. stud spacing is standard to provide an evenly spaced nailing surface for typical 48 × 96 in. (1200 × 2400 mm) exterior sheathing and interior drywall. 2 × 6 studs spaced 16 or 24 in. on center are commonly used to frame exterior walls, providing adequate depth to meet insulation code requirements. Keeping stud spacing at the maximum 24 in. on center helps increase wall insulation value by decreasing the amount of wood used in the framing. Wood does not have good insulation value. However, the use of 24 in. stud spacing requires that structural sheathing or siding be rated for 24 in. stud spacing and must be confirmed by structural engineering. 2 × 4 studs spaced 16 or 24 in. on center can be used for framing interior partitions where there are no insulation code requirements. Interior partitions containing plumbing are usually framed using 2 × 6 or 2 × 8 studs to accommodate the diameter of the plumbing pipes, such as the partition behind a toilet. **Jack stud** is a partial stud nailed next to **king studs** to support the header at door or window openings. The king stud is nailed to the header and the jack stud is nailed to the full king stud. A jack stud used at a **rough** door and window **opening.** A rough opening (RO) is any unfinished opening that is framed to specific measurements to accommodate the finish product. The rough opening for a window is given by the window manufacturer for each specific window. For example, one manufacturer specifies framing the rough opening 72 × 36 in. for their 71-1/2 × 35-1/2-in. window frame. This allows for 1/4 in. of clearance on each side of the window, making it very important to frame the rough opening perfectly **plumb, level,** and **square.** Plumb is the term used to describe true vertical. Level is the term used to describe true horizontal. Used in construction, the term **square** is any four-sided shape with four straight sides, four right angles, and equal diagonal measurements, as shown in Figure 8.22. King studs, shown in Figure 8.22, are studs used to support and trim both ends of a header, and run from the sole plate to the top plate. A **cripple** is a short stud framed above a door rough opening, or above and below a window rough opening between the header and a top plate, or between the sill and sole plate, as shown in Figure 8.22.

When the building you are framing has a standard floor to ceiling height, then pre-cut studs can be ordered and delivered to the construction site by the lumber company. This saves construction time by not having to cut every stud to the correct length before framing. Stud length is determined by the dimension from the top of **underlayment** to the finish ceiling. This dimension is 8 ft for a typical building but can vary depending on the architectural design. The pre-cut stud length is calculated to provide 8 ft plus 1/8 in. tolerance from the top of 1/2 in. underlayment to the bottom of 1/2 in. finish ceiling drywall, and by subtracting the top and sole plate thickness like this and as shown in Figure 8.23:

96-1/8: 8 ft plus 1/8 in. tolerance

−1/2: underlayment

−1/2: ceiling drywall

−1-1/2: sole plate

−3: top plate

92-5/8: inch pre-cut stud lengths

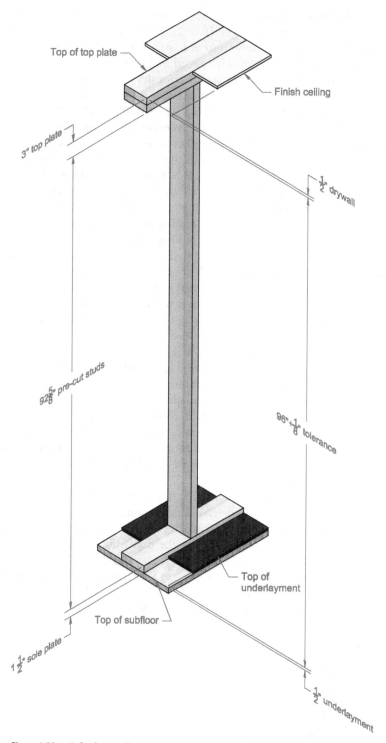

Top of top plate

Finish ceiling

3" top plate

$\frac{1}{2}$" drywall

$92\frac{5}{8}$" pre-cut studs

$96" + \frac{1}{8}$" tolerance

Top of underlayment

Top of subfloor

$1\frac{1}{2}$" sole plate

$\frac{1}{2}$" underlayment

Figure 8.23 Calculating the length of standard pre-cut studs for a typical 8-ft floor-to-ceiling height. (*Previously published in Modern Residential Construction Practices, Routledge, 2017 and reproduced here with permission.*)

Underlayment, described in Chapter 12, *Finish Work and Materials*, is construction material used over subflooring and under the finish floor material to provide a base for finish floor material.

The terms **sole plate, sill plate,** or **bottom plate** are all used to describe the single bottom horizontal member of a wall or partition to which the vertical members are attached, as shown in Figure 8.22. The sole plate is the same size material as the intersecting studs. There is no sole plate at a door opening. A sole plate used on a concrete slab must be pressure-treated lumber or a vapor barrier must be used under the slab. Confirm this requirement with the plans and specifications.

The **top plate,** shown in Figure 8.22, is a framing member on top of a stud wall on which joists rest to support an additional floor or to form a ceiling, or upon which rafters rest to form a roof. The top plate is generally constructed using double stacked dimensional lumber that matches the stud size. Initially, the first top plate lumber is nailed to the ends of the studs and then the second top plate lumber is nailed to the first. Splices in the double top plate must be staggered at least 24 in. (600 mm) to help maintain the strength of the entire top plate. See Figure 8.24a. A single top plate can be used if permitted by the structural requirements. The single top plate normally has a 6-in. (150-mm) 20-gauge galvanized steel plate centered over each lineal splice and fastened with six 8d nails on each side of the splice, as shown in Figure 8.24b. Figure 8.24c shows a 3 × 6-in. 20-gauge steel plates nailed with 3-8d nails on each side for single plate corners and intersecting walls.

A **header,** found in the wall framing system, is a horizontal structural member that supports the load over an opening such as a door, window, or any other opening. Headers framed in a **load-bearing** wall are either solid dimensional lumber, glue-laminated member, or two pieces of dimensional lumber separated by rigid insulation. A header can be a steel beam if required by structural engineering. A load-bearing wall, also called a **bearing** wall, is a wall that supports the weight of the structure resting on it from above, and by transferring the weight to the foundation structure. The size and type of header construction depends on the opening span and the structural load. This information is detailed and specified on the plans. Modern insulation typically requires the header to be insulated. Figure 8.25a shows a solid header and Figure 8.25b shows an insulated header over rough openings in a load-bearing wall. The insulated header is constructed using two 2× dimensional lumber or engineered wood separated by rigid insulation. The specifications for the dimensional lumber or engineered wood are found on the plans. The header framing shown in Figure 8.24b uses a header framed with a king stud and a jack stud to support the ends. Removing unnecessary wood framing within walls can increase energy efficiency of the wall system. A method used to reduce wood members in header construction is to replace the jack studs with steel header hangers, as shown in Figure 8.24c. Notice in this example that the single top plate technique is used to decrease additional wood framing. Structural engineering is needed to support these applications.

A headers framed in a **non-load-bearing wall** uses less wood, because it does not have the structural requirements of headers framed in a load-bearing wall. A non-load-bearing wall is a wall that supports only its own weight and does not support structural weight from above. Less wood used in the framing allows for more insulation to be used and helps reduce hot and cold spots in the wall framing system. A common framing technique for a non-bearing header is to use a single piece of dimensional lumber placed flat and framed

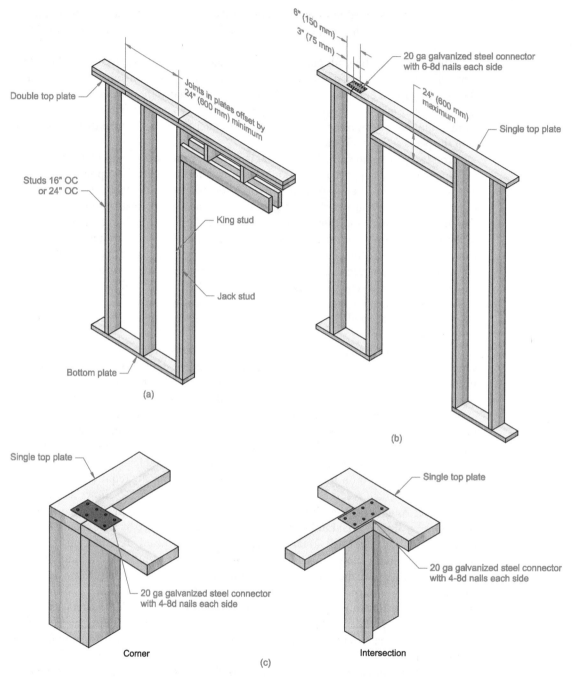

Figure 8.24 (*a*) Splices in the double top plate must be staggered at least 24 in. (600 mm) to help maintain the strength of the entire top plate. (*b*) The single top plate normally has a 6-in. (150-mm) 20-gauge galvanized steel plate centered over each lineal splice and fastened with six 8d nails on each side of the splice. (*c*) A 3 × 6-in. 20-gauge steel plates nailed with 3-8d nails on each side for single plate corners and intersecting walls. (*Previously published in Modern Residential Construction Practices, Routledge, 2017 and reproduced here with permission.*)

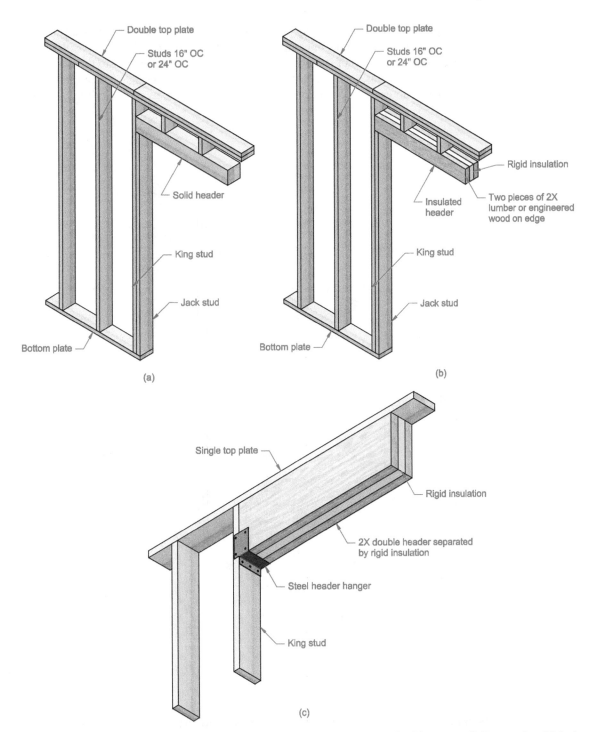

Figure 8.25   (*a*) A solid header. (*b*) An insulated header over rough openings in a load-bearing wall. (*Previously published in Modern Residential Construction Practices, Routledge, 2017 and reproduced here with permission.*)

with cripples connecting it to the top plate, as shown in Figure 8.26*a*. A king stud and jack stud are used to strengthen the opening and support the header in this example. When the opening is 8 ft (2400 mm) or less wide and the distance from the header to the top plate is 24 in. (600 mm) or less, then the jack stud can be omitted and a single flat 2× can be used as a header in an exterior or interior non-load-bearing wall, as shown in Figure 8.26*b*.

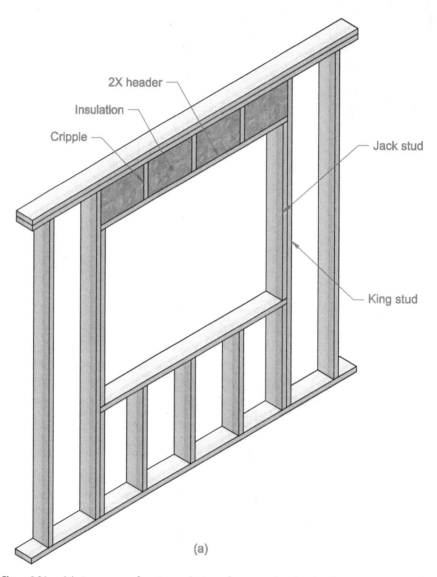

(a)

Figure 8.26   (*a*) A common framing technique for a non-bearing header is to use a single piece of dimensional lumber placed flat and framed with cripples connecting it to the top plate. (*b*) When the opening is 8 ft (2400 mm) or less wide and the distance from the header to the top plate is 24 in. (600 mm) or less, then the jack stud can be omitted and a single flat 2× can be used as a header in an exterior or interior non-load-bearing wall. (*Previously published in Modern Residential Construction Practices, Routledge, 2017 and reproduced here with permission.*)

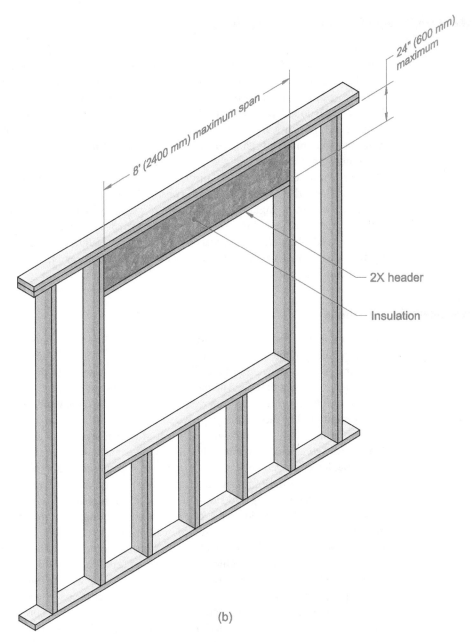

Figure 8.26   (*Continued*)

Cripples are not required above the header using this method and the full cavity is insulated on the exterior walls.

Rough window openings are framed with a header on top and a **sill** on the bottom, as shown in Figure 8.22. The word window sill is also spelled windowsill but is referred to as sill in the construction industry. The sill is the bottom member of the window framing rough opening and is the same size material as used for the studs.

Wall Framing with Engineered Wood Products

Engineered wood products include oriented strand board (OSB), glued-laminated beams (glulam), laminated veneer lumber (LVL), I-joists, plywood, and rim board. Engineered wood offers cost, environmental, and quality advantages over dimensional lumber. The initial cost of engineered wood studs can be more than or equal to dimensional lumber, but savings is found in less waste and reduced labor costs, especially when preassembled framing components are used. Use of engineered wood studs helps protect the environment by using material made from young trees and different tree species than trees used for dimensional lumber. Engineered wood studs provide improved appearance, strength, stability, wind resistance, and uniformity with less defects than dimensional lumber. Engineered wood studs can be used to construct any wall framing system but are especially useful when framing high walls with long studs. Quality long dimensional lumber studs are difficult or impossible to buy and can be very expensive if available. Long engineered wood studs are available at an economical cost and provide straight and uniform framing materials with exceptional strength. The wall framing methods previously described throughout this chapter apply equally to using engineered wood products.

Wall Sheathing and Siding

Sheathing is fastened to the exterior wall studs to reinforce, support, and strengthen the structure and provide a backing for finish materials. Wall sheathing is generally 1/2 in. structural exterior plywood or OSB. Sheathing is often fastened to the wall framing using 8d common or ring shank nails. The nail spacing can be specified on the plans or in the specifications. Light wood framing, referred to as **double wall construction,** uses sheathing that is commonly C-D grade exterior plywood or OSB with a **vapor barrier** placed over the sheathing and then covered with a finish **siding** material, as shown in Figure 8.27. Double wall construction is typically used on high-quality buildings. Double wall construction provides improved appearance, insulation, vapor barrier, and noise reduction. Siding is an exterior covering of finish material on a building. Vapor barriers, a layer of material used between the framing and siding to help prevent moisture from transferring from the outside through the exterior walls, can reduce possible moisture-related mold and mildew growth or even block gases that can be a health risk in the building. Historically, **tar paper** was used as a vapor barrier, but technically advanced products called **housewrap** are used today. Tar paper is a thick product that is manufactured in rolls and is made by impregnating paper with tar, producing a waterproof material. Housewrap is the term that describes a variety of synthetic products that have replaced tar paper for use as a vapor barrier. Housewrap is light weight and manufactured in wide rolls, liquid form, spray foam, or insulated sheathing for easy use in construction. Housewrap provides a weather-resistant barrier that prevents rain from getting into the wall, while allowing water vapor to pass to the exterior. A few of the many brand names for housewrap are Tyvek®, Typar®, WEATHERMATE™, Grip-Rite, BuildingGuard, or other products that can be found by searching the Internet for housewrap products. Siding can be wood, metal, plastic, composite materials, stucco, or masonry. The type of siding can be used to define or enhance the architectural style. Siding is selected for its architectural style, appearance, and durability. Additional siding information is provided in Chapter 13.

Single wall construction, not typically used in commercial construction, uses structural plywood or OSB siding over a vapor barrier placed over the exterior studs. The structural plywood or OSB siding serves as the sheathing and siding all in one layer. Plywood or OSB siding

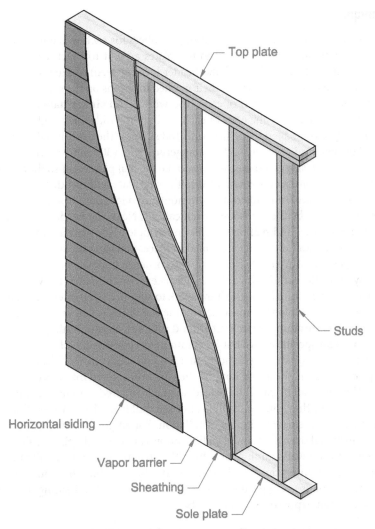

Top plate

Studs

Horizontal siding

Vapor barrier

Sheathing

Sole plate

Figure 8.27   Double wall construction uses sheathing that is commonly C-D grade exterior plywood or OSB with a vapor barrier placed over the sheathing and then covered with a finish siding material. (*Previously published in Modern Residential Construction Practices, Routledge, 2017 and reproduced here with permission.*)

is manufactured in 4 × 8-ft (1200 × 2400 mm) sheets. The most common structural plywood siding for single wall construction is **T1-11 siding** that has a textured finish exterior veneer that provides the appearance of traditional solid-wood siding. Vertical groves are cut into the face of the sheet to provide an additional siding appearance. T1-11 siding can be ordered with 4, 6, 8, or 12 in. on center vertical groves depending on the manufacturer and supplier. T1-11 siding can also be ordered without groves when it is desirable to install 1 × 2 equally spaced vertical boards to give the appearance of board and bat siding. Similar OSB products are available with a rough sawn appearance and vertical grooves. The OSB siding has a hardboard exterior face that protects and provides a durable outer surface. These products can also be used as the finish siding over double wall construction.

Wall Framing Techniques

An excellent place to layout the framing members for each wall and partition is on the flat framed subfloor. The sole plate and top plates are cut to exact lengths, laid out on their edges, and marked for the stud and header locations. All of the full-length studs are then placed on edge and aligned with their marked locations on the sole and top plates. The sole plate is fastened to the ends of each stud with **2-16d sinker nails.** The first board of the top plate is fastened to the ends of each stud with 2-16d sinker nails, and then the second top plate board is nailed to the first with 16d sinkers. The exact number of nails used can be based on standard practice, such as **staggered** 12 in. (300 mm) apart, or as specified on the plans. The term **staggered** means to alternate the nail placement so they are not in a straight row. Jack studs are placed and nailed to the sole plate and the king studs after all full studs are nailed in place. 16d sinkers are commonly used for framing applications, but other nails can be required by the plans and specifications. Next, the headers are nailed in place, followed by locating and nailing the sills and cripples. The wall framing is carefully squared after all of the wall members are nailed in place. Squaring the wall frame is done my making diagonal measurements from each top corner to opposite bottom corner, and adjusting the layout until each diagonal dimension is exactly the same. Now, the wall sheathing or structural siding is nailed to the framework to support, strengthen, and maintain the square structure. For double wall construction, the sheathing is laid on the framework and nailed in place. The sheathing is cut to fit around rough openings. Nailing can vary depending on the plans and specifications, but 8d common or ring shank nails are generally spaced 6 in. (150 mm) apart on the edges and 12 in. (300 mm) apart in the fields. The vapor barrier can be placed over the sheathing now, or after the wall is tilted into place.

Windows can be carefully placed, aligned, squared, and fastened to the wall framing at this time, or they can be installed after the wall is in place. Always follow manufacturer instructions for correct window installation. For single wall construction, the vapor barrier is installed over the framed wall, windows are carefully placed, aligned, squared, and fastened to the wall framing, and then the structural siding is nailed to the framework in the same manner described for installing the sheathing on double wall construction. Galvanized nails are typically used to fasten the structural siding, because they do not corrode when exposed to the elements.

> **NOTE**
>
> A sinker nail is a type of nail typically used in wood-frame construction. A sinker nail is thinner than a **common nail,** and coated with adhesive for easier driving and to enhance holding power. A sinker nail has a funnel-shaped head with a grid stamped on the head. A common nail has a smooth uncoated shank less than one-third the diameter of the nail head and is most commonly used framing. A 16d nail is commonly used in framing and is 3-1/2 in. long. **d** in a nail specification refers to penny, which is an old term still used to specify nail sizes today. The nail size gets larger as the number gets larger.

Locations for hold-down anchors are prepared by drilling or other means as needed to later fasten the anchor systems. The wall is tilted in place after the wall is framed, squared, and the sheathing or structural siding is installed. A long wall can be very heavy and jacks must be used, for safety, to tilt the wall in place. The wall is properly aligned in its location and the sole plate is nailed to the structure below with 16d sinkers placed staggered at least 12 in. (300 mm) apart. Confirm nailing with structural engineering on plans and specifications. The wall is then plumbed and braced with long equally spaced boards that are temporarily nailed near the top of studs and placed at an angle down to the floor where they are temporarily nailed to the subfloor. This wall bracing is very important to keep the wall standing until the entire wall system is framed in place and ceiling joists are installed to secure the structure. Hold-down anchors are properly fastened based on manufacturer instructions.

Standard height straight walls can normally be framed as previously described, but custom wall designs are often framed in place one board at a time. Walls can be framed with or without installing the sheathing and windows depending on the complexity of the structure or the type of windows.

FRAMING CORNERS

Exterior wall corners are framed to create a strong connection between intersecting walls and to provide a nailing surfaces for exterior sheathing, siding, and interior **drywall** or other interior finish materials. This is done using a minimum number of corner studs to allow for maximum insulation at the corners of the building. **Drywall** is also called **gypsum board, gyprock, plasterboard, sheetrock,** or **wallboard,** is a plaster panel made of **gypsum** pressed between two thick sheets of paper. Gypsum is a soft white or gray mineral consisting of hydrated calcium sulfate and is used to make plaster, which is mixed with water and allowed to harden for various applications such as making drywall. Drywall is available in sheets ranging from 4 × 8 to 4 × 16 ft. Other interior wall materials can include solid wood boards, metal, plastic, or plywood **paneling** with an outer hardwood veneer. Paneling is material used to cover an interior wall, and usually manufactured in 4 × 8-ft (1200 × 2400-mm) sheets. Many different types of paneling are available with a variety of surfaces or patterns for a desired appearance. Plywood paneling is made with hardwood or softwood veneer. Hardwood paneling can be selected with one of many attractive and exotic hardwood veneers for quality architectural appearance. One of the three following different exterior corner framing methods is commonly used in building construction.

The exterior **two-stud corner,** shown in Figure 8.28, has the least number of studs and allows for the maximum amount of insulation in the corner. This corner framing has two

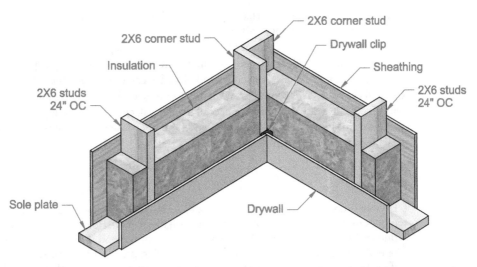

Figure 8.28   Exterior two-stud corner. (*Previously published in Modern Residential Construction Practices, Routledge, 2017 and reproduced here with permission.*)

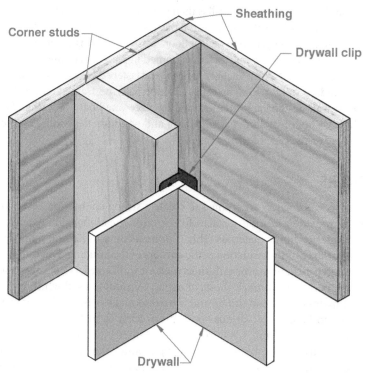

Figure 8.29   Drywall clips or other nailing brackets are used to secure the other drywall material edge at a corner that has no framing support. Drywall clips are placed on the edge of the drywall sheets and fastened to the stud with a nailing strip that is fastened to the studs with nails, staples, or screws and placed every 16 in. along the stud. (*Previously published in Modern Residential Construction Practices, Routledge, 2017 and reproduced here with permission.*)

studs that provide nailing surfaces for the exterior sheathing and siding and for one surface of the interior drywall. **Drywall clips** or other nailing brackets are used to secure the other drywall material edge at a corner that has no framing support. Drywall clips are placed on the edge of the drywall sheets and fastened to the stud with a nailing strip that is fastened to the studs with nails, staples, or screws and placed every 16 in. along the stud. See Figure 8.29. Drywall clips can also be used to secure the top edge of drywall along the top plate. An alternate practice uses a 1 × 6 **nailer** in place of the drywall clip when framing the two-stud corner, as shown in Figure 8.30. A nailer is a wood member fastened to the structure and used for attaching other wood members or finish materials.

NOTE

Drywall is described here because it relates to wall construction, and specific framing techniques must be used to accommodate drywall installation. However, drywall is not installed during the framing process. Drywall is installed near the end of the construction process as one of the last applications to complete the building project. Drywall installation is normally done by a subcontractor who specializes in professional drywall installation and finish. Additional information is provided in Chapter 12.

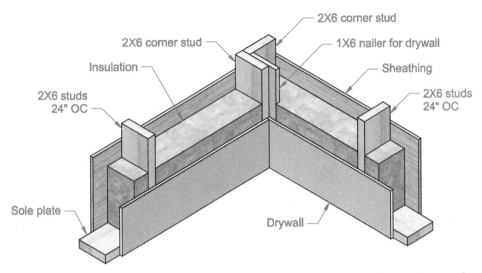

Figure 8.30   A 1 × 6 nailer in place of the drywall clip when framing the two-stud corner. (*Previously published in Modern Residential Construction Practices, Routledge, 2017 and reproduced here with permission.*)

The **three-stud corner** uses three studs to secure and strengthen the exterior corner and provide a nailing surface for sheathing, siding, and interior drywall or other interior finish materials. There are two basic three-stud corner systems that only differ by the way the studs are placed, as shown in Figure 8.31. The corner framing in Figure 8.31*a* uses three studs, with two studs placed to provide a nailing surface for the interior drywall or other finish material. This three-stud corner framing method is called a California corner or open corner in case you hear this terminology used in the industry. This is commonly used, because it allows for good insulation installation and does not require drywall clips for interior surface fastening. Figure 8.31*b* uses three studs, with two studs placed for exterior sheathing and siding nailing, and requires drywall clips to be used to fasten the edge of one sheet at the inside corner.

Interior partitions usually do not require insulation unless insulation is used to reduce sound between rooms. Interior corner framing can be done without the concern for providing maximum insulation as important with exterior corner framing. A common interior partition corner framing method uses three studs, as shown in Figure 8.32*a*. A traditionally common interior corner framing technique uses three studs with two of the studs separated by blocking to reduce material, as shown in Figure 8.32*b*.

FRAMING INTERSECTING PARTITIONS

A **T-intersection** occurs when an interior partition intersects an exterior wall and the number of studs used needs to be kept to a minimum for the maximum amount of insulation to be used. One framing option for the T-intersection is the use of 1 × 6 or 2 × 6 horizontal blocking placed 24 in. (600 mm) on center behind the intersecting interior partition, as shown in Figure 8.33*a*. Another T-intersection option is the use of a full length vertical 1 × 6 or 2 × 6 behind the intersecting interior partition, as shown in Figure 8.33*b*.

One of the previously described framing methods can be used when interior partitions intersect, or a four-stud intersection can be used, as shown in Figure 8.34. The commonly used four-stud intersection has four studs next to each other and is easy to frame.

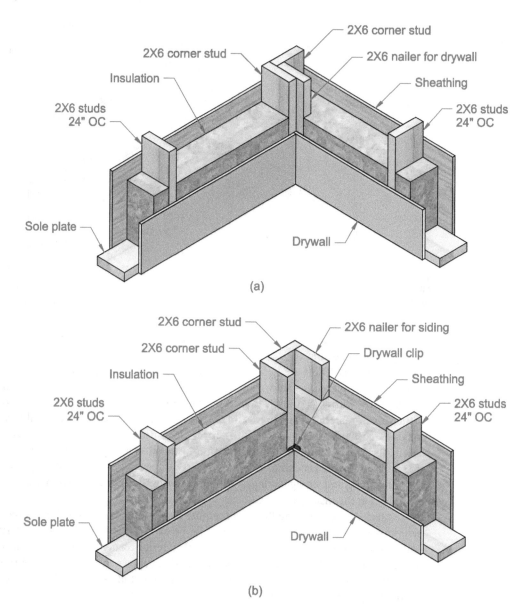

Figure 8.31 Two basic three-stud corner systems. (*a*) This corner framing uses three studs, with two studs placed to provide a nailing surface for the interior drywall or other finish material. (*b*) This corner framing uses three studs, with two studs placed for exterior sheathing and siding nailing, and requires drywall clips to be used to fasten the edge of one sheet at the inside corner. (*Previously published in Modern Residential Construction Practices, Routledge, 2017 and reproduced here with permission.*)

FRAMING SHEAR WALLS

Framing shear walls is an important part of building construction, with information repeated here for your review and understanding. A shear wall, also called a braced wall line, is a wall made of a shear panel to oppose the effects of lateral loads acting on a structure. A shear panel, also called a braced panel, is typically part of a wood frame stud wall

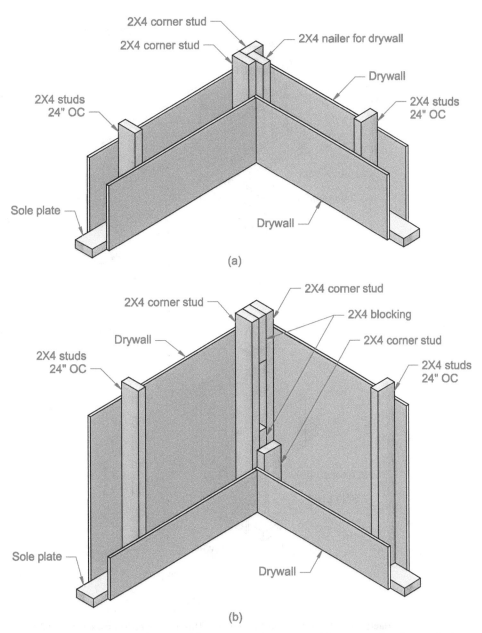

Figure 8.32 Interior corner framing. (*a*) A common interior partition corner framing method uses three studs. (*b*) A traditionally common interior corner framing technique uses three studs with two of the studs separated by blocking to reduce material. (*Previously published in Modern Residential Construction Practices, Routledge, 2017 and reproduced here with permission.*)

that is covered with structural sheathing such as plywood, but other materials such as steel and bracing systems can be used. Hold-down anchors, also called hold-downs or holdowns, are an important part of shear wall construction. Hold-down anchors are fastening systems that are embedded in concrete foundation walls and attach to wood structural walls above, or attach wood structural walls between floors, and especially at shear walls to control uplift

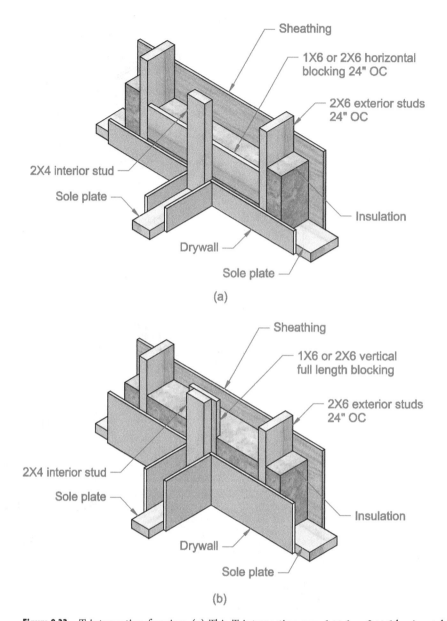

Figure 8.33 T-intersection framing. (*a*) This T-intersection uses 1 × 6 or 2 × 6 horizontal blocking placed 24 in. (600 mm) on center behind the intersecting interior partition. (*b*) This T-intersection option uses a full length vertical 1 × 6 or 2 × 6 behind the intersecting interior partition. (*Previously published in Modern Residential Construction Practices, Routledge, 2017 and reproduced here with permission.*)

in the wall system. The hold-down is connected to the concrete foundation or structural slab by an embedded or adhesive-installed anchor bolt at the bottom. The hold-down is connected to a wood post with screws, nails, or bolts at the top, as shown in Figure 8.35. In structural engineering, a shear wall, also called a braced wall line, is a wall composed of shear panels, also known as braced panels, to counter the effects of **lateral load** acting on a structure.

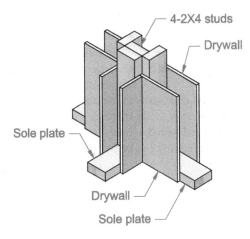

Figure 8.34 This commonly used four-stud intersection has four studs next to each other and is easy to frame. (*Previously published in Modern Residential Construction Practices, Routledge, 2017 and reproduced here with permission.*)

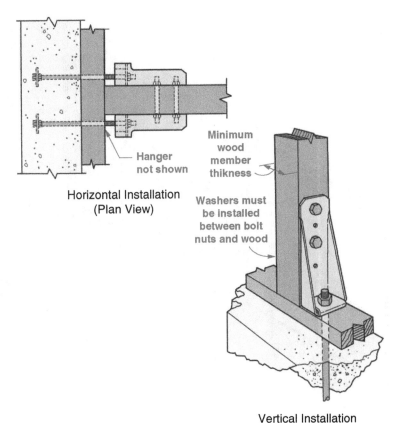

Figure 8.35 The hold-down is connected to a wood post with screws, nails, or bolts at the top. (*©Simpson Strong-Tie Inc. and/or its affiliates. All Rights Reserved.*)

A shear panel is typically part of a wood frame stud wall that is covered with structural sheathing such as plywood, but other materials such as diagonal bracing systems, sheet steel, and steel-backed shear panels can be used. All exterior walls and some interior walls have braced panels depending on structural engineering requirements. Interior shear walls can be bearing or non-bearing. The plywood sheathing on a shear panel is securely nailed to the stud wall framing at the edges and in the field with nails of a specific size and spacing based on structural engineering. An engineered shear wall has the strength and stiffness to make the structure safe against **lateral loads.** The construction of a structural plywood-reinforced shear wall is shown in Figure 8.36. A lateral load is a force working on a structure applied parallel to the ground, and diagonally to the structure. Wind and **seismic** loads are the most common lateral loads. Seismic loads are caused by earthquakes.

A shear wall can also be constructed with diagonal wood or metal T-bracing by cutting a diagonal slot in the studs where the bracing can be inserted and screwed or nailed to the

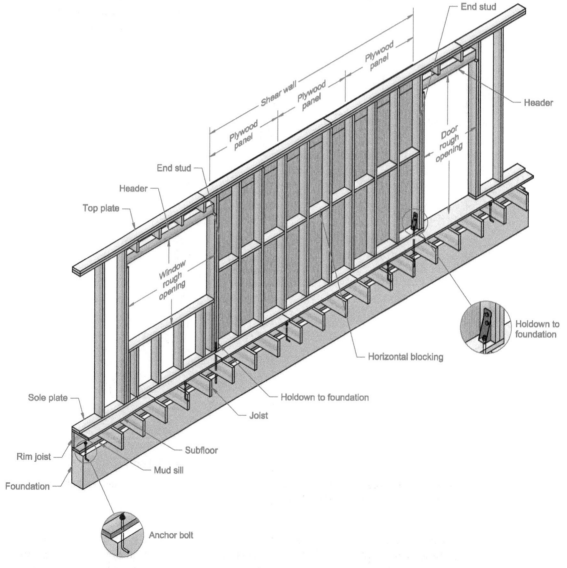

Figure 8.36   The construction of a structural plywood-reinforced shear wall. (*Previously published in Modern Residential Construction Practices, Routledge, 2017 and reproduced here with permission.*)

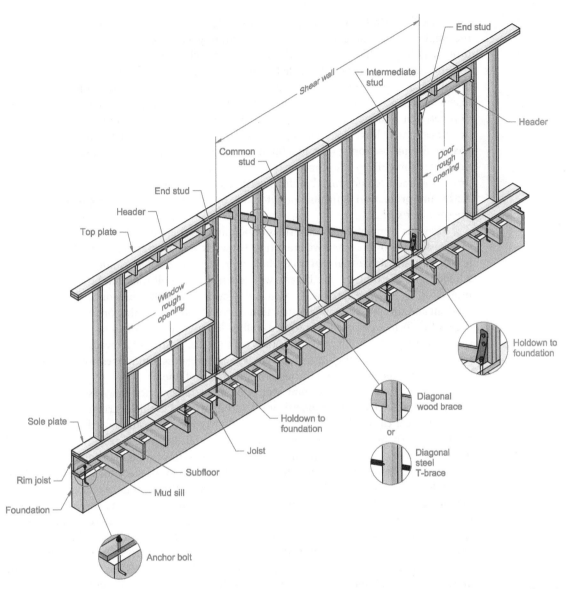

Figure 8.37   A shear wall constructed with diagonal wood or metal T-bracing. (*Previously published in Modern Residential Construction Practices, Routledge, 2017 and reproduced here with permission.*)

studs, as shown in Figure 8.37. The structural engineer determines the system to be used and provides drawings and specifications included on the set of plans.

FIRE AND SOUND WALLS

Multifamily buildings such as apartments, condominiums, and townhomes require fire-resistant-rated **building assemblies** on **common walls** and between adjacent units to prevent the spread of fire, smoke, and heat through walls and floors from one unit to another for a specific period of time. Fire-resistant standards also allow the structural integrity of the building to be maintained during a fire. **Building assembly** is all the parts of the structure, including the floor, walls, and roof including all layers of material used in the construction. A common wall, also called a party wall, is a structural **fire-rated wall** and floors dividing partition between two adjoining living units or buildings shared by the separate occupants of each residence or business. This type of wall is usually required for construction between a garage and living space, and at exit corridors and stairways to make sure people can safely leave the building in the event of a fire. A fire-rated wall is a feature of a **passive fire protection** system in a building designed to meet code requirements. Passive fire protection (PFP) is the combination of all construction components of structural fire protection and fire safety in a building, designed to contain fires or slow the spread of fire. **Active fire protection** is manual or automatic fire detection and fire suppression, such as fire sprinkler systems.

Firewalls completely separate sides of buildings and provide stable structural support. Two basic types of fire-rated walls are **true firewalls** and **fire barrier walls.** A true firewall is a structurally stable wall that is certified to prevent fire from spreading from one side of a building to the other. A fire barrier wall is a nonstructural wall that provides fire-rated protection. A true fire-rated wall extends downward through the floor to the foundation mud sill and through the roof construction under the roof sheathing or continues through the roof to form a parapet. A fire-rated wall is designed and certified to prevent fire from spreading through a building if a fire starts on one side. A fire barrier wall, also called a fire partition, helps to contain fire and prevent it from spreading, but it does not provide structural support, does not extend through the floor or roof line.

The fire-resistance of floor and wall wood-frame assemblies depends on the gypsum board, also called drywall, wallboard, or plasterboard, used to shield structural wood members from heat caused by fire. There are different types of gypsum board on the market, but the construction of fire-rated wood-frame assemblies requires specially manufactured fire-rated sheets. Fire-rated gypsum has glass fibers that improve dimensional stability and nail-head pull-through resistance, allowing them to remain in place for longer periods of time when exposed to fire. Specific codes and construction design practices are used to maximize the length of time gypsum board remains in place during a fire. Fire-rated gypsum is installed with the goal of creating a fire separation that is a requirement design for specific locations in building codes. Two types of fire-rated gypsum are **Type X** and **Type C**. Type X gypsum is 5/8 in. (16 mm) thick and installed on each side of a wood framed wall to create the fire separation. Type C gypsum is made 1/2 in. (13 mm) and 5/8 in. thick with more glass fiber reinforcement and other ingredients that make it an improved fire separation material than Type X. Fire-resistant ratings are given in the number of minutes or hours a structure can withstand a fire simulation test. A common example is a 1-hour rating that indicates a wall constructed can contain flames and high temperatures, and support its full load, for at least 1 hour after the fire begins. A typical 1-hour fire-rated wall is built with 2×4 studs spaced 16 in. (400 mm) on center, covered with 5/8 in. Type X gypsum, as shown in Figure 8.38a. Two-hour firewall constructed can contain flames and high temperatures, and support its full load, for at least 2 hours after the fire begins. There is more than one construction method used to build a 2-hour firewall. One method shown in Figure 8.38b uses a double wood-frame wall construction with Type X gypsum on each side of each framing. Another option for an area separation firewall, as shown in Figure 8.39, is designed for the construction to collapse on the fire-exposed side, without collapse of the entire wall. This construction uses aluminum breakaway clips attaching the separation wall to the adjacent framing. When one side of the separation wall is exposed to fire, the clips are designed

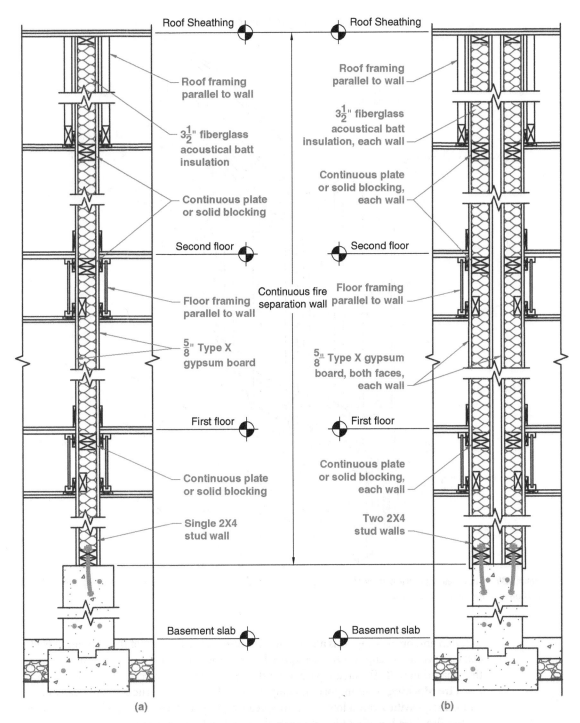

Figure 8.38   (*a*) One-hour fire-rated wall construction. (*b*) Two-hour fire-rated wall construction.

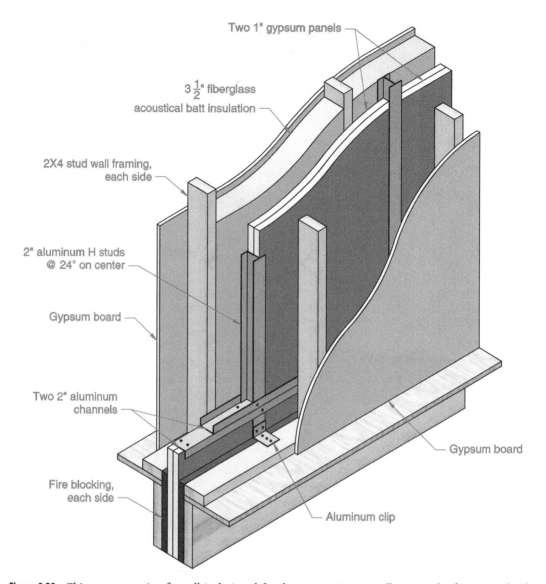

Two 1" gypsum panels

$3\frac{1}{2}$" fiberglass acoustical batt insulation

2X4 stud wall framing, each side

2" aluminum H studs @ 24" on center

Gypsum board

Two 2" aluminum channels

Fire blocking, each side

Aluminum clip

Gypsum board

Figure 8.39 This area separation firewall is designed for the construction to collapse on the fire-exposed side, without collapse of the entire wall.

to soften and break away. This allows the structure on the fire side of the separation wall to collapse, while the clips on the unexposed side of the separation wall continue to support the separation wall. The area separation wall remains intact, protecting the adjacent unit.

Fire blocking is an important component of firewall construction. Any gaps in wall and ceiling cavities can allow a fire to spread rapidly, because they provide airflow that feeds a fire. Fire blocking fills and seals the construction cavity to prevent or slow down the spread of fire. Specific fire blocking is required in firewall construction and must be confirmed with building codes.

Another consideration is openings in fire walls, such as doors. Automatically activated fire doors or shutters are required to secure the opening in the event of a fire. Further, opening sizes are restricted by building codes and need to be confirmed for passive and active fire protection.

This fire protection content has focused on wood-frame construction for light commercial and multifamily projects. Concrete and masonry firewalls are also an option for passive fire protection but more commonly used in other commercial construction projects. Excellent fire-resistant construction materials and systems described in previous chapters include precast concrete, concrete masonry units, and other masonry products such as bricks.

Materials used in firewalls and fire separation construction practices can improve sound dampening control between dwelling units. **Sound transmission class (STC)** is a rating of how well a building partition reduces airborne sound, which is any sound transmitted by the air. The STC rating reflects the **decibel** reduction in noise that a partition can provide and a measurement of how much noise is stopped. Decibel is a unit used to measure the intensity of a sound. STC ratings are used for windows, doors, walls, and most building materials. STC ratings range from 18 to 38 for windows. STC ratings are the only way to accurately compare various noise reduction products.

A construction assembly can increase its STC rating by doubling the gypsum board layers on each side of framing, installing extra sound reduction fiberglass insulation in the wall cavity and including an air space, which helps eliminate **structure-borne sound.** Structure-borne sound is the sound that spreads in solid bodies such as walls and floors, and pipes by tremors conducted by vibrations. A construction example is a double stud walls with a single layer of laminated noise reducing gypsum to dampen sound can give a higher STC rating.

ENERGY-EFFICIENT WALL FRAMING METHODS

The following discussions and examples represent advanced framing practices that are becoming an industry standard and used to meet current and future energy code requirements. These framing techniques help increase energy efficiency without compromising structural integrity by eliminating unnecessary framing members. The following reviews and expands on the previous information and is demonstrated in Figure 8.40.

Buildings can be designed using **24-in. (600-mm) modules**, which means that every dimension is a 24-in. increment, such as 30×48 ft ($9000 \times 14,600$ mm). The use of 2×6 exterior studs 24 in. (600 mm) on center rather than 2×4 or 2×6 exterior studs 16 in. (400 mm) on center reduces the amount of wood used and helps increase the amount of insulation that can be placed in the exterior walls. Additional insulation increases the energy efficiency of the building. Insulation and barriers used in building construction to achieve quality energy efficiency is described in Chapter 10, *Insulation and Barriers, and Indoor Air Quality and Safety*.

When structural engineering permits the use 2×6 studs 24 in. on center, the framing members below and above are also framed 24 in. on center to align with the studs. This practice helps increase energy efficiency between walls, floors, and ceilings. This structural alignment method also improves structural effectiveness, decreases drywall cracking, and reduces material waste, as shown in Figure 8.40.

Use of the exterior corners and T-intersections between interior partitions and exterior walls as described in this chapter helps reduce the amount of wood used in framing and maximizes the amount of insulation that can be used in the exterior walls, as shown in Figure 8.40.

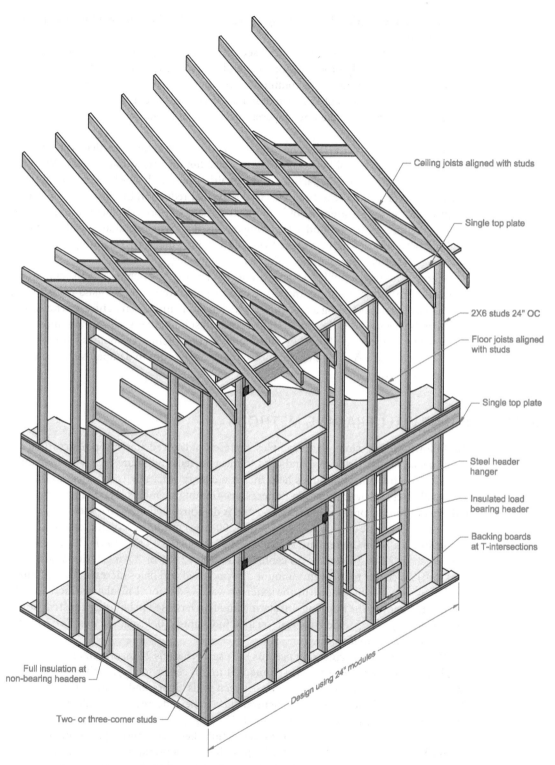

Figure 8.40 Advanced framing methods used to reduce wood framing members and increase insulation. (*Previously published in Modern Residential Construction Practices, Routledge, 2017 and reproduced here with permission.*)

Advanced header framing techniques previously described allow for partial insulation to be used at headers for exterior bearing walls. Headers used in non-bearing exterior walls can be framed to allow for full insulation, as shown in Figure 8.40.

Using Rigid Insulation Sheathing

Building design and construction continues to move toward the development of highly energy-efficient **sustainable** building systems. The word sustainable is commonly used today to describe anything that is capable of being continued or maintained with minimum or no long-term effect on the environment. Sustainable buildings are efficient, durable, and economically practical and are able to maintain their desired function into the future.

Advances in wall construction use rigid insulation with sheathing or in the place of sheathing to help increase energy efficiency. Exterior rigid insulation sheathing is manufactured in a variety of dimensions from 3/4- to 2-1/2-in. thick and 4 × 8-ft (1200 × 2400 mm) sheets. Additional information about insulation is described in Chapter 1. When properly installed, rigid insulation improves energy efficiency and reduces the possibility of **condensation** within exterior wall systems. Condensation is water that collects when humid air comes in contact with a cold surface and can cause many problems within a wall, such as mold, mildew, and rot. There are several options for using rigid insulation as sheathing. Plywood or OSB sheathing can be installed over the stud framing, followed by housewrap, followed by rigid insulation, and finally the finish siding installed, as shown in Figure 8.41a. Rigid insulation can be used over housewrap that is installed directly on the stud framing, as shown in Figure 8.41b. Rigid insulation can be installed over the stud framing, followed by housewrap, as shown in Figure 8.41c. Rigid insulation sheathing is available that has an integral vapor barrier or housewrap connected by the manufacturer and can be installed directly over the stud framing, as shown in Figure 8.41d. The options shown in Figure 8.41b, c, and d eliminate the need for structural plywood or OSB sheathing. However, the structure must be engineered to accept one of these applications. These rigid insulation system used with advanced framing practices can provide cost savings by reducing building materials, such as less studs, less top plate material, and the elimination of plywood or OSB sheathing. These practices can also save in labor costs and reduce waste in the construction project. Always confirm local building code approval of these practices and follow manufacturer instructions and specifications.

> **NOTE**
>
> Rigid insulation sheets are available with interlocking edges that provide a tight seal between sheets. It is very important to install housewrap and rigid insulation carefully following manufacturer instructions in an effort to maintain quality and performance, and to make sure the wall system has no air or water leaks. Also, use manufacturer specific seam tape and follow application instructions.

Structural Insulated Panels

A **structural insulated panel (SIP)**, also called a **panelized wall system,** is a highly energy-efficient building system for light commercial construction that is custom manufactured for each building and delivered to the project site for installation. The panels are made with a rigid insulating core between two structural outer sheets, as shown in Figure 8.42. The outer sheets can be sheet metal, plywood, cement board, or typically OSB. SIPs are manufactured

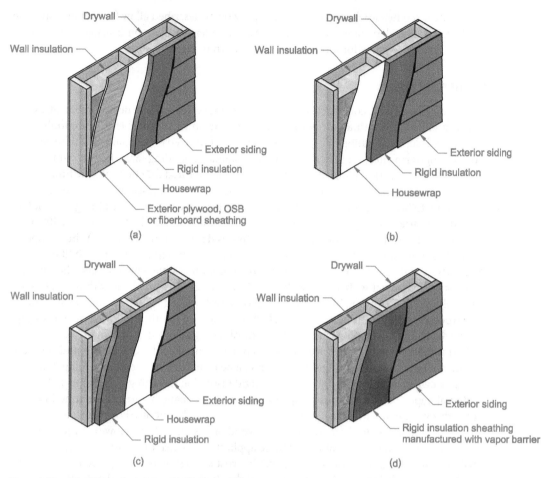

Figure 8.41 (*a*) Plywood or OSB sheathing installed over the stud framing, followed by housewrap, followed by rigid insulation, and finally the finish siding. (*b*) Rigid insulation used over housewrap that is installed directly on the stud framing. (*c*) Rigid insulation installed over the stud framing, followed by housewrap. (*d*) Rigid insulation sheathing is available that has an integral vapor barrier or housewrap connected by the manufacturer, and can be installed directly over the stud framing. (*Previously published in Modern Residential Construction Practices, Routledge, 2017 and reproduced here with permission.*)

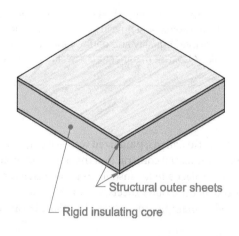

Figure 8.42 Structural insulated panels (SIP), also called a panelized wall system, are made with a rigid insulating core between two structural outer sheets.

Figure 8.43 Structural insulated panels being installed at a building construction project.

under factory-controlled conditions and can be fabricated to fit nearly any building design. The result is a building system that is extremely strong, energy efficient, and cost effective, while reducing sound between rooms. Building with SIPs generally costs about the same as building with wood-frame construction when you consider the labor savings resulting from shorter construction time and less construction waste. Long-term saving is also found with energy savings when a building is constructed using SIPs. The SIP system has no studs. Plates, or headers, reduce energy efficiency through the wood in a conventional stud framed wall. The result is a wall system with twice the insulation value and less possibility of air leaking as with a traditional exterior wall. SIP products vary between manufacturers, but custom panels can be made up to 8 × 24 ft (2400 × 7000 mm) with T&G edges for easy and tight assembly. Panel thickness is the same as traditional stud wall framing thickness, allowing for the use of standard doors and windows. SIPs are manufactured to match the building design, making panels ready to install when delivered to the project site. At the jobsite, each panel is laid out using an arrangement based on manufacturer instructions. The panels are tilted into place, with all joints sealed and properly fastened. The walls for the new building are now ready for finish work. Figure 8.43 shows a structural insulated panel being installed at a building construction project.

> **NOTE**
>
> When using SIPs on a construction project, a location needs to be prepared for delivery. SIPs must be protected from exposure to the elements and must not be stored directly on the ground. SIPs are bulky and heavy. Manpower is enough to move small panels, but larger panels often require a forklift or crane.

INTRODUCTION TO LIGHT WOOD ROOF FRAMING

Commercial roof construction was covered in Chapter 7 without light wood roof construction content. Light wood roof construction materials and practices are better covered here with continuity related to the previous wood construction practices in this chapter.

The following introduces you to **roof** framing materials and their applications, along with common roof framing construction practices. Framing techniques that help improve energy efficiency is also described in this chapter. Roof framing starts after wall framing is complete. A roof is the supporting structure and exterior surface on top of a building. The roof does more than cover and protect the building. The roof also represents an architectural style of the building. There are many different roof types and combinations that define the architectural style of the building.

ROOF STYLES

There are many different roof styles for use on a building. The roof style selected by the architect or building designer is important to establish the architectural style of a building. You can see different roof styles on buildings as you drive around your local community. The roof styles you see are often determined by the age of the buildings in a specific area. Buildings built in the 1930s and 1940s likely have **traditional architecture** with steep gable and hip roof styles. Buildings built in the 1950s and 1960s probably have low **slope** hip, gable, or flat roofs. Buildings constructed in the 1970s and 1980s often have steeper roof pitches in a shed style. As architectural age approaches 2000 and beyond, architecture has commonly evolved to a **Craftsman style,** where roofs again take on a historically traditional style with steep hip and gable roofs. These observations are generalities that do not always hold true, because you also see **contemporary** and **modern architecture,** where roof style is unique and as different as the architectural creativity. The point is that architectural style changes throughout the years and roof styles change to represent the trends. Roof slope, also called pitch, is the amount of rise a roof has compared to a horizontal measurement called the run. Figure 8.44 shows a 5/12 roof slope, which means that the roof rises 5 in. for every 12 in. of run. The roof slope symbol in Figure 8.44 is typically displayed throughout a set of drawings wherever the roof slope is shown. The term **traditional architecture** used here refers to architectural styles that evolved from the early American influence and from

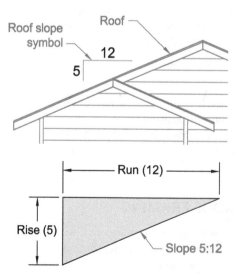

Figure 8.44   Roof pitch example. (*Previously published in Modern Residential Construction Practices, Routledge, 2017 and reproduced here with permission.*)

regions around the world. The following gives a few common architectural style names. Use these names to search the Internet for you to see example of how these styles look. Styles often associated with traditional architecture are Colonial, Cape Cod, Tutor, and Victorian. Craftsman style architecture is influenced by the use of natural materials and colors, heavy posts and wood beams, trim boards, masonry veneer accents, and wide porches. Modern architecture refers to a unique architectural style that is different from other styles and can be seen in any time period. The term **contemporary architecture** is the modern architecture of today. Modern and contemporary architectural examples are wide and varied because of their unique characteristics. What these buildings have in common is irregular shapes with frequent flat vertical surfaces with extensive large windows, and flat or shed roofs. Contemporary buildings of the twenty-first century often incorporate green design with an effort to fit into the environment without distraction in the surroundings. The following provides additional roof terminology:

A **flat roof** for a building is constructed almost horizontal or level with a slight slope that allows for rainwater drainage. The flat roof is often used in modern and contemporary architecture.

A **shed roof** has a single slope that is often used in modern and contemporary architecture.

A **gable roof** is a roof system that slopes downward both ways from a central horizontal ridge so as to leave a **gable** at each end. A gable is the vertical triangular wall built on each end of a building with a gable roof. The gable roof style is one of the most popular used in building design and construction.

The **A-frame roof** is an architectural building style with steep roof planes that also function as walls and meet at a central horizontal ridge. The two roof slopes commonly start at or near the foundation line and meet at the ridge, forming A-shaped gable ends. A-frame buildings became an option for mountain and seaside buildings.

A **hip roof** has the ends and sides sloping with the same pitch. The hip roof style is one of the most popular used in building design and construction.

The **Dutch hip roof** is a combination of hip and gable roof, with the gable portion at the top.

A **Gambrel roof** has a double slope on each side, where the lower slope is normally steeper than the upper slope. The Gambrel roof style is sometimes used in building design, but you will often see a Gambrel roof used on agricultural buildings.

The upper story of a **mansard roof** building is generally designed under the lower slope of the roof. A mansard roof has a double-hip roof on each face with a steeper lower portion. The roof above the double hip can be flat.

A **dormer** is a small structure extending out from the slope of the main roof of a building. The dormer is generally designed and built to provide additional space and a window for a room framed under the main roof slope. Dormers have walls and a roof of their own. The front face generally has a window, and the roof can be a gable or hip, designed to match the architectural style of the building. Dormer construction is not as common in commercial architecture as in residential, because of the limited space provide as related to the construction cost.

A **solarium** is a room with walls of glass and often have roofs made with extensive glass. A solarium can extend the architectural area by providing an outdoor-style atmosphere where plants grow well all year. A thermal mass can also be used in the solarium to store heat from the sun. Heat from the solarium can be circulated throughout the entire building by natural convection or through a **forced-air system.** Forced-air refers to a system that used a mechanical blower to circulate air throughout the building. Forced-air systems are described in Chapter 13, *Mechanical, Plumbing, and Electrical Systems*. Solariums and solar practices were described in Chapter 2, *Sustainable Technology*.

A **green roof** is also known as a **rooftop garden** has plants over the roof structure to help reduce building temperatures, filter pollution, and lessen water runoff. A green roof reduces the **urban heat island effect** and helps temper heating and cooling loads in the building. The urban heat island effect means that city areas are warmer than suburbs or rural areas due to less vegetation, more land coverage, and other infrastructure. The green roof also slows stormwater runoff from the roof and lessens the load on the public wastewater system. Green roof practices were described in Chapter 2.

The number of floors in a building also influences the architectural style. A **one-story building** has one level with no stairs. The roof over a one-story building is generally over an unusable attic space, and the rafters are typically fairly low pitch but can be any desired pitch to match the style. A **two-story building** has two floors with two full-height walls built with the second floor framed above the first floor, and there is a stairs for access between floors. A **one and a half story building** has one level on the main floor with one full-height wall, and with a second floor under the rafters, not generally used in commercial design because of the limited space for the cost.

ROOF FRAMING MATERIALS AND TERMINOLOGY

Dimensional lumber is the most commonly used material for light wood-frame roof construction, but light steel materials are an option for designers and contractors using steel products for building construction. It is important for you to learn roof framing terminology and at the same time you can discover the many options and techniques used in roof framing applications.

The most basic components of the roof framing system are the **rafters, ridge, joists,** and **collar ties.** Look at Figure 8.45 as you discover the following roof framing members and terminology. The construction practiced using individual roof members is called **stick framing,** because one board or *stick* is used at a time to assemble the structure. A rafter is the sloped structural members of a roof system used to support the roof loads and connect from the wall top plate to the ridge. There are different types of rafters explained and shown throughout this discussion. The main rafter in the roof system is called a **common rafter.** Common rafters are typically dimensional lumber such as 2 × 6 or 2 × 8 spaced 16 or 24 in. (400 to 600 mm) on center that run at the desired roof pitch from the wall top plate at the bottom to the ridge at the top. The ridge is the very top of the roof and is framed with a ridge board that runs horizontally and provides support for the upper ends of the rafters. The ridge is generally 2× dimensional lumber that is 2 in. (50 mm) deeper than the rafters. This allows the full end of rafters to rest on the ridge board, where the rafters are nailed to the ridge. The bottom of the rafters, called the **tails,** generally extend past the exterior wall a desired distance to create an **eave** or **overhang.** The terms **eave** and **overhang,** also called **cornice,** refer to the lower portion of the roof construction beyond the exterior wall. The rafter ends can be cut different ways depending on the desired finish, which is described in more detail later. A notch is cut out of the rafter at the location where the rafter meets the

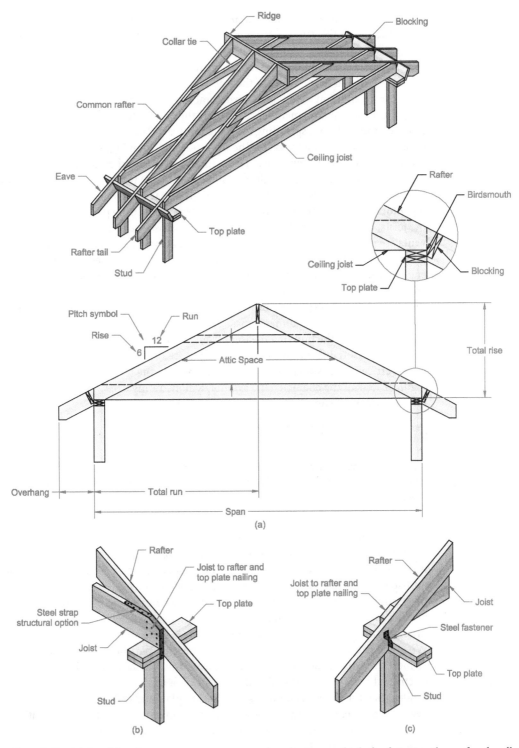

Figure 8.45 (*a*) Roof framing system components and anchoring methods for fastening the roof and wall system. (*b*) Steel strap anchor roof system to wall system. (*c*) Steel fasteners tie the roof system to wall system. (*Previously published in Modern Residential Construction Practices, Routledge, 2017 and reproduced here with permission.*)

wall top plate. This notch, called a **birdsmouth,** allows the rafter to fit exactly on the top plate. The top of the rafter is cut at an angle to match perfectly with the face of the ridge. Another rafter is fastened to the ridge directly across, and this rafter runs down to the opposite exterior wall. Ceiling joists are generally 2× dimensional lumber or engineered wood products fastened to the raters at the wall top plates commonly extending across the entire structure. In addition to providing structural support for the ceiling below, the ceiling joists are critical construction members used to tie the outside walls together and keep downward forces on the roof from spreading the structure apart. The ceiling joists are nailed to the rafters and toe nailed to the top plate. The steel strap shown in Figure 8.45*b*, or the steel fastener shown in Figure 8.45*c* can be used if required by structural engineering. The ceiling joists act as support for the ceiling material and tie the outside walls together when the area under the rafters is an unused **attic** space. An attic is a space or room under the roof of a building. **Blocking** is used between the ceiling joists at the bearing partition top plate to stabilize the joists, and at the exterior wall top plate to enclose the space between rafters. **Screened vents** are provided in the blocking to help ventilate the attic space. Ventilation options are described in more detail later in this chapter. Collar ties are used depending on structural engineering requirements to help tie the rafters together and support the ridge. Collar ties are typically 1× or 2× dimensional lumber fastened to the rafters in the upper one-third of the attic space at every, every other, or every third set of rafters. Additional roof framing terms relate to structure dimensions, as shown in Figure 8.45. The **span** is the horizontal dimension across the building measure from the outside of top plates. The **total run** is half the span measured from the outside of a top plate to the center of the ridge. The **total rise** is the vertical dimension from the top of the plate to the top of the ridge. Pitch is shown in Figure 8.45 with the characteristic symbol demonstrating a 6/12 pitch, which is 6 in. of rise for every 12 in. of run.

NOTE

The roof framing system rafters, ridge, joists, and collar tie sizes are determined by structural engineering based on roof loads and other engineering factors with dimensions given on the plans.

The example in Figure 8.45 shows the ceiling joists running all the way across between exterior walls without mid-span support. This application is possible in some cases, but it is common to have one or more interior bearing partitions supporting the ceiling joists at mid-span, as shown in Figure 8.46*a*. In this example, the joists are either continuous across the partition or join at the center of the partition when two members meet. Structural engineering can require a steel strap or other steel **gusset,** where two members meet at the partition. A gusset is a steel or other material plate used to strengthen a joint in the structure. When a header supports an opening in the bearing partition, the ceiling joists can cross or meet over the header, as shown in Figure 8.46*b*. The header in this example is exposed or covered by drywall and visible from below. If a flush ceiling is desired with no exposed header, then the header is framed above the top plate and ceiling joists are connected to the header with steel joist hangers allowing the bottom of the joists to be even with the bottom of the header, as shown in Figure 8.46*c*. This option extends the header into the attic space, which can be acceptable.

Ceiling joists typically run parallel to the rafters, as shown in Figure 8.46, which is especially true for gable roof construction. However, in some cases, especially in hip roof

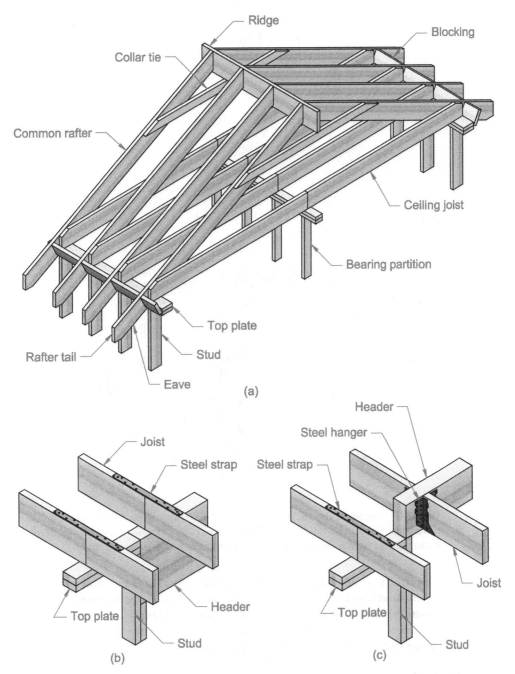

Figure 8.46 (*a*) It is common to have one or more interior bearing partitions supporting the ceiling joists at mid-span. (*b*) When a header supports an opening in the bearing partition, the ceiling joists can cross or meet over the header. (*c*) When no exposed header is desired, then the header is framed above the top plate and ceiling joists are connected to the header with steel joist hangers allowing the bottom of the joists to be even with the bottom of the header. (*Previously published in Modern Residential Construction Practices, Routledge, 2017 and reproduced here with permission.*)

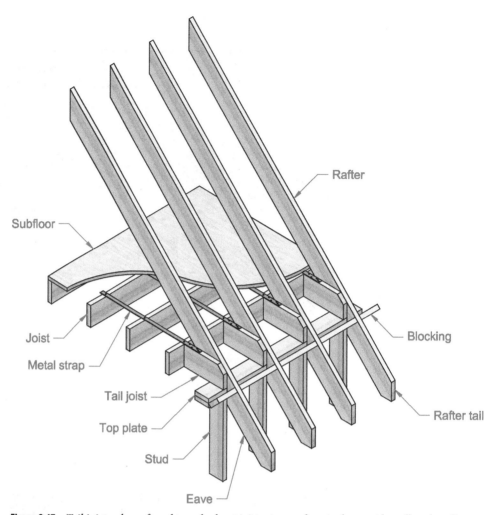

Figure 8.47 Tail joists, also referred to as lookout joists, are used to tie the outside wall to the adjacent joists when the main joists run perpendicular to the rafters. (*Previously published in Modern Residential Construction Practices, Routledge, 2017 and reproduced here with permission.*)

construction, ceiling joists can run perpendicular to the rafters. **Tail joists,** also referred to as **lookout joists,** and are used to tie the outside wall to the adjacent joists when the main joists run perpendicular to the rafters, as shown in Figure 8.47. Plywood or OSB subflooring, steel straps, steel joist hangers, or a combination can be used to tie the tail joists to the main joists depending on structural requirements.

Framing a Hip Roof

Hip roof framing is more complex than framing a gable roof. A hip roof system has **hip rafters, valley rafters,** and **jack rafters.** Look at Figure 8.48 as you discover the following hip roof framing members and terminology. If a hip roof is longer than it is wide, then part of the roof system is the same as framing a gable roof with common rafters that connect the top plate to the ridge. The hip roof framing starts when the roof slopes to an inside or outside corner of the building. On a hip roof, the **hip** is the corner of an external angle formed

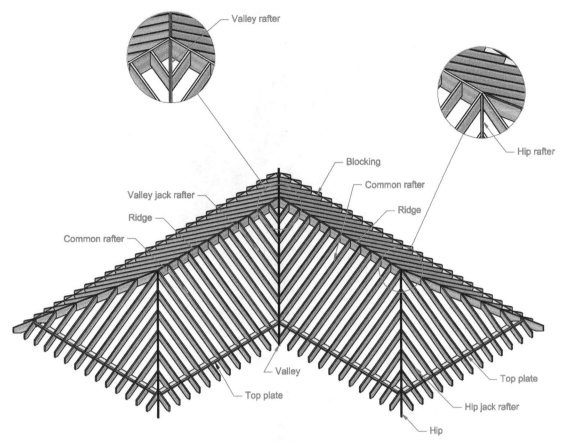

Figure 8.48 Hip roof framing members and terminology. (*Previously published in Modern Residential Construction Practices, Routledge, 2017 and reproduced here with permission.*)

by two intersecting exterior roof surfaces. The hip rafter is the rafter that forms the hip and runs from the ridge to the top plate. On a hip roof, the **valley** is the corner of an interior angle formed by two intersecting interior roof surfaces. The valley rafter is the rafter that forms the valley and runs from the ridge to the top plate. A jack rafter is any rafter that is shorter than a common rafter and connecting to a hip rafter or a valley rafter. The jack rafter that connects to the hip is also called a **hip jack rafter,** and the jack rafter that connects to a valley is also called a **valley jack rafter.**

Framing the Eave

There are a variety of options for framing the eaves depending on the desired results and the architectural style. Modern architecture often has a large overhang, while traditional architecture usually has a short overhang. Long overhangs are also used to protect the exterior walls from weather and provide shade from the summer sun. The eave can be open or enclosed. An open eave is generally found on buildings where it is important to reduce cost, and enclosed eaves are common on high-quality buildings where appearance and quality is more important than cost. An open eave, also called an open **soffit,** is shown in Figure 8.49. Soffit is the term used to describe the underside of any architectural feature, such as a beam, arch, ceiling, overhang, or vault. The term **cornice** can also be associated with overhangs, but cornice also has

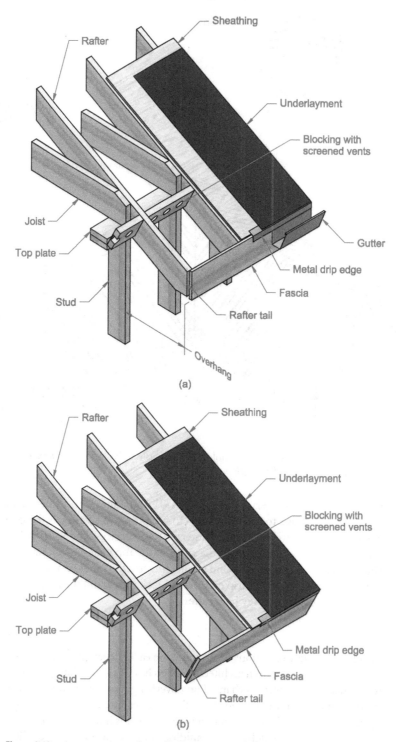

Figure 8.49 An open eave, also called an open soffit. (*a*) Construction of an open soffit where the rafter tails are cut for a fascia and gutter to be installed. (*b*) Rafter tails cut at an angle with a fascia installed and no gutter used.

varied meanings in architectural applications. A cornice is generally any horizontal decorative molding that is commonly found over a door or window, or along the corner where a wall meets the ceiling. This kind of decorative cornice is described in Chapter 13. The term **cornice** in this discussion is a projection out from the building to protect the building from weather, and generally has decorative moldings used in the finish construction.

Figure 8.49*a* shows the construction of an open soffit, where the rafter tails are cut for a **fascia** and gutter to be installed. A fascia is usually 1× dimensional lumber, plastic, or other material installed over the ends of rafters as a finish or trim. The fascia is optional on an open soffit. A **gutter** is normally installed directly over the rafter ends when there is no fascia used, or over the fascia. A gutter is a metal or plastic trough that runs horizontally along the edge of the roof overhang to catch rainwater and divert the water to **downspouts.** The gutter also has an attached flashing that goes under the roofing to keep rainwater from entering the roof structure. Downspouts are vertical square, rectangular, or round pipes that connect to the bottom of gutters for taking rainwater down to a system of underground rain drain pipes that take the water away from the building and generally into a stormwater system. Figure 8.49*b* shows an option with rafter tails cut at an angle with a fascia installed and no gutter used. Gutters are typically required, but some locations do not require gutters. The roof sheathing is generally 1/2 in CDX plywood or OSB down to the blocking and ACX plywood or OSB over the exposed open soffit. The A grade veneer is fastened face down at the soffit to provide the best appearance. CDX refers to C grade veneer on one side, D grade veneer on the other side and the X indicates exterior application.

An enclosed soffit is generally used on high-quality building construction, where it is desired to hide the rafter tails and provide a better finished appearance. Figure 8.50 shows typical enclosed soffit options. In Figure 8.50*a*, the enclosed soffit construction uses a 2 × 4 horizontal **ledger** fastened to the siding directly across from the bottom of the rafter tails. The ledger is used to support lookout joists that connect to each rafter tail. The ledger and lookout system is used to attach the enclosed soffit material. There are a variety of soffit materials that can be used, such as 1 × 4 tongue and grove wood, plastic, or metal. There is normally a continuous **screened vent** placed near the center of the soffit for providing ventilation to the attic or joist space. Continuous screened soffit vents are part of a roof ventilation system that circulates cool air in from the soffit vents into the attic space and moves hot air out through vents high in the roof and in gable ends in some cases. The continuous soffit vent is usually a prefabricated aluminum, plastic, or steel perforated product that runs the full length of the soffit. Additional information about roof ventilation is provided later in this chapter. Blocking is provided at the top plate between the rafters to support the rafters. The blocking material is 2× with a width that allows 2 in. (50 mm) of air spaced between the top of the blocking and the roof sheathing. A half-inch CDX plywood or OSB sheathing is used over the entire rafter system, because the appearance of ACX is not needed for an enclosed soffit. A horizontal molding is generally placed along the seam, where the soffit meets the wall to finish the installation and seal the corner. Caulking can also be used at the joint for additional sealing. Any type of prefabricate decorative molding or dimensional lumber can be used depending on the desired appearance.

The soffit option shown in Figure 8.50*b* is used when a small overhang is desired as commonly seen in traditional architecture. In this example, the soffit material is fastened directly to the cut ends of the rafter tails and no lookout joists are used. Everything else is constructed as described and shown in Figure 8.49*a*.

Framing the overhang at a gable end is typically done, as shown in Figure 8.51. The last common rafter, just before the gable end, supports **lookout rafters** that extend out over the gable top plate the desired distance of the overhang. The lookout rafters support the

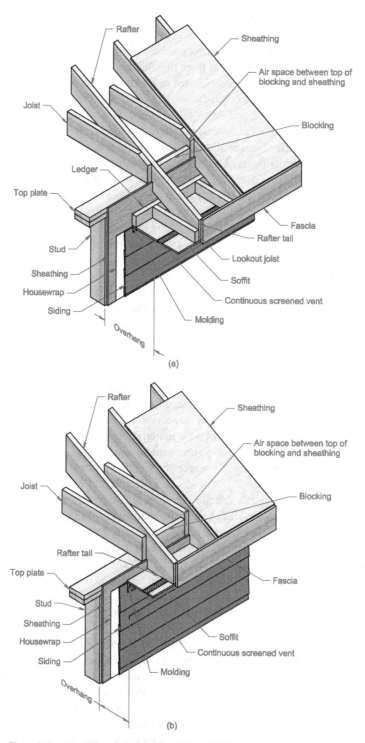

Figure 8.50   Typical enclosed soffit options. (*a*) This enclosed soffit construction uses a 2 × 4 horizontal ledger fastened to the siding directly across from the bottom of the rafter tails. (*b*) This soffit option is used when a small overhang is desired as commonly seen in traditional architecture. (*Previously published in Modern Residential Construction Practices, Routledge, 2017 and reproduced here with permission.*)

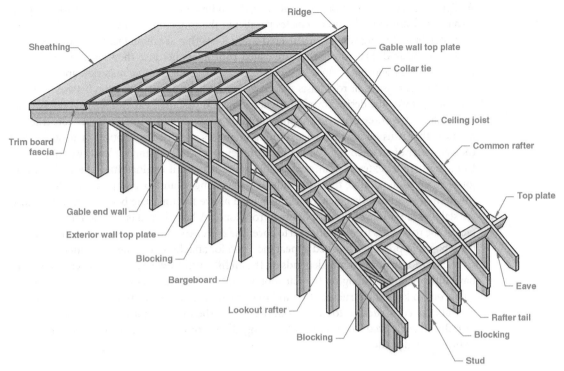

Figure 8.51 Framing the overhang at a gable end. (*Previously published in Modern Residential Construction Practices, Routledge, 2017 and reproduced here with permission.*)

bargeboard and roof sheathing at the **rake.** A bargeboard is attached to the ends of the lookout rafters and used to trim the gable end overhang. The bargeboard is often dimensional lumber one size larger than the lookout rafters. There is often an additional trim board placed at the top of the bargeboard used to cover the exposed roof sheathing. This trim board is called a **fascia, rake board,** or **rake molding.** The bargeboard can also add ornate detail to the gable end as is historically the purpose of a bargeboard in traditional architecture. The term **rake** refers to the slant edge of a gable roof at the end wall of the house. Blocking is placed between the lookout rafters to stabilize the structure and block the opening. ACX plywood or OSB sheathing is commonly used to provide a quality appearance. The gable end overhang can be open with the lookout rafters exposed as described here, or soffit material can be attached to the bottom of the lookout rafters to enclose the overhanging structure.

As you look at Figure 8.51, also notice how the gable end wall is framed above the exterior end wall of the building. The last ceiling joist is framed next to the inside face of the gable end wall. The top plate of the gable end wall runs at the same angle as the roof pitch and the top surface is exactly at the bottom of the adjacent common rafters. The studs for the gable end wall are individually cut and nailed at the exterior wall top plate and the gable wall top plate.

Framing a Flat Roof

Flat roofs are used on buildings most commonly in dry climates where rain and weather changes are minimal, because the flat roof is more difficult to insulate and construct weathertight than a building with a sloped roof. Flat roofs are not generally framed perfectly flat.

A slight roof slope of at least 1/4 in./ft is used to allow for drainage. The slight roof slope is created by framing walls and top plates parallel to the roof slope, or by framing all walls the same height and using sloping engineered wood joists that are manufactured for the specific roof slope desired. Flat roof rafters are called **roof joists** because they serve as rafters and ceiling joist for the rooms below. **Figure 8.52** shows flat roof framing materials and methods. Continuous roof joists or roof joists spliced at interior bearing partitions run between exterior walls and provide the desired overhang where they run past their exterior walls. A double roof joist called a **trimmer** is used to support outlook roof joists. A trimmer is any construction member that runs parallel to other framing members and used as support or to strengthen the perimeter of an opening. The double trimmer is generally framed at least two joist spaces from the parallel exterior wall as determined by structural engineering. The lookout roof joists fasten to the double trimmer with steel joist hangers and extend past the exterior wall the desired overhang distance. Blocking is used above the top plate between each roof joist for stability and to seal the space between roof joists. A bargeboard is attached to the outside face of the roof joists for support, and to cover the ends. An additional fascia is placed over the bargeboard to provide a finished appearance, and for attaching flashing and gutter.

The amount of overhang depends on the architectural style. Some architectural designs have little or no overhang on the flat roof. Quality eave construction is very important on a flat roof to keep the structure dry and to ventilate the roof joist space. **Figure 8.53** shows typical eave construction for a flat roof overhang. The overhang distance can be increased or reduced as desired. A minimum overhang allows for only the continuous screened vent to be installed at the soffit.

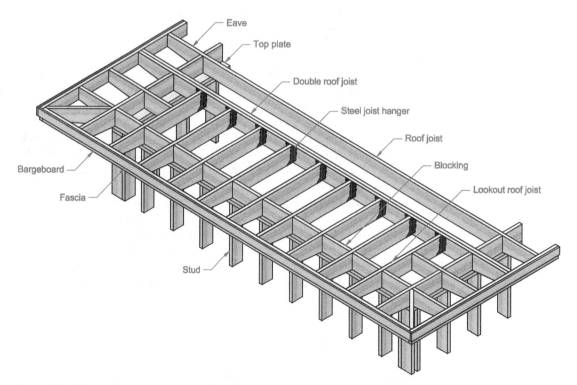

Figure 8.52 Flat roof framing materials and methods. (*Previously published in Modern Residential Construction Practices, Routledge, 2017 and reproduced here with permission.*)

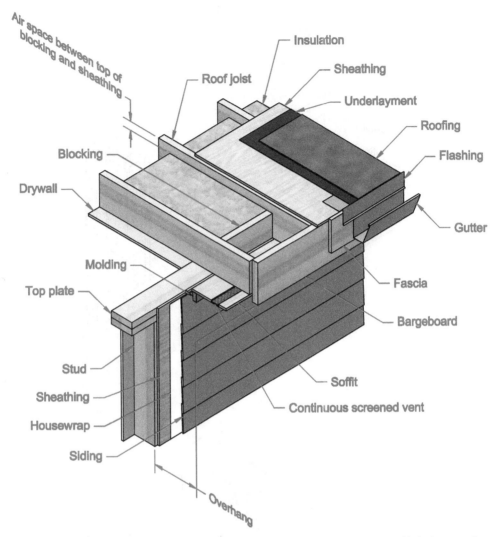

Figure 8.53 Typical eave construction for a flat roof overhang. (*Previously published in Modern Residential Construction Practices, Routledge, 2017 and reproduced here with permission.*)

Some flat roof designs use a **parapet** constructed around the outside perimeter of the roof. A parapet is a protective wall along the edge of a roof, or other structure. A parapet is often designed on a flat roof to maintain a clean vertical appearance in the architecture. Parapet construction is shown in Figure 8.54. Figure 8.54*a* shows parapet construction along the top of the slope and sides of the flat roof, and Figure 8.54*b* shows the parapet construction at the bottom slope of the flat roof, where gutter outlets and a gutter are installed for the roof drainage system.

Framing a Roof Opening

An opening through the roof framing is required around a chimney or for framing a skylight rough opening. Figure 8.55 shows how to frame an opening through the roof for a chimney. Any opening through a roof system can be framed by using double trimmers

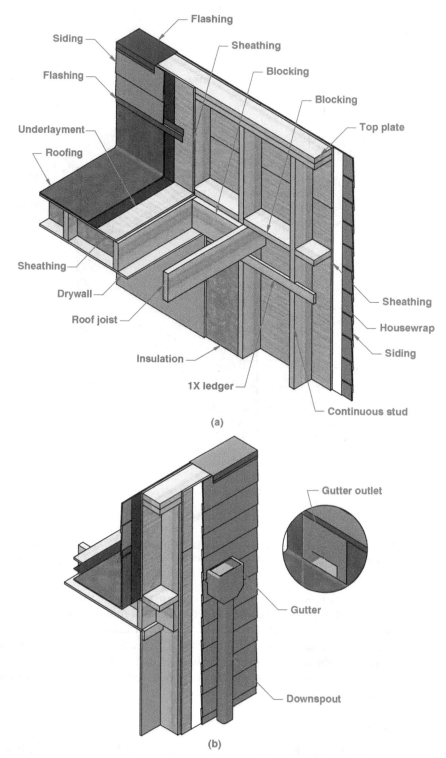

Figure 8.54 Parapet construction. (*Previously published in Modern Residential Construction Practices, Routledge, 2017 and reproduced here with permission.*)

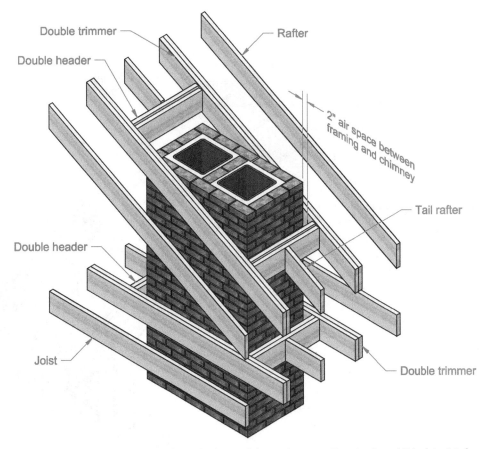

Figure 8.55 Frame an opening through the roof for a chimney. (*Previously published in Modern Residential Construction Practices, Routledge, 2017 and reproduced here with permission.*)

to establish the width of the opening. Double headers are used at the top and bottom of the opening to establish the opening depth. The double headers are framed plumb with the inside face of the headers parallel to the chimney surface. The measurements between the double trimmers and the double headers are the rough opening dimensions. A 2 in. (100-mm) air space is required between the chimney and wood framing, making the rough opening: chimney width plus 4 in. (200 mm) X chimney depth plus 4 in. The chimney rough opening is also framed in the same manner at the ceiling structure below.

Framing a Roof for a Vaulted Ceiling

A **vaulted ceiling** can have unequal sloping sides, a single sloping side, an arch shape, or other design possibilities. Vaulted ceilings can have the same pitch or a different pitch than the roof. A vaulted ceiling can angle upward from the plate to the ridge, or stop at a flat ceiling part way to the ridge. A **cathedral ceiling** has equal sloping sides that meet at a ridge, or stop at a flat ceiling part way to the ridge.

A vaulted ceiling can be framed with only roof joists, as shown in Figure 8.56. This is done by using a ridge beam that is solid lumber or a glu-lam beam. The ridge beam is

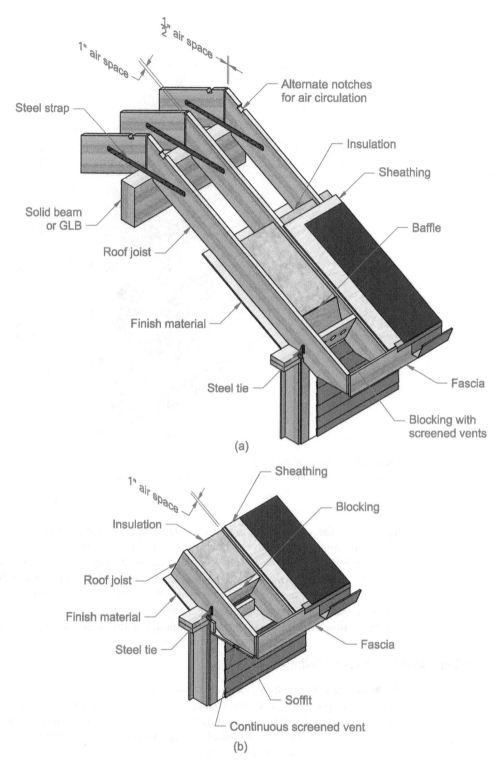

$\frac{1}{2}$" air space

1" air space

Steel strap

Alternate notches
for air circulation

Insulation

Sheathing

Solid beam
or GLB

Baffle

Roof joist

Finish material

Steel tie

Fascia

Blocking with
screened vents

(a)

Sheathing

1" air space

Blocking

Insulation

Roof joist

Finish material

Steel tie

Fascia

Soffit

Continuous screened vent

(b)

Figure 8.56   A vaulted ceiling framed with roof joists. (*Previously published in Modern Residential Construction Practices, Routledge, 2017 and reproduced here with permission.*)

supported at each end with a post that distributes the load to the foundation system. A solid dimensional lumber beam can be used if the desire is a natural wood appearance, and an option is a **rough sawn** beam for a rustic appearance. The term **rough sawn lumber** refers to nominal size lumber, also referred to as rough lumber, which is the size before it is planed. Glu-lam beams are more common, because of their structural preference, quality, and availability. The ridge beam supports the roof structure and keeps the roof joists from pushing the exterior walls outward. Roof joists are generally 2× dimensional lumber or engineered wood products spaced 16 or 24 in. (400 to 600 mm) on center. The roof joists need to be sized big enough to support the roof load and to provide enough space for insulation requirements. The roof joists have a birdsmouth at the exterior wall top plate where they are anchored to the top plate with steel anchors or straps as determined by structural engineering. The eaves can be constructed with an open or closed soffit as previously described. A screened vent block is placed at each roof joist space, or a continuous soffit vent is used to provide ventilation to the roof system. Ventilation is critical to keep the structure dry. Insulation is installed in each roof joist space leaving a minimum 1-in. (25-mm) air space between the top of the insulation and the sheathing. It is very important for the insulation to be properly installed without blocking the air space. A plywood or OSB **baffle** is placed at an angle from the base of each eave block to the bottom of the air space at each roof joist space. The baffle is used to keep the insulation from plugging the screen vents. The roof joists are cut at the top to fit over the ridge beam, and each pair of roof joists is framed directly across from each other. A 1/2-in. (13-mm) air space is provided between the roof joist ends allowing for air to circulate from the eave block vents through the air space and around the roof joists. Alternately, a notch can be cut near the top of each roof joist to provide ventilation, depending on engineering specifications. A steel strap is typically used to tie the roof joists together and strengthen the structure, depending on engineering specifications. Structural plywood or OSB sheathing is used over the roof joists. The underside of the roof joists can be finished with any material such as drywall, or something like 1 × 6 T&G cedar or other desired wood can be used for a natural appearance.

A cathedral ceiling can be framed to provide an architectural style with a combination of sloped and flat ceiling in one or more rooms of the building. Figure 8.57 shows cathedral ceiling framing where rafters run from the top plate to the ridge with sloped exposure up to collar ties fastened to the rafters near the lower one-third of the structure. Structural engineering for the collar tie placement and fastening is typically detailed in the plans. Open or closed soffits can be framed as previously described.

FRAMING WITH ROOF TRUSSES

A **roof truss** is a manufactured structural support for the roof system. The roof framing practices previously described throughout this chapter are referred to as stick framing, because one board or stick is used at a time to assemble the structure. A roof truss combines all of the individual components together into one structural unit. Figure 8.58 shows a typical building construction roof truss and related terminology. Manufactured roof trusses are structurally engineered for each specific building design and are fabricated using quality materials providing uniform pitch and size. A complete building roof structural system is delivered to the project on a crane truck and each roof truss is placed directly on the framed walls where they are quickly assembled to create the roof structure. Trusses are commonly made using 2 × 4 dimensional lumber and steel truss plates that act as gussets to strengthen and support each joint. The use of manufactured roof trusses can save about 25 percent in

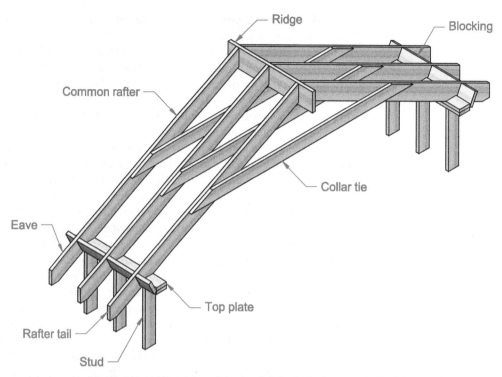

Figure 8.57 Cathedral ceiling framing where rafters run from the top plate to the ridge with sloped exposure up to collar ties fastened to the rafters near the lower one-third of the structure. (*Previously published in Modern Residential Construction Practices, Routledge, 2017 and reproduced here with permission.*)

construction time. There are many different truss designs available, as shown in Figure 8.59. The **fink truss** and **scissors truss** are the most common. The fink truss provides the typical pitch roof and flat ceiling. The scissors truss provides a pitch roof and sloped ceiling for a cathedral design. Custom trusses can be manufactured, but they become more expensive and the cost should be compared with stick framing as an option.

When using roof trusses for building construction, the walls are framed and ready for the roof system before the trusses are delivered to the site. Trusses are delivered on a truck with a crane that lifts trusses onto the top plates individually or in groups of three or four, as shown in Figure 8.60. After all trusses are placed by the crane, the framer sets each truss in place and supports the trusses with **continuous lateral bracing,** as shown in Figure 8.61. Continuous lateral bracing allows the trusses to be accurately spaced and stabilizes the roof system. The truss manufacturer delivers the complete truss system as a kit that includes everything needed for the entire roof framing system, including a gable end truss, gable end lookout rafters, and barge board.

TIMBER CONSTRUCTION

You were introduced to the pier foundation in Chapter 4. A pier foundation is used to support post and beam, timber-frame, or pole construction. The term **post** and **beam** was used earlier with the post and beam foundations system, and has similar characteristics as timber-frame construction, described in this chapter.

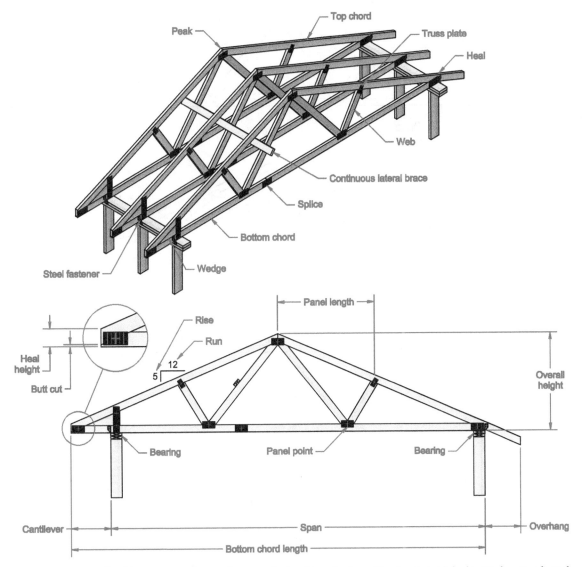

Figure 8.58   Typical building construction roof truss and related terminology. (*Previously published in Modern Residential Construction Practices, Routledge, 2017 and reproduced here with permission.*)

TIMBER-FRAME CONSTRUCTION

Timber-frame construction uses generally large posts and beams or timbers for the members of the structural system. Timber-frame buildings are constructed using a combination of post and beam, timber framing, and conventional stud framing depending on the requirements of the specific building. The vertical timbers or posts are used as the building supports to which the horizontal framing beams are fastened. Timber-frame construction is a general term for building with heavy timbers rather than the dimensional lumber, described previously, for use in light wood-frame construction. Dimensional lumber is lumber that is cut and planed to standard width and depth specified in inches, and lengths

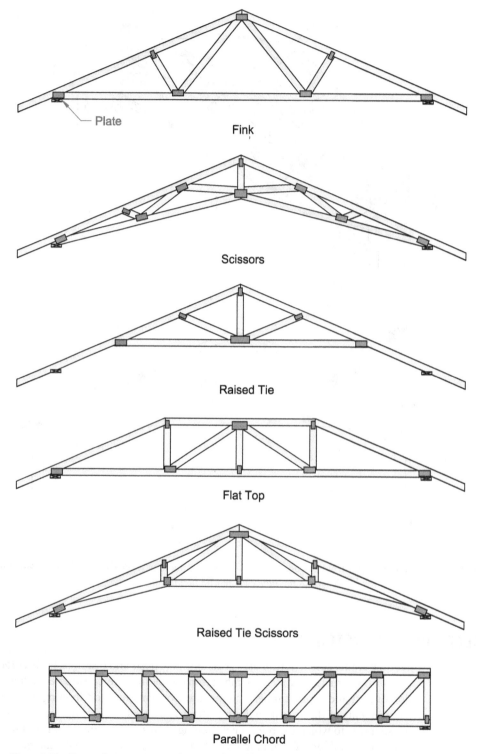

Plate

Fink

Scissors

Raised Tie

Flat Top

Raised Tie Scissors

Parallel Chord

Figure 8.59 Truss design options. (*Previously published in Modern Residential Construction Practices, Routledge, 2017 and reproduced here with permission.*)

Figure 8.60   Trusses are delivered on a truck with a crane that lifts trusses onto the top plates individually or in groups of three or four.

Figure 8.61   After all trusses are placed by the crane, the framer sets each truss in place and supports the trusses with continuous lateral bracing.

specified in feet. A 6 × 12 dimensional beam, for example, is not actually 6 × 12 in. A 6 × 12 dimensional beam is planned to finish measurements of 5-1/2 × 11-1/2 in. The lumber used in timber-frame construction can be **rough sawn (RS)** lumber. When logs are cut into lumber, they are sawn to their nominal size. Nominal size lumber is the size before it is planned, which is also referred to as rough sawn lumber. So, the 6 × 12-in. beam example is 6 × 12 in. when rough sawn. The surface of rough sawn lumber is rough, while the surface of dimensional lumber is smooth. Timber-frame construction often uses rough sawn lumber for the structural components because of the desire to show the natural characteristics of the wood. Figure 8.62 shows the difference between rough sawn lumber and dimensional lumber.

Timber-frame construction practices go back thousands of years when buildings were built using heavy carefully fitted **mortise and tenon** joined timbers with joints secured by

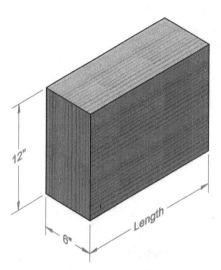

Rough Sawn Lumber

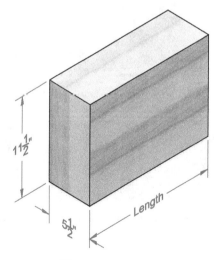

Dimensional Lumber

Figure 8.62 The difference between rough sawn lumber and dimensional lumber. (*Previously published in Modern Residential Construction Practices, Routledge, 2017 and reproduced here with permission.*)

large wooden **pegs.** Historically, the timbers were handmade square or rectangular from logs, or the logs were stripped of bark and used in their round shape, especially for the vertical supports. This practice is still used today in the construction of **pole buildings.** Pole buildings are similar to timber-frame buildings except the vertical supports are used directly from the round tree with bark removed and finished to a desired diameter. Timber-frame construction practices allow craftsman to demonstrate the beauty of exposed wood in the building. The timber-frame structure can also be built without any interior supports, allowing the vertical posts, bracing, and rafter system to support the entire building. A mortise and tenon joint is commonly used when construction members connect at a 90° angle. Also used in furniture construction, the mortise and tenon joint is simple and strong. There are many joint variations, but the basic characteristics are the mortise, which is a rectangular hole and the tenon, which is cut to fit the mortise hole exactly, as shown in Figure 8.63a. A shoulder can be made that allows the tenon additional support when inserted and resting in the mortise. Authentic timber-frame construction uses wooden pegs to fasten joints, as shown in Figure 8.63b. Pegs are cylindrical wooden fasteners driven into a hole that connects between two or more construction members.

Timber-Frame Floor Framing

There are many possible methods for framing the floor system in timber-frame construction. The basic application is to have horizontal girders connect between vertical square or round posts that are placed at building corners and mid-support locations. The vertical posts run all the way from the foundation footings or piers to the roof structure. The girders can be joined to the posts with mortise and tenon joints, other joints, or by connecting with large bolts or other steel connectors. Bracing is often used to strengthen the connection between the posts and girders and reduce lateral movement in the structure. Floor beams are the placed over the girders and are spaced apart a distance such as 32 in. on center to support subfloor material above. Figure 8.64a shows girders connected to posts using a joint system such as mortise and tenon. Figure 8.64b shows double girders connected on each side of posts using a **lap joint** and bolt fasteners. The bolt fastener is generally a large hexagon head bolt with a washer at the head and at the hexagon nut on the other side. A lap joint is basically a joint where one or both members have a grove cut through that allows the other member to fit exactly. This method is commonly used in pole building construction, because it is easier to use a lap joint when connecting girders to round posts.

TIMBER-FRAME ROOF CONSTRUCTION

Timber-frame construction offers the natural beauty of exposed wood. A timber-frame roof system can be one of the most impressive forms of construction found in the building industry. Classic timber-frame structures are fabricated using wood joints such as mortise and tenon described earlier. When this is done, roof construction members are joined together with the precision used in furniture construction, as shown in Figure 8.65a, and structures are often self-supporting, as shown in Figure 8.65b. Alternately, connections between framing members can be done with steel brackets, hangers, and straps that are structurally engineered for each application. There are a variety of construction options for the timber-frame roof system. Figure 8.66 shows a common method and terminology used to build a timber-frame roof truss. The posts run from the foundation to the roof system. Horizontal headers connect between posts to support **tie beams** and roof beams.

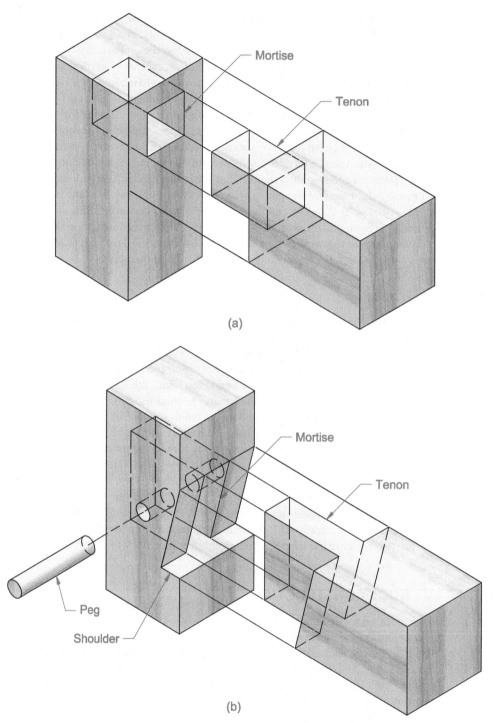

(a)

(b)

Figure 8.63   (*a*) The basic characteristics are the mortise, which is a rectangular hole and the tenon, which is cut to fit the mortise hole exactly. (*b*) Authentic timber-frame construction uses wooden pegs to fasten joints. (*Previously published in Modern Residential Construction Practices, Routledge, 2017 and reproduced here with permission.*)

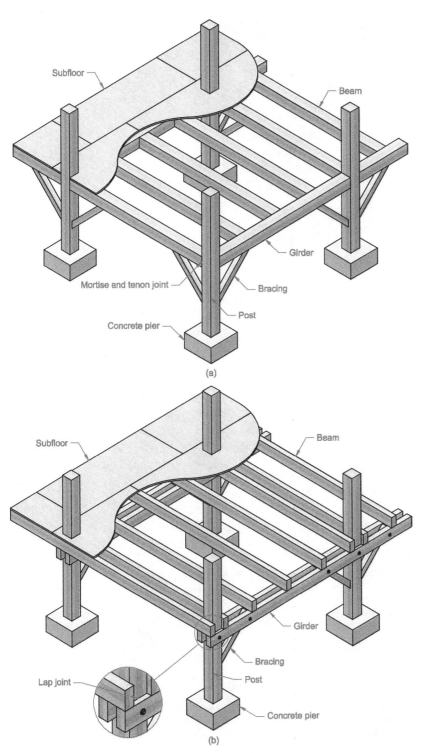

Figure 8.64   (*a*) Girders connected to posts using a joint system such as mortise and tenon. (*b*) Double girders connected on each side of posts using a lap joint and bolt fasteners. The bolt fastener is generally a large hexagon head bolt with a washer at the head and at the hexagon nut on the other side. (*Previously published in Modern Residential Construction Practices, Routledge, 2017 and reproduced here with permission.*)

(a)

(b)

Figure 8.65 (*a*) Timber-frame roof construction members can be joined together as shown here. (*b*) Timber-frame roof structures are often self-supporting.

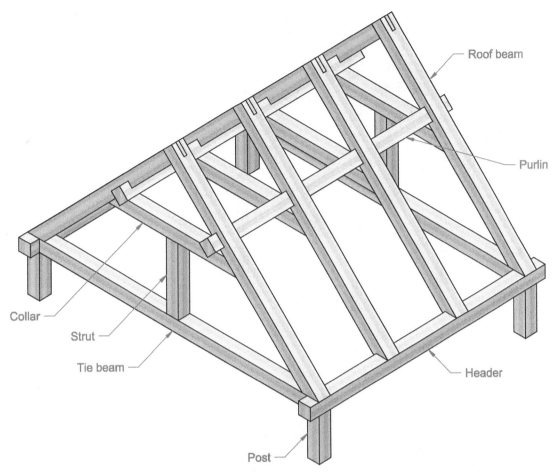

Figure 8.66   Common method and terminology used to build a timber-frame roof truss. (*Previously published in Modern Residential Construction Practices, Routledge, 2017 and reproduced here with permission.*)

Tie beams connect horizontally across the structure between posts. The roof beams are rafters that connect from the headers to the ridge where they are fastened together with a wood joint or steel connector. Spacing between roof beams depends on structural engineering, but 4 or 8 ft (1200 to 2400 mm) is common. Collars are attached to each set of roof beams to strengthen the structure and keep the roof beams from pushing outward at the headers. **Purlins** are horizontal lateral members that extend the length of a roof, used to support and tie roof beams together and can be used to attach roof decking depending on the installation practice. Another timber-frame roof construction method is shown in Figure 8.67, where posts support headers and roof beams at the eaves. Posts also support the ends of a ridge beam that runs between gable ends. Additional interior posts can be used as needed to support the ridge beam to the foundation. Roof decking is quality T&G lumber placed perpendicular over the roof beams, as shown in Figure 8.67*a*, or perpendicular to purlins, as shown in Figure 8.67*b*. T&G wood decking is selected for appearance, structural requirements, where it is designed to span between roof beams or purlins, and wood species such as cedar, pine, fir, or hemlock. Standard 2 × 6 V-joint T&G decking is shown in Figure 8.68*a*, and 3 × 6 or 4 × 6 V-joint T&G decking is shown in Figure 8.68*b*.

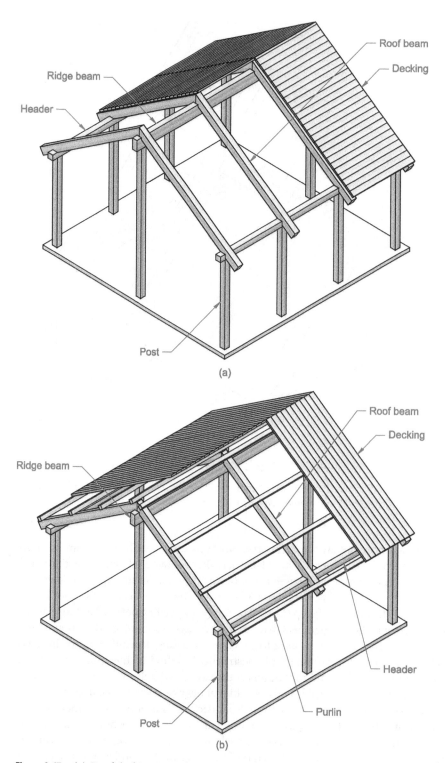

Figure 8.67   (*a*) Roof decking is quality tongue and grove (T&G) lumber placed perpendicular over the roof beams. (*b*) Roof decking perpendicular to purlins. (*Previously published in Modern Residential Construction Practices, Routledge, 2017 and reproduced here with permission.*)

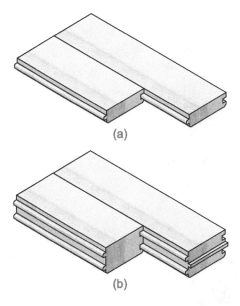

(a)

(b)

Figure 8.68   (*a*) Standard 2 × 6 V-joint T&G decking. (*b*) 3 × 6 or 4 × 6 V-joint T&G decking. (*Previously published in Modern Residential Construction Practices, Routledge, 2017 and reproduced here with permission.*)

A variety of steel fabrications systems can be used in timber-frame construction. Prefabricated products are available for use in nearly every application, but custom steel fastening systems are often used and painted black to match an old-world appearance. Figure 8.69*a* shows the use of a steel strap connecting over the top of roof beams and a steel plate connecting to each side of the roof beams above the ridge beam. Figure 8.69*b* shows roof beams connected above a ridge beam by drilling a through hole with a **counterbore** on each side to insert a bolt with washer and nut. A **spotface** is cut where the through hole meets the inside face between the intersection of the ridge beams. A round **shear plate** is inserted in each spotface at this location. A counterbore is a flat-bottomed cylindrical enlargement of the mouth of a hole with enough depth to hide the bolt head or washer and nut below the surface. A spotface is a flat-bottomed cylindrical enlargement of the mouth of a hole with slight depth to insert a washer or shear plate below the surface. A shear plate is a special round plate inserted in the face of a timber to improve shear resistance in the wood-to-wood joint. Figure 8.70 shows basic options for connecting ridge beams to posts using steel fasteners. Figure 8.70*a* shows a steel angle bracket bolted on each side of a roof beam connecting to the header and the same used to connect the post to the header below. Figure 8.70*b* shows a steel strap used to connect between a post, header, and roof beam to secure the structure and protect against uplift in windy conditions.

MASS TIMBER CONSTRUCTION

Mass timber is a term used to describe a variety of large engineered wood products commonly manufactured by compressing multiple layers of wood together to create solid members or panels of wood or a combination of materials such as concrete and wood. You were previously introduced to engineered wood products that are a combination of smaller components used to make structural products that have been engineered for specific applications, fabricated in a manufacturing facility, and delivered to the construction site. The most notable engineered wood product previously described and one of the mass timber construction components is **glue-laminated timber,** called glulam, which is an engineered

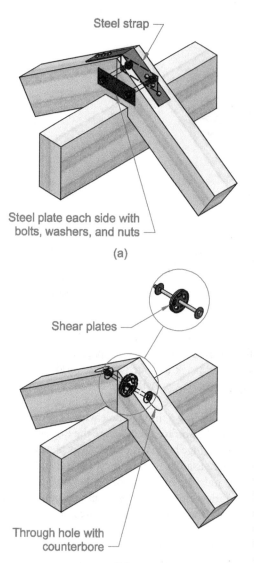

Figure 8.69   (*a*) Use of a steel strap connecting over the top of roof beams and a steel plate connecting to each side of the roof beams above the ridge beam. (*b*) Roof beams connected above a ridge beam by drilling a through hole with a counterbore on each side to insert a bolt with washer and nut. (*Previously published in Modern Residential Construction Practices, Routledge, 2017 and reproduced here with permission.*)

wood used for products include beams, columns, and roof joists that can carry heavier loads than equally sized conventional timbers. The wood used in making glulams and other mass timber products is called **lamstock.** Lamstock is quality high-grade dimensional lumber with low-moisture content and minimal defects, designed for specific strength requirements for use in mass timber products. Mass timber products are generally manufactured by pressing together dimensional lumber using moisture resistant formaldehyde-free polyurethane glue applied using heat and pressure in a factory.

Mass timber products offer sustainable construction by reducing the impact that the manufacturing of building materials has on the environment. Mass timber products use wood that is the only primary building product that grows naturally and is renewable. Compared to other common traditional building materials such as concrete, masonry, and steel, timber contains relatively low amounts of **embodied energy** and also has a lighter

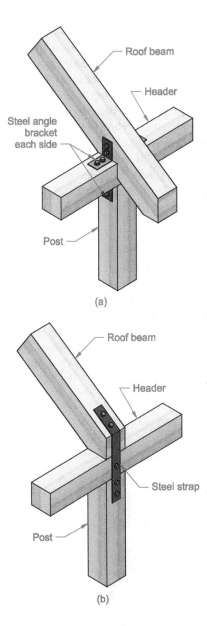

Steel angle
bracket
each side

Roof beam

Header

Post

(a)

Roof beam

Header

Steel strap

Post

(b)

Figure 8.70 Basic options for connecting ridge beams to posts using steel fasteners. (*a*) A steel angle bracket bolted on each side of a roof beam connecting to the header and the same used to connect the post to the header below. (*b*) A steel strap used to connect between a post, header, and roof beam to secure the structure and protect against uplift in windy conditions. (*Previously published in Modern Residential Construction Practices, Routledge, 2017 and reproduced here with permission.*)

carbon footprint with less air and water pollution during manufacture. **Embodied energy** is the combination of all energy required to produce a product or service, as if the energy is in the product.

Mass timber construction allows for lighter structures, potentially decreasing the foundation requirements, because wood is about five times lighter than concrete and about 15 times lighter than steel. Additional advantages over concrete and steel include thermal performance, acoustic, fire, and seismic performance. Wood has good thermal performance when compared with other equally sized construction materials. Thick mass timber panels require less insulation to achieve improved insulated value than other construction materials. Mass timber panels are solid wood and manufactured to close tolerances with tight-sealed joints that results in energy efficiency, because there is minimal airflow through the structure. The same mass timber materials and construction practices for floor and wall panels provide good sound insulation between flows and rooms. Thick mass timber construction provides fire resistance because mass timber panels burn slowly and in a controlled and predictable manner. For example, steel softens and loses its structural integrity during a fire more than a wood structure. Mass timber interior panels can be used to design spaces that reduce the ability for a fire to spread undetected. The dimensional stability and rigidity of mass timber panels create an effective lateral load resisting system for earth quake construction engineering.

This introduction gives you basic general information about mass timber construction. Specific types of mass timber construction practices are described in the following, including cross-laminated timber (CLT), dowel-laminated timber (DLT), nail-laminated timber (NLT), and timber-concrete composite construction.

Cross-Laminated Timber Construction

Cross-laminated timber (CLT), also called **X-lam,** is an engineered wood product used for floor, wall, and roof structures. CLT is prefabricated using several layers of kiln-dried lumber, laid flat with each layer placed alternating 45° or 90° with each other, and glued together on their wide faces. See Figure 8.71. Completed CLT units are called panels. Panels are always manufactured with an odd number of layers typically consist of three, five, seven, or nine alternating layers of quality high-grade dimensional lumber with low-moisture content. The CLT lumber used is typically spruce, fir, poplar, and sometimes cedar. Other soft and hardwoods can also be used. The alternating directions of the CLT laminations provide

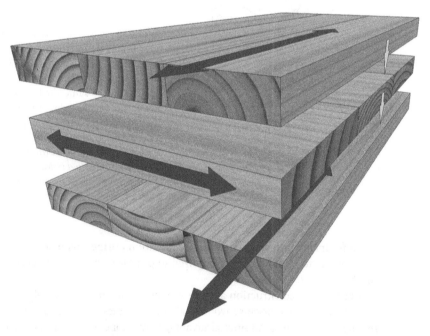

Figure 8.71   Cross-laminated timber (CLT), also called X-lam, is an engineered wood product used for floor, wall, and roof structures. CLT is prefabricated using several layers of kiln-dried lumber, laid flat with each layer placed alternating 45° or 90° with each other, and glued together on their wide faces. (*Courtesy of Reid Middleton, Inc.*)

high dimensional stability, and high strength to weight ratio, along with structural, fire, thermal and acoustic performance advantages. Panel thicknesses usually range between 4 and 12 in. (100 and 300 mm), but panels as thick as 20 in. (500 mm) can be made. Panel sizes are controlled by the manufacturing machinery and shipping regulations, but can range from 4 to 10 ft (1.2 to 3 m) in width and 16 to 64 ft (5 to 19.5 m) in length with lengths up to 98 ft (30 m). Each piece of lumber in the panel length can be **finger-jointed** at random lengths to create continuous boards throughout the panel length. Lumber crossing in the panel width are solid wood. A finger joint is a woodworking joint made by cutting a set of complementary, interlocking profiles in two pieces of wood, which are then assembled together with glue. All the layers in one direction are commonly manufactured using the same grade and species of lumber, but adjacent layers can be different thickness and made of alternative grades or species. CLT panels are manufactured for specific applications, resulting in minimum waste. Manufacturing scraps are used for other construction products, or for biofuel. CLT panels can be finished a variety of different ways, including staining and painting, but it is common to leave the beauty of natural wood exposed or sealed with a transparent product. CLT layers are face-glued and then pressed together, followed by planning and sanding as desired for the finished panels. The CLT panels are then processed using **computerized numerical control (CNC)** machinery, where the panels can be custom fabricated to create openings, compound angles and unique features requiring complex geometry to meet specific architectural and construction applications. CNC is the automation of machine tools through the use of software in a microcomputer attached to the machinery. The CNC equipment also makes panels with high tolerance dimensions for accurate fit when assembled at the construction site. Prefabricated CLT panels are delivered to the construction site where erection time is greatly reduced, allowing for improved construction efficiency, lower construction costs, and faster occupancy. Prefabricated CLT panes can include exact openings for doors and windows, along with assembled stairs, service channels, and ducts. Required insulation and finishes can also be applied at the factory and ready for installation when delivered to the construction site. These assemblies can help save time and cost by reducing subcontractor requirements on site. Figure 8.72 shows a crane placing two CLT panels. The CLT floor panel on the right shows the size of CLT panels and demonstrates the speed and efficiency of construction. The panel on the left shows the accuracy and quality of CNC window cutouts.

Dowell-Laminated Timber Construction

Dowel-laminated timber (DLT) is a structural and economic mass timber panel that is used for floor, wall, and roof construction. DLT panels are the only 100 percent wood, using no glue as in CLT manufacture. DLT panels can be made from the same wood species used for CLT panels. Lumber for DLT panels is stacked on edge next to each other. Mating boards are drilled with a pilot hole and hardwood dowels are press-fit into the holes holding the boards tightly together, as shown in Figure 8.73. The lumber is finger-jointed end-to-end and the dowels hold each board side-by-side to form a rigid and strong connection. DLT panels are processed using CNC machinery, just like CLT panels. CNC manufacturing creates a close tolerance panel that can enclose pre-integrated electrical conduit, fire sprinklers, and other services. DLT panels are prefabricated in sizes of up to 12 ft (3600 mm) wide and 60 ft (18 m) long. Each panel is put through a panel planer to establish dimensional accurate and smooth surfaces. DLT has many of the same characteristics found in CLT with additional architectural values. The DLT panel laminations can be milled with different profiles, including notches and curves. Curved DLT panels can be made by milling

Figure 8.72 A crane placing two CLT panels. The CLT floor panel on the right shows the size of CLT panels and demonstrates the speed and efficiency of construction. The panel on the left shows the accuracy and quality of CNC window cutouts. (*Courtesy of KATERRA.*)

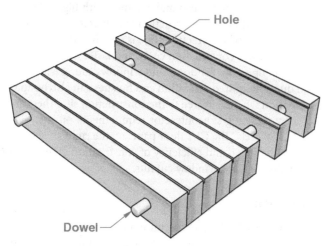

Figure 8.73 Lumber for DLT panels is stacked on edge next to each other. Mating boards are drilled with a pilot hole and hardwood dowels are press-fit into the holes holding the boards tightly together.

custom profiles into each lamination, creating curves perpendicular to the panel length. Curved panels are formed into shape at the job site.

Nail-Laminated Timber Construction

Nail-laminated timber (NLT) is the oldest mass timber product used in construction for over 150 years and commonly used to build warehouse and large industrial buildings. NLT is made using 3-, 4-, 6-, 8-, 10-, or 12-in. (76 to 300 mm) dimensional lumber placed on edge and fastened together with nails, as shown in Figure 8.74. NLT is commonly used for floor

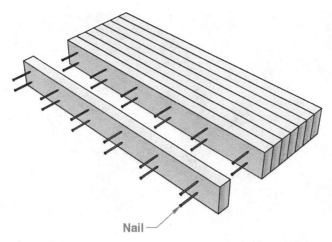

Nail

Figure 8.74 Nail laminated timber (NLT) is made using 3-, 4-, 6-, 8-, 10-, or 12-in. (76 to 300 mm) dimensional lumber placed on edge and fastened together with nails.

and roof panels with plywood or OSB sheathing is often added to the top side to provide a structural diaphragm. NLT cannot be processed with CNC machinery because of the nails, making DLT less uniform in appearance than CLT. NLT panels can span in only a single direction, providing less strength than CLT. NLT does provide a cost-effective alternative to CLT and its uneven rustic appearance offers an aesthetic reminder of past construction practices.

Laminated Veneer Lumber

Laminated veneer lumber (LVL) is an engineered wood product that uses multiple layers of wood veneer assembled with adhesives. LVL is typically used for headers, beams, rim joist, and edge-forming material. LVL is stronger, straighter, and more uniform than most dimensional lumber. Due to its composite nature, LVL is less likely to warp, twist, bow, or shrink than dimensional lumber. LVL is a type of structural composite lumber, comparable to glulams but with a higher allowable stress.

Timber-Concrete Composite Construction

The term **composite construction** refers to two or more different materials securely bonded together so the two or more individual materials act together as a single structural unit. **Composite materials** are engineered for use for a variety of reasons, including to increase strength over one material alone, to improve appearance, and for possible environmental sustainability. **Timber-concrete composite construction** uses the combination of timber beams or panels connected to a reinforced concrete slab using connectors that secure the concrete to the wood, as shown in Figure 8.75. This composite relationship takes advantage of concrete compression strength and timber tension strength, plus providing advantages of timber and concrete slabs. Timber panel advantages are increased stiffness, improved load capacity, better sound insulation, reduced vibrations, less possible horizontal bracing of the structure Compared to a concrete slab alone, the advantages are reduced dead load, improved sustainability of wood, increase of prefabrication for faster structure erection, reduced volume of concrete, which has many construction advantages.

Figure 8.75 Timber-concrete-composite construction uses the combination of timber beams or panels connected to a reinforced concrete slab using connectors that secure the concrete to the wood. (*Courtesy of SFS intec AG—Division Construction.*)

COMMERCIAL CONSTRUCTION MASS TIMBER APPLICATION

This commercial construction application taken, in part, from a Nordic Structure project is an example of the universal characteristics of mass timber construction systems, combining glulams, CLT, timber concrete-composite systems, steel-concrete, and steel-wood all in one project. The project is the John W. Olver Design Building at the University of Massachusetts shown in Figure 8.76. This project achieved the 2018 Wood Design Award, and the Jury's Choice for Wood Innovation.

Introduction

Prior to the construction of the John W. Olver Design Building, the design and construction related education programs were scattered around the University of Massachusetts (UMass) campus, in Amherst. The intent of the Design Building was to unite the Department of Architecture, the Department of Landscape Architecture and Regional Planning, and the Building Construction Technologies (BCT) of the Environmental Protection Department under the same roof to facilitate communication and collaboration between the students of the different disciplines.

Figure 8.76 The John W. Olver Design Building at the University of Massachusetts Nordic Structures project is an example of the universal characteristics of mass timber construction systems, combining glulams, CLT, timber concrete-composite systems, steel-concrete, and steel-wood all in one project. (*Courtesy of Nordic Structures.*)

The building is strategically located in the center of the campus. The east side of the building, three storeys high, is in harmony with the lower-rise historic buildings of the campus. In contrast, the more imposing west side, four storeys high, reflects the larger scale of the modern campus buildings.

The architectural program is located on these four floors which cover an area of 87,500 ft² (8130 m²). There is a total mass timber volume of 72,467 ft³ (2052 m³) in the slabs, walls, posts, and beams of the structure.

Architectural Design

The architectural concept was developed by Leers Weinzapfel Associates. The building forms a volume around an interior courtyard serving as a space for gatherings and teaching, and which rises up and becomes a green roof featuring native plants. The building opens and tilts up on one side to reveal the fully glazed ground floor. This architectural feature allows a fluid circulation on the ground level under the upper floor. The glazing combined with the natural wood beauty provides an abundance of natural light, and adds transparency and a connection to the exterior environment, as shown in Figure 8.77.

In this widely forested region of Massachusetts, the landscape is dotted with many brown wood barns that were historically used to dry tobacco leaves. The vertical brown metallic cladding covering the pavilion reminds of these "tobacco barns" which are an important element of the area. Like the bark on a tree trunk, the building's dark cladding envelops the natural light wood color of the interior structure. The choice of an exposed wood structural system provides an ever-present teaching demonstration for the architecture and building technology students.

Figure 8.77 The glazing combined with the natural wood beauty provides an abundance of natural light, and adds transparency and a connection to the exterior environment. (*Courtesy of Nordic Structures.*)

Structural Design

At its highest point, the UMass integrated design building reaches a height of 82 ft (25 m). Composed of three hybrid structural elements including wood-concrete, steel-concrete, and steel-wood, the school incorporates the latest technologies in the field of structural building design. The project contains a total of 20,129 ft³ (570 m³) and 65,721 ft³ (1861 m³) of glulam and CLT, respectively. Although primarily a wood structure, the building does include various structural steel elements.

The primary gravity force resisting system consists of a glulam and steel post-and-beam structure that supports wood-concrete composite floor slabs. Typically, the 5-ply 175-mm CLT slabs and glulam beams have slots to insert and adhere shear connectors shown in Figure 8.78, forming a composite system with the 4-in. (102-mm) concrete slab. Likewise, the steel beams typically have shear studs welded to the top flange to create a steel-concrete composite system. Besides this main structure, the center portion of the building consists of a roof that is supported by **zipper trusses,** shown in Figure 8.79. Zipper trusses use compression and tension members to transfer gravity loads to the timber structural system. This structural element is a wood-steel composite system supported throughout with **Besista steel tension rods** and glulam compression members, forming a trapezoidal structural truss which has a depth of 27.9 ft (8.5 m) and a span of 84 ft (25.6 m) along the west side of the roof. Besista steel tension rods is a patented system that consists of steel bars called tension rods with external threads that are connected to each other and to the structure by special connecting devices.

Glulam bracing and CLT shear walls were used in combination with a CLT-concrete composite diaphragm. The bracing system is composed of **steel knife plates** with tight-fit pins as the main connection. A steel knife plate is a flat plate steel to timber post connector with rebar embedded in the concrete foundation, and a post base attached to the timber post through the knife plate and held together with a steel pin. The shear wall hold-downs and shear connectors consist of steel-to-wood **Holz-Stahl-Komposit (HSK)™** plates that

Figure 8.78 The 5-ply 175 mm CLT slabs and glulam beams have slots to insert and adhere shear connectors as shown here. (*Courtesy of Nordic Structures.*)

Figure 8.79 The center portion of the building consists of a roof that is supported by zipper trusses that use compression and tension members to transfer gravity loads to the timber structural system. (*Courtesy of Nordic Structures.*)

were installed at the factory. For the panel-to-panel connections, steel plates with annular ring nails were preferred. Holz-Stahl-Komposit (HSK)™ plates are high-capacity hold-downs for mass timber construction.

Collaborative Design

All parties demonstrated flexibility and openness to the complexities of this project, including training and supervision of site crews previously unfamiliar with mass timber construction. A key aspect of the project was the wood-concrete composite floor, a proprietary system developed with the support of Peggi L. Clouston, Ph.D. and Alexander C. Schreyer, M.A.Sc., Dipl.-Ing., both from UMass. The integration of this first in the United-States unique composite system had a significant impact during manufacturing, shipping, handling, and installation. Due to seasonal challenges, the assembly of metal components to the wood elements was done in the factory and required a unique quality control process.

The success of the project was due, in part, to the crucial and extensive collaboration between Nordic engineering staff and the extended team of designers, engineers, and the general contractor. All parties collaborated to ensure that the building would meet, or exceed, the intent of the school: to provide visionary and welcoming spaces that serve as meeting places, exposition halls, and workshops that demonstrate mass timber effectiveness as a visual, structural, and sustainable construction product.

Nordic's position as a leader and essential partner in the design, fabrication, and installation of large-scale mass timber projects was confirmed and solidified with the successful delivery of this ground-breaking project, with its many design, engineering, and execution challenges.

Go to Downloads & Resources tab at **www.mhprofessional.com/Commercial Building Construction-Ancillaries** to access the images provided in this chapter and correlate with the content of this chapter. Here you can look at the chapter figures on your computer screen to pan and zoom in and out for better observation, because full drawings reduced to fit on a textbook page are often difficult to read. The image bank also includes the complete set of plans for the Brookings South Main Fire Station used as examples throughout this textbook.

Go to **Downloads & Resources tab at www.mhprofessional.com/Commercial Building Construction-Ancillaries** to access the test and creative thinking problems for this chapter. The chapter test can be used for review or to evaluate content knowledge depending on your course objectives. Creative thinking problems are provided to help expand your knowledge by researching given subjects.

Print reading problems have questions that ask you to find information on the Brookings South Main Fire Station plans. The print reading problems can help you become familiar with the format of construction documents and reinforce your ability to seek specific information. This is an important skill for you to master as you prepare to enter the construction industry. Go to **Downloads & Resources tab at www.mhprofessional.com/Commercial Building Construction-Ancillaries** to access the complete set of plans for the Brookings South Main Fire Station.

CHAPTER 9

Doors, Windows, and Installations

INTRODUCTION

This chapter introduces you to commercial doors, windows, skylights, curtain wall, storefronts, and cladding.

COMMERCIAL DOORS

Commercial doors are specifically manufactured for use in business and industry-related activities in buildings such as offices, stores, schools, hospitals, and manufacturing industries to name a few. Commercial doors are manufactured for extra durability over residential doors. Commercial doors are designed and manufactured endure weather changes and adverse use. The following describes commercial door types, materials used to make doors, and commercial door products. There are too many topics to show an image or photo of each, so it is suggested that you search the Internet using the key term to see additional reference.

Types of Commercial Doors

There are many commercial door products that generally fit into the following types:

Roll-up doors are typically galvanized steel doors manufactured in different shapes, sizes, and configurations of rolled formed slats that interconnect with each other to form a door curtain that rolls up into a coil above interior or exterior building openings.

Fire-rated doors are manufactured with fire-resistant materials and designed to prevent fire from spreading throughout the building. Steel fire doors are rated by time that a door can withstand exposure to fire-test conditions. Hourly ratings include 3-hours, 2-hours, 1-1/2-hours, 1-hour, 3/4-hour, and 1/3-hour. Certain buildings containing fire hazardous materials, and specific building locations require fire-rated doors. Confirm regulations with building codes.

Scissor gates, also called **security gates,** made in the form of two or more vertical hinged leaves that can be folded one against another. Scissor gates are made from high-quality steel to provide maximum security for use by commercial institutions like banks, lending firms, and other finance-related businesses.

Overhead doors are conventional garage door that rotates on a horizontal axis and is operated and supported by rollers running in a horizontal track when opening and closing. Overhead doors are preferably suitable for warehouse and garage openings. Overhead doors save space when compared to swinging doors.

Commercial Door Manufacturing Materials

Commercial doors are often manufactured with one type of material, but the doors can be made using a variety of materials. The following describes materials commonly used to manufacture commercial doors:

Steel doors provide a high level of security and durability due to the material sturdiness and thickness. Steel is one of the most common materials used for commercial doors. These doors are manufactured from steel sheets wrapped around honeycombed core or insulation.

Wood doors are manufactured using solid wood for exterior installations, or wood veneer sheets that are wrapped around a composite core made from solid lumber, particle board, or pressed mineral products. These interior commercial door applications are not as durable as the other types of materials.

Aluminum doors are commonly used in commercial businesses where a stylish, and clean appearance is required. Aluminum doors can have the natural material polished or unpolished for a desired appearance, or the doors can be **anodized** or painted for any color combination. Aluminum doors are often combined with glass. Anodize is a coat, especially for aluminum, that provides a protective color oxide layer by an electrolytic process.

Fiberglass doors are manufactured with **fiberglass,** which is a reinforced plastic material composed of glass fibers embedded in a resin matrix. Fiberglass doors are durable and can be made in any color and with or without glass.

Full glass doors are used for interior locations were an open environment is desired between areas. Glass doors are made with **safety glass,** which is glass that has been toughened or laminated so that it is less likely to splinter when broken.

Overhead doors are manufactured from steel sheets or sections that fold or rotate on a horizontal axis and is supported horizontally when open or rolled into a cylinder above the opening. Overhead doors are commonly used in garages, storage facilities, warehouses, docks, and other locations due to its large door openings and space saving design.

Commercial Doors Products

There are many commercial door products and you can search the Internet to experience the options available for specific applications. The following describes commonly available commercial door products identified in general terms rather than specific products:

High-performance doors are a strong and durable class of doors for high-use conditions.

Weather-edge doors are manufactured to have an improved resistance to dust, air, and water infiltration.

Insulated commercial doors combine panels with weather seals and foam insulation on the guides, hood, bottom bar, for doors that provide excellent temperature and climate control.

Ventilated doors are typically used for loading docks, warehouses, and retail storefronts. Ventilated doors provide security while offering ventilation and visibility.

Counter service doors are designed for small openings over a counter with reduced clearances and attractive appearance. Counter service doors have applications in commercial, retail, corporate and professional environments, designed as metal curtain, wood curtain, and integral frame and sill.

Fire doors meet fire safety standards with the design described earlier, and automatic closing, specialized construction, and auto-reset or easy-reset functions, causing the door to close automatically when a fire alarm is activated.

High-performance grilles are commonly used for facilities such as parking garages where there is a need for numerous use and high-speed operation. Available doors are designed to minimize operating noise and vibration.

Side-folding closures are slide to the side of small openings with low headroom, stack compactly with easy operation, and are generally made with light-weight aluminum.

Folding gates, also called collapsible, expanding, accordion, or scissor gates are fixed or portable, and either bi-passing or single sliding.

Boarding bridge door are roll up doors used at an airline board bridge for passengers to board the aircraft, and can have side curtains that serve as walls.

PRE-HUNG DOORS

Pre-hung doors are commonly used for interior doors in commercial construction and exterior doors in light wood frame or steel frame in light commercial projects. A pre-hung door is a pre-manufactured door assembly with jambs and door mounted and ready for installation in the project. The components of a pre-hung door assembly are shown in Figure 9.1. The following briefly defines each component of the pre-hung door assembly and also apply to other doors:

A **jamb** is a vertical-side and horizontal-top surrounding frame for a door to open from or close into. This is also called side jamb and head jamb. The **hinge jamb** is part of the door frame where the hinges are attached. The **strike jamb,** or **latch jamb,** is the vertical side of the door frame on the lock side of the door. The **head jamb** is the horizontal part of the door frame above the door. A **stop molding** is a small rectangular-shaped molding positioned on the jamb, slightly past the width of the door, and is used to stop the door in the jamb at a specific location. A **face plate** is a rectangular plate on the edge of the door around the **latch bolt.** A **strike plate** is a metal plate attached to the door frame that the lock engages. The strike plate has a curved leading edge that allows the lock to smoothly engage the strike plate. The strike plate also has a rectangular cutout with a **mortise** cut in the jamb behind to accept the latch bolt and to keep the door closed. A mortise for the strike plate is a square hole, pocket, or relief cut into the jamb to accept the latch bolt and to keep the door closed. The **lockset bore** is a hole cut through the door where the **lockset** is installed. A lockset is complete locking system, including knobs, plates, **latch bolt,** and a

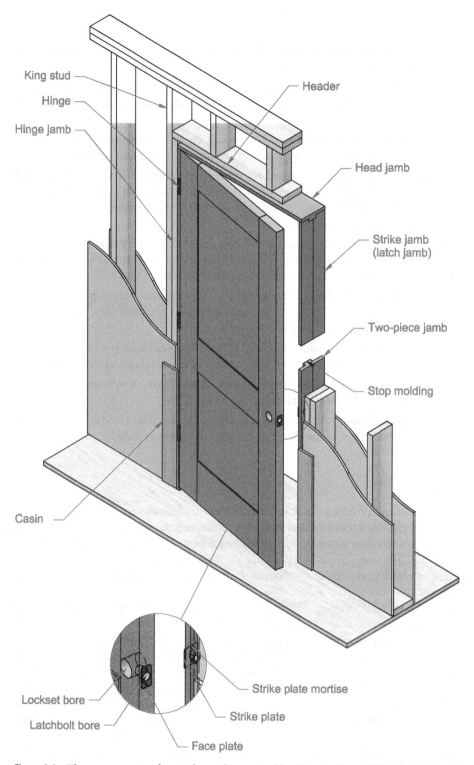

Figure 9.1 The components of a pre-hung door assembly. (*Previously published in Modern Residential Construction Practices, Routledge, 2017 and reproduced here with permission.*)

locking mechanism. The lockset is normally not part of the pre-hung door assembly and is purchased separately. The **latch bolt bore** is a hole cut in from the latching side of the door that intersects the center of the lockset bore, and is where the latch bolt slides. The latch bolt is a spring-operated cylinder assembly in the edge of a door that keeps the door closed, and is operated by turning the door handle. The latch bolt is normally not part of the pre-hung door assembly and is purchased with the lockset. **Casing** is the molding placed around the door frame to cover the joint between the frame and the wall. The casing on the outside of an exterior door is called a **brick mold,** which is part of the pre-hung exterior door assembly and used to trim a door installation around the exterior top and sides. A pre-hung exterior door also comes with a **threshold** that is the metal, plastic, or wood sill.

Specifying Door Swing

The direction doors' swing is very important and is carefully considered by architects when designing a building. As work in the commercial construction industry, you can read the floor plan door symbols to see the direction each door is placed and the related swing. When ordering the doors it is important to specify the correct swing for each door, so they open as intended when set in each location. All doors are designated by the direction of their swing, referred to as **hand.** This must be determined to be sure that the door you order opens in the correct direction for its location in the building. There are different ways to determine door swing, so this is why you need to be sure you and the manufacturer or supplier are communicating door swing in the same manner. Figure 9.2 shows four ways to specify door swing based on when you are standing looking at the door and if the door is left hand, right hand, inswing, or outswing. Figure 9.2a shows a left-hand (LH) inswing, when the door swings

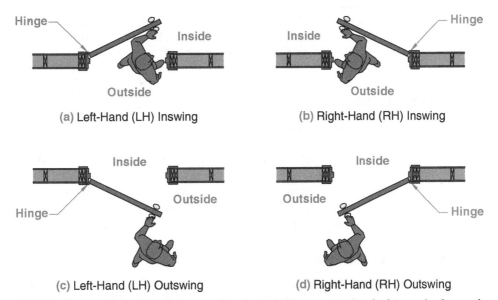

(a) Left-Hand (LH) Inswing   (b) Right-Hand (RH) Inswing

(c) Left-Hand (LH) Outswing   (d) Right-Hand (RH) Outswing

Figure 9.2   Four ways to specify door swing based on when you are standing looking at the door and if the door is left hand, right hand, inswing, or out swing. (*a*) Left-hand (LH) inswing, when the door swings in away from you and to the left. (*b*) Right-hand (RH) inswing, when the door swings in away from you and to the right. (*c*) Left-hand outswing, when the door swings out toward you and to the left. (*d*) Right-hand outswing, when the door swings out toward you and to the right.

in away from you and to the left. Figure 9.2*b* shows a right-hand (RH) inswing, when the door swings in away from you and to the right. Figure 9.2*c* shows a left-hand outswing, when the door swings out toward you and to the left. This is also called left-hand reverse (LHR), because you have to backup or move in reverse when opening the door. Figure 9.2*d* shows a right-hand outswing, when the door swings out toward you and to the right. This is also called right-hand reverse (RHR). You can say reverse hand door if you are standing facing the outside of the door and you have to backup when opening the door. Also, if the hinges are on the left side of the door, it is an LHR, and hinges on the right side of the door is RHR.

GLASS SLIDING DOORS

Glass sliding doors have two or more glass panels that slide horizontally on a track, where one panels slides next to another panel to open and close. Glass sliding doors are made with wood, metal, or vinyl frames and **tempered glass** for safety. Tempered glass is a safety glass that is four to five times stronger than standard glass and shatters into small oval-shaped pebbles when broken. Glass sliding doors can be found in multifamily commercial buildings. These doors are used to provide glass areas and are excellent for access to a patio or deck. Glass sliding doors typically range in width from 5 ft (1500 mm) through 12 ft (3600 mm) at 1 ft (300 mm) intervals. Common sizes are 6 ft and 8 ft (1800 and 2400 mm).

> ### NOTE
>
> Metric door sizes vary from 600 to 1000 mm depending on use and location. Standard metric door sizes for all large openings such as sliding glass or double French doors vary from 1500 to 3600 mm.
>
> The Americans with Disabilities Act (ADA) specifies a minimum opening 36 in. (915 mm) wide for wheelchair access. This is a clear, unobstructed dimension measured to the edge of the door at 90°, if the door does not open 180°.

COMMERCIAL WINDOWS

Commercial windows must meet specific ratings set by the North American **Fenestration** Standard. Standards for commercial windows include heavy-duty frame manufacturing and reinforced glazing designed to withstand high wind pressure and structural loads. The term **fenestration** refers to any opening in a building envelope, including windows, doors, and skylights. Commercial windows are designed to let in extra light for larger spaces, such as offices and schools. Commercial windows are built to last for many years, because they are expensive to buy and install, and their replacement is even more costly.

Commercial Window Materials

The following briefly describes materials generally used in the manufacture of commercial windows. The term **glazing** refers to the glass panes that make up a window. There are single, double, triple, and higher glazes. The greater the number of window panes, the more heat and noise insulation is provided. Manufacturers also produce multiple pane glazing that uses a combination of glass and plastic. These windows can be less expensive and easier to install than all-glass windows.

Tempered Glass

Tempered glass is widely used in commercial windows and required by code for specific installations. Tempered glass, also called **toughened glass,** is a type of **safety glass** made by controlled thermal or chemical treatments to increase the glass strength. Tempered glass is four to six times stronger than standard glass, and shatters in small harmless pieces when broken, as compared to standard glass that breaks into large sharp pieces. Safety glass is a general term that refers to glass that has been toughened or laminated so that it is less likely to splinter when broken.

Laminated Glass

Laminated glass is a type of safety glass that holds together when shattered. In the event of breaking, it is held in place by an interlayer, typically of **polyvinyl butyral** (PVB) or ethylene-vinyl acetate (EVA), between its two or more layers of glass. The interlayer keeps the layers of glass bonded even when broken, and its high strength prevents the glass from breaking up into large sharp pieces. This increases protection in the event of a natural or man-made disaster by making the window stronger and causing broken glass to stick to the plastic center rather than shatter outwardly. Laminated glass is normally used when there is a possibility of human impact or where the glass could fall if shattered and also for architectural applications. Skylight glazing and automobile windshields typically use laminated glass. In geographical areas requiring hurricane-resistant construction, laminated glass is often used in exterior storefronts, curtain walls, and windows.

Laminated glass is also used to increase the sound insulation rating of a window, where it significantly improves sound reduction compared to standard glass panes of the same thickness. For this purpose, a special acoustic PVB compound is used for the interlayer.

Tinted Windows

Window tint is special film or coating applied to windows for a variety of reasons, most commonly to prevent a specific spectrum or amount of sunlight from passing through the glass, but also for one of a variety of other reasons such as create privacy, eliminate glare, and reduce the amount of heat absorbed from sunlight. Many of the possible window tinting applications are briefly described in the following.

Standard Tint
Commonly used bronze and gray tints reduce glare but also decrease the illumination benefit of daylight and are less effective at blocking sun's heat.

High-Performance Tint
Spectrally selective glazing is a high-performance tint that allows more illumination than standard tints but also blocks much of the heat absorption from the sun. Spectrally selective glazing allows as much daylight as possible, while preventing transmission of as much solar heat as possible.

Reflective Glass
Reflective glass has a coating that acts as a mirror and provides privacy and heat reduction by reflecting light. Reflective glass coating is common for large commercial windows in areas that have very sunny climates.

Low-Emissive Glass

Low-emissive glass, also called **low-e glass,** is a technology that improves window energy efficiency. Low-e glass has a transparent coating that acts as a thermal mirror, which increases insulating value, blocks heat from the sun, and reduces fading of objects inside the building. Low-e window tint works as a window insulation and blocks heat. Low-e tint holds indoor cool air during the summer and heat during the winter for improved energy conservation. Another energy saving window tint product reduces heat during peak sun conditions while acting as a clear insulator during colder periods.

Sun Blocking

Sun blocking window film can be used to block harmful UV rays and moderate excessive heat gain from the sun.

A specific sun blocking window tint option can be used exclusively to block harmful UV rays while allowing natural light to enter. The UV controlling tint helps reduce hot spots and protects indoor items from becoming faded.

If window walls are desired, but excess sun exposure is a concern, then sub block window tinting can be used to provide a comfortable indoor environment with less heat and glare while preserving the view.

Sun blocking window tint can also be used specifically to reduce glare from the sun and block harmful UV rays.

Daylight redirecting window tint allows natural light to enter without glare, hot spots, or UV exposure, and optically redirects sunlight up toward the ceiling.

Photochromic window tinting was described earlier in this chapter, also called transitional window tint, this product provides anti-glare and UV protection during the day, with clear transparency at night.

Skylight and sloped window glass commonly use clear glass for natural light, but window tinting can be used to help reduce energy loss and UV exposure through the glass.

Privacy Window Tint

A variety of privacy window tint options are available to produce different effects. Frosted window tint, also called two-way tint, provides privacy inside while blocking the view from outside, while offering natural light at the same time. Mirrored window tint is a privacy option that provides a dramatic mirror appearance from the outside, while delivering one-way vision and privacy during bright conditions, along with sun and heat control.

Safety and Security Window Film

Safety and security window film products are used to reinforce glass and improve the glass ability to withstand impact from attempted break in or impact from projectiles. This type of window protection can be used on specific glass locations such as entry doors and ground level windows. Specific bullet impact resistant glass film, referred to as **ballistics film** is used on glass where this severe protection is a required. These products can also help reduce glass from shattering when hit by flying objects.

Window Graphics

Different from the window tint and film products previously described, window graphics are generally a film or vinyl product with standard or custom printed graphics. These products can be used for aesthetes with a landscape scene or used for advertising with a promotion graphic.

Window Frames

Commercial window frame are available in metal, wood, plastic, fiberglass, or other material. Metal frames provide the least amount of insulation. A **thermal break** is typically used in the window frame to provide insulation. Thermal break system is a longitudinal channel, longitudinal flange, or side walls that create a hollow center in which a thermal barrier is integrally formed.

Insulated Windows

Windows can be a big source of energy loss by heat escaping from the building during the winter and allowing extra heat to enter the building during the summer. Doors can also contribute to loss of energy efficiency in the building. However, doors are generally more energy efficient than windows. There are a variety of ways to insulate windows. A common method to insulate windows is to use two or more panes of glass with a sealed space between, as shown in Figure 9.3. This is referred to as **insulating glazing (IG)** or **insulated glass units.** Another technique is to fill the sealed space with argon, which is a colorless gas that raises the window insulating value. Low-emissive glass, also called low-e glass, introduced earlier, is an additional technology that improves window energy efficiency. Low-e glass has a transparent coating that acts as a thermal mirror, which increases insulating value, blocks heat from the sun, and reduces fading of objects inside the building. Window frames made of wood or a combination of materials that have a thermal brake installed provide an energy efficiency advantage.

Insulation is rated by **R-value,** and windows are rated by **U-value.** R is the reciprocal of the U-value: 1/U = R. R-value of a material is a measure of thermal resistance to heat flow. U-value is the coefficient of heat transfer expressed as **Btuh** sq ft/°F of surface area. Btuh is heat loss calculated in Btu's per hour. **Btu** means British thermal unit, a measure of heat. The lower the U-value, the more the insulating value. The U-value of windows can also be identified by window class (CL). For example, a window with a value of U = 0.40 is CL40.

THERMOCHROMIC GLASS

The following content is provided, in part, by Suntuitive Dynamic Glass. Dynamic glass is based on **thermochromic technology.** Thermochromic technology is the property of substances that change color due to a change in temperature. Thermochromic elements are the dynamic principle of glass activate by the heat from direct sunlight causing the window to tint as necessary. These elements are embedded in a PVB (polyvinyl butyl) interlayer, which is then laminated between two pieces of glass. This laminate is then incorporated into the final product, a dynamic insulated glass unit (IGU), as shown in Figure 9.4.

Throughout the day whenever the sun shines directly on the window the sunlight heats the outboard laminate of the window. The PVB interlayer of the laminate becomes warm and darkens, creating a tinting effect over the window. Low-e coating in the IGU help reject that heat to the outside. As the sunlight recedes, the glass and interlayer cool, returning the glass to a clear state.

Passive and Sustainable

Thermochromic glass is a passive technology that does not require a power source, other than solar energy from direct sunlight, and requires no additional energy input.

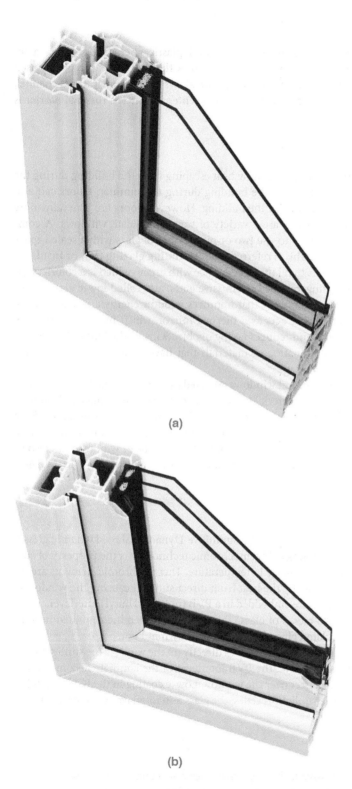

(a)

(b)

Figure 9.3   A common method to insulate windows is to use two or more panes of glass with a sealed space between. (*a*) A double pane window example. (*b*) A triple pane window example.

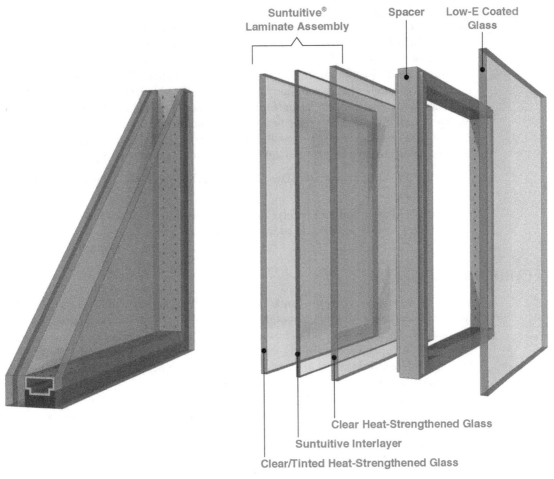

Suntuitive® Laminate Assembly   Spacer   Low-E Coated Glass

Clear Heat-Strengthened Glass

Suntuitive Interlayer

Clear/Tinted Heat-Strengthened Glass

Figure 9.4   A dynamic insulated glass unit (IGU). (*Courtesy of Suntuitive Dynamic Glass.*)

Thermochromic Glass Performance

The following are thermochromic glass performance advantages over standard clear glass:

- **Visible Light Transmittance (VLT):** Thermochromic dynamic glass constantly adapts to changing sunlight conditions. Throughout the year, this adaptive technology attenuates the amount of visible light and glare let into a building by dynamically tinting when called for by various weather conditions.

- **Solar Heat Gain Coefficient (SHGC):** As thermochromic technology adjusts to direct sunlight throughout the day, it optimizes solar control and minimizes solar heat gain.

- **U-Value:** Suntuitive Dynamic Glass has all the advantages of dynamic glass while at the same time retaining the benefits of a conventional high-efficiency IGU using the best low-e coatings to give low U-values to prevent heat loss during cold nights.

- **Energy Savings:** Using thermochromic dynamic glass reduces the heat load of a building and can save up to 20 percent on energy.

Maximize Energy Savings

Suntuitive Dynamic Glass technology is incorporated in high-efficiency IGUs, featuring both excellent U-values, daylighting and **SGHC** properties, saving energy and lowering greenhouse emissions. SHGC is the fraction of incident solar radiation admitted through a window, both directly transmitted and absorbed and later released inward. SHGC is expressed as a number between 0 and 1. The lower a window's solar heat gain coefficient, the less solar heat it transmits. Independent research has shown that using Suntuitive Energy Saving Dynamic Glass reduces the heat load of a building and minimizes energy use by reducing heat loads in winter and cooling loads in summer. Retaining daylight independence, reducing the need for artificial lighting, can result in further energy savings.

- **UV Protection:** Suntuitive Dynamic Glass technology blocks 99 percent of harmful UV light.
- **Noise Reduction and Safety:** Suntuitive Dynamic Glass laminated construction offers noise reduction and added safety benefits.

ELECTROCHROMIC GLASS

Electrochromic glass, also called **smart glass** or **electronically switchable glass,** is a technologically advanced building glass that can be used to create partitions, windows, and skylights that change opacity from clear to a desired shade using electricity. Electrochromic glass allows occupants to control the amount of heat or light that passes through the glass by switching electrical current on or off, giving them the ability to regulate temperatures or create privacy as desired.

Electrochromic glass uses the principle of **electrochromism** that allows certain materials to change color or opacity when an electrical charge is applied. A small amount of electricity is required for changing the glass opacity, but no electricity is needed to maintain a desired shade once the change has been made. Electrochromic glass can be transformed from clear to opaque for having light and bright spaces or shaded private enclosures. Transition from clear glass to opaque and back again can be controlled by wall switches, remote controls, movement sensors, light sensors, or timers.

NATIONAL FENESTRATION RATING COUNCIL

The following is taken in part from the NFRC website. The **National Fenestration Rating Council (NFRC)** is a nonprofit organization that administers the only uniform, independent rating, and labeling system for the energy performance of windows, doors, skylights, and attachment products. The term **fenestration** is any opening in a building envelope, including windows, doors, and skylights.

NFRC Label

The NFRC energy performance label, as shown in Figure 9.5, can help you determine how well a product performs the functions of helping to cool your building in the summer, warm your building in the winter, keep out wind, and resist condensation. By using the information contained on the label, building designers, builders, and consumers can reliably compare one product with another and make informed decisions about the windows, doors, and skylights they specify and buy.

Figure 9.5 The NFRC energy performance label. (*Courtesy of the National Fenestration Rating Council (NFRC).*)

The NFRC label lists the manufacturer, describes the product, provides a source for additional information, and includes ratings for one or more energy performance characteristics. All energy performance values on the label represent the rating of windows and doors as whole systems, which includes glazing and frame. Look at Figure 9.5 as you review the following information found on the NFRC label.

U-Factor

U-factor measures how well a product prevents heat from escaping. The rate of heat loss is indicated in terms of the U-factor (U-value) of a window assembly. U-factor ratings generally fall between 0.20 and 1.20. The insulating value is indicated by the R-value which is the inverse of the U-value. The lower the U-value, the greater a window's resistance to heat flow and the better its insulating value.

Solar Heat Gain Coefficient

Solar heat gain coefficient (SHGC) measures how well a product blocks heat caused by sunlight. The SHGC is the fraction of directly transmitted and absorbed solar radiation admitted through a window and then released inward. SHGC is expressed as a number between 0 and 1. The lower a window's solar heat gain coefficient, the less solar heat it transmits in the house.

Visible Transmittance

Visible transmittance (VT) measures how much light comes through a product. The visible transmittance is an optical property that indicates the amount of visible light transmitted. VT is expressed as a number between 0 and 1. The higher the VT, the more light is transmitted.

Air Leakage

Air leakage (AL) is indicated by an air leakage rating expressed as the equivalent cubic feet of air passing through a square foot of window area (cfm/sq ft). Heat loss and gain occur by infiltration through cracks in the window assembly. The lower the AL, the less air passes through cracks in the window assembly.

Condensation Resistance **Condensation resistance (CR)** measures the ability of a product to resist the formation of condensation on the interior surface of the product. The higher the CR rating, the better the product is at resisting condensation formation. While this rating cannot predict condensation, it can provide a credible method of comparing the potential of various products for condensation formation. CR is expressed as a number between 0 and 100.

NOTE

Fenestration products, especially windows and skylights, have a unique challenge of providing adequate thermal performance while also providing daylight and views. While system manufactures work to develop improved frame designs to lower U-factor ratings, glass manufacturers are constantly advancing low-e technology to block the non-visible light spectrum to reduce the greenhouse effect by lowering the SHGC value, while maintaining the highest visible transmittance possible. Glass manufacturers list thermal performances to aid in glass selection, which are center of glass values (COG). COG values are always significantly better than full product NFRC-rated values. COG values are measured from a single optimal center point, whereas NFRC-rated values measure the entire assembly performance. Also note that skylights are tested in a 20° orientation and on a wooden curb, whereas windows and doors are tested vertically and mounted in wood framing. This causes skylight values to often seem less efficient due to the varied tested procedures.

WINDOW TYPES AND STYLES

Windows are available in a variety of types and styles with fixed glass that is not operable, or with one of several different opening and closing mechanisms. Operable windows usually have a screen over the opening to keep insects, birds, and some pollutants out of the building when the window is open. Windows are often designed to match the architecture where contemporary windows look much different from traditional windows. Windows are also selected to function best in their environment, where certain types of windows are better for cold wet climates than other windows. Window frames are typically made from aluminum, vinyl, or wood. Wood-frame windows can be aluminum- or vinyl-clad on the outside for weather protection and wood exposed on the inside for desired finishes or painting, as shown in Figure 9.6.

Windows are typically manufactured in widths ranging from 1 ft through 12 ft at intervals of about 6 in., depending on the manufacturer and style. Vinyl- and aluminum-frame windows generally fall within the range of these nominal sizes, but sizes of wood-frame windows are often different and should be confirmed with manufacturer specifications. Metric window sizes also vary and should be confirmed. The top of windows is typically 6'-8" height from the floor in standard multifamily homes, but can be higher in special rooms with higher than normal ceilings.

The location of a window and the way the window opens has an effect on the window size. A window between 6 and 12 ft wide is common for a living, dining, or family room. To let the occupants take advantage of a view while sitting, and windows in these rooms are normally between 4 and 5 ft high.

Windows in bedrooms are typically 3 to 6 ft wide. The height often ranges from 3'-6" to 4'-0". The type of window used in the bedroom is important because of emergency egress requirements in most codes, which typically states that a bedroom window sill can be no higher than 44 in. from the floor.

Figure 9.6 Wood-frame windows can be aluminum- or vinyl-clad on the outside for weather protection and wood exposed on the inside for desired finishes or painting.

Kitchen windows are often between 3 and 5 ft wide and between 3′-0″ and 3′-6″ tall. Wide windows are nice to have in a kitchen for the added light they provide, but the number of upper cabinets is reduced to accommodate a wider window. If the tops of windows are at the normal 6′-8″ height, windows deeper than 3′-6″ can interfere with the countertops.

Bathroom windows often range between 2 and 3 ft wide, with an equal height. A wider window with less height is often specified if the window is to be located in a shower area. Most bathroom windows have **opaque glass.** Opaque glass is made translucent, which is not clear, instead of transparent, which is clear.

> **NOTE**
>
> This window type content is related to windows commonly used in multifamily homes such as apartments, condominiums, and townhomes. Other commercial buildings typically use windows that are not operable.

Sliding Windows

The **sliding window** is a popular 50 percent operable window when there are two **sashes** in the window. Sliding windows commonly have two side-by-side sashes where one sash slides horizontally past the other. Large sliding windows can have three sashes, with the center sash fixed between two smaller side sashes that each open by sliding horizontally. The term **sash**, plural sashes, is a framework that holds the panes of a window in the window frame. Sliding windows are typically aluminum or vinyl framed, but wood frames are available.

Casement Windows

Casement windows can be 100 percent operable and are best used where extreme weather conditions require a tight seal when the window is closed, although these windows are in common use everywhere. Casement windows open from the outside vertical edge, generally using a crank mechanism. A variety of casement window combinations are available. Single casement windows have one sash that opens outward. Casement windows can have two or more connected sashes with any combination fixed or operable.

Double-Hung Windows

The traditional **double-hung window** has two sashes with one above the other. Both sashes open to any desired extent by sliding vertically. Double-hung windows allow for natural ventilation control, you can let the warm air in that rises to the ceiling by lowering the top sash, or raise the bottom sash to let cool air flow inside. A double-hung window operates by means of two sashes that slide vertically past each other. Double-hung wood-frame windows are designed for energy efficiency and are commonly used in traditional architectural designs. Double-hung windows are typically higher than they are wide. A double-hung window usually ranges in width from 1'-6" through 4'-0". It is common to have double-hung windows grouped together in pairs of two or more. A **single-hung window** has two sashes of glass, with the top sash stationary and the bottom sash movable.

Fixed Windows

Fixed windows are popular when a large, unobstructed area of glass is required to take advantage of a view or to allow for solar heat gain. A fixed window has a single non-operable glass pane, or a group of large windows each separated by a vertical support.

SKYLIGHTS AND SLOPED GLAZING

A **skylight** is a window in a roof used to allow sunlight to enter. An individual skylight typically used in building design and construction is referred to as a **unit skylight.** A unit skylight is a complete factory-assembled glass- or plastic-glazed opening consisting of not more than one panel of glass or plastic installed in a sloped or horizontal orientation primarily for natural **daylighting.** Daylighting is the use of various design techniques to enhance the use of natural light in a building. A skylight is used when additional daylight is desirable in a room, or to let natural light enter an interior room through the ceiling. Skylights are available fixed or operable. Skylights are made of plastic in a dome shape or flat tempered glass. Tempered double-pane insulated skylights are energy efficient, do not cause any distortion of view, and generally are not more expensive than plastic skylights.

The term **skylight** generally refers to both unit skylights and **sloped glazing.** Sloped glazing is a glass and framing assembly that is sloped more than 15° from the vertical, forms the entire roof of the structure, and is generally a single-slope construction, as shown in Figure 9.7. Skylights have been historically restricted in the amount of roof area in which they can cover, due to the concern of thermal heat loss in cold climates and excessive solar heat gain in hot climates. Advancement of low-e glass technologies has made skylights significantly more energy efficient. Green energy codes recognize the benefits of providing natural daylight into the interior spaces far outweigh the minimal thermal losses. Now, codes require a minimum skylight area rather than limiting skylights to a maximum allowable area as in the past.

Figure 9.7   Sloped glazing is a glass and framing assembly that is sloped more than 15° from the vertical, forms the entire roof of the structure, and is generally a single-slope construction. (*Courtesy of CrystalLite, Inc.*)

Tubular Daylighting Devices

It is important to determine accurate and adequate lighting needs for a building. Daylighting is often a major consideration. Natural light in addition to or in place of artificial lighting is a decision made based on safety considerations, aesthetic preferences, environmental practices, cost efficiency, health benefits, and expense. Natural light is not always readily available within every space, especially in small rooms, corridors, and multiple-story buildings.

A natural lighting option that is often selected for its compact size, technologically advanced abilities and flexibility are **tubular daylighting devices (TDD).** Figure 9.8 shows an example of a tubular daylighting device. A standard TDD system is a transparent, roof-mounted dome with self-flashing curb, reflective tubing, and a ceiling-level diffuser assembly that transmits the daylight into interior spaces.

DOOR AND WINDOW SCHEDULES

Numbered symbols used on the floor plan key specific items to charts known as schedules. A **schedule** is a grouping of related items with corresponding individual features, with a heading and a minimum of three columns of related information. A schedule formats information into rows and columns in order to more easily present design information.

Figure 9.8 An example of a tubular daylighting device (TDD). (*Courtesy of CrystalLite, Inc.*)

Schedules are used to describe items such as doors, windows, appliances, materials, fixtures, hardware, and finishes. A typical door schedule is shown in Figure 9.9. Schedules help keep plans clear of unnecessary details and notes. Schedules typically include the following kind of information about the product:

Manufacturer's name	Quantity
Product name	Size
Model number	Rough opening size
Type	Color

Schedule Key

Doors and windows described in a schedule are keyed from the floor plan to the schedule. For example, the key can label doors with a number and windows with a letter. The key letters and numbers can be enclosed in different geometric figures to help make them more obvious when looking at the floor plans. Another option for the key is the use of a divided circle with the letter D for door or W for window above the dividing line, and the number of the door or window, using consecutive numbers, below the line. Figure 9.10 shows how the door and window floor plan symbols key to the schedules.

> **NOTE**
>
> The schedule key examples provided here demonstrate light commercial residential applications, such as in apartments, condominiums, and townhomes. Other commercial door and window installations are shown on the plans in a similar manner using drawings with schedule keys and schedules, although schedule descriptions can be more detailed.

DOOR SCHEDULE

DOOR NUMBER	ROOM NAME	SIZE			DOORS				FRAME				DETAILS			RATING	HW SET	NOTES
		WIDTH	HEIGHT	THK	MATL.	TYPE	FINISH	GLZ	MATL.	TYPE	FINISH	GLZ	JAMB	HEAD	SILL			
FIRST FLOOR																		
100A	DRIVE-THRU BAYS	3'-0"	7'-0"	1 3/4"	ALUM	F	PREFIN.	-	ALUM	2	PREFIN	-	5C/A200	5C/A200	4C/A200	-	SEE SPEC	
100B	DRIVE-THRU BAYS	3'-0"	7'-0"	1 3/4"	ALUM	F	PREFIN.	-	ALUM	2	PREFIN	-	5C/A200	5C/A200	4C/A200	-	SEE SPEC	
100C	DRIVE-THRU BAYS	3'-0"	7'-0"	1 3/4"	ALUM	F	PREFIN.	-	ALUM	2	PREFIN	-	5C/A200	5C/A200	4C/A200	-	SEE SPEC	GLASS
100D	DRIVE-THRU BAYS	14'-0"	14'-0"	2"	ALUM	OH Glass Door - Panel	PREFIN.	INSULATED TEMP/INSULATED	SL	SL	-	-	6D/A301	5D/A301		-	SEE SPEC	GLASS
100E	DRIVE-THRU BAYS	14'-0"	14'-0"	2"	ALUM	OH Glass Door - Panel	PREFIN.	INSULATED TEMP/INSULATED	SL	SL	-	-	6D/A301	5D/A301		-	SEE SPEC	SOLID
100F	DRIVE-THRU BAYS	14'-0"	14'-0"	2"	ALUM	OH Glass Door - Panel	PREFIN.	INSULATED TEMP/INSULATED	SL	SL	-	-	6D/A301	5D/A301		-	SEE SPEC	GLASS
100G	DRIVE-THRU BAYS	14'-0"	14'-0"	2"	ALUM	OHS	PREFIN.	-	SL	SL	-	-	6D/A301 (SIM.)	5D/A301 (SIM.)		-	SEE SPEC	SOLID
100H	DRIVE-THRU BAYS	14'-0"	14'-0"	2"	ALUM	OHS	PREFIN.	-	SL	SL	-	-	6D/A301 (SIM.)	5D/A301 (SIM.)		-	SEE SPEC	SOLID
100J	DRIVE-THRU BAYS	14'-0"	14'-0"	2"	ALUM	OHS	PREFIN.	-	SL	SL	-	-	6D/A301 (SIM.)	5D/A301 (SIM.)		-	SEE SPEC	SOLID
100K	DRIVE-THRU BAYS	3'-0"	7'-0"	1 3/4"	HM	F	FIELD PAINT	-	HM	1	PT	-	1C/A200	1C/A200	-	90MIN	SEE SPEC	SIGN 'B'
102	DRIVE-THRU BAYS	3'-0"	7'-0"	1 3/4"	HM	F	FIELD PAINT	-	HM	1	PT	-	1C/A200	1C/A200	-	90MIN	SEE SPEC	SIGN 'D'
103	MAINT/MECH																	

Figure 9.9 Door schedule. (*JLG Architects, Alexandria, MN.*)

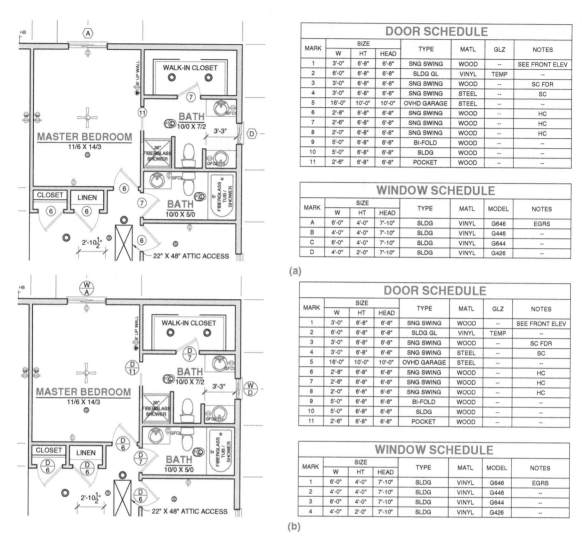

DOOR SCHEDULE

MARK	SIZE			TYPE	MATL	GLZ	NOTES
	W	HT	HEAD				
1	3'-0"	6'-8"	6'-8"	SNG SWING	WOOD	--	SEE FRONT ELEV
2	6'-0"	6'-8"	6'-8"	SLDG GL	VINYL	TEMP	--
3	3'-0"	6'-8"	6'-8"	SNG SWING	WOOD	--	SC FDR
4	3'-0"	6'-8"	6'-8"	SNG SWING	STEEL	--	SC
5	16'-0"	10'-0"	10'-0"	OVHD GARAGE	STEEL	--	--
6	2'-8"	6'-8"	6'-8"	SNG SWING	WOOD	--	HC
7	2'-6"	6'-8"	6'-8"	SNG SWING	WOOD	--	HC
8	2'-0"	6'-8"	6'-8"	SNG SWING	WOOD	--	HC
9	5'-0"	6'-8"	6'-8"	BI-FOLD	WOOD	--	--
10	5'-0"	6'-8"	6'-8"	SLDG	WOOD	--	--
11	2'-6"	6'-8"	6'-8"	POCKET	WOOD	--	--

WINDOW SCHEDULE

MARK	SIZE			TYPE	MATL	MODEL	NOTES
	W	HT	HEAD				
A	6'-0"	4'-0"	7'-10"	SLDG	VINYL	G646	EGRS
B	4'-0"	4'-0"	7'-10"	SLDG	VINYL	G446	--
C	6'-0"	4'-0"	7'-10"	SLDG	VINYL	G644	--
D	4'-0"	2'-0"	7'-10"	SLDG	VINYL	G426	--

(a)

DOOR SCHEDULE

MARK	SIZE			TYPE	MATL	GLZ	NOTES
	W	HT	HEAD				
1	3'-0"	6'-8"	6'-8"	SNG SWING	WOOD	--	SEE FRONT ELEV
2	6'-0"	6'-8"	6'-8"	SLDG GL	VINYL	TEMP	--
3	3'-0"	6'-8"	6'-8"	SNG SWING	WOOD	--	SC FDR
4	3'-0"	6'-8"	6'-8"	SNG SWING	STEEL	--	SC
5	16'-0"	10'-0"	10'-0"	OVHD GARAGE	STEEL	--	--
6	2'-8"	6'-8"	6'-8"	SNG SWING	WOOD	--	HC
7	2'-6"	6'-8"	6'-8"	SNG SWING	WOOD	--	HC
8	2'-0"	6'-8"	6'-8"	SNG SWING	WOOD	--	HC
9	5'-0"	6'-8"	6'-8"	BI-FOLD	WOOD	--	--
10	5'-0"	6'-8"	6'-8"	SLDG	WOOD	--	--
11	2'-6"	6'-8"	6'-8"	POCKET	WOOD	--	--

WINDOW SCHEDULE

MARK	SIZE			TYPE	MATL	MODEL	NOTES
	W	HT	HEAD				
1	6'-0"	4'-0"	7'-10"	SLDG	VINYL	G646	EGRS
2	4'-0"	4'-0"	7'-10"	SLDG	VINYL	G446	--
3	6'-0"	4'-0"	7'-10"	SLDG	VINYL	G644	--
4	4'-0"	2'-0"	7'-10"	SLDG	VINYL	G426	--

(b)

Figure 9.10 Door and window floor plan symbols key to the schedules. (*Previously published in Modern Residential Construction Practices, Routledge, 2017 and reproduced here with permission.*)

NOTE

Windows must be installed square and plumb to allow for proper operation. Proper window installation is critical to ensure there is no air infiltration around the frame, proper insulation, no water leakage, and proper function of operable windows. Always confirm correct window installation with manufacturer specifications and requirements.

Window manufacturers typically have very specific installation instructions for their windows. While there are similarities in window installation methods, some windows require specific installation practices. You need to follow window manufacturer installation instruction to help make sure the windows you install do not leak, are as energy efficient as possible, and to protect the manufacturer warranty on the windows.

Skylights require special instructions for proper installation. Confirm manufacturer instructions and if there is a required placement direction. Skylights can have a specific top and bottom with weep holes.

Skylights should never be set flat on the roof prior to installation. Set the skylight on blocks to provide air circulation when you take it to the roof prior to installation. Skylights placed flat on the roof can develop high temperatures that can damage the glazing in a short period of time.

Skylight installations described in this textbook are general suggestions. Always use manufacturer provided preparation and installation instructions and confirm plans with local or national codes.

The American Architectural Manufacturers Association (AAMA) publishes the Installation Guidelines for Skylights that provides detailed installation methods for each skylight style, and for all roofing types currently used in the construction industry.

WINDOW WALL SYSTEMS

There are several different types of window wall systems used on commercial buildings. The most often used are storefronts, curtain walls, window walls, punched windows, and strip windows. The following briefly describes these window wall systems.

Storefront

A **storefront** is a nonresidential, non-load-bearing assembly for a commercial building entrance system using metal framing and windows. Storefronts are typically installed at street level between the ground floor and the structure above, as shown in Figure 9.11. Storefronts are designed and constructed for high use and strength, and are generally not more than 10 ft (3 m) high. While a storefront is commonly used at the ground floor of a building, they can be used up to the second or third floor when the architectural design requires a high glass wall inside the building for a lobby, showroom, or other architectural design.

Figure 9.11   Storefronts are typically installed at street level between the ground floor and the structure above. (©lbarn–Can Stock Photo Inc.)

Storefronts have been historically used as an architectural statement that provides a dramatic building entrance and matches the surrounding area historic architecture.

Punched Windows

A **punched window** refers to a construction practice where a hole is punched in the exterior wall of the building and filled with a window, as shown in Figure 9.12. The term **punched** sounds nontechnical, but the actual construction practice provides an exact opening with required **rough opening (RO)** dimensions to accommodate a specific window in the opening. RO is any unfinished opening that is framed to specific measurements to accommodate the finish product. The punched window architectural arrangement and composition can vary greatly in its application, including manufactured as a round window, arched window, or custom forms.

Punched windows referred to as **Schüco AWS 75.SI+** are highly thermal insulated aluminum-frame window systems with improved thermal insulation, security, and fabrication for increased energy efficiency performance. Schüco is a modular system designed for the basic construction framing depth of 3 in. (75 mm), with impressive function, energy efficiency, and architectural characteristics.

Strip Windows

Strip windows are a row of horizontal windows placed next to each other at each floor of a building, as shown in Figure 9.13. The strip of windows can be along a wall or wrapped

Figure 9.12   A punched window refers to a construction practice where a hole is punched in the exterior wall of the building and filled with a window. (*©meinzahn–Can Stock Photo Inc.*)

Figure 9.13 Strip windows are a row of horizontal windows placed next to each other at each floor of a building. (©*PaulMatthew–Can Stock Photo Inc.*)

around a corner. The windows can be any height between floor slabs. Strip windows are usually installed between concrete panels, aluminum or steel frames, or other common construction materials. Strip windows are a cost-effective way to maximize daylighting without having the cost of curtain wall systems, described next.

Window Walls

Window walls are framed window panels mounted in between the concrete floor slabs from the top of the floor slab at the bottom of the window wall to the bottom of the next floor slab above attaching the top of the window wall, as shown in Figure 9.14. Window walls are non-load bearing. Window wall panels are manufactured and glazed at a manufacturing facility and then delivered to the construction site. Window wall premanufacturing can help to improve quality control and speed manufacturing and installation time for faster project completion. Window walls are constructed between concrete slab floors, which reduces sound transfer between floors.

Curtain Wall

A basic **curtain wall** definition is any non-load-bearing exterior wall building facade that hangs like a curtain from the floor slabs, as shown in Figure 9.15. A curtain wall is designed and constructed to resist air and water infiltration, horizontal wind and seismic forces acting on the building. A curtain wall can be 25 ft (7.6 m) or higher, depending on the architectural design. Curtain walls are commonly designed and built using aluminum or steel **stick framing.** Stick framing is when one member, referred to as a stick, is used

Figure 9.14   Window walls are framed window panels mounted in between the concrete floor slabs from the top of the floor slab at the bottom of the window wall to the bottom of the next floor slab above attaching the top of the window wall. (*©fabian19–Can Stock Photo Inc.*)

Figure 9.15   A curtain wall definition is any non-load-bearing exterior wall building facade that hangs like a curtain from the floor slabs. (*©eldadcarin–Can Stock Photo Inc.*)

at a time to assemble the structure. Insulated glass panels are installed between the completed framework. The frame and glass are combined to provide the architectural style and desired visual appearance with sunlight for the interior. Curtain walls are different from storefront systems because they are designed and constructed to span multiple floors. Curtain walls are typically more expensive to build than window walls, mainly because of the additional labor required. However, the higher costs provide advantages such as improved structural engineering. Curtain walls offer improved resistance to moisture, wind, earthquakes, and heat transfer. Curtain walls provide the following architectural and engineering requirements:

- Thermal expansion and contraction
- Building sway and movement
- Water infiltration
- Thermal energy efficiency for cost-effective heating and cooling
- Natural lighting

There are a few types of curtain wall systems that are different by how each system is built. A curtain wall system commonly incorporates different sized **mullions** that make up the panel framing. Mullions are horizontal or vertical divides between sections of a window. The mullion framing is anchored to the concrete or mass timber slabs. Glazing is then attached to the mullions onsite. Curtain wall window panels can also be prefabricated in a manufacturing facility and then delivered to the construction site where they are anchored to the slabs. The most common types of curtain wall systems include the following.

The stick system, previously introduced, uses vertical mullions anchored to the floor slabs and glazed at the construction site. **Muntins** and **spandrel panels** can be attached after the vertical mullions have been installed, depending on the architectural style. Muntins are horizontal or vertical dividers within a section of a window. Spandrel panels are a preassembled structural panel that can have a variety of textures and colors, and used to replace the appearance of a masonry wall.

Panel systems are more economical than stick systems, because they require much less onsite labor than stick framing. The panels are prefabricated at a manufacturing plant, delivered to the construction site, and then attached to anchors within the building structure.

The unit-and-mullion system is a hybrid panel system that uses mullions anchored to the structure, and then prefabricated glass-framed units are attached to the mullions.

Column cover and spandrel are a modified stick framing system that used large cladding components that use column and spandrel covers that function as sticks. Column cover and spandrel systems join the building frame by aligning mullions to structural columns. Preassembled or field-assembled infill units of framed glass panels are fitted between the column covers.

Go to Downloads & Resources tab at **www.mhprofessional.com/Commercial Building Construction-Ancillaries** to access the images provided in this chapter and correlate with the content of this chapter. Here you can look at the chapter figures on your computer screen to pan and zoom in and out for better observation, because full drawings reduced to fit on a textbook page are often difficult to read. The image bank also includes the complete set of plans for the Brookings South Main Fire Station used as examples throughout this textbook.

Go to **Downloads & Resources tab at www.mhprofessional.com/Commercial Building Construction-Ancillaries** to access the test and creative thinking problems for this chapter. The chapter test can be used for review or to evaluate content knowledge depending on your course objectives. Creative thinking problems are provided to help expand your knowledge by researching given subjects.

Print reading problems have questions that ask you to find information on the Brookings South Main Fire Station plans. The print reading problems can help you become familiar with the format of construction documents and reinforce your ability to seek specific information. This is an important skill for you to master as you prepare to enter the construction industry. Go to **Downloads & Resources tab at www.mhprofessional.com/Commercial Building Construction-Ancillaries** to access the complete set of plans for the Brookings South Main Fire Station.

CHAPTER 10

Insulation and Barriers, and Indoor Air Quality and Safety

INTRODUCTION

This chapter introduces you to energy-efficient construction practices that can help decrease dependency on the heating and cooling system and result in energy savings. These materials and practices are commonly used in building today and involve a variety of applications working together to build a quality insulated and airtight building. The combination of practices includes framing techniques, insulation, radiant barriers, vapor barriers, and caulking. These practices apply to commercial and residential construction. Much of the content focuses on applications, technologies, and terminology related to light commercial construction that includes wood- and steel-frame office buildings, multifamily apartments, townhomes, and condominiums.

This chapter also introduces you to indoor air quality and occupant safety factors and applications that need to be considered as part of standard and energy-efficient construction practices. Government energy agencies, architects, designers, and contractors around the country and in Canada have been evaluating construction methods that are designed to reduce energy consumption. Some of the tests have produced super-insulated, vapor-barrier-lined, airtight buildings. The result has been a dramatic reduction in heating and cooling costs. However, the air quality in these buildings can be significantly reduced and can be harmful to health. The problem is that the building is constructed tightly, and stale air and pollutants have no place to escape. The solution is mechanical ventilation, which is an important part of designing and building an energy-efficient building.

INSULATION

Insulation is material used to restrict the flow of heat, cold, or sound, saves energy costs and helps make the building comfortable. A properly insulated building helps maintain a uniform temperature throughout the building. Properly installed insulation can also reduce noise. Various types of insulation are available, but any type must be installed properly for the best efficiency. Figure 10.1 shows locations where insulation is typically installed in a building.

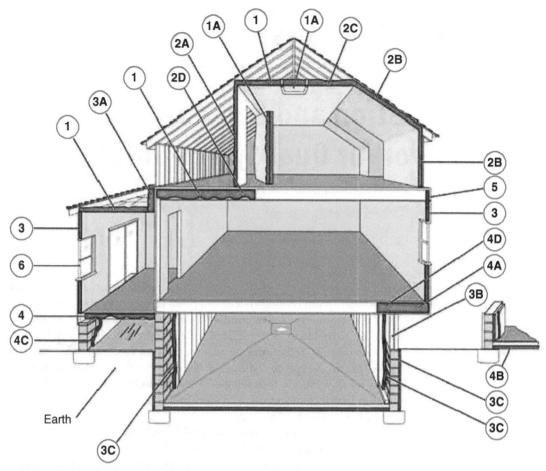

1. In unfinished attic spaces, insulate between and over the floor joists to seal off living spaces below. **(1A)** attic access door.
2. In finished attic rooms with or without dormer, insulate **(2A)** between the studs of "knee" walls, **(2B)** between the studs and rafters of exterior walls and roof, **(2C)** and ceilings with cold spaces above. **(2D)** Extend insulation into joist space to reduce air flows.
3. All exterior walls, including **(3A)** walls between living spaces and unheated garages, shed roofs, or storage areas; **(3B)** foundation walls above ground level; **(3C)** foundation walls in heated basements, full wall either interior or exterior.
4. Floors above cold spaces, such as vented crawl spaces and unheated garages. Also insulate **(4A)** any portion of the floor in a room that is cantilevered beyond the exterior wall below; **(4B)** slab floors built directly on the ground; **(4C)** as an alternative to floor insulation, foundation walls of unvented crawl spaces. **(4D)** Extend insulation into joist space to reduce air flows.
5. Band joists.
6. Replacement or storm windows and caulk and seal around all windows and doors.

Figure 10.1 Examples of locations where insulation is typically installed in a home. (*Courtesy U.S. Department of Energy | Oak Ridge National Laboratory.*)

Insulation reduces the amount of heat lost through walls, ceilings, and floors during the winter and helps keep heat from entering the building during the summer. Insulation is rated by its **R-value,** which determines its **thermal resistance.** The R-value of a material is a measure of thermal resistance to heat flow. Thermal resistance is the ability of material to slow heat transfer. The higher the R-value assigned to a material, the greater the insulating ability. Insulation is critical in helping reduce heat loss, but it must be combined with proper construction methods, caulking, and vapor barriers.

How much insulation a building should have depends on local and national codes, climate, energy costs, budget, and personal preference. The International Energy Conservation Code (IECC) provides minimum required insulation levels for building design and construction, and the U.S. Department of Energy (DOE) provides recommended insulation levels, by zip codes, for walls, ceilings, floors, and foundations. Recommended R-values in the ceiling range from R-30 to R-60 depending on the type of heating to be provided and the climate zone where the building is located. Vaulted ceilings can vary from R-22 to R-60. R-13 to R-30 insulation is recommended for use in floors although buildings built in some warm climates do not require floor insulation. Wall insulation levels vary between R-13 and R-21 depending on the climate zone where the building is located. In general, energy efficiency can be increased by using more properly installed insulation. The amount and type of insulation is found on the plans and specifications for the construction. Energy-efficient construction methods include advanced framing practices and double-wall construction described in Chapter 8, *Wood Construction*. These practices allow for added insulation to help reduce air infiltration and heat loss.

Types of Insulation

Common insulation materials include fiberglass, rock wool, wool, cellulose, urethane foam, and recycled cotton. Insulation is available in the form of loose fill, blanket, rigid foam, and expanding spray foam. Regardless of the type of insulation used, insulation must be professionally installed to completely fill all voids in quality framing and used with properly installed vapor barriers and caulking.

Loose-Fill Insulation
Loose-fill insulation, shown in Figure 10.2, is fibers or granules made from cellulose, fiberglass, rock wool, cotton, or other materials. Loose-fill insulation is normally blown into areas and cavities with special equipment. A benefit of this insulation is that it conforms to and completely fills the space where it is applied. This helps eliminate air spaces and increases efficiency. Some applications mix loose-fill fibers with a spray adhesive to cover areas before drywall is installed. This is a special advantage for irregular and hard-to-reach areas. Some applications use a vapor barrier, a net, or a temporary frame over the studs to hold the insulation while it is blown into the wall **cavity.** The word cavity refers to the blank area or void between construction members such as studs in walls or joists in ceilings where insulation is placed. Loose-fill insulation must be installed correctly, without excess air in the cavity. Some manufactures guarantee that the insulation has a specific R-value when installed to a given thickness. Loose-fill insulation settles over a period of time, leaving an uninsulated area at the top of wall cavities. Settling is reduced or avoided by installing higher-than-normal density. This technique, called **dense-pack**, also reduces air leakage.

Figure 10.2   Loose-fill insulation is normally blown into areas and cavities with special equipment and long hoses to reach all attic areas. (©*photovs—Can Stock Photo Inc.*)

Blanket Insulation

Blanket insulation is the most common and widely used type of insulation that comes in the form of batts or rolls. **Batts** are insulation strips that are 103 in. (2350 mm) long to fit in the wall space between studs of a typical 8-ft (2400-mm) wall. **Roll insulation** is the same as batts except it comes in approximately 40 ft (12,000 mm) rolls rather than strips. Blanket insulation is commonly manufactured with flexible fiberglass fibers but is also made from mineral wool, plastic fibers, cotton, and sheep wool. Blanket insulation is available 15 or 23 in. (380 or 580 mm) wide to fit between standard 16 or 24 in. (400 or 600 mm) on center stud spacing. Figure 10.3*a* shows batt insulation used in between wall framing. Figure 10.3*b* shows batt insulation used between ceiling joist framing. When batt insulation is used between floor beams or floor joists, a mesh or string system is placed under the beams or joists and a vapor barrier sheet is placed over the beams or floor joists. This is used to hold the batt insulation in place during construction and to protect the insulation from moisture during and after construction is complete. The thickness of the insulation determines its R-value. Batts are available with or without vapor-retarded facings. Careful installation of batts and blanket insulation is important to avoid gaps that reduce their effectiveness. Be sure the insulation is tight to the framing members and is not compacted.

The inside of concrete walls can be insulated with batts or blanket by building a stud-frame, called **furring,** next to the concrete wall. Furring is pressure-treated 2 × 4 studs fastened to the concrete wall. The insulation is placed between the framing of the furring members.

(a)

(b)

Figure 10.3 Batt insulation is strips, blankets, and roll insulation made of fiberglass or cotton fibers. (*a*) Batt insulation used in between studs in wall framing. (*©Kadmy—Adobe Stock*.) (*b*) Batt insulation used between ceiling joist framing. (*©Jeanette Dietl—Adobe Stock*.)

Figure 10.4   A worker installing rigid insulation on a wall. (©*Simin Zoran—Adobe Stock.*)

Rigid Insulation
Rigid insulation is made from fibrous material or plastic foams and pressed or extruded into sheets of varying thickness. Figure 10.4 shows a worker installing rigid insulation on a wall. Rigid insulation provides high R-values per inch of thickness, sound insulation, some structural strength, and air sealing. Rigid foam insulation is used in foundations, under siding to increase R-value, and in tight places such as vaulted ceilings where other materials are too thick to be effective. Foam insulation must be covered with finish material for fire safety. A common use is under concrete slabs and on the outside of concrete walls. Any insulation exposed aboveground must be covered with protective material such as plastic, treated wood, plaster, or concrete.

Rigid Fiber Board Insulation
Rigid fiber board insulation contains fiberglass or mineral wool material primarily used for **insulating ducts,** and for insulation that can withstand high-temperature applications. Ducts are the pipes used to move hot or cold air in a heating, air-conditioning, and ventilation system. Rigid fiber board insulation products come in a range of thicknesses from 1 to 2.5 in. (25 to 65 mm), and provide about R-4 insulation value per inch (25 mm) of thickness. Duct insulation is done by mechanical contractors, as described in Chapter 13, *Mechanical, Plumbing, and Electrical Systems.*

Structural Insulated Panels

Structural insulated panels (SIP), described in Chapter 8, are a high energy-efficient building system for residential and light commercial construction that uses rigid foam insulation core between a structural plywood or OSB sheet on each side.

Insulating Concrete Forms

Insulating concrete forms (ICF), described in Chapter 4, *Concrete Construction and Foundation Systems*, and are rigid foam forms that hold concrete in place during pouring and curing. These ICF walls use rigid foam blocks that are stacked and filled with concrete and reinforcing steel. These foundations provide structural qualities and excellent insulation.

Foam-in-Place Insulation

Foam-in-place, also referred to as **spray-in foam** insulation, delivers high R-values per inch of insulation and provides excellent sealing against air infiltration. These materials require special equipment to spray or extrude into place. They are sprayed into open wall and ceiling cavities and hard-to-access areas, where they expand to fill the space. Foam-in-place insulation is also used to fill small voids in framing, around door and window frames, and at any penetration through sills and plates to stop air infiltration at those locations. Any excess foam is trimmed away when it hardens. Foam-in-place insulation is used in energy-efficient construction to get high R-values and airtight seals. Foam-in-place insulation also provides low toxicity for a healthier environment.

VAPOR BARRIER

For an energy-efficient system, airtight construction is critical. The ability to eliminate air infiltration through small cracks is necessary to minimize heat loss. A **vapor barrier** or **vapor diffusion retarder** is a material that reduces the rate at which water vapor can move through a material. Vapor barriers help prevent moisture from transferring from the ground through the slab or from the inside and outside through the exterior walls, can reduce possible moisture-related mold and mildew growth, and block gases that can be a health risk in the building. Vapor barriers are a very effective method of decreasing heat loss. Most building codes require 6-**mil**-thick plastic to be placed over the ground in the crawl space. Mil is a unit of measure equal to 1/1000 (.001) inch used especially in measuring material thickness. Many energy-efficient construction methods add a continuous vapor barrier to the walls. This added vapor barrier is designed to keep moisture from the walls and insulation. Figure 10.5 shows the effect of water vapor on insulated walls with and without a vapor barrier.

Vapor barriers are available as **membranes** or coatings. Membranes are generally thin, flexible materials that come in rolls or as parts of building materials. These vapor barriers include polyethylene sheeting and aluminum- or paper-faced blanket insulation. Drywall is also available with a foil-backing that acts as the vapor barrier. Painting products can also have vapor barrier properties that can be confirmed by checking the manufacturer specifications. Thicker construction materials such as rigid foam insulation are used for insulation and as vapor barriers. These types of vapor diffusion retarders are usually mechanically fastened to the structure and sealed at the joints with special adhesive or special tape available from the manufacturer.

House wrap is the term that describes a variety of synthetic products that have replaced tar paper for use as a vapor barrier. House wrap is usually placed over the studs or over the sheathing on the exterior side of a building. House wrap is designed to be permeable, which means that it allows water vapor to pass through but stops rain water or other hard moisture from passing through the material. House wrap also reduces air movement through exterior walls, making it especially important to use in windy locations.

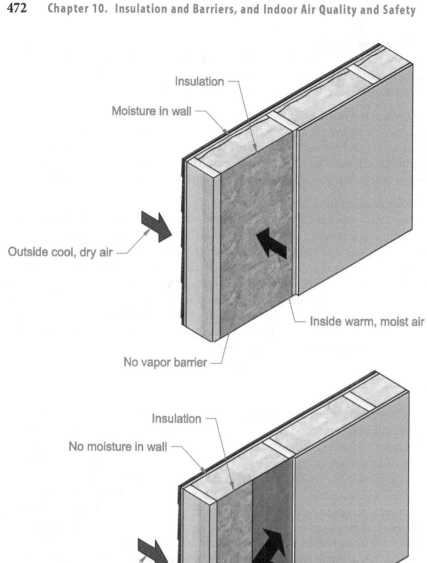

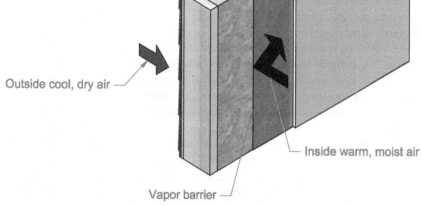

Figure 10.5 The effect of water vapor on insulated walls with and without a vapor barrier. (*Previously published in Modern Residential Construction Practices, Routledge, 2017 and reproduced here with permission.*)

> **NOTE**
>
> House wraps and vapor barriers function to control moisture and air infiltration. Some house wraps work as vapor barriers and some vapor barriers act as house wraps. The use of house wraps and vapor barriers is important, but they must be used correctly and in accordance with local building code requirements and manufacturer installation instructions. A few of the many brand names for house wrap are Tyvek®, Typar®, WEATHERMATE™, Grip-Rite, Building-Guard, or other products that can be found by searching the Internet for house wrap products.

Insulated Structural Sheathing

Insulated structural sheathing combines engineered wood structural sheathing, added insulation, air-infiltration barrier, and moisture-resistant barrier in exterior wall and roof sheathing panels. Standard 4×8 to 4×10 ft $(1200 \times 2400 - 1200 \times 3000$ mm) sheets are available for installation over wood- or steel-frame construction. Thicknesses and R-values range from 1 in. (25 mm) with R-3.6 to 2-1/2 in. (64 mm) with R-12.6 insulation value.

Vapor Barrier Installation

Vapor barriers work best when installed close to the exterior in hot and wet climates, such as between the sheathing and siding. In cold climates, the vapor barrier should be installed close to the inside, such as between the studs and the drywall. Painting products and plaster wall coatings containing vapor barrier properties that can be used in mild climates to control water vapor movement through the material. Building construction in extreme climates requires more extensive vapor barrier applications. Under normal conditions, wall and ceiling insulation with a foil face on the interior side allows small amounts of air to leak in at each seam and at construction members. A continuous vapor barrier can be installed to reduce this leakage. Vapor barrier installation should be continuous and as close to perfect as possible. It is very important to completely seal tears, openings, or punctures that happen during construction. To be effective, the vapor barrier must be lapped and sealed to keep air from penetrating through the seams in the material. The vapor barrier can be installed in the ceiling, walls, and floor system for effective air control. Figure 10.6 shows three different vapor barrier ceiling applications. Each is designed to help prevent small amounts of heated or cooled air from escaping to the attic. All the effort required to keep the vapor barrier intact at the seams must be continued wherever an opening in the wall or ceiling is required. The cost of materials for a vapor barrier is low compared with the overall cost of the project. The expense of the vapor barrier comes in the labor to install and to maintain the seal during construction. Great care is required by the entire construction crew to maintain the barrier during construction.

> **NOTE**
>
> Caution should be used when building a completely airtight building with increased insulation and vapor barriers. This type of construction can cause problems with the air quality inside the building. These potential problems can be countered with the installation of an air-to-air exchanger. A complete discussion of air contaminants and air-to-air exchangers is found in Chapter 13.

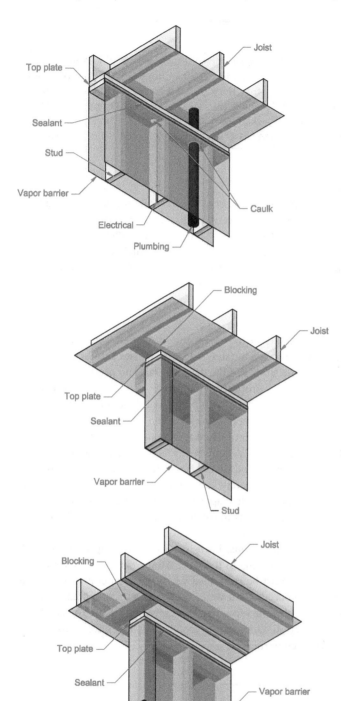

Figure 10.6 Ceiling vapor barrier applications. (*Previously published in Modern Residential Construction Practices, Routledge, 2017 and reproduced here with permission.*)

RADIANT BARRIERS

Radiant barriers stop heat from radiating through the attic and help reduce attic temperatures in hot climates. When the hot sun heats the roof, the heat is transferred from the roof structure into the attic space and continues to move from the warm attic to the cool living space below. Radiant barriers work by reflecting this radiant heat away from living spaces below. Insulation placed in the ceiling between the attic and the living space slows heat transfer but does not reduce heat enough in warm climates without the addition of a radiant barrier. Radiant barriers should be used in hot climates where they are more effective than in cool climates. Increasing ceiling insulation without using a radiant barrier is more effective for increasing building energy efficiency in cool climates.

Radiant Barrier Installation

Radiant barriers have a highly reflective material applied to one or both sides of insulation or other building products, such as drywall, plywood, and oriented strand board. A radiant barrier is most effective if it can radiate heat away from the structure before the heat enters the attic. This is done by placing a rolled-foil radiant barrier foil-face down over the roof rafters just before the roof sheathing is installed, or by stapling the material to the bottom of the rafters after the roof sheathing is installed. When installing the radiant barrier over the rafters during construction, the material should hang down at least 1 in. (25 mm) between the rafters to make an air space between the product and the roof sheathing.

A common method for insulating and providing a radiant barrier at the same time is by using **reflective insulation.** Reflective insulation includes radiant barriers that are typically highly reflective aluminum foil backing on one or both sides of the insulation. A radiant barrier should not be installed on top of attic insulation, because efficiency can be reduced by dust accumulation, and condensation moisture can be trapped in the insulation.

Radiation-control roof coatings are also available to reduce roof temperature in hot climates.

CAULKING

Insulation in walls, ceilings, and floors reduces heat lost during the winter and helps keep heat from entering the building during the summer. Insulation alone does not stop air infiltration due to leaks through the structural system. Air leakage happens when outside air enters and conditioned air leaves the building uncontrollably through cracks, gaps, and openings. Air leakage can also cause **moisture** problems that can deteriorate the structure and affect occupant health. The most noticeable benefit of sealing cracks is improving comfort in the building by eliminating drafts and creating a uniformly heated or cooled environment. Professional **caulking** products and applications are used to stop air infiltration. **Caulking** is a waterproof product used to reduce air infiltration by filling gaps, openings, seams, cracks, and voids to make the structure air- and watertight. Caulking is typically purchased in tubes that are used in a manual caulking gun to extract the product in a bead at the desired location, as shown in Figure 10.7. Cordless electric caulking guns are also available with variable speed to adjust the discharge rate to control the bead of caulk for high-volume professional projects. **Expanding foam insulation** is a spray-applied insulating foam in a pressurized container that is installed from a nozzle as a liquid and then expands many times its original size until it hardens in its final form upon contact with the air.

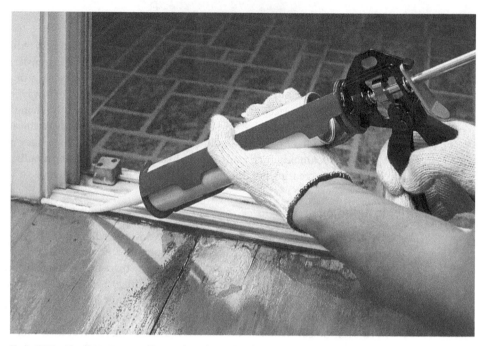

Figure 10.7   Caulking is typically purchased in tubes that are used in a manual caulking gun to extract the product in a bead at the desired location. (©*okinawakasawa—Adobe Stock*.)

Expanding foam insulation is used as a highly effective insulation when applied between framing members during construction, as shown in Figure 10.8*a*. Expanding foam insulation is often used as a fast and effective way to seal and add insulation value in framing gaps, around door and window frames, and holes or other penetrations through framing members, especially sills and top plates. Figure 10.8*b* shows a typical expanding foam insulation application for filling minor cracks and holes in construction. Professional spray applications are also available for filling entire wall or ceiling cavities for complete sealing and effective thermal and sound insulation.

Figure 10.9 shows the areas of the building most likely to have air leakage where caulking should be applied.

SOURCES OF POLLUTANTS

Air pollution in a building is a main reason for controlling indoor air quality. The following briefly describes sources that can contribute to an unhealthy environment within a building.

Moisture in the form of **relative humidity** can cause structural damage, respiratory problems, and other health concerns. Relative humidity is a percentage ratio of the amount of water vapor in the air at a specific temperature to the maximum amount that the air can hold at that temperature. Sources of relative humidity include the atmosphere, especially in warm humid climates. Steam from cooking, showers, and steam from other sources can increase humidity. Each occupant can also produce up to 1 gal (3.8 L) of water vapor per day.

(a)

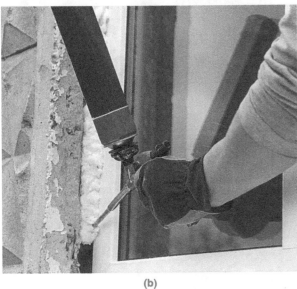

(b)

Figure 10.8 (*a*) Expanding foam insulation applied between framing members during construction. (©*Adobe Stock.*) (*b*) A typical expanding foam insulation application for filling minor cracks and holes in construction. (©*sociopat_empat—Adobe Stock.*)

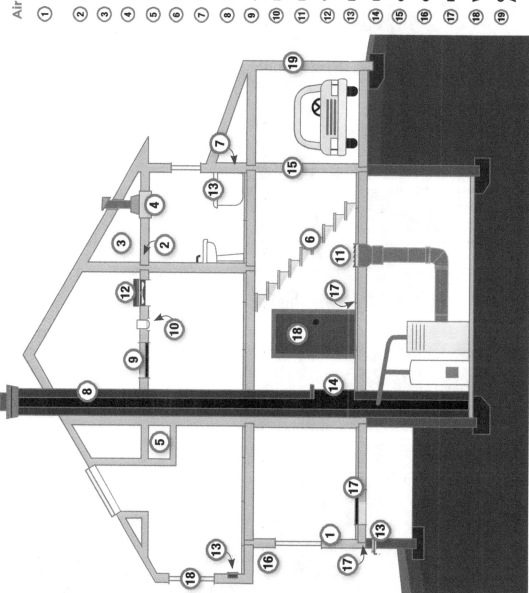

Air Sealing Trouble Spots

1. Air Barrier and Thermal Barrier Alignment
2. Attic Air Sealing
3. Attic Kneewalls
4. Shaft for Piping or Ducts
5. Dropped Ceiling/Soffit
6. Staircase Framing at Exterior Wall
7. Porch Roof
8. Flue or Chimney Shaft
9. Attic Access
10. Recessed Lighting
11. Ducts
12. Whole-House Fan
13. Exterior Wall Penetrations
14. Fireplace Wall
15. Garage/Living Space Walls
16. Cantilevered Floor
17. Rim Joists, Sill Plate, Foundation, Floor
18. Windows & Doors
19. Common Walls Between Attached Dwelling Units

Figure 10.9 The areas of the building most likely to have air leakage where caulking should be applied. (*Courtesy of the Department of Energy.*)

Incomplete combustion from gas-fired appliances and fireplaces can generate a variety of pollutants, including **carbon monoxide (CO), aldehydes,** and **soot.** Carbon monoxide, has the chemical formula CO, and is a colorless, odorless, poisonous gas, produced by incomplete burning of carbon-based fuels. Aldehydes are reactive organic compounds that contribute to ozone production. Soot is finely divided carbon deposited from flames during the incomplete combustion of organic substances such as coal.

Pollutants from humans and pets can transmit bacterial and viral diseases through the air.

Tobacco smoke adds chemical compounds to the air that can adversely affect all occupants. Household products such as those available in aerosol spray cans and craft materials such as glues and paints can contribute a number of toxic pollutants.

Products and materials containing **formaldehyde** can be a factor in the cause of eye irritation, certain diseases, and respiratory problems. Formaldehyde is a chemical found in disinfectant, preservative, carpets, furniture, and the glue used in construction materials, such as plywood and particle board, as well as some insulation products.

Volatile organic compounds (VOC) are chemicals contained in the items used in home construction, which can emit pollutants throughout the lifespan of the product. It is important to specify and use products containing low or no VOC. Products that should have low or no VOC include paint, caulks, sealants, cabinet materials, plastics, and carpets. Reducing the amount of VOCs improves indoor air quality, and reduces exposure to toxic compounds. A healthier alternative can be achieved with the same color, texture, sheen and quality of finish using VOC-free paint. Look for the **GreenSeal**™ and ask your paint supplier about the VOC content when purchasing paint. Green Seal is a nonprofit organization that uses science-based programs to empower consumers, purchasers, and companies to create a more sustainable world.

Polyvinyl chloride (PVC) is a widely used building material found in products such as window frames, flooring, pipes, bathtubs, and shower units. Despite appearing to be an ideal building material, studies have shown that PVC can have environmental and health risks. Architects and contractors should consider alternatives such as wood framed or aluminum windows, natural flooring materials, non-PVC plumbing pipes, and tiled showers.

When buying or building cabinets, specify materials that contain a urea-formaldehyde-free binder. Urea-formaldehyde, found in many engineered wood products, has been classified as a toxic air contaminant by the California Air Resources Board, due to its potential to cause cancer. Water-based wood finishes, such as waterborne urethane or acrylic, have decreased toxic compounds, while still providing comparable durability.

Radon is a naturally occurring radioactive gas that breaks down into compounds that can cause cancer when large quantities are inhaled over a long period of time. Radon is invisible to sight, smell, and taste. Radon can be more apparent in a building containing a large amount of concrete, or radon enters from the ground, with specific geological areas carrying different levels of contamination. Radon can be monitored scientifically at a nominal cost, and barriers can be built that help reduce concern about radon contamination. The basic methods for radon buildup prevention include a gas permeable layer placed under the concrete slab or flooring system to allow the soil gas to move freely beneath the structure. In many cases, the material used is a 4-in. (100 mm) layer of clean gravel. Plastic sheeting on top of the gas permeable layer prevents the soil gas from entering the building. Radon gas can also be directed away from the building by using a 3- or 4-in. (75-100 mm) gastight vent pipe that runs from the gas permeable layer through the building and out through the roof or exterior wall. This safely vents radon and other soil gases above the building.

Additional radon ventilation can be provided by using an electric exhaust fan connected to the vent system. Options for radon detection are professional digital radon detectors that can require annual calibration for accuracy, a radon test kit available at hardware stores for a one-time test, and radon detectors designed to monitor and measure radon levels.

> **NOTE**
>
> The following content describes practices and technologies that can be used to help maintain a clean and healthy indoor environment. Part of the discussion refers to heating, ventilation, and air-conditioning (HVAC). HVAC is the terminology used to refer to the industry that deals with the heating and air-conditioning equipment and systems found in a building. This HVAC content is related to systems that exchange inside air with fresh outside air. Additional HVAC information is provided in Chapter 13.

HEAT RECOVERY AND VENTILATION

Air-to-air heat exchanger technology has emerged from a need to properly ventilate highly energy-efficient airtight buildings. In the past, the air in a building was exchanged by leakage through walls, floors, ceilings, and around openings. This random leakage created a certain amount of heat loss and provided no assurance that the building was properly ventilated. With the concern for energy conservation, and air quality, it is clear that internal air quality cannot be left to chance. An air-to-air heat exchanger is a heat recovery and ventilation device that pulls polluted, stale warm air from the living space and transfers the heat in that air to fresh, cold air being pulled into the building. Heat exchangers exchange heat from one airstream to another, but they do not produce heat. The heat transfer takes place in the core of the heat exchanger, which is designed to avoid mixing the two airstreams and makes sure that indoor pollutants are removed. Moisture in the stale air condenses in the core and is drained from the unit. Figure 10.10 shows the function and basic components of an air-to-air heat exchanger.

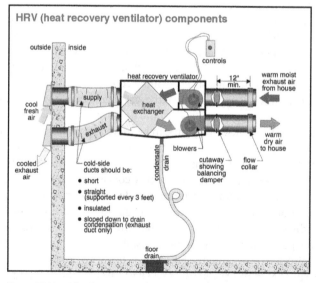

Figure 10.10   The function and basic components of an air-to-air heat exchanger. (©*Carson Dunlop 2013. www.carsondunlop.com. All Rights Reserved.*)

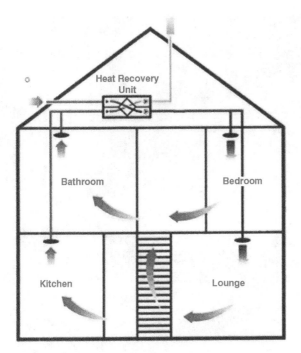

Figure 10.11   A complete heat recovery and ventilation system in a building. (*Courtesy of Systemair Ltd.*)

A **heat recovery and ventilation (HRV)** system is the complete system that is operated by a **heat exchanger, air exchanger,** or air-to-air heat exchanger. A heat recovery ventilation system is a ventilation system that uses a counter-flow heat exchanger between the inbound and outbound air flow. The HRV provides fresh air and improved climate control, while also saving energy by reducing the heating or cooling needs of the building. An **energy recovery ventilator (ERV)** system is closely related to a HRV, except the ERV also transfers the humidity level of the exhaust air to the intake air. HRV and ERV systems are part of the HVAC system that require mechanical engineering for proper design and function. Figure 10.11 shows a complete HRV system in a building.

The recommended minimum effective air change rate is 0.5 air change per hour (ACH). Codes in some areas of the country have established a rate of 0.7 ACH. The American Society of Heating, Refrigeration, and Air-Conditioning Engineers, Inc. (ASHRAE) recommends ventilation levels based on the amount of air entering a room. The recommended amount of air entering most rooms is 10 cubic feet per minute (cfm). The rate for a kitchens is 100 cfm, and bathrooms is 50 cfm. Mechanical exhaust devices vented to outside air should be added to kitchens and baths to maintain the recommended air exchange rate.

EXHAUST SYSTEMS

Exhaust systems are required to remove odors, steam, moisture, and pollutants from inside a building, and generally located in specific rooms such as bathrooms, laundry rooms, and kitchens where excess moisture and pollutants can be found. The following are basic general requirements that are considered during the design process and detailed in plans and specifications.

Figure 10.12   This range hood has outer wood style to match the kitchen cabinets and a metal insert with lights and exhaust fan. (©*Ambient Ideas, LLC—Adobe Stock.*)

Range hoods and **range down-draft systems** must be vented to the outside by a minimum 0.016-in.-thick (0.4-mm) galvanized, stainless steel, or copper duct. A range hood is a metal hood over a range that has lights and an exhaust fan that vents to the outside. The range hood can be utilitarian or decorative to enhance the kitchen. The range hood can have outer wood style to match the kitchen cabinets and a metal insert, as shown in Figure 10.12. A down-draft range exhaust system is generally directly behind the range and has an exhaust fan that ventilates the range downward to the outside in duct work generally through the crawl space. The range hood or range down-draft systems duct has a smooth inner surface, is airtight, and has a **back draft damper.** A back draft damper is a movable plate that regulates the draft or air flow in a chimney or vent pipe with blades that are activated by gravity, permitting air to pass through them in one direction only. The range hood cannot be any closer than 24 in. (610 mm) to the range.

> **NOTE**
>
> If length dimensions are required, they are given at the end of the callout in feet and inches except for plates, which should be given only in inches.

The laundry room is another place where specific exhaust venting is required. Clothes dryer vents must be independent of all other systems and carry the moisture outside. The vent cannot be connected with screws that extend into the vent. Clothes dryer exhaust vents require a full opening back draft exhaust damper at the outside. The vent should also have a termination cap to keep vermin out of the building. There should not be a screen over the opening that can trap lint and cause a fire. The vent duct should be rigid metal 0.016 in. (0.4 mm) thick with a smooth inside and joints running in the direction of the air-flow. Approved flexible duct can be used but cannot be concealed within the construction. A typical close dryer vent duct is 4 in. (100 mm) in diameter, and cannot be longer than 25 ft (7620 mm) from the dryer to the wall or roof vent. The total length should be reduced by 30 in. (762 mm) for each 45° bend in the duct, and reduced 60 in. (1524 mm) for each 90° bend in the duct. The vent outlet should be at least 3 ft from any other opening. A gas clothes dryer must be approved by the manufacturer for placement in a closet, and no other fuel-burning appliance can be in the same closet.

Bathrooms and spas also have specific ventilation requirements due to the amount of moisture that can be created in these areas. Exhaust fans that run continuously must be rated at a maximum of 1.0 **sone.** A sone is a sound rating, where the lower the number the quieter the sound. The exhaust system that does not run continuously must be controlled by a **humidistat**, timer, or other automatic control. A humidistat device automatically regulates the humidity of the air in a room or building. Exhaust fans that run intermittently, by switch or timer, must be rated at a maximum of 3.0 sones. Fans that are located 4 ft (1200 mm) or more from the air inlet grill are exempt from sone requirements. Continuous fans must circulate air at 20 **cubic feet per minute** (cu ft/min), abbreviated CFM (34 cubic meters per hour). CFM is a measurement of the velocity at which air flows into or out of a space. Intermittent fans must circulate air at 80 CFM (136 cubic meters per hour). **Half bathrooms** must have fans that circulate 50 cfm (85 cubic meters per hour). A half bathroom is a bathroom without a shower or bathtub.

CLEANING THE AIR

There are a variety of construction alternatives and environmental applications that can be used to improve indoor air quality. This discussion includes features that can be used to achieve better air quality, and daily applications that occupants can use to continue to improve the environment.

Ventilation and Air Filtration

Air movement in and out of the building needs to be controlled for maximum efficiency. In order to achieve desired air movement control, it is important to seal the **envelope** tightly and control ventilation properly. An envelope, related to building design and construction, is the entire exterior of the building that includes exterior walls, the roof, the foundation, and doors and windows. An HVAC professional contractor can properly calculate the amount of air required for ventilation in the building design. This professional takes into account all the variables associated with properly sizing ventilation equipment, which includes the number of bathroom and kitchen fans, regional air pressure, size of indoor spaces, and proper heating and cooling equipment selection. To maintain adequate fresh air, the ventilation system can include a heat recovery ventilation system to regularly change the air within the building. This makes sure that you always breathe fresh air while maintaining a high level of energy efficiency by exchanging energy from outgoing air to incoming air.

The air used to ventilate a building comes from the outside, and is subjected to external pollutants. The air should run through a filtering system in order to protect the occupants from airborne pollutants. There are various types of air filters including **Minimum Efficiency Reporting Value (MERV)** and **high-efficiency particulate air (HEPA)** filters. MERV refers to the filtration efficiency of an air filter. MERV performance is determined by comparing airborne particle counts upstream and downstream of the air filter. HEPA filters are designed to be 99.97% effective in capturing particles as small as 0.3 **micrometer.** A micrometer (μm) is a unit of measure equal to one millionth of a meter. When designing the HVAC system, it is important to confirm that all equipment is compatible.

Electrostatic Air Filters

Electrostatic air filters clean the air flowing through the HVAC system by using static electricity in a unit placed in the return air duct directly next to the furnace or air handler of the system. The electrostatic air filter works by an electrostatic charge generated by air flowing through a network of static fibers. Airborne particles are attracted and held by the static charge until released by removing the filter and washing with water. After the filter is washed, it is placed back in the unit where it can continue collecting pollutants until the next cleaning.

Ultraviolet Light Systems

Specific **ultraviolet (UV)** light can be beneficial for improving indoor air quality. Ultraviolet energy is measured in **nanometers.** A nanometer (nm) is one billionth of a meter. UV wavelengths are shorter than visible light, and are invisible to the human eye. One of the wavelengths of UV light is the C bandwidth. This desirable **UV-C light** has a wavelength of 253.7 **nanometer (nm),** which is called the **germicidal bandwidth**. A nanometer is a **unit** of **measure** in the **metric system**, equal to one billionth of a **meter** (0.000000001 m). The germicidal bandwidth measures from 100 to 280 nm, which is a wavelength, the distance between the crest of two waves, measured in nanometers. The building can have germs, molds, bacteria, toxins, and gases that cause unhealthy air conditions. These contaminants can pass through the forced air heating and cooling system several times per hour, making the furnace a collection place for recirculating air pollutants. This makes the duct work, near the furnace, an ideal place to install an indoor air cleaning system to control harmful levels of indoor air pollutants. UV-C lights are mounted inside the duct work at the return air duct just before the air cleaner, or in the conditioned air supply duct immediately after the HVAC system. See Figure 10.13. The furnace fan circulates air across the UV-C light, neutralizing up to 98 percent the microorganisms. Exposure to UV-C light can destroy or damage microorganisms by disrupting their reproductive code (DNA) and inhibiting them from reproducing. UV radiation in wavelengths of 253.7 nm is capable of destroying nearly all known forms of microorganisms and viruses. While there are varying opinions about the effectiveness of UV light systems for pollutant control, you should do additional research and consult with HVAC professionals to determine their use in your construction projects.

Controlling Dust

While preventing the buildup of dust may not completely reduce asthmatic or allergic symptoms, controlling dust is one of the most proactive measures you can take in keeping the interior environment healthy. The following are a few steps that can be taken to control the amount of dust that accumulates in a building:

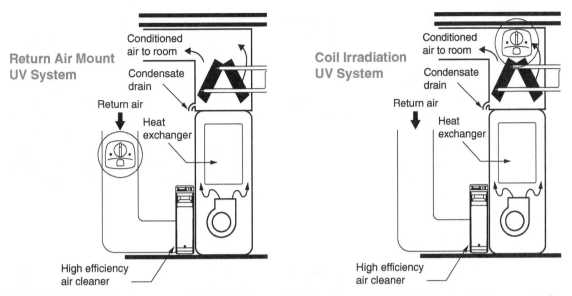

Figure 10.13 UV-C lights are mounted inside the duct work at the return air duct just before the air cleaner, or in the conditioned air supply duct immediately after the HVAC system. (*Courtesy of National Trade Supply.*)

Provide shoe storage at entrances. Removing your shoes at the door prevents tracking dust further into the building.

Install a special HEPA filter in the air conditioner or air purifier. This can help remove some allergens such as pollen or animal dander, and tobacco smoke from the air.

Limit carpet, upholstered furniture, and heavy drapes that collect dust. Upholstery and drapes should be made from a tight weave fabric that keeps out dust. Use hard-surfaced furniture that you can wipe clean.

Avoid wall-to-wall carpeting. Use smaller rugs made of cotton or wool that you can clean. Use roll-down shades or washable curtains.

Damp-mop tile and hard surfaced floors rather than sweeping. Use a washable cloth, instead of disposable wipe.

Install a central vacuum system that vents directly outdoors. If a central vacuum is not an option, use a vacuum cleaner with a HEPA filter or a special double-thickness bag that collects dust-mite particles and pollen. Consider wet-vacuum cleaning when possible. This can help remove allergens from carpeting as it washes the carpet. Consider periodic steam-cleaning of carpets when possible. In addition to cleaning the carpet, the heat of the steam kills dust mites.

Mold Prevention

Exposure to **mold** can cause or worsen conditions such as asthma, hay fever, or other allergies. The most common symptoms of mold overexposure are cough, congestion, runny nose, eye irritation, and aggravation of asthma. Depending on the amount of exposure and individual sensitivity, more serious health effects can occur, such as fever and breathing problems. Mold is a superficial growth produced especially on damp or decaying organic matter or on living organisms by a fungus.

Mold is most likely to grow where moisture is trapped in improperly ventilated spaces such as damp crawlspaces, misty bathrooms, and defective wall, eave, and roof construction systems. Because mold increases with humidity, it is important to be aware of areas that can trap humidity and have inadequate ventilation.

The key to preventing mold from forming in the building is to prevent weather and moisture from entering the building envelope and by controlling the humidity inside. In the design and construction for energy efficiency, the exterior building envelope must be air- and watertight, and air movement must be controlled. To make sure that all parts of the building remain dry, it is important to include systems to supply proper ventilation. Examples of systems used to help prevent water damage and mold growth include house wrap behind exterior finishes, properly sealed building envelopes, the use of moisture-resistant building materials, heat recovery ventilation, and ventilation fans in interior spaces.

SEALING OFF THE GARAGE

Many multifamily architectural designs include an attached garage. An attached garage keeps you dry as you come and allows for ease of moving items from the car into the house. An attached garage can allow for a more compact building footprint in places where space is limited. While an attached garage has advantages, there are items inside the garage that need to be isolated from the living space. Chemicals, paints, oils, and cleaning agents stored in the garage can attribute to air quality concerns if they leak. The automobile emits dangerous carbon monoxide (CO) gas when the engine is running. It is important to prevent dangerous substances, gases, and toxins from entering the living area. An easy solution is to seal gaps in walls that shares with the garage. Install a CO_2 monitor in any room adjacent to the garage and consider installing a fan attached to the garage door opener to vent the garage when you leave with the automobile.

Mechanical Systems and the Garage

Mechanical systems and ductwork placed in the garage can pull air from the garage into the home. When possible, make sure to avoid the placement of ductwork or mechanical equipment in the garage to prevent toxins from entering the HVAC system. Placing the mechanical system and related duct work in an insulated area other than the garage is required for new construction and also saves energy, because any loss from the system remains indoors.

Detached Garages

Multifamily homes such as apartments and condominiums often have detached garages. Designing and building a detached garage is the safest way to stop gas and chemicals found in a garage from entering the living space. Many factors need to be considered when designing a with a detached garage, such as architectural style, personal preference, and available space on the building site. A disadvantage of a detached garage is that you have to go outside when transporting items from the vehicle. This may not be as much of an issue in moderate and warm climates as it is in cold climates. An option for providing some protection from weather is to design and build an open covered structure between the garage and living area.

Go to Downloads & Resources tab at **www.mhprofessional.com/Commercial Building Construction-Ancillaries** to access the images provided in this chapter and correlate with the content of this chapter. Here you can look at the chapter figures on your computer screen to pan and zoom in and out for better observation, because full drawings reduced to fit on a textbook page are often difficult to read. The image bank also includes the complete set of plans for the Brookings South Main Fire Station used as examples throughout this textbook.

Go to **Downloads & Resources tab at www.mhprofessional.com/Commercial Building Construction-Ancillaries** to access the test and creative thinking problems for this chapter. The chapter test can be used for review or to evaluate content knowledge depending on your course objectives. Creative thinking problems are provided to help expand your knowledge by researching given subjects.

Print reading problems have questions that ask you to find information on the Brookings South Main Fire Station plans. The print reading problems can help you become familiar with the format of construction documents and reinforce your ability to seek specific information. This is an important skill for you to master as you prepare to enter the construction industry. Go to **Downloads & Resources tab at www.mhprofessional.com/Commercial Building Construction-Ancillaries** to access the complete set of plans for the Brookings South Main Fire Station.

CHAPTER 11

Stair Construction

INTRODUCTION

The **staircase** can be very simple and hidden in a location where it only provides access from one floor level to another or to multiple floors for emergency use, or the staircase can be a main focal point, providing beautiful design and craftsmanship in a ballroom, lobby, or living area, as shown in Figure 11.1. A staircase is the entire stair construction including framing the stairwell opening, adjacent walls, and the railing systems. This chapter describes **stair** design and construction with basic information about stair codes and different types of stairs. Stair terminology is provided so you can communicate in the construction industry. You will discover how to make calculations for building a **stairs** in a specific location and how to set up a stair layout. A stair, commonly referred to as stairs, is a set of steps leading from one floor of a building to another. Stairs are typically inside the building, but they can be outside. Stairs are also called a **flight**, staircase, or **stairway.** A flight is a set of steps between one floor or **landing** and the next. There can be a series of flights of steps connecting separate levels. A landing is a platform between flights of stairs, at the end of a flight of stairs, or used when stairs change direction. A stairs has a **handrail** or **guardrail** that provides a support that can protect you from falling when going up or down the stairs. A handrail is generally a single rail attached to a wall or **balusters** following the rise of the stairs and used for hand support and safety. A guardrail, also called a **banister,** is a handrail and its supporting posts made up of **newel posts** and balusters. A newel post is the large vertical support for the handrail at the ends of each flight of stairs, and often placed at regular intervals when additional support is needed. A baluster, also called a **spindle** or **picket,** is one of a series of vertical handrail supports placed equally spaced between newel posts. Figure 11.2 shows the basic stair parts described here. Additional stair terminology is described and illustrated throughout this chapter.

TYPES OF STAIRS

Stair planning can be one of the most complex characteristics of building design, because stairs require a significant amount of floor space. Fortunately, stairs can be designed a variety of ways to help take advantage of available space. Figure 11.3 shows the plan view of the following stair types.

Figure 11.1 The staircase can be a main focal point, providing beautiful design and craftsmanship. (©*pics721—Adobe Stock.*)

Figure 11.3*a* is a **straight flight** of stairs with a landing at the bottom and top. A straight flight of stairs is a flight between landings without any turns. Figure 11.3*b* shows a straight flight of stairs with a landing at the bottom, middle, and top. This is sometime done to provide a landing or stopping place when the distance between two floors is greater than normal. The middle landing is called a **platform.** A platform is any intermediate landing in a stairway and is also an extension of the floor landing, which is often used as the top tread of a spiral stairs. Figure 11.3*c* is an **L-shaped stairs** with a landing between flights. An L-shaped stairs makes a 90° turn at a landing or with **winders** between flights, as shown in Figure 11.3*d.* A winder is a special type of **tread** used for making a turn in a staircase at a mid-staircase landing. The tread is the horizontal portion of a stair where you step when going up or down the stairs.

Figure 11.3*e* is a **U-shaped stairs** with a landing between flights. A U-shaped stairs makes a 180° turn at a landing. The landing can extend all the way across between the flights, or an additional step can be placed in the landing, as shown in Figure 11.3*f.*

Figure 11.3*g* is a **spiral stairs,** also called a **circular stairs** that has treads winding around a center newel.

STAIR FLOOR PLAN REPRESENTATIONS

The first place you see the stairs is on the floor plans when reading a set of plans. Stairs are displayed on floor plans by the width of the tread, the direction and number of **risers,** and the lengths of handrails or guardrails, as shown in Figure 11.4. A stair **riser** is the vertical

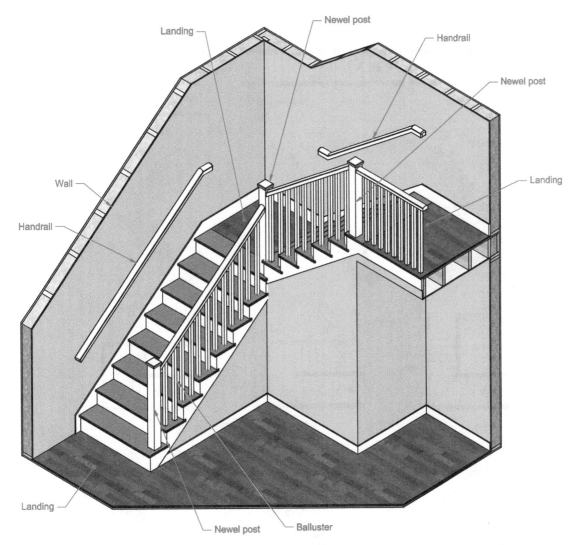

Figure 11.2   Basic stair parts. Additional stair terminology is described and illustrated throughout this chapter. (*Previously published in Modern Residential Construction Practices, Routledge, 2017 and reproduced here with permission.*)

part, forming the space between each step. The number of steps in a flight of stairs is the number of risers. An arrow is shown on the floor plan indicating if the stairs are going up or down. The abbreviations UP for up, DN for down, and R for the number of risers are placed next to the arrow. The note 14R means there are 14 risers in the flight of stairs. Partial stairs are commonly shown on the floor plans. The stairs in Figure 11.4 are displayed as partial flights with a **long break line** at approximately mid-height in the flight of stairs. A long break line is a graphic symbol used to break away a portion of a drawing that is not shown.

Figure 11.5 shows a common straight stair layout with a wall on one side and a guardrail on the other side. Some stair drawings omit the break line and display the full flight.

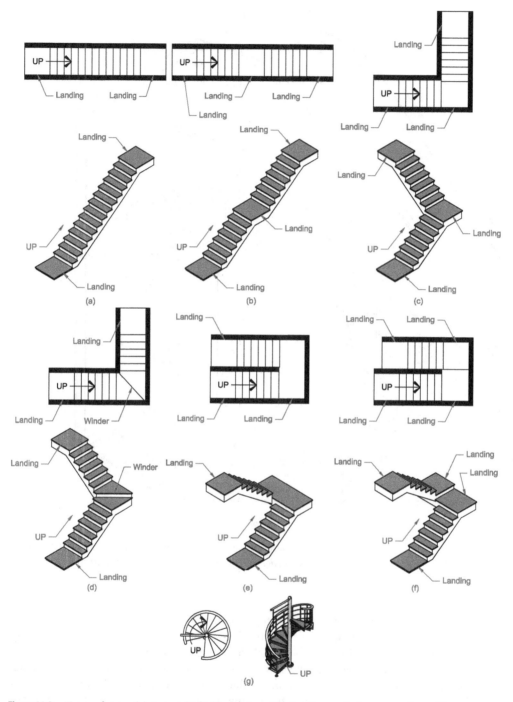

Figure 11.3 Types of stairs. (*a*) A straight flight of stairs with a landing at the bottom and top. (*b*) A straight flight of stairs with a landing at the bottom, middle, and top. (*c*) An L-shaped stairs with a landing between flights. (*d*) An L-shaped stairs with a winder in the landing between flights. (*e*) A U-shaped stairs with a landing between flights. (*f*) A U-shaped stairs with a step in the landing between flights. (*g*) A spiral stairs, also called a circular stairs, has treads winding around a center newel. (*Previously published in Modern Residential Construction Practices, Routledge, 2017 and reproduced here with permission.*)

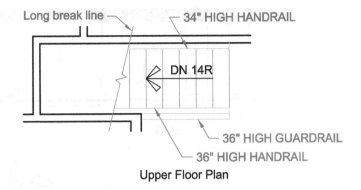

Upper Floor Plan

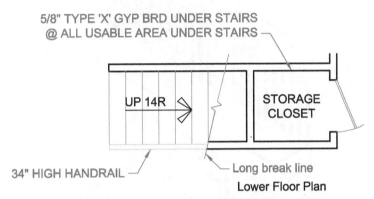

Lower Floor Plan

Figure 11.4 Stairs are shown on floor plans by the width of the tread, the direction and number of risers, and the lengths of handrails or guardrails. The long break line is a graphic symbol used to break away a portion of a drawing that is not shown. (*Previously published in Modern Residential Construction Practices, Routledge, 2017 and reproduced here with permission.*)

Figure 11.6 shows a flight of stairs with guardrails all around at the top level and a handrail running down the stairs. Figure 11.7 shows stairs between two walls. Figure 11.8 shows stacked stairs with one flight going up and the other down. This position is common when access from the main floor to both the second floor and basement or second and third floors is designed for the same area.

The plan view of **winder** and **spiral stairs** is shown in Figure 11.9. Winder and spiral stairs are often designed to save space and as an attractive alternative to straight and square-shaped stairs. There are a variety of styles of winder stairs that include **circular stairs** and **helical stairs,** which are any stairway with treads that do not have parallel edges. A **circular stairs** has circular construction with treads in a spiral pattern around a central support post. A helical stairs is also called a curved stairs that forms an attractive curve without the use of a center post. Helical stairs are often used as an architectural feature in an office building or retail space where they attract visitors or shoppers gracefully between floors. Winder stairs have certain requirements, such as the minimum tread width on the inside radius can be no less than 6 in. (150 mm), and the tread depth can be no less than 10 in. measured at the **walk line**, which is12 in. (300 mm) from the inside tread edge. Winder stairs are also required to have a handrail on the inside and outside radius.

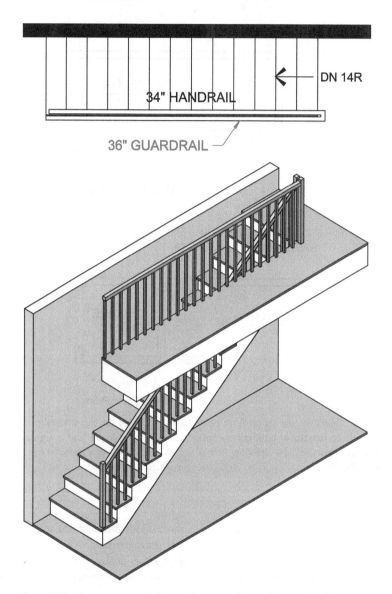

Figure 11.5   A common straight stair layout with a wall on one side and a guardrail on the other side. Some stair drawings omit the break line and display the full flight. (*Previously published in Modern Residential Construction Practices, Routledge, 2017 and reproduced here with permission.*)

A flight of stairs is considered to have at least three steps. A sunken or raised room, such as a living room or family room is often a popular design feature. When a room is either sunken or raised, there is at least one step into the room. The number of steps is noted with an arrow, as shown in Figure 11.10. A few steps up or down into a sunken or raised room do not require a handrail unless each floor level is more than 30 in. (760 mm) apart. Even though not required, a guardrail can be used for decorative and safety purposes.

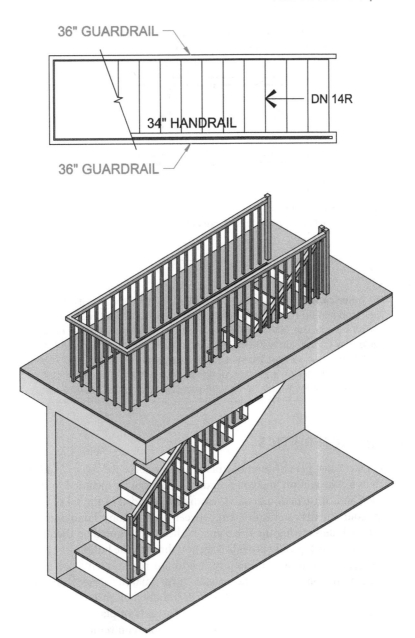

36" GUARDRAIL

34" HANDRAIL

DN 14R

36" GUARDRAIL

Figure 11.6 A flight of stairs with guardrails all around at the top level and a handrail running down the stairs. (*Previously published in Modern Residential Construction Practices, Routledge, 2017 and reproduced here with permission.*)

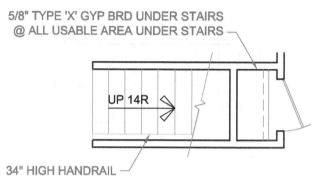

5/8" TYPE 'X' GYP BRD UNDER STAIRS
@ ALL USABLE AREA UNDER STAIRS

UP 14R

34" HIGH HANDRAIL

Figure 11.7   Stairs between two walls. (*Previously published in Modern Residential Construction Practices, Routledge, 2017 and reproduced here with permission.*)

NOTE

The **Americans with Disabilities Act (ADA)** specifies the use of a ramp for wheelchair access. The ramp should have a slope of 1:12 for 30 in. (762 mm) maximum rise and 30 ft (9 m) of horizontal run. The ADA of 1990 prohibits discrimination and ensures equal opportunity for persons with disabilities in employment, state and local government services, public accommodations, commercial facilities, and transportation. Although, ADA regulations are not required in the design and construction of private homes, the regulations do provide excellent guidelines for designing a home with access and facilities for homeowners with disabilities.

BASIC STAIR CHARACTERISTICS

Stairs are a key part of the floor plans as described previously. Additional stair drawings such as a stair section and construction details can be included in the set of plans. A stair section and construction details is normally provided in the set of commercial plans. Stair plan requirements can be different between jurisdictions. Fundamental minimum requirements include showing the rise, run, and headroom noted on the floor plans and in a section. Architects typically provide a section through the building at the stairs, to graphically show the rise, run, and minimum headroom requirements are met. Plan requirements can be more complicated when a stair turns, making it necessary to provide additional stair sections and construction details. Carefully look through the set of plans to find the related stair drawings and information. Most sets of plans have general notes pages that also describe stair construction including railing requirements. The stair section shown in Figure 11.11 is a typical example and provides additional stair terminology and definitions.

Stair layout depends on the amount of space available, comfortable use of the stairs, and code requirements. The stair layout and specifications are already created by the architect or designer when you see the plans, but you should be familiar with stair characteristics. The following outlines fundamental stair design and construction characteristics:

- There is always one less **run** than rise in a flight of stairs. For example, the stairs in Figure 11.11 with 14 risers have 13 runs. Run is the horizontal dimension one tread, measured from the face of one riser to the face of the next riser, or from the edge of one **nosing** to the edge of the next nosing. Rise, previously defined, is the vertical distance between one tread and another. Nosing is the edge part of the tread that projects from the face of the riser below, as shown in Figure 11.11.

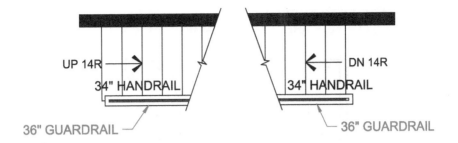

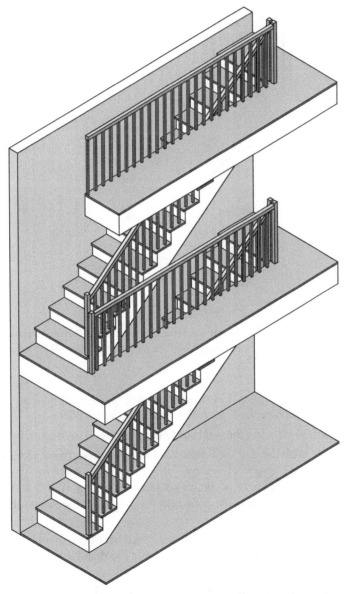

Figure 11.8 Stacked stairs with one flight going up and the other down. This position is common when access from the main floor to both the second floor and basement or second and third floors is designed for the same area. (*Previously published in Modern Residential Construction Practices, Routledge, 2017 and reproduced here with permission.*)

5/8" TYPE 'X' GYP BRD UNDER STAIRS
@ ALL USABLE AREA UNDER STAIRS

UP 14R

34" HIGH HANDRAIL

Stairs with Winders

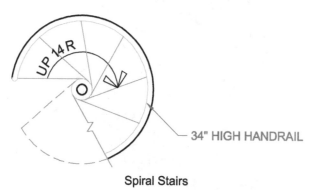

UP 14R

34" HIGH HANDRAIL

Spiral Stairs

Figure 11.9 The plan view of winder and spiral stairs. (*Previously published in Modern Residential Construction Practices, Routledge, 2017 and reproduced here with permission.*)

- The average set of stairs for an 8-ft (2400-mm) floor to ceiling, plus construction members, have 14 to 15 risers between two floors.
- To calculate the **total run,** multiply the length of a run by the number of runs (R).
- With the stairs in Figure 11.11 as an example, use the formula: Number of runs (R) × dimension of one run = total run. If the stairs in Figure 11.11 has 13 runs and each run is 11 in., then 11 × 13R = 143 in., or 11'-11" is the total run. The total run of the stairs is the horizontal distance from the face of the first riser to the face of the last riser.
- Landings take up additional space, and they should be as deep as the stairs are wide. For example, if a 36-in.-wide (900 mm) stairway has a minimum landing that measures 36 × 36 in. (900 × 900 mm).
- Minimum stair width is 36 in. (900 mm) but more width, such as 48 to 60 in. (1200 to 1500 mm), is preferred if space is available.
- Stair tread depth should be 10 to 12 in. (250 to 300 mm) with 10 in. minimum.

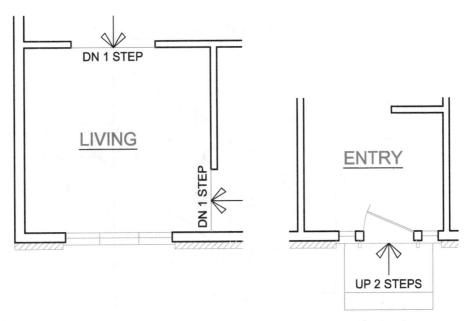

Figure 11.10   The number of steps into a sunken or raised room are noted with an arrow. (*Previously published in Modern Residential Construction Practices, Routledge, 2017 and reproduced here with permission.*)

- Individual risers can range between 6 and 9 in. (150 and 220 mm) in height. A comfortable riser height is 7-1/4 in. Each riser height measurement should not vary more than 3/8 in. (10 mm).
- The stair angle should be between 30° and 35° when measured from the floor.
- A clear height of 6′-8″ is the minimum amount of headroom required for the length of the stairs. A headroom clearance height of 7′-0″ is preferred.
- Stairs with over three risers require handrails. Stairs must have a handrail that measures between 34 and 38 in. (860 and 965 mm) above the nosing, as shown in Figure 11.11. Handrails are placed at a blank wall along a flight of stairs to be used for support and should run the entire length of the stairs, even though the stairs are enclosed between two walls. Handrails must extend 6 in. (150 mm) past the edge of the first and last step.
- Guardrails are placed where there is no protective wall. They serve the dual purpose of protecting people from falling off the edge of the stairs and providing a rail for support. Guardrails at landings above stairs, at balconies, at lofts, or at any area above another floor must be at least 36 in. (914 mm) high above the floor and have openings no greater than 4 in. (102 mm) between railings and balusters.

Stair Framing

Continue to use Figure 11.11 as an example as you discover the following stair framing applications and terminology. The stairs in Figure 11.11 provide a typical framing example, but many alternatives and options are possible. The stairs in Figure 11.11 are framed between

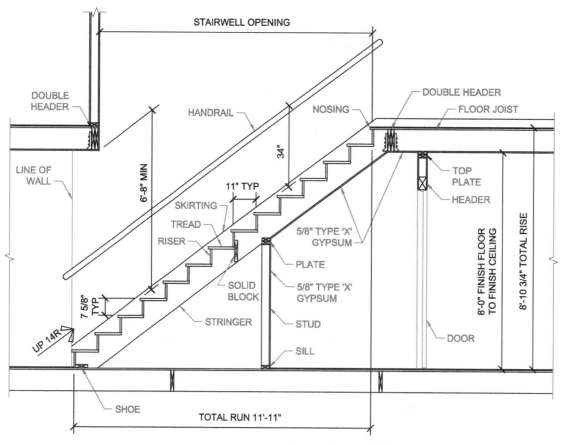

STAIRWELL OPENING

DOUBLE HEADER

HANDRAIL

NOSING

34"

DOUBLE HEADER

FLOOR JOIST

LINE OF WALL

6'-8" MIN

11" TYP

SKIRTING

TREAD

RISER

5/8" TYPE 'X' GYPSUM

TOP PLATE

HEADER

PLATE

SOLID BLOCK

5/8" TYPE 'X' GYPSUM

8'-0" FINISH FLOOR TO FINISH CEILING

8'-10 3/4" TOTAL RISE

7 5/8" TYP

STRINGER

STUD

UP 14R

SILL

DOOR

SHOE

TOTAL RUN 11'-11"

Figure 11.11 A typical stair section with stair construction terminology labeled. (*Previously published in Modern Residential Construction Practices, Routledge, 2017 and reproduced here with permission.*)

two floors with a closet space provided under the stairs at the first floor. This is a common design, allowing the space under the stairs to be used for storage or a closet, such as an entry coat closet. When this design is used, the stair and adjacent wall framing is generally covered with drywall, and a door is added to complete the enclosed area. Another common stair design practice is to provide access to another stairway directly below for access to a lower floor. This allows one stairway going to an upper floor and another stairway going to a lower floor occupying the same space with one above the other.

The main structural supports for a stairs are called **stringers.** Stringers, also called **carriage** or **stair horse,** are the supporting member running the length of a stair incline on which treads and risers are mounted. Stringers need to be made with lumber that provides substantial support. 2 × 12 dimensional lumber is commonly used, with at least three stringers used to frame the stairs. One stringer is placed at each side and one in the middle of the stair for a 36-in.-wide (900 mm) stairs. Additional stringers or double stringers are used for wider stairs or when the stairs are designed to carry additional weight. Solid blocking is generally placed mid-height between the stringers to stabilize the stringers and provide for **fire blocking,** as shown in Figure 10.11. Fire blocking is used in a variety of concealed or hollow construction places in the framing of a building, such as between stringers, between studs in walls, between joist in floors and ceilings, and between rafters in specific roof framing locations. The fire blocking fills and seals the construction cavity to prevent or slow the spread of fire.

A stub partition is framed near the lower end of the stringers to provide a wall at the back of the closet in Figure 11.11. This stub partition can be framed at any desired location depending on how much space is used under the stairs. A doorway is framed on the right side, providing access to the closet under the stairs. The framing under the stairs is covered with **Type X gypsum.** Type X gypsum is 5/8-in.-thick drywall, manufactured for use in locations where building codes require a fire resistance rating in construction.

The required handrail for the stairs in Figure 11.11 is installed at a height of 34 to 38 in. (900 mm) above the nosing. This stairway is enclosed by walls on both sides, making a handrail required along one side.

Stringer layout and cutting is generally done by an experienced framer. Some contractors have the rough framing done by the framers, and the stair construction performed by a **stair builder.** A stair builder is an especially skilled finish carpenter who designs and builds stairways, and guardrail systems. Figure 11.12 shows stair builders at work on a set of stairs. Exceptional attention to accuracy and detail is required when building and finishing stairs. Stringers need to be accurately laid out with the length and size of each rise and run carefully determined, as shown in Figure 11.13. The stairwell opening is framed with double headers and double trimmers. Notice the double headers on each side of the stairwell opening in Figure 10.13. An interior second floor partition is framed directly above the left header in this example, and the upper ends of the stringers are fastened to the right header. The lower ends of the stringers are cut into a 2 × 4 or 2 × 6 **shoe** that is nailed through the subfloor and into floor joists below. A shoe is a blocking used to reinforce and stabilize the ends of stringers at the floor.

A **skirt board** is optional but commonly used against the wall as a trim. A skirt board is a nonstructural fascia used to trim the sides of stairs to which the treads and risers are fitted.

Figure 11.12 Stair builders at work on a set of stairs. Exceptional attention to accuracy and detail is required when building and finishing stairs. (©*Agence DER—Adobe Stock.*)

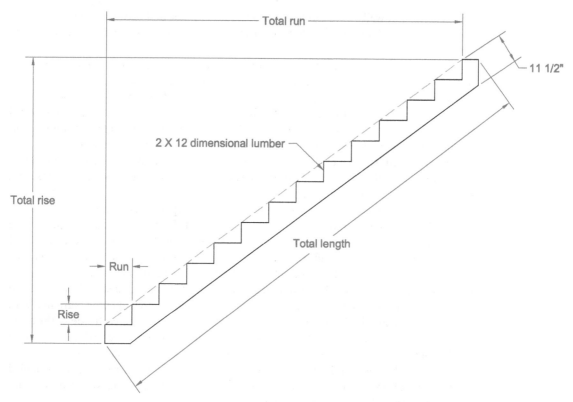

Figure 11.13   Stringers need to be accurately laid out with the length and size of each rise and run carefully determined. (*Previously published in Modern Residential Construction Practices, Routledge, 2017 and reproduced here with permission.*)

Riser and tread material is typically fastened to the stringers with construction adhesive and screws to secure and help keep the steps from squeaking when used. Risers and treads can be made with plywood when the finish material is carpet pad and carpet. Carpeting can be wrapped around the nosing or the nosing can be flush with the riser and cut at a 45° angle for a smooth transition of carpeting from the tread to the riser, as shown in Figure 11.14. The stair treads and risers are often made from fine **hardwood** to demonstrate the architectural beauty of wood as shown in Figure 11.15*a*. Carpet is commonly used on the stairs for quiet comfort and beauty as shown in Figure 11.15*b*. Hardwood is wood from a broadleaf tree, but more specifically one of the fine hardwoods typically used in finish architectural applications, such as cherry, maple, oak, and walnut. These woods have beautiful grain and provide a durable surface. **Softwood** can be used for treads and risers when a more rustic look is desired; however, softwoods are not a durable as hardwoods. Softwood is wood that comes from a coniferous tree, such as fir, hemlock, and pine. The finish hardwood or softwood treads and risers are usually attached to a base plywood tread and riser using construction adhesive, as shown in Figure 11.16.

When the stairs are open on one side, the stringer is generally covered with hardwood or painted wood trim called a **face stringer,** and balusters are attached to the treads, as shown in Figure 11.17*a*, or to a **bottom rail** or **shoe rail.** A face stringer is an exposed stringer on the open side of stairs. The bottom rail is a short wall built up above a stringer to enclose the ends of the treads and risers and used to attach the balusters, as shown in Figure 11.17*b*. A shoe rail is used to receive the square bottom end of balusters when they

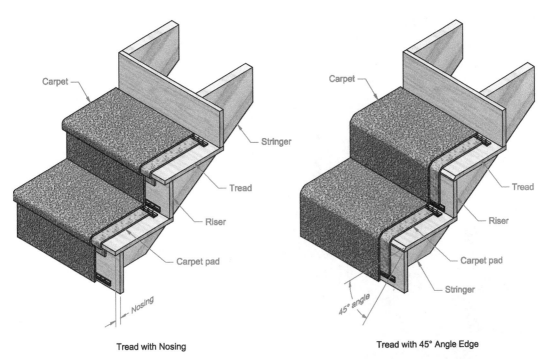

Figure 11.14   Plywood tread and risers used for finish carpeting. (*Previously published in Modern Residential Construction Practices, Routledge, 2017 and reproduced here with permission.*)

are not connected directly to the treads. An alternative to fastening the balusters to the treads or bottom rail is to attach the balusters vertically to the side of the face stringer, as shown in Figure 11.17c. The balusters can be attached to the face stringer using screws that have a counterbore into the balusters and a **wood plug** used to cover the hole. Alternately, a pilot hole can be drilled without a counterbore when the fastener heads are exposed and selected for their desired appearance. A wood plug is a cylindrical flat or rounded wooden plug of the same diameter as the counterbore, used to plug the holes bored for concealed fasteners. Wood plugs can be the same wood species to match the balusters when a natural wood appearance is desired, or **paint grade** when the balusters are painted. Paint grade is any wood or other material that looks best when paint painted, and it is less expensive than **stain grade.** Stain grade wood generally matches the wood species where it is assembled and is used with stain and clear finish to accent the natural wood beauty.

A **housed stringer,** also called a **routed stringer,** can be used on high-quality hand-crafted stairs. A housed stringer is a closed stringer that has the ends of the treads and risers recessed into channels cut into the stringer, as shown in Figure 11.18.

> **NOTE**
>
> The previous content introduced you to stair design, construction, and terminology related to wood frame construction for stairs in light commercial buildings such as apartments, town-homes, condominiums, and offices. The same terminology is used for any stair design and construction, with materials such as glulams, steel, or concrete typically used in other types of commercial construction projects.

(a)

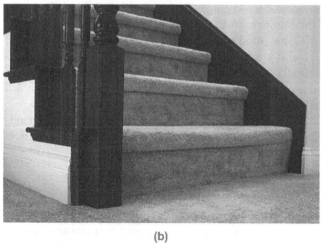

(b)

Figure 11.15 (*a*) The stair treads and risers are often made from fine hardwood to demonstrate the architectural beauty of wood. (*©Photographee.eu—Adobe Stock.*) (*b*) Carpet is commonly used on the stairs for quiet comfort and beauty. (*©Inna Efimchik—Adobe Stock.*)

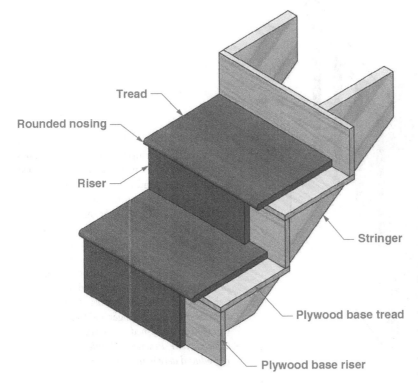

Tread

Rounded nosing

Riser

Stringer

Plywood base tread

Plywood base riser

Figure 11.16 Using finish hardwood for the treads and risers on a stairs. (*©alexandre zveiger—Adobe Stock*.) Model *Previously published in Modern Residential Construction Practices, Routledge, 2017 and reproduced here with permission.*)

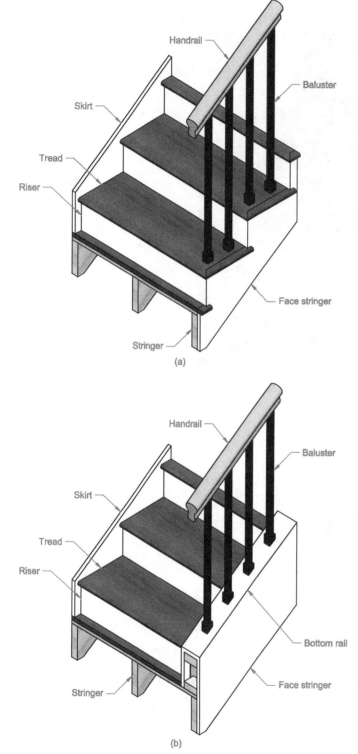

(a)

(b)

Figure 11.17 When the stairs are open on one side, the stringer is generally covered with hardwood or painted wood trim called a face stringer, and balusters are attached to the treads or to a bottom rail or shoe rail. (*a*) A face stringer is a stringer on the open side of stairs generally covered with hardwood or painted wood trim, and balusters are attached to the treads. (*b*) The bottom rail is a short wall built above a stringer to enclose the ends of the treads and risers and used to attach the balusters. (*c*) Balusters can also be fastened to the side of the face stringer and to the handrail. (*Previously published in Modern Residential Construction Practices, Routledge, 2017 and reproduced here with permission.*)

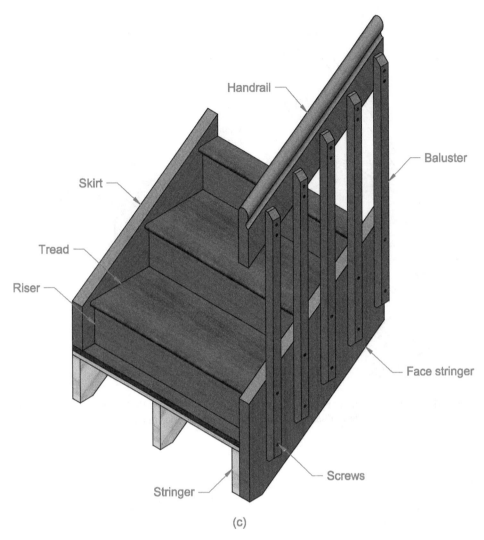

(c)

Figure 11.17   (*Continued*)

Building Freestanding Stairs

A **freestanding stairs,** also called **open stairs,** has stringers or carriages that are unsupported by walls, and attached only to the floor at the bottom and the structural framing at the top. Freestanding stairs are often used in contemporary architecture, as shown in Figure 11.19, because of the simple style, and used in rustic architecture with natural exposed wood stringers and treads. Freestanding stairs are also often used as a basic and inexpensive way to build outside stairs. This is commonly done by using the stringers with heavy boards, such as 2 × 12s, for treads, and using no risers, as shown in Figure 11.20. A concrete slab can be used for a landing at the base of the stairs, and the bottom of the stringers is placed on a tar pad or pressure treated wood to minimize decay.

There are a variety of methods that can be used to build freestanding stairs, with two typical techniques shown in Figure 11.21. To maintain the best open stairs design,

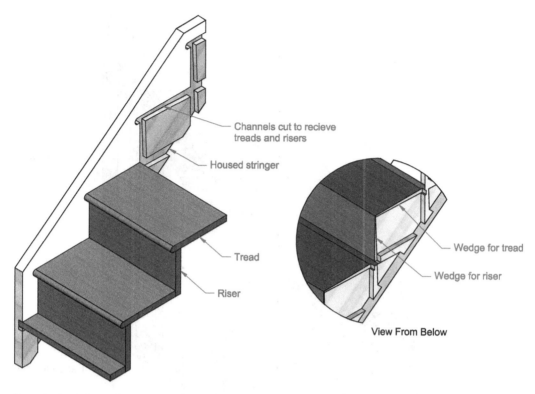

Channels cut to recieve
treads and risers

Housed stringer

Tread

Riser

Wedge for tread

Wedge for riser

View From Below

Figure 11.18 A housed stringer is a closed stringer that has the ends of the treads and risers recessed into channels cut into the stringer. (*Previously published in Modern Residential Construction Practices, Routledge, 2017 and reproduced here with permission.*)

Figure 11.19 Freestanding stairs are often used in contemporary architecture. The guardrails are not yet installed on this stairs. (*©Photographee.eu—Adobe Stock.*)

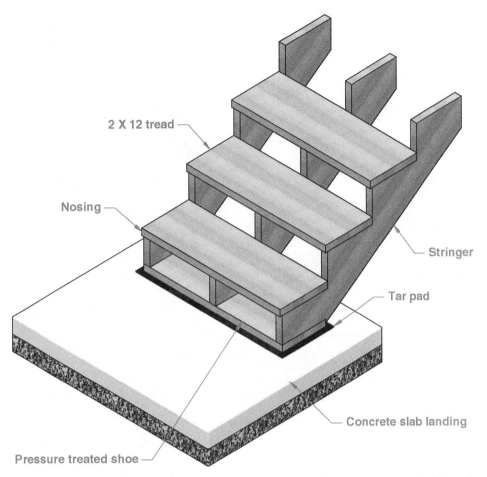

2 X 12 tread

Nosing

Stringer

Tar pad

Concrete slab landing

Pressure treated shoe

Figure 11.20 Freestanding stairs are also often used as a basic and inexpensive way to build outside stairs. This is commonly done by using the stringers with heavy boards such as 2 × 12s for treads with no risers. A concrete slab can be used for a landing at the base of the stairs, and the bottom of the stringers are placed on a tar pad or pressure treated wood to minimize decay. (*Previously published in Modern Residential Construction Practices, Routledge, 2017 and reproduced here with permission.*)

one heavy carriage is used on each side of the stair. The term **carriage** is substituted for stringer in this application, because the carriage is generally considered more heavy duty than typical stringers. The carriage material is generally selected for strength and appearance. High-quality 4 × 12, 6 × 12, 4 × 14, or 6 × 14 dimensional or rough sawn lumber is an option, or glulams can be used for a more contemporary look and for the availability of quality material. The treads are also selected for strength and appearance, with 3× or 4× dimensional lumber or glulam material used. Hardwood lumber can also be selected for desired appearance. Figure 11.21*a* shows the carriages attached to the floor structure using steel angles and lag bolts. The lag bolts can fasten directly to floor joists below or to solid blocking between floor joists. The treads are fastened to the carriages with steel angles and lag bolts. The steel angles and bolt heads are generally painted or made with stainless steel for appearance. Figure 11.21*b* shows an application without using steel angles. In this

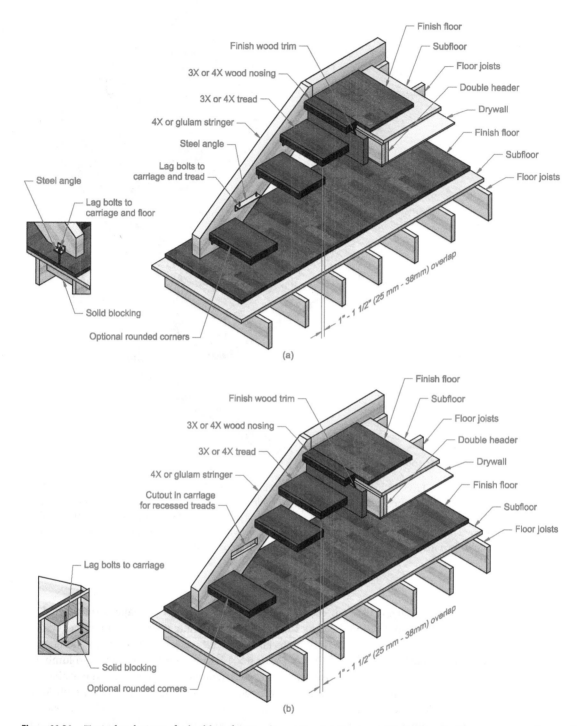

Figure 11.21   Typical techniques for building freestanding stairs. (*a*) Carriages attached to the floor structure using steel angles and lag bolts. Treads are fastened to the carriages with steel angles and lag bolts. (*b*) In this example, a cutout is made in each carriage where the treads are recessed into the cutouts. (*Previously published in Modern Residential Construction Practices, Routledge, 2017 and reproduced here with permission.*)

example, a cutout is made in each carriage where the treads are recessed into the cutouts and secured with construction adhesive. Each cutout depth is made at least half the thickness of the carriage material.

Manufacturers provide freestanding stair kits that can be purchased to fit in nearly any design and stair type. Products include wood or steel stair kits or a combination of materials and styles. The kits are generally ordered based on specific dimensions for total run and rise, and stair configuration such as straight, L shape, or U shape. Figure 11.22 shows a straight contemporary open stair design.

Building Spiral Stairs

Spiral stair manufacturers provide spiral stair kits that can be used for almost any design application and to meet local and national building codes. Spiral stair kits are available in a variety of materials including metal and wood. Figure 10.23 shows a spiral stair installation.

Concrete Stairs

Concrete stairs are common in commercial and industrial construction projects and used for access to the garage or to a concrete porch or patio. The distance from the floor inside the building to the floor of the entry, garage, or patio can vary depending on the type of construction. For example, post and beam construction allows for a minimum distance from the floor inside to the garage where only one step is needed, as shown in Figure 11.24a. The design and construction using no steps is becoming more popular with an aging population and when disability considerations are important. Construction with a concrete slab

Figure 11.22   A straight contemporary open stair design. (©*musa_263—Adobe Stock.*)

Figure 11.23   A spiral stair installation. (©*Mazur Travel—Adobe Stock.*)

foundation can easily accommodate no step construction with the entry, garage, and patio slabs at the same elevation as the inside floor. Alternately, flashing can be used to protect wood structures when built next to concrete for a no step design. Construction for Americans with Disabilities Act (ADA) requirements are thresholds cannot exceed 3/4 in. (20 mm) in height for exterior sliding doors or 1/2 in. (13 mm) in height for other types of doors. Changes in elevation up to 1/4 in. (6 mm) can be vertical and do not require a beveled edge.

Typical floor joist construction can require two or three steps from the floor inside to the garage floor depending on the sizes of the framing members. Figure 11.24*b* shows concrete stairs with three steps from the inside floor to the garage floor. The concrete slab under the stairs and over the foundation footing should be expanded to fill the area. Steel reinforcing bars can be used to minimize cracks in the concrete and depending on structural engineering specifications.

> **NOTE**
>
> The concrete stair construction examples used here are basic and minimal, but concrete stairs for commercial projects have the same or similar characteristics. Commercial concrete stairs are designed and built using poured-in-place reinforced concrete, or steel reinforced precast concrete based on structural engineering.

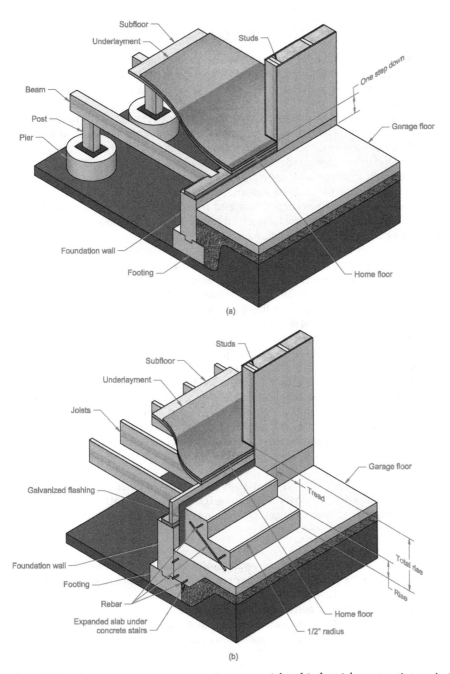

Figure 11.24   Concrete stairs are common in commercial and industrial construction projects, and used for access to the garage and to a concrete porch or patio. The distance from the floor inside to the floor of the garage or patio can vary depending on the type of construction used. (*a*) For example, post and beam construction allows for a minimum distance from the floor inside to the garage where only one step is needed. Typical floor joist construction can require two or three steps from the floor inside to the garage floor depending on the sizes of the framing members. Additional information is provided in the content. A small wood stairs can be built for this application or a concrete stairs can be constructed. (*b*) Concrete stairs with three steps from the inside floor to the garage floor. The concrete slab under the stairs and over the foundation footing should be expanded to fill the area. Steel reinforcing bars can be used to minimize cracks in the concrete and depending on structural engineering specifications. (*Previously published in Modern Residential Construction Practices, Routledge, 2017 and reproduced here with permission.*)

COMMERCIAL BUILDING CONSTRUCTION STAIR APPLICATION

BUILD recently completed this project in the Ballard neighborhood of Seattle. The upper grade is a full 14 ft above the street, while the hillside slopes parallel to the street. In order to produce an efficient and usable stair project, the design process focused heavily on the stair design strategy, both inside and out.

In order to navigate from the sidewalk to the entry, two types of stairs were implemented. A cast-in-place, concrete slab on grade stair ascends to the top of the hill and is lit at every fourth step by a recessed tread mount exterior grade light. Because the height between the top of the stairs and the adjacent grade is always less than 30 in., a guardrail is not required but a handrail is required. In this case, we use a slender galvanized steel rectangular cross section that attaches to the side of the concrete stair and extends up 36 in. From the top of grade to the main floor entry, a wood-framed stair is wrapped in **ipe** wood, while the galvanized hand-rail geometry is outfitted with horizontal steel rods every 3-1/2" to double-act as a guardrail. Ipe, also called Brazilian walnut, is an exotic hardwood from South America, used on structures that are hard, strong, and naturally resistant to rot, abrasion, weather, and insects.

Once inside, the interior circulation stacks with the stairs to the upper floor directly above the stairs to the lower floor in order to consume as little square footage as possible. A solid stair leads to the lower level, while an open riser stair extends to the common areas of the upper floor, as shown in **Figure 11.25**. At the upper stair, 3/8-in.-thick steel plates extend from blocking at the wall to a blackened steel channel at the open side. A blackened steel guardrail attaches to the channel while the handrail is located at the wall side to keep the guardrail uncluttered. Solid oak boards then wrap the steel plates creating a series of monolithic treads. All interior stairs are solid oak with an ebonized stain to match the hardwood floors.

To best optimize the view and natural light, the steel stair minimizes obstructions at the common space. A thin steel guardrail encompasses the stair opening from below while a set of blackened steel channels reaches to the roof above. A roof deck less than 200 ft² in areas does not require a code compliant stair, allowing fewer limitations around the rise and run of the treads as well as the guardrails and handrails. With minimal obstruction, the roof stair delivers inhabitants and visitors to a sweeping view of Ballard, the Olympic Mountains, and Seattle beyond.

One of the important design intents of the stairs was to create a series of experiences with smart, crafted stairs that make navigating a pleasure. The nearly 40 vertical feet of stairs in this project do their job effectively without getting in the way of the experience or the views.

Go to Downloads & Resources tab at **www.mhprofessional.com/Commercial Building Construction-Ancillaries** to access the images provided in this chapter and correlate with the content of this chapter. Here you can look at the chapter figures on your computer screen to pan and zoom in and out for better observation, because full drawings reduced to fit on a textbook page are often difficult to read. The image bank also includes the complete set of plans for the Brookings South Main Fire Station used as examples throughout this textbook.

Go to **Downloads & Resources tab at www.mhprofessional.com/Commercial Building Construction-Ancillaries** to access the test and creative thinking problems for this chapter. The chapter test can be used for review or to evaluate content knowledge depending on your course objectives. Creative thinking problems are provided to help expand your knowledge by researching given subjects.

Figure 11.25   Commercial building construction stairs application provides an open riser stair extending to the common areas of the upper floor. (*Courtesy of BUILD LLC.*)

Print reading problems have questions that ask you to find information on the Brookings South Main Fire Station plans. The print reading problems can help you become familiar with the format of construction documents and reinforce your ability to seek specific information. This is an important skill for you to master as you prepare to enter the construction industry. Go to **Downloads & Resources tab at www.mhprofessional.com/Commercial Building Construction-Ancillaries** to access the complete set of plans for the Brookings South Main Fire Station.

CHAPTER 12

Finish Work and Materials

INTRODUCTION

This chapter describes and illustrates the materials and practices used to finish a construction project. These applications include the installation of drywall, subfloor, siding, wood work, and finish flooring. **Painting** and other surface finishes are also described in this chapter as part of the completion process for the home. Painting is the process of using a solid, semitransparent, or transparent surface covering as a decoration and protective coating.

DRYWALL INSTALLATION AND FINISH

Drywall or other wall covering materials are put on the walls and ceilings before finish millwork is installed. For review, drywall is also called gypsum board, gyprock, plasterboard, sheetrock, or wallboard, is a plaster panel made of gypsum pressed between two thick sheets of paper. Standard drywall used to cover most walls and ceilings is 1/2 in. thick. A product called Type X gypsum is 5/8 in. thick drywall, manufactured for use in locations where building codes require a fire-resistance rating in construction. Green board is a drywall product with green-colored paper covering containing an oil-based additive that provides moisture resistance. It is commonly used in bathrooms and other areas where elevated levels of humidity can exist. Blue board drywall has blue face paper with a coat of plaster finish providing water and mold resistance.

Specific framing techniques must be used in corners before drywall is installed. These framing practices were described in Chapter 8, *Wood Construction*. Drywall is typically installed by a drywall subcontractor who specializes in the professional hanging and finish of the material.

Drywall sheets are manufactured in 48 in. (1200 mm) widths and a variety of lengths starting with 8 ft (2400 mm) and increasing in 2 ft (600 mm) modules, such as 10, 12, and 14 ft long. These dimensions allow the drywall to be fastened equally spaced over 12, 16, or 24 in. (300 to 600 mm) on center studs and ceiling joists, which is the common spacing for framing members. Drywall sheets are typically hung on the walls and ceilings with the sheet length perpendicular to the framing members. The drywall installer marks framing member locations on the drywall to make sure the fasteners attach the drywall into the framing. Drywall is fastened using special drywall nails or screws, as shown in Figure 12.1.

Figure 12.1 Drywall is fastened using special drywall screws. (*©Coprid Imagery—Adobe Stock.*)

A special drywall hammer is used to set the nail head slightly below the surface when nailing drywall. A cordless portable electric drill is used to drive the drywall screws into place, as shown in Figure 12.2. A special attachment is used in the drill that automatically sets the screw head slight below the surface of the drywall. The nail and screw heads need to be recessed slightly below the drywall surface so the locations can be filled with **drywall joint compound** during the finishing process. Drywall joint compound is described later in this discussion. Drywall installers often place a bead of construction adhesive at the center of each framing member before installing the drywall and screwing it in place. The use of construction adhesive helps secure the drywall and reduces the possibility of screw heads popping out through the surface after installation. Drywall is typically cut using a straight edge and a utility knife. Drywall has a gypsum core separated by a paper layer over each surface. The utility knife is used to cut through the paper on the face side of the drywall and then the installer quickly snaps the drywall on one side of the cut to break the gypsum along the cut line. The drywall is then folded back along the cut line, exposing the paper on the other side through the fold. The installer then uses the utility knife again to cut through the break in the drywall cutting the paper on the other side to separate the two pieces of drywall. This process takes some practice but is fairly easy to learn. Holes need to be cut in the drywall at locations where there are electrical outlets or other projections in the wall or ceiling so the drywall can fit over the projections. This is done by carefully measuring the exact location of each projection, then transferring and marking the measurements on the drywall surface. The hole is then cut out using a keyhole saw. A **keyhole saw** is a small handsaw with a long narrow blade used for cutting holes, short radius curves, and other small features.

The finishing process can start after the drywall is nailed or screwed in place. Drywall is finished using **drywall tape** and **joint compound.** Drywall tape is a paper, vinyl, or mesh product used to cover drywall joints and corners. Drywall tape is purchased in large rolls that range from 25 to 200 ft (7600 to 60000 mm) long. Paper drywall tape is typically used on inside corners and can be used over flat wall joints. Vinyl drywall tape is commonly used

Figure 12.2   A cordless electric drill is used to drive drywall screws into place using a special attachment in the drill that automatically sets the screw head slight below the surface of the drywall. (©*gwimages—Adobe Stock.*)

on outside corners where durability is needed to minimize corner damage. Mesh drywall tape is typically used on flat wall seams and has a self-adhesive backing for easy installation. A roll of mesh drywall tape is shown in Figure 12.3. Premanufactured outside drywall corners are available to simplify corner construction and to make durable and strong 90° and rounded corners. Joint compound, commonly called **mud,** is used to finish drywall joints, drywall tape, corners, and screw locations, and is used as a **skim coating.** Skim coating is the application of a thin layer of joint compound over drywall to smooth out walls and ceilings. Joint compound is a combination of several technical ingredients and is available premixed for immediate use. Joint compound has a creamy texture that spreads easily onto drywall surfaces. Joint compound is applied with a **trowel** or **drywall knife** using three or four thin coats to achieve the desired coverage. A trowel is a flat-bladed hand tool used for spreading, leveling, or shaping joint compound in drywall finishing applications. A drywall knife is a small hand tool with a blade used for applying joint compound when finishing drywall. Drywall knives are available in several blade widths and handle lengths. Figure 12.4*a* shows a drywall finisher using a drywall knife to spread joint compound over drywall. When installing drywall tape over a joint, the drywall finisher first applies a coat of joint compound and then rolls out drywall tape over the joint at that location. Drywall finishers typically use a tool-belt tape holders that allow them to easily dispense the tape. Then they lay a piece of tape over the joint and smooth it with a drywall knife or trowel. Drywall tape bonds with the drywall paper surface during this process, making a continuous smooth surface between adjoining sheets of drywall. Properly taped and finished drywall joints are not visible. Figure 12.4*a* shows a drywall finisher applying joint compound over a tape seam on a wall, and Figure 12.4*b* shows drywall joint compound applied over tape seams on ceilings and walls.

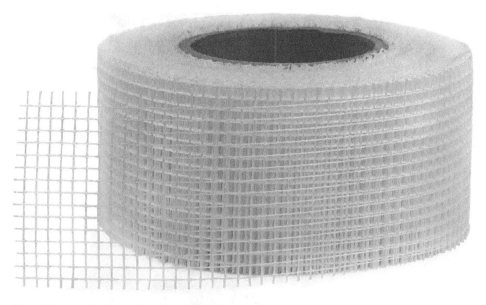

Figure 12.3 A roll of mesh drywall tape. (*©Coprid Imagery—Adobe Stock.*)

After the compounds are dry, the drywall surface is sanded to smooth out imperfections after the drywall is taped and the desired coats of joint compound is used to seal and level the joints, corners, and screw locations. A coat of primer paint can be applied to the drywall to seal all surfaces, provide bonding and uniformity for **texturing,** and to allow the texturing to dry evenly. The drywall can be left smooth or texturing can be applied. Smooth drywall requires special care with applying drywall tape, using joint compound, and doing quality surface sanding to make sure there are no irregularities. Smooth drywall surfaces are commonly used in bathrooms and kitchens, making it easier for cleaning than textured surfaces. Smooth drywall is also used under **wall paper.** Wall paper is a paper or vinyl product that usually has printed decorative patterns and colors, used for pasting in vertical strips over the walls and ceilings to provide a decorative or textured surface. Texturing is applied to drywall to make the surface look even and cover any surface irregularities. There are many different styles of texturing. It is also common to use one type of texturing on walls and another on ceilings for contrasting appearance. A commonly used wall texturing style is called **orange peel,** as shown in Figure 12.5. Orange peel texturing makes a surface finish that looks similar to an orange and provides a delicate uniform appearance that is easy to paint and maintain. Orange peel texturing is applied using a sprayer called a **hopper gun.** The hopper gun has a container with a spray nozzle that is operated by an air compressor. The container or hopper is filled with joint compound that is sprayed out through the nozzle in a consistent pattern on the drywall. A desired texture from light to heavy can be controlled by adjusting the hole size in the nozzle or by increasing or reducing the air pressure. **Knockdown** texturing is commonly used on walls and ceilings where a more artistic appearance is desired than orange peel. It is also common to use knockdown texture on the ceiling and orange peel on the walls providing a contrasting appearance and easier to maintain walls. Knockdown texturing is created by spraying texturing material on the drywall and then using a trowel to flatten the high places while still wet. Varying degrees of surface relief can be achieved by the amount of spray on material used before troweling. Figure 12.6 shows surface relief created using the knockdown texturing process.

(a)

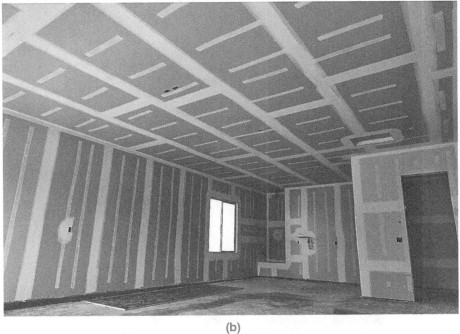

(b)

Figure 12.4 (*a*) A drywall finisher using a drywall knife to spread joint compound over a tape seam on a wall. (©*tinabelle—Adobe Stock*.) (*b*) Drywall joint compound applied over tape seams and fasteners on ceilings and walls. (©*1jaimages—Adobe Stock*.)

Figure 12.5 Orange peel texturing makes a surface finish look similar to an orange, and provides a delicate uniform appearance that is easy to paint and maintain. (©*travelview—Adobe Stock.*)

Figure 12.6 Surface relief created using the knockdown texturing process. (©*gio—Adobe Stock.*)

An interesting effect can be achieved by first using a primer paint followed by a desired paint color before using the knockdown texturing application. The painted surface can be seen behind the areas where the trowel skips over the texturing material creating an interesting two-color and three-dimensional appearance. After texturing has been completed and allowed to dry, the surfaces can be primed and painted as desired.

MILLWORK

Finish materials and the craftsmanship it takes to install can be one of the most exciting phases of construction, because of the beauty and creativity possible with quality **millwork.** Figure 12.7 shows two examples of the beauty found in quality millwork. Finish work happens near the end of the project after the drywall or other wall covering material has been installed. Millwork is finished woodwork that has been manufactured in a **milling plant,** and is anything that is considered finish trim or finish woodwork, including cabinets. A milling plant is an industry that manufactures millwork. Millwork is manufactured in many wood species, shapes, and styles for use in a variety of applications. Wood millwork components are shaped using a **milling** process. Milling is the use of a rotating cutting tool where the tool forms the desired shape of the millwork. Plastic, plaster, and ceramic products are also available that duplicate the traditional wood millwork. Plastic, plaster, and ceramic millwork is generally **extruded** or molded into the desired shape. Extrusion is a manufacturing process used to create long objects of a fixed cross-sectional profile, where material is pressed or pushed under pressure through a desired shaped opening.

Complete millwork details are often not required in a set of plans, but written requirements can be found in general notes and specifications. The practice of creating millwork representations depends on the specific requirements of the project. For example, some commercial building plans can provide very detailed and specific drawings of the finish woodwork in the form of plan views, elevations, construction details, and written specifications. Typical plans often show the cabinets on the floor plans, with specifications, description of materials, and contracts that provide information about materials and desired workmanship. The requirements of specific lenders and local codes can also determine the extent of millwork representations on a set of plans.

TYPES OF MILLWORK

Millwork can be designed for appearance and for function. When it is designed for appearance, ornate, and decorative millwork can be created with a group of shaped wood forms placed together to capture a style of architecture, as shown in Figure 12.8. Shaped millwork is available in as many styles as the builder can imagine. There is also a wide variety of prefabricated millwork moldings available at less cost than custom designs.

Functional millwork can be very plain in appearance. This type of millwork is also less expensive than shaped forms. In some situations, natural wood millwork can be replaced with plastic, ceramic, or rubber products. Millwork is also manufactured out of engineered wood products. A laundry room can use a plastic or rubber base strip to protect the wall at the floor, for example. This material stands up to abuse better than wood. There are as many possible options as there are design ideas. The following discussion provides definitions and example of many available finish materials and products.

(a)

(b)

Figure 12.7 Examples of the beauty found in quality millwork. (*a*) Millwork used at the wall base, ceiling cove, fireplace mantel, and window and doorway trim. (©*Wollwerth Imagery—Adobe Stock.*) (*b*) Millwork used to detail this stairs, stair railings, and surrounding areas in the home. (©*Rodenberg—Adobe Stock.*)

Figure 12.8 When millwork is designed for appearance, ornate, and decorative designs can be created with a group of shaped wood forms placed together to capture a style of architecture. (*©Wollwerth Imagery—Adobe Stock.*)

Baseboards

Baseboards are placed at the intersection of walls and floors and are generally used for appearance and to protect the wall from damage, as shown in Figure 12.9. There are a large variety of baseboard shapes, sizes, and styles. Baseboards can be as ornate or as plain as the specific design or location requires. In some designs, baseboards are the same shape as other millwork members, such as trim around doors and windows, as shown in Figure 12.10. Baseboard material is cut at a 45° angle at inside and outside square corners, as shown in Figure 12.11*a*, and formed around rounded drywall corners by making multiple cuts to fit the corner or by using a premanufactured molded corner, as shown in Figure 12.11*b*.

Wainscot

A **wainscot** is any wall finish where the material on the bottom portion of the wall is different from the upper portion. The lower portion is called the wainscot, and the material used is called wainscoting. Exterior wainscoting is often brick or stone veneer. Interior wainscoting can be any material that is used to divide walls into two visual sections. For example, wood paneling, plaster texture, ceramic tile, wallpaper, or masonry can be used as wainscoting. Figure 12.12 shows handcrafted wood paneling used for wainscoting. A more contemporary or rustic appearance can be achieved by using horizontal or vertical tongue and grove lumber for wainscot paneling, as shown in Figure 12.13. A fairly inexpensive wood wainscot can be built with plywood panels. The plywood panel can have a hardwood

Figure 12.9   Baseboards are placed at the intersection of walls and floors and are generally used for appearance and to protect the wall from damage. (*©S_E—Adobe Stock.*)

Figure 12.10   Baseboards can be the same shape as other millwork members, such as trim around doors and windows. (*©peshkova—Adobe Stock.*)

outer **veneer** to match the surrounding hardwood material. Veneer is thin sheets of wood glued together to form plywood or glued to a wood base material. When used for millwork, the outer face can be any desired hardwood. The plywood panels are cut to any desired size and evenly spaced along the length of the wall. Matching hardwood is used to trim around the plywood panels like a picture frame and also used between, above,

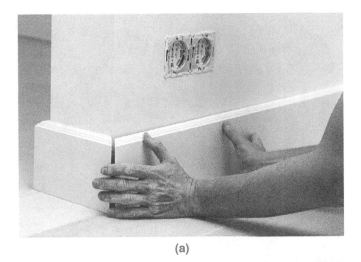

(a)

(b)

Figure 12.11   Baseboard outside corner applications. (*a*) Baseboard material is cut at a 45° angle at inside and outside square corners. (©*Photographee.eu—Adobe Stock.*) (*b*) Baseboard formed around rounded drywall corners by making multiple cuts to fit the corner or by using a premanufactured molded corner as shown in this example. (©*Andy Dean—Adobe Stock.*)

Figure 12.12 Handcrafted wood paneling used for wainscoting. (*©nikolarakic—Adobe Stock.*)

Figure 12.13 A more contemporary or rustic appearance achieved using horizontal or vertical tongue and grove lumber for wainscot paneling. (*©Photographee.eu—Adobe Stock.*)

and below the panels to build the complete wainscot system on the wall. **Paint grade** engineered wood products can also be used in the same manner to construct an attractive but less expensive wainscot. Paint grade refers to a wood product that is more suitable for painting than for a clear finish where the actual wood grain is visible.

Chair Rail

A **chair rail** is traditionally placed horizontally on the wall at a height where chair backs would otherwise damage the wall, as shown in Figure 12.14. Chair rails are usually found in dining rooms, dens, offices, meeting rooms, or other areas where chairs are frequently moved against a wall. Chair rails can be used individually for appearance or together with wainscoting, as shown in Figure 12.15. In some applications the chair rail is an excellent division between two different materials or wall textures, such as wall paper on the bottom below the chair rail and textured painted above the chair rail.

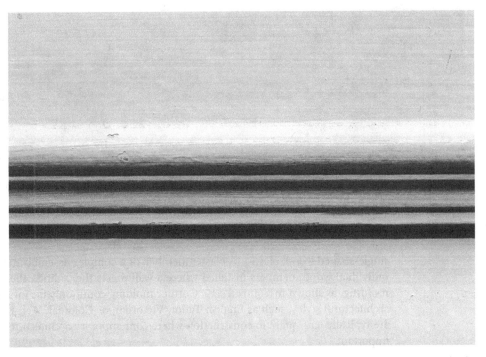

Figure 12.14   Chair rail is traditionally placed horizontally on the wall at a height where chair backs would otherwise damage the wall. (©*weedezign—Adobe Stock.*)

Figure 12.15   Chair rails can be used individually for appearance or together with wainscoting. (©*fraismedia—Adobe Stock.*)

Cornice

A **cornice** is generally any horizontal decorative molding that is commonly found over a door or window, or along the corner where a wall meets the ceiling, and is a projection out from the building to protect the wall from weather. A cornice generally has decorative moldings used in the finish construction, as shown in Figure 12.16. A cornice can be a

Figure 12.16 A cornice is any horizontal decorative molding that is commonly used along the corner where a wall meets the ceiling. (*©Wollwerth Imagery—Adobe Stock.*)

single-shaped wood, plastic, or plaster member, or a complex assembly made up of several individual wood members installed where a wall meets the ceiling called **cove** or **crown molding,** as shown in Figure 12.17. Cornice molding commonly fits into specific types of architectural styles, such as English Tudor, Victorian, or Colonial. Wall-to-ceiling corners are typically left square in construction where contemporary architecture or cost saving is important.

Figure 12.17 A cornice can be a single-shaped wood member called cove or crown molding, or a more complex assembly made up of several individual wood members as shown here. (*©Wollwerth—Can Stock Photo Inc.*)

Casing

Casing is the molding or trim used around doors and windows. Casing is attached around the edge of the door or window jamb and to the adjacent wall. Casing can be decorative to match other moldings or plain to serve the functional purpose of covering the space between the door jamb or window jamb and the wall. The baseboard is often wider than the casing, but baseboard and casing can be the same style of molding to match, as shown in Figure 12.18. The casing around doors is usually installed like a picture frame with 45° mitered corners, but **corner blocks** can also be used. Corner blocks are square decorative prefabricated blocks that are traditionally used at the top corners of doors for ornamental appearance, and to simplify casing installation. A cornice can also be used as a horizontal decorative molding over a door or window as shown over the window in Figure 12.19. The cornice design can be as creative as the molding styles available. The cornice over the window in Figure 12.19 has a **keystone** set in the center for a unique architectural style. Keystones are wedge-shaped blocks used in the center of curved and straight arches for a decorative appearance or to match Italian architecture. Keystones are commonly used in masonry arches for appearance and to lock the other stones into place, but they can also be used in millwork applications as shown in this example. Window casing can be installed all around the window using 45° mitered corners like a picture frame, as shown in Figure 12.20a, or with casing at the sides and top joined with 45° mitered corners and a sill projecting out at the bottom, as shown in Figure 12.20b.

Mantels

The **mantel** is an ornamental shelf or structure built above a fireplace opening. Mantel designs vary with individual preference. Mantels can be made of masonry as part of the fireplace structure, or ornate decorative wood moldings, or even a rough-sawn length of

Figure 12.18 Casing is the molding or trim used around windows and doors. The baseboard is often wider than the casing, but baseboard and casing can be the same style. (©*Iriana Shiyan—Adobe Stock.*)

Figure 12.19 A cornice can also be used as a horizontal or arched decorative molding over a door or window. This cornice has a keystone set in the center for a unique architectural style. (*©josunshine—Adobe Stock.*)

lumber bolted to the fireplace face. Figure 12.21 shows a traditional decorative wood mantel application. Figure 12.22 shows a wood mantel attached to a masonry fireplace structure.

Bookshelves

Bookshelves and display shelves can have simple construction with metal brackets and metal or wood shelving, or they can be built and detailed the same as fine furniture. Bookshelves are commonly found in the den, library, office, or living room. Shelves are also used for functional purposes in storage rooms, linen closets, laundry rooms, and any other location where additional storage is needed. Figure 12.23 shows bookshelves and display shelves built into the woodwork in an office.

Handrails and Guardrails

Stair construction was described in Chapter 11, *Stair Construction*, with information about the construction of stairs and components such as **handrails** or **guardrails.** A stairs has a handrail or guardrail that provides a support and protects you from falling when going up or down the stairs. A handrail is generally a single rail attached to a wall or **balusters** following the rise of the stairs and used for hand support and safety. A guardrail, also called a **banister,** is a handrail with supporting posts made up of **newel posts** and balusters. A newel post is the large vertical support for the handrail at the ends of each flight of stairs, and often placed at regular intervals when additional support is needed.

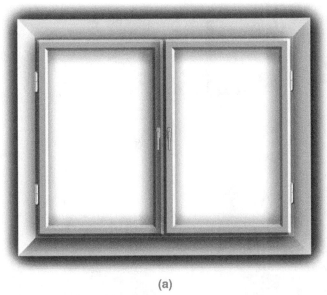

(a)

(b)

Figure 12.20   (*a*) Window casing installed all around the window using 45° mitered corners like a picture frame. (©*Iscatel—Adobe Stock*.) (*b*) Window casing installed with casing at the sides and top joined with 45° mitered corners and a sill projecting out at the bottom. (©*Iriana Shiyan—Adobe Stock*.)

A baluster, also called a **spindle** or **picket,** is one of a series of vertical handrail supports placed equally spaced between newel posts. Guardrails are used for safety at stairs, landings, decks, and open balconies, and are required at any rise over 30 in. (760 mm) or three or more stair risers in height. Verify the size and requirements with local codes. Guardrails can also be used as decorative room dividers or for special accents. Guardrails can be built

Figure 12.21 A traditional decorative wood mantel. (*©Iriana Shiyan—Adobe Stock.*)

Figure 12.22 A wood mantel attached to a masonry fireplace structure. (*©pics721—Adobe Stock.*)

Figure 12.23   Bookshelves and display shelves built into the woodwork in a den. (©*Iriana Shiyan—Adobe Stock.*)

enclosed or open and constructed of wood or metal. Enclosed rails are often the least expensive to build because they require less detailed components and labor than open railing systems. A decorative wood cap is often used to trim the top of enclosed railings.

Open railing systems can be one of the most attractive elements of interior home design. Open railings can be as detailed as the designer's or craftsperson's imagination. Detailed open railings built of exotic hardwoods can be one of the most expensive and most impressive features in the building. Figure 12.24 shows the beauty of a custom stair and balcony guardrail system.

CABINETS

Cabinets are generally a storage or display system with drawers, and doors with shelves behind. The quality of cabinetry can vary greatly. Cabinet designs can reflect individual taste, with a variety of styles available. Cabinets are used for storage and as furniture. Cabinets are commonly found in the kitchen and bath but are also used in the laundry room, family room, walk-in closet, garage, and other storage locations.

The design and arrangement of kitchen cabinets has been the object of many studies. Over the years design ideas have resulted in attractive and functional kitchens. Kitchen cabinets have two basic elements: the **base cabinet** and the **upper cabinet,** as shown in Figure 12.25. A base cabinet is built under the countertop of a kitchen or bathroom. Base cabinets are the largest storage area of a cabinet system. Kitchen base cabinets are typically 36 in. (900 mm) high and 24 in. (600 mm) deep, while bathroom base cabinets are typically 30 in. (760 mm) high by 22 in. (550 mm) deep. Drawers, shelves, cutting boards,

Figure 12.24   The beauty of a stair and balcony guardrail system. (©*pics721—Adobe Stock*.)

Figure 12.25   Cabinets have two basic elements: the base cabinet and the upper cabinet. (©*la_photog—Adobe Stock*.)

pantries, and appliance locations must be carefully considered for convenient use. Upper cabinets, also called wall-mounted or wall cabinets. Cabinets that are hung on a wall are referred to as upper cabinets, because they are generally built above the base cabinets. Upper cabinets are commonly 12 in. (300 mm) deep and vary in height. In addition to the cabinets, the kitchen has appliances such as the range, range hood, refrigerator, oven, and microwave, as shown in Figure 12.26. It is also common for a custom kitchen to have an island, as shown in Figure 12.27, with cabinets below a work preparation area, a built-in range, or a sink.

Figure 12.26 In addition to the cabinets, the kitchen has appliances such as the range, range hood, oven, microwave, and refrigerator. (*©pics721—Adobe Stock.*)

Figure 12.27 It is common for a kitchen to have an island with cabinets below a work preparation area, a built-in range, or a sink. (*©Wollwerth Imagery—Adobe Stock.*)

Figure 12.28   A bathroom with a cabinet on the left with vanity and mirror, and a built-in tub and separate shower on the right. (*©pics721—Adobe Stock.*)

Bathroom cabinets are called **vanities,** linen cabinets, and medicine cabinets. Figure 12.28 shows a bathroom with a cabinet on the left with vanity, mirror, and medicine cabinet above, and a built-in tub on the right. Other cabinetry is found throughout a house, as in utility or laundry rooms, for storage. For example, a workspace cabinet and storage cabinets above a washer and dryer are common, as shown in Figure 12.29.

Cabinet Types

There are as many cabinet styles and designs as you can imagine. However, there are two general types of cabinets, based on their method of construction. These are **modular** or **prefabricated cabinets,** and **custom cabinets.**

Modular Cabinets

Modular cabinets are designed and built at a manufacturing company and delivered to the project for installation during construction. There are many manufacturers of modular cabinets, providing a wide variety of styles and design alternatives. The term **modular cabinets** refers to prefabricated cabinets because they are constructed in specific sizes called modules. Modular cabinets are used by placing a group of modules side by side in a given space. If a little more space is available, then pieces of wood, called **filler,** are spliced between or at the ends modules. Many brands of modular cabinets are well crafted and very attractive. Most modular cabinet manufacturers offer different door styles, wood species, and finish colors. Modular cabinets are sized in relationship to standard or typical applications, although many modular cabinet manufacturers can make components that fit nearly every design situation. Modular cabinets are often manufactured nationally and delivered in modules. Good places to see modular cabinets on display are your local multiproduct

Figure 12.29 Cabinetry is used for storage throughout, such as in utility or laundry rooms. (*©Iriana Shiyan—Adobe Stock.*)

lumber company, Home Depot, or Lowe's Home Improvement Center. The suppliers of modular cabinets often have detailed displays showing the available door designs and other available features. These suppliers typically have a cabinet designer on staff that can create a computer-aided design of your desired cabinets from building plans. The computer-aided design commonly shows the floor plan, cabinet elevations, three-dimensional model, and provides a cost analysis based on your desired features. Figure 12.30 shows several very basic modular cabinet options. Modular cabinets are typically prefinished by the manufacturer, providing a variety of colors and surface options.

Custom Cabinets

Custom cabinets are usually designed by the architect, or custom cabinet shop, and are generally built at a local cabinet shop and delivered to the project for installation in large sections. One of the advantages of custom cabinets is their design is limited only by the imagination of the architect and the cabinet shop. Custom cabinets can be designed and built for any situation, such as for any height, space, type of exotic hardwood, type of hardware, any geometric shape, or for any other design requirements.

Custom cabinet shops estimate a price based on the architectural drawings. The actual cabinets are constructed from measurements taken at the job site after framing. In most cases, if cabinets are ordered as soon as framing is completed, the cabinets should be delivered on time for installation as one of the last phases of construction. Most custom cabinets are designed and built for the average home, but custom cabinets can be as decorative as the design, shown in Figure 12.31. Custom cabinets can be prefinished at the cabinet shop before installation, but custom cabinet builders often prefer to finish the cabinets on the job after the cabinets are installed.

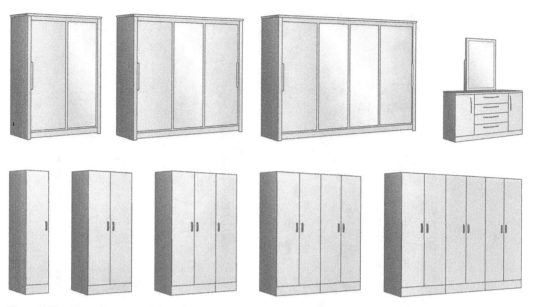

Figure 12.30 Several basic modular cabinet options. (©*Ana Vasileva—Adobe Stock.*)

Figure 12.31 Custom cabinets can be as decorative as the design shown in this example. (©*Melanie DeFazio—Adobe Stock.*)

Cabinet Substrate and Finish Tops

Base cabinets are the lower cabinet system typically covered with a durable hard surface for working and appearance. Typical finish surfaces include **laminate, melamine,** tile, and granite. Laminate, also referred to as a laminated structure or material, especially one made of layers fixed together to form a hard, flat, or flexible material. Melamine is a plastic-based product used to cover cabinet surfaces, or build drawers and shelves. Each finish material requires specific installation methods, skills, and the installation of proper base material. Cabinet top base material is typically called **backer board** or **substrate.** Backer board or substrate is used as a base cabinet top upon which the surface finish material is installed. Substrate for laminate tops is usually 3/4 in. plywood or OSB. Substrate for tile and granite tops can be 3/4 in. plywood but cement board or cement board over 3/4 in. plywood is preferred.

Cabinet Options

Cabinet designs available from either custom or modular cabinet manufacturers include the following:

- Various door styles, materials, and finishes
- Self-closing hinges
- A variety of drawer slides, rollers, and hardware
- Glass cabinet fronts for a traditional appearance where desired
- Wood or metal range hoods
- Specially designed pantries, appliance hutches, and a lazy Susan or other corner cabinet design for efficient storage
- Bath, linen storage, and kitchen specialties

APPLIANCES AND FIXTURES

An important part of the kitchen, bath, and laundry design and construction includes the **major appliances** and fixtures. A major appliance is usually a large product that is usually built-in as a permanent part of the cabinets or self-standing and used for daily activities, such as cooking, or food storage, cleaning, and disposal. Major appliances are generally powered or produce heat generated by electricity, natural gas, or propane. Major appliances commonly found in the kitchen include cooktop range, range and oven combination, range hood exhaust system, oven, microwave oven, refrigerator, trash compactor, garbage disposal. Laundry room major appliances are the clothes washer and clothes dryer. Fixtures are found in the kitchen, bathroom, and laundry and are used to supply and contain water, and discharge waste. Typical fixture examples are the kitchen sink, bathroom lavatory, laundry tray, shower, and bath tub, along with the water heater.

In addition to selecting the desired product, the most important thing to know about major appliances and fixtures as a builder is the installation dimensions and instructions. Each major appliance and fixture has installation specifications that you need to correlate during construction. Determine the exact installation dimensions with the appliance manufacturer and then provide that information to the cabinet builder. The preinstallation preparation for fixtures such as sinks and lavatories is done after the base cabinet or vanity substrate is installed. The sink or lavatory manufacturer usually ships a **cutout template** with the product. The cutout template is a paper outline of the exact substrate cutout required for the specific fixture. Place the template on the substrate surface in the exact location where the fixture will be installed. Use a pencil to mark the surface with the template outline as a guide. Finally, use a jig saw to cutout the substrate material at the marked location.

READING CABINET ELEVATIONS

Cabinet elevations are exterior views of the cabinets developed directly from the floor plan drawings. The purpose of the cabinet elevations is to show how cabinet exteriors look when completed and to give general dimensions, notes, and specifications. Cabinet elevations can be as detailed as the architect or designer desires and as detailed as needed for the cabinet builder. Figure 12.32 shows quality cabinet elevation drawings and the three-dimensional cabinet representation created by an architectural design software program for a custom home.

Keying Cabinet Elevations to Floor Plans

When you read plans you can notice that cabinet elevations are keyed to the floor plans, allowing you to correlate each cabinet elevation with its location on the floor plan. Several methods can be used to key the cabinet elevations to the floor plan. A basic method has cabinet elevations keyed to the floor plans with room titles such as KITCHEN ELEVATION or BATH ELEVATION. Another method uses an arrow with a letter inside to correlate the elevation to the floor plan. For example, the A arrow in Figure 12.33 points to a specific group of cabinets on the kitchen floor plan and the same letter is displayed below the cabinet elevations for the group of cabinets. Other techniques can be used, and various degrees of detailed representation can be provided for millwork, depending on complexity and the design specifications.

SURFACE FINISHES

Painting is the process of using a solid, semitransparent, or transparent surface covering as a decoration and protective coating. Painting is typically done to finish walls and can also be used on millwork when a solid color finish is desired. **Stain** and a clear finish coating can also be used on millwork when it is desired to preserve or enhance the natural beauty of the wood. Stain is a penetrating liquid used to preserve and tint woodwork for a desired appearance. Typically, the inside walls are painted before the millwork is installed. This practice avoids the need to **mask** the millwork before painting the walls. The word mask means to cover, usually with special tape, plastic, or paper to keep specific areas from getting painted. Millwork can be **prefinished** or finished after installation. Prefinished cabinets and woodwork is finished at the manufacturer and then delivered to the project for installation.

UNDERLAYMENT

Underlayment is construction material used over floor sheathing and under the finish floor material to provide a base for finish floor material. The underlayment is usually glued and screwed to the subfloor. Underlayment is generally 1/2 in. 4 × 8 ft sheets of **particleboard**

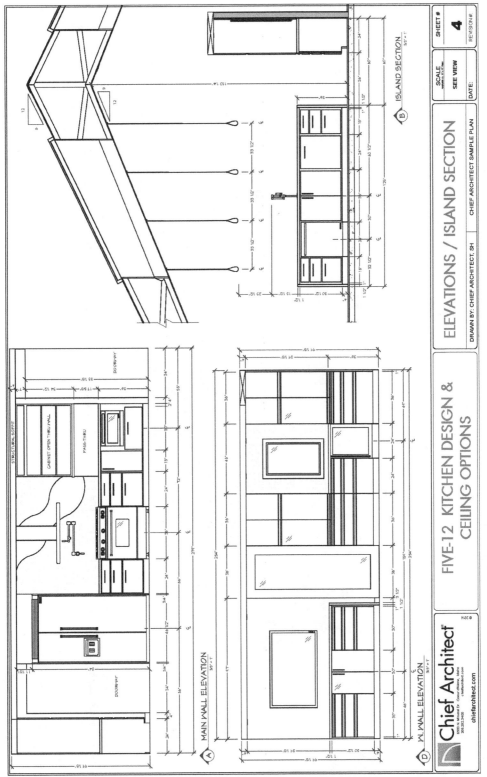

Figure 12.32 (a) Quality cabinet elevation drawings for custom cabinets. (b) Three-dimensional (3D) cabinet drawing. (c) 3D cabinet rendering. (*Courtesy of Chief Architect® Software.*)

543

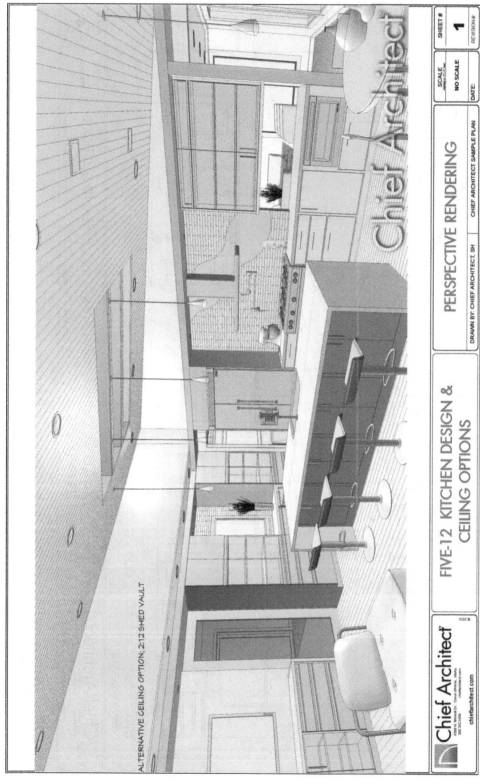

Figure 12.32 *(Continued)*

Figure 12.32   (*Continued*)

when used under finish floor such as carpet or wood. Particleboard is wood fibers that are glued and pressed into 4 × 8 ft sheets. Underlayment used under finish floor material such as tile is usually **concrete board,** which is concrete pressed into sheets.

Underlayment Material

Underlayment is material that is normally manufactured in sheets or rolls and is placed between the structural subfloor and the finish flooring material. This section describes the use of underlayment as a prefinish material. Underlayment is generally one of the last construction materials to be installed, after all framing is complete, rough utilities are installed, and the roof is finished. The building must be clean and dry before installing underlayment to protect the quality and performance. Finish floor material is generally installed shortly after the underlayment. Underlayment is used to smooth out imperfections and roughness that can exist on subfloors, provides a hard surface to support the finish flooring material, can provide insulation, and can reduce noise. Each type of finish floor material requires specific underlayment material and application. Proper selection and installation of underlayment is important because it can affect how the finish flooring feels, looks, and wears. The type of underlayment used for a flooring project should be based on the installation environment, such as moisture, and type of finish flooring. Always follow finish flooring manufacturer instructions about the type of underlayment to use, how to install the underlayment, and how to prepare the underlayment surface before installing the finish flooring. Improper installation can shorten the finish floor life and can void the manufacturer warranty. The most commonly used underlayment materials are **particleboard,** plywood, OSB, **cement board,** and **technologically advanced materials** for specific applications. The following describes a variety of underlayment materials and their applications.

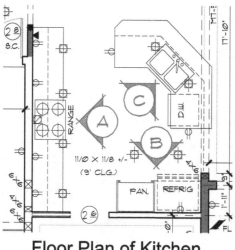

Floor Plan of Kitchen

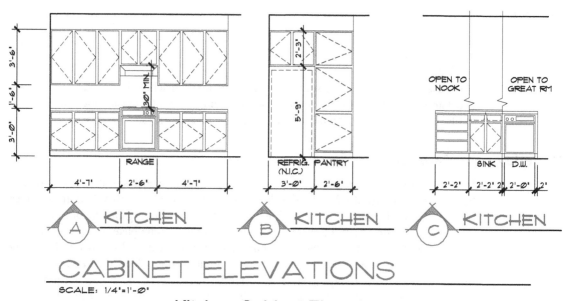

Kitchen Cabinet Elevations

Figure 12.33 Correlating cabinet elevations with the floor plan. The A arrow points to a specific group of cabinets on the floor plan and the same letter is displayed below the cabinet elevations for the group of cabinets. (*Previously published in Modern Residential Construction Practices, Routledge, 2017 and reproduced here with permission.*)

Particleboard Underlayment

Particleboard is a nonstructural hard, smooth material made in rigid sheets or panels from compressed wood chips and resin. Particleboard is commonly used as an underlayment for **carpet,** and **resilient flooring.** Particleboard should be avoided in areas where moisture can occur. Particleboard edges can swell and seams separate when moisture is present. Carpet is a finish floor material made from woven fabric, such as nylon or wool. Carpet is popular in living areas and bedrooms, because it is soft, quiet, and warm. Resilient flooring, also called vinyl flooring, is either sheet or tile vinyl material that is firm and flexible. Resilient flooring is a good economical hard surface that is commonly used in kitchens and utility rooms.

Oriented Strand Board Underlayment

Oriented strand board (OSB) looks similar to particleboard, but OSB is a structural underlayment that is fire and impact resistant, and it helps control temperatures and sound. OSB is also moisture resistant but not waterproof. OSB has advantages over particleboard for underlayment, but it is more expensive than particleboard.

Plywood Underlayment

Plywood is an excellent underlayment material for use under resilient flooring and carpet. Exterior plywood can be used in areas where moisture can be present. Plywood is not usually necessary under carpet unless moisture is expected.

Foam Underlayment

Foam underlayment is commonly used under wood and laminate floors. **Wood flooring** is any finish flooring product manufactured from wood. Wood flooring provides a natural beauty, durability, and restorability. **Laminate flooring,** also called floating wood floor, is a multilayer synthetic flooring product fused together with a lamination process. Laminate flooring simulates wood, or other natural materials, with a photographic layer under a clear protective layer. Standard foam is thin foam placed under the wood or laminate finish flooring, and is the least expensive underlayment for this application. Standard foam underlayment should be avoided in areas where moisture can be present.

Combination film and combustion foam is an underlayment with a sheet of moisture barrier attached and is used on projects where moisture can rise from concrete.

Cork Underlayment

Cork underlayment, also referred to as acoustic underlayment, is used under wood and laminate floors when noise reduction is desired between floors. Acoustic underlayment can be cork, recycled rubber, or a combination of cork and recycled rubber and foam.

Cement Board Underlayment

Cement board, also called **backer board,** is a combination of cement and reinforcing fibers formed into 1/4 to 1/2 in. thick 4 × 8 ft (1200 × 2400 mm) sheets, or 3 × 5 ft sheets (900 × 1500 mm), that are typically used as a tile underlayment. Cement board can be nailed or screwed to wood or steel joists to create an underlayment for tile floors, or fastened to studs for vertical tile applications.

Advanced technology underlayment for tile and stone is available in rolls for lightweight shipping and handling. A product such as Schluter-DITRA provides underlayment support and load distribution for tile flooring, with waterproofing and vapor control. This is a polyethylene product with a grid structure of square cavities, and an anchoring fleece laminated to the base, and installed over a layer of **thin-set mortar.** Thin-set mortar, also called thin-set or thin-bed mortar, is used to adhere tile to the floor or wall, with a very thin layer of cement often containing other additives, such as acrylic for strength. Mortar is a mixture of sand, cement, and water, and can contain lime. The Schluter-DITRA anchoring fleece provides a mechanical bond to the thin-set. Tile is installed over the product using another thin-set layer that becomes mechanically anchored in the square cavities of the underlayment.

Moisture Barriers Used with Underlayment

A **moisture barrier,** also called a vapor barrier, is waterproof material used to protect the finish flooring from moisture coming up from the subfloor. A moisture barrier helps prevent moisture from transferring to the flooring, can reduce possible moisture-related mold and mildew growth or even block gases that can be a health risk. Installing a moisture barrier over underlayment can be required by the flooring manufacturer. A variety of moisture barrier materials are available, including common 6-mil polyethylene sheeting. Check the specification of the moisture barrier products to find the material that best fit your requirements. Moisture barrier products are generally nontoxic plastic sheeting or recycled plastic sheeting.

Carpet Pad

Carpet pad is not an underlayment that fits the descriptions previously given, but it is described here because it is installed under carpet as the finish floor material. The carpet pad is installed between the underlayment and the carpet. Selecting the proper carpet pad for the carpet being used is important to provide desired cushion, comfort, and prolong carpet life. The following briefly describes the types of carpet pad commonly used. **Rebond carpet pad** is made from high-density foam that is bonded together to make a variety of densities and thicknesses. **Frothed foam carpet pad** is made of densely packed urethane, and designed for use under any carpet, and wears under heavy traffic and weight. **Foam carpet pad** is made from urethane that is not designed for use in high traffic areas. **Waffle rubber carpet pad** is rubber material pressed into a waffle shape that provides a soft cushion that is not designed for use in high traffic areas. **Slab rubber carpet pad** is a single thickness of rubber material that stands up under pressure for a long period of time, and is used in high traffic areas. **Fiber carpet pad** is made from a nylon that provides a high-density product.

Installing Underlayment

Most underlayment such as particleboard, OSB, and plywood are manufactured in 4×8 ft (1200×2400 mm) sheets that are commonly installed over the same size plywood or OSB subfloor. If this is the case, you should stagger the underlayment covering the seams of the subfloor, and also stagger underlayment end joints at the center of adjacent sheets. The underlayment should be glued to the subfloor and use ring-shank nails or wood screws to fasten the underlayment to the subfloor. Nail or screw spacing should be 6 to 8 in. (150 to 200 mm) at the edges and 10 to 12 in. (250 to 300 mm) in the field. Nail or screw heads must be flush with or recessed below the surface of the underlayment. Always confirm underlayment installation with finish flooring manufacturer instructions, because different finish floor materials require alternate applications. Advanced technology products typically have very specific installations instructions and requirements.

EXTERIOR SIDING

Siding is described here because it relates to the finish material used over wood-frame wall construction. Siding is material used to surface the exterior of a building to protect against exposure to the elements, prevent heat loss, and provide a desired appearance. Siding is commonly installed near the end of the construction process as one of the last applications to complete the building project. Siding installation is normally done by a subcontractor who specializes in professional siding installation.

Sheathing, described in Chapter 8, *Wood Construction*, is fastened to the exterior wall studs to reinforce, support, and strengthen the structure and provide a backing for finish materials. Sheathing is commonly C-D grade exterior plywood or OSB with a **vapor barrier** placed over and then covered with a finish **siding** material. Siding is an exterior covering of finish material. Vapor barriers are a layer of material used between the framing and siding to help prevent moisture from transferring from the outside through the exterior walls, can reduce possible moisture-related mold and mildew growth or even block gases that can be a health risk. Housewrap, described earlier, is lightweight and manufactured in wide rolls or other materials for easy use in construction. Housewrap provides a weather-resistant barrier that prevents rain from getting into the wall, while allowing water vapor to pass to the exterior. A few of the many brand names for housewrap are Tyvek', Typar', WEATHERMATE™, Grip-Rite, HomeGuard, or others available by searching the Internet for housewrap. Siding can be wood, metal, plastic, composite materials, stucco, or masonry. The type of siding can be used to define or enhance the architectural style. Siding is selected for its architectural style, appearance, and durability. A few of the large variety of siding materials are shown in **Figure 12.34** and described in the following:

- **Bevel siding** is bevel-shaped wood, metal, or plastic material applied with each horizontal row overlapping the previous row. Figure 12.34*a* shows horizontal bevel wood siding used on a home. Figure 12.34*b* shows metal horizontal bevel siding being installed on a home. The metal siding is durable, requires less maintenance than wood siding, and often has an insulation backing.

- **Lap siding** has a notch on one side that allows each piece to lap over the previous, as shown in Figure 12.34*c*.

- **Board and bat** siding is installed with wide vertical boards placed next to each other with a small narrow board fastened over each joint between the large boards, providing a rustic appearance.

- **Shingle siding** is installed using wood shingles, which are thin, tapered pieces of wood placed side by side in rows with each row overlapping the row below, as shown in Figure 12.34*d*. Shingle siding is often used on coastal or rustic architecture.

- **Stucco siding** is composed of cement, sand, and lime, and is applied while wet. The stucco dries to a hard durable material for exterior and interior walls, and is commonly used on Southwest architecture and other dry climate areas, as shown in Figure 12.34*e*.

- **Masonry** is one of the most durable, long-lasting, and maintenance-free sidings available. The material is referred to as masonry units, which are laid next to each other and bound together by mortar. Common masonry materials for masonry construction are brick, stone, marble, granite, travertine, limestone, cast stone, concrete block, glass block, and tile. Figure 12.34*f* shows a home constructed using a combination of brick and stone masonry. Masonry wall construction is described in detail later in Chapter 5, *Masonry Construction*.

Figure 12.34 Siding materials. (*a*) Horizontal bevel siding. (©*Elenathewise—Adobe Stock.*) (*b*) Metal horizontal bevel siding being installed. (©*Wendy Kaveney—Adobe Stock.*) (*c*) Lap siding. (©*smailik—Adobe Stock.*) (*d*) Shingle siding. (©*2tun—Adobe Stock.*) (*e*) Stucco siding. (©*Jeffrey Banke—Adobe Stock.*) (*f*) A combination of brick and stone masonry. (©*PhotoSerg—Adobe Stock.*)

Go to Downloads & Resources tab at **www.mhprofessional.com/Commercial Building Construction-Ancillaries** to access the images provided in this chapter and correlate with the content of this chapter. Here you can look at the chapter figures on your computer screen to pan and zoom in and out for better observation, because full drawings reduced to fit on a textbook page are often difficult to read. The image bank also includes the complete set of plans for the Brookings South Main Fire Station used as examples throughout this textbook.

Go to **Downloads & Resources tab at www.mhprofessional.com/Commercial Building Construction-Ancillaries** to access the test and creative thinking problems for this chapter. The chapter test can be used for review or to evaluate content knowledge depending on your course objectives. Creative thinking problems are provided to help expand your knowledge by researching given subjects.

Print reading problems have questions that ask you to find information on the Brookings South Main Fire Station plans. The print reading problems can help you become familiar with the format of construction documents and reinforce your ability to seek specific information. This is an important skill for you to master as you prepare to enter the construction industry. Go to **Downloads & Resources tab at www.mhprofessional.com/Commercial Building Construction-Ancillaries** to access the complete set of plans for the Brookings South Main Fire Station.

CHAPTER 13

Mechanical, Plumbing, and Electrical Systems

INTRODUCTION

This chapter introduces you to three construction phases that include the **mechanical system,** the **plumbing system,** and the **electrical system.** The mechanical system provides heat, fresh or **conditioned air,** and ventilation. Conditioned air refers to a heated or cooled space within the home. Controlling air quality was introduced in Chapter 10, Insulation and Barriers, Indoor Air Quality and Safety where the mechanical systems needed for improving air quality was described. The plumbing system includes all of the pipes, tanks, fittings, and fixtures required for the water supply, water heating, and sanitation. The electric system is a network of wiring, outlets, and fixtures used to transmit and supply the electrical needs. The project is **dried-in** before the mechanical, plumbing, and electrical systems can be installed, because the building needs to be dry and not subject to any further wetness from the weather. The term **dried-in** was described in Chapter 7, *Roof Construction and Materials.* The only related construction phase done before dry-in is the installation of mechanical equipment chimneys and vents, and plumbing roof vents that are part of the roofing process. Mechanical equipment chimneys and vents and plumbing vents are describe later in this chapter.

The **National Energy Conservation Code** regulates the design and construction of the **exterior envelope** and selection of heating, ventilating, and air-conditioning (HVAC); service water heating; electrical distribution and illuminating systems; and equipment required for effective use of energy in buildings for human occupancy. The exterior envelope is made up of elements of a building that enclose conditioned spaces through which thermal energy transfers to or from the exterior.

INTRODUCTION TO MECHANICAL SYSTEMS

Heating, ventilating, and air-conditioning (HVAC) is the terminology used to refer to the industry that deals with the heating and air-conditioning equipment and systems found in a building. The HVAC system is used to regulate temperature and filter the air in a building. This includes the furnace, air-conditioning equipment, and **duct systems,** which ensure the uniform transfer of the cold, hot, and filtered air throughout the building. The duct system is square, rectangular, and round sheet metal or plastic pipes used to conduct hot or cold air of the HVAC system.

Central Forced-Air Systems

Central forced air is one of the most common systems for heating and air-conditioning by circulating the air to and from occupied spaces through or around heating or cooling devices. A fan forces the air into **ducts,** which connect to openings called **air supply registers.** An air supply register is a grill with moving parts that can be opened, closed, and can direct the air flow. The placement and size of registers is important for HVAC efficiency. Warm air or cold air passes through the ducts and registers to enter the rooms and provides heating or cooling as needed. Air then flows from the room through another opening into the **return duct,** or return air register. The return air register and return duct directs the air from the rooms over the heating or cooling device. If warm air is required, the return air is passed over the surface of either a combustion chamber, which is the part of a furnace where fuel is burned, or a heating coil. If cool air is required, the return air passes over the surface of a cooling coil. Finally, the conditioned air is picked up again by the fan and the air cycle is repeated. Figure 13.1 shows the air cycle in a central forced-air system. Conditioned air is air that has been heated, cooled, humidified, or dehumidified to maintain an interior space within the desired **comfort zone.** The comfort zone is the range of temperatures, humidity, and air velocities at which people generally feel comfortable.

Heating Cycle

If the air cycle previously described is used for heating, the heat is generated in a **furnace.** Furnaces for heating produce heat by burning fuel oil or natural gas, or by using electric heating coils or heat pumps. If the heat comes from burning fuel oil or natural gas, the

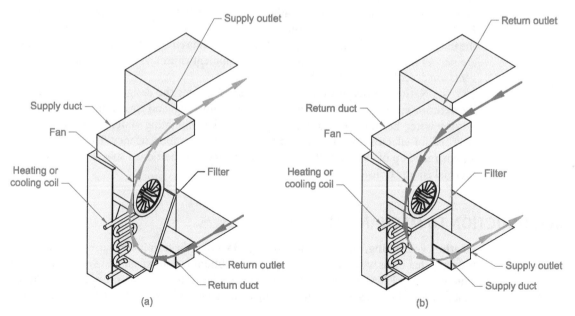

Figure 13.1   The air cycle in a central forced-air system. (*Previously published in Modern Residential Construction Practices, Routledge, 2017 and reproduced here with permission.*)

combustion takes place inside a combustion chamber. The air to be heated does not enter the combustion chamber but absorbs heat from the outer surface of the chamber. The gases given off by combustion are vented out through the roof in a vent pipe or chimney. In an electric furnace, the air to be heated is passed directly over the heating coils. This type of furnace does not require a vent pipe or chimney.

Cooling Cycle

If the air from the room is to be cooled, it is passed over a cooling coil, which is a **refrigeration system.** The principal parts of a refrigeration system are the cooling coil called an **evaporator,** the **compressor,** the **condenser,** and the **expansion valve.** Figure 13.2 shows a diagram of the cooling cycle. The cooling cycle operates when warm air from the ducts is passed over the evaporator. As the cold liquid refrigerant moves through the evaporator coil, it picks up heat from the warm air. As the liquid picks up heat, it converts into vapor. The heated refrigerant vapor is then drawn into the compressor, where it is put under high pressure. This pressure raises the temperature of the vapor even more. The high-temperature, high-pressure vapor passes to the condenser, where the heat is removed. This is done by blowing air over the coils of the condenser. The high-pressure vapor converts into liquid as the condenser removes heat. From the condenser, the refrigerant flows to the expansion valve. As the liquid refrigerant passes through the valve, the pressure is reduced, lowering the temperature of the liquid still further, so it picks up more heat. The cold low-pressure liquid then moves to the evaporator. The pressure in the evaporator is low enough to allow the refrigerant to boil again and absorb more heat from the air passing over the coil of the evaporator. The evaporator is a device in which a liquid refrigerant is vaporized. The compressor maintains adequate pressure to cause refrigerant to condense and flow in sufficient quantities to meet the cooling requirements of the system. The condenser is the portion of a refrigeration system where the compression and condensation of refrigerant is accomplished. The expansion valve controls the amount of refrigerant flow into the evaporator.

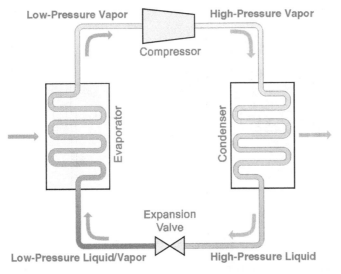

Figure 13.2   The cooling cycle of a refrigeration system. (©*mrhighsky—Adobe Stock.*)

Forced-Air Heating Plans

Complete plans for the heating system are normally needed when applying for a building permit or a mortgage loan, depending on the requirements of the local building department or the lending agency. Plans for the HVAC system show the size and location of all equipment, ductwork, and components with accurate symbols, specifications, notes, and schedules that form the basis of contract requirements for construction. **Specifications** are documents that accompany the drawings and contain all written information related to the HVAC system. When forced-air electric, gas, or oil heating systems are used, the warm-air outlets and return air locations can be seen on the plan, as shown in Figure 13.3. Notice the warm air registers are normally placed in front of a window so warm air is circulated next to the coldest part of the room. As the warm air rises, a circulation action is created as the air goes through the complete heating cycle. Cold-air returns are often placed in the ceiling or floor at a central location.

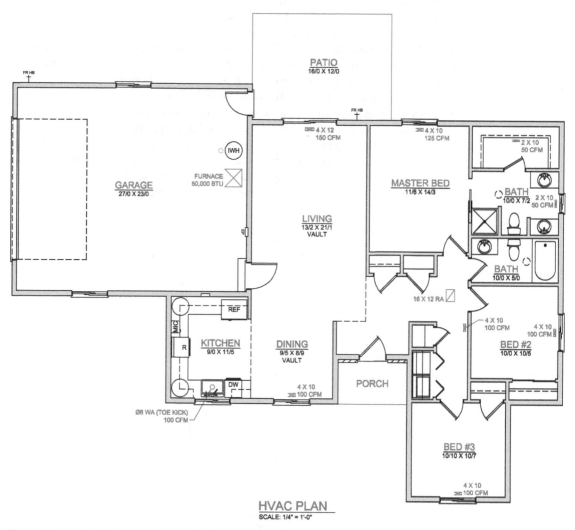

Figure 13.3 A simplified HVAC plan with the size and air flow shown by each warm air and return air symbol on the floor plan. (*Courtesy Engineering Drafting & Design, Inc.*)

A complete forced-air heating plan shows the size, location, and number of **BTU** dispersed to the rooms from the warm-air supplies. Btu stands for British thermal unit, which is a measure of heat. The location and size of the cold-air return and the location, type, and output of the furnace are also shown. Warm-air registers are sized in inches such as 4 × 12, or millimeters. The size of the duct is also given, as shown in Figure 13.4. The note 20/24 on a duct means a 20 × 24 in. size register. A number 8 next to a duct means an 8 in. diameter duct. Millimeters are used for metric values. This same system can be used as a central cooling system when cool air is forced from an air conditioner through the ducts and into the rooms. **WA** means warm air, and **RA** is return air. **CFM** is cubic feet per minute, which is the rate of airflow.

HVAC Schedules   Numbered symbols can be used on the HVAC plan to key specific items to schedules. HVAC schedules describe items such as ceiling outlets, supply and exhaust grills, hardware, and equipment. HVAC schedules include size, description, quantity used,

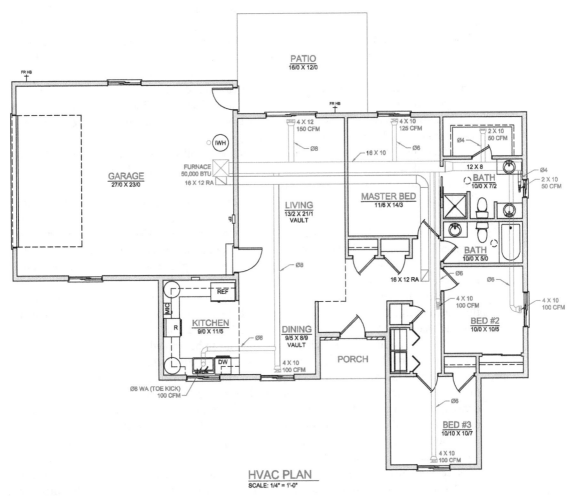

Figure 13.4   A detailed HVAC plan with the size and air flow shown by each warm air and return air symbol, and duct runs with size on the floor plan. (*Courtesy Engineering Drafting & Design, Inc.*)

capacity, location, manufacturer specifications, and any other information needed to construct or finish the system. Schedules keep the plans clear of unnecessary notes. Schedules are generally placed in any convenient area of the drawing field or on a separate sheet. Items on the plan can be keyed to schedules by using a letter and number combination, such as C-1 for CEILING OUTLET NO. 1, E-1 for EXHAUST GRILL NO. 1, or ACU-1 for EQUIPMENT UNIT NO. 1.

Providing Duct Space

Ducts are placed in a crawl space, attic, or above a drop ceiling when possible. When ducts cannot be confined to a crawl space, attic, or above a drop ceiling they must be run inside the occupied areas. Ducts are typically hidden when they must be placed in locations where they can be seen. There are several ways to conceal ducts. When ducts run parallel to structural members, the ducts can be placed within the space created by the structural members. This is referred to as running the ducts in the joist space, rafter space, or stud space. These terms identify the type of construction members used to provide duct space. This works when the duct size is equal to or smaller than the size of the space between construction members. In the case of return air ducts, the construction members and enclosing materials can be used as the duct **plenum.** A plenum is a chamber that can serve as a distribution area for heating or cooling systems, generally between a false ceiling and the actual ceiling, the crawl space area, or between construction members. When a duct can be in a space created between construction members, no extra framing needs to be done to conceal the ducts. When ducts must be exposed to occupied areas, they are normally enclosed in a **chase.** A chase is a continuous recessed or expanded area built to carry or conceal ducts, pipes, or other construction products or utilities. Ducts can be left exposed when run in an area such as an unfinished basement, or for convenience or architectural design as in the ceiling of a restaurant or warehouse.

When possible, ducts are run in the ceiling of a hallway, because they can be framed and finished with a lowered ceiling. The minimum ceiling height in hallways can be 7 ft (2134 mm). Bathrooms and kitchens can also have a minimum ceiling height of 7 ft (2134 mm), though this is normally undesirable. The typical ceiling height in habitable rooms is approximately 8 ft (2440 mm), but a ceiling can be as low as 7'-6" (2286 mm) for at least 50 percent of the room, with no portion less than 7 ft (2134 mm). This information is valuable, because it tells you how low ceilings can be framed down to provide space for ducts when needed. It is normally preferred to frame higher walls to accommodate space for ductwork when necessary.

When ducts run vertically between floors, they need to be in an easily concealed location, such as in a closet or in the stud space. The stud space can be used for ducts that are 3-1/2 in. deep for 2 × 4 studs, 5-1/2 in. deep for 2 × 6 studs, or 7-1/2 in. deep for 2 × 8 studs. If the duct can be run up through a closet, then it can be framed in and covered in a chase. If the duct cannot be located in a convenient place for concealment, then it might need to be framed into the corner of a room; although, this is normally not preferred. The framing for a chase in Figure 13.5 is shown on the floor plan as a wall surrounding the duct to be concealed, and it is located in a portion of a bedroom wardrobe closet. A note is usually provided that indicates the use, such as CHASE FOR 24 × 24 RETURN DUCT.

Commercial and industrial construction projects often place ducts in the groundwork during excavation, as shown in Figure 13.6. This photo shows excavation with two ducts in place. Look near the center of the photo where you can see concrete poured in forms to build a reinforced concrete collar around the ducts to hold them in place. The forms are removed after the concrete sets. The excavation is then carefully backfilled with clean compacted soil or gravel based on engineering specifications.

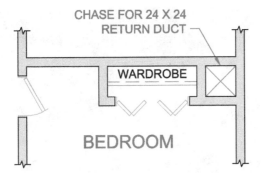

CHASE FOR 24 X 24
RETURN DUCT

WARDROBE

BEDROOM

Figure 13.5 The framing for a chase is shown on the floor plan as a wall surrounding the duct to be concealed, and is located in a portion of a bedroom wardrobe closet. (*Previously published in Modern Residential Construction Practices, Routledge, 2017 and reproduced here with permission.*)

Figure 13.6 Commercial and industrial construction projects often place ducts in the groundwork during excavation. (*©blinow61—Can Stock Photo Inc.*)

Ductwork in commercial and industrial buildings are often left exposed for convenient access and economy, as shown in Figure 13.7. There is no need to conceal ducts in this environment, because they function well when exposed and there is no advantage to covering the ducts. Ductwork is even left exposed in some commercial buildings such as museums, showrooms, restaurants, factories, and warehouses for convenience and as an architectural design. This type of industrial architecture is commonly used when an aesthetic or decorative statement is intended or easy access is desired. This type of industrial architectural influence often leaves the steel and wood trusses or beams exposed for the same effect. The ductwork might even be painted a bright color to provide additional emphasis.

Figure 13.7 Ductwork in commercial and industrial buildings are often left exposed. (*©edu1971— Can Stock Photo Inc.*)

HOT WATER HEATING SYSTEMS

In a **hot water system,** the water is heated in an oil- or gas-fired **boiler** and then circulated through pipes to radiators or convectors in the rooms. A boiler is a fuel-burning container for heating water. The boiler is supplied with water from the fresh water supply. The water is circulated around the combustion chamber, where it absorbs heat. A **one pipe hot water system** has one pipe that leaves the boiler and runs through the rooms of the building and back to the boiler. In the one-pipe system, the heated water leaves the supply, is circulated through the outlet, and is returned to the same pipe. A **two-pipe hot water system** has two pipes running throughout the home. One pipe supplies heated water to all of the outlets. The other is a return pipe, which carries the water back to the boiler for reheating, as shown in Figure 13.8.

Hot water systems use a pump, called a **circulator,** to move the water through the system. The water is kept at a temperature between 150 and 180°F (65.5 to 82°C) in the boiler. When heat is needed, the **thermostat** starts the circulator, which supplies hot water to the convectors in the rooms. A thermostat is an automatic temperature control mechanism for regulating the amount of heating or cooling in a room or building area. The thermostat is described later in this chapter.

HEAT PUMP SYSTEMS

A **heat pump** is a forced-air central heating and cooling system that operates using a compressor and a circulating refrigerant system. Heat is extracted from the outside air and pumped inside the building. The heat pump supplies up to three times as much heat per

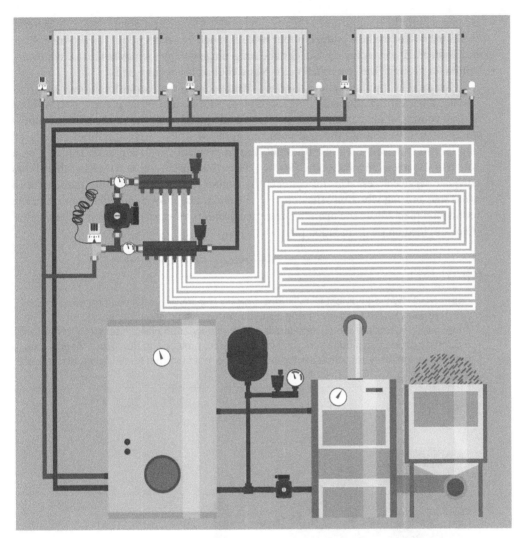

Figure 13.8   A two-pipe hot water system has two pipes running throughout the building. (©*sivvector—Adobe Stock.*)

year for the same amount of electrical consumption as a standard electric forced-air heating system. In comparison, this can result in a 30 to 50 percent annual energy saving. In the summer the cycle is reversed, and the unit operates as an air conditioner. In this mode, the heat is extracted from the inside air and pumped outside. On the cooling cycle the heat pump also acts as a dehumidifier.

Heat pumps vary in size from 2 to 5 **tons.** Some vendors carry two-stage heat pumps that alternate between 3 and 5 tons, for example. During minimal demand, the more efficient 3-ton phase is used, and the 5-ton phase is operable during peak demand. Each ton of rating removes approximately 12,000 Btu per hour (Btuh) of heat.

Large square footage buildings or complex floor plans can require a split system with two or more compressors. The advantage of a split system is that two smaller units more effectively control the needs of two zones within the building. Other features, such as an air cleaner, a humidifier, or an air freshener, can be added to the heat pump system. In general, the initial cost of the heat pump is about twice as much as a conventional forced-air electric heat system. However, the advantages and long-range energy savings make it a significant option for heating. The cost difference is less if cooling is a requirement.

The total heat pump system uses an outside compressor, an inside blower to circulate air, a backup heating coil, and a complete duct system. Heat pump systems move a large volume of air, making it important to adequately size the return air and supply ducts. Figure 13.9 shows the heat pump compressor. The compressor should be placed in a location where some noise does not cause a problem. The compressor is placed on a concrete slab about 36 × 48 × 6 in. (900 × 1200 × 150 mm) in size in a location that allows adequate service access. Do not connect the concrete slab to the building foundation to avoid transmitting vibration. The concrete slab should be shown and dimensioned on the foundation plan. Verify the manufacturer specifications for exact concrete slab dimensions if not specified on the foundation plan.

NOTE

Commercial HVAC equipment installations typically require structural engineered reinforced concrete foundations with anchoring systems and structural steel framework for support. These requirements are shown and detailed on the foundation plans. Alternately, HVAC roof mounted systems also require structural engineering to provide additional roof-member support and fastening systems that can be found on the roof plan, roof framing plan, and related details.

Figure 13.9　A heat pump compressor. (©*GalinaSt—Adobe Stock.*)

Heat pumps may not be as efficient in some areas of the country as in other areas because of annual low or high temperatures. Verify the product efficiency with local heating and cooling contractors.

The heat pump system described previously is an air-to-air system. A heat pump system can also be a water system where heat or cooling is extracted from a water source such as a water well, pond, or other water source. A heat pump system can also be a geothermal system where heating or cooling is extracted from the ground. Geothermal and water systems are describe later in this chapter.

Ductless Heat Pump Systems

Ductless heat pumps, also called **mini-split-system heat pumps,** are heat pump systems that heat or cool directly into the rooms without the use of ductwork. Ductless heat pumps are effectively used in energy-efficient new buildings that require only a small space conditioning system, and for remodeling additions where extending or installing ductwork is not practical. Multiple ductless systems can be used for larger spaces. Ductless heat pump systems have the two main components found in the standard heat pump system that includes one or more outdoor compressor and condenser units and one or more indoor air-handling units. The outdoor unit is much smaller than a standard heat pump and is mounted on a smaller concrete pad. The indoor units are small and can be recessed or suspended from a ceiling or wall. A conduit containing the power cable, refrigerant tubing, suction tubing, and a condensate drain is connected between the outdoor and indoor units. Ductless heat pumps provide small size and flexibility for **zone heating** and cooling individual rooms. Zone heating is described in detail in the next section. Most ductless heat pump systems allow for up to four indoor air-handling units that can be designed for four zones or rooms connected to one outdoor unit. A separate thermostat is used to control the heating and cooling needs of each zone. Ductless heat pumps have no ducts, so they avoid the energy losses associated with the ductwork of central forced-air systems.

ZONE CONTROL SYSTEMS

A **zone heating system** requires one heater and one thermostat per room or area. No ductwork is required, and only the heaters in occupied rooms need to be used as desired.

One of the major differences between a zone and a central system is flexibility. A zone heating system allows the occupant to determine how many rooms are heated, how much energy is used, and how much money is spent on heat. A zone system allows the heating to occupant needs, while a central system requires using all the heat produced. If the airflow is restricted, the efficiency of the central system is reduced. There is also a 10 to 15 percent heat loss through ductwork in central systems.

Regardless of building square footage, occupants normally use less than 40 percent of the entire area on a regular basis. A zone system is very adaptable to heating the 40 percent that is occupied. Automatic controls allow for night and day settings, and unheated areas controlled as needed. As much as 60 percent on energy costs can be saved through controlled zone heating systems.

There are typically two types of zone heaters: baseboard and fan. **Baseboard heaters** are placed along the bottom of a wall. They are used in many different climates and under various operating conditions. No ducts, motors, or fans are required. Baseboard heaters have an electric heating element, which causes a **convection current** as the air around the

Figure 13.10 A wall-mounted fan heater. (©*magraphics—Adobe Stock.*)

unit is heated. A convection current is a natural condition where warm air rises and cool air falls. The heated air rises into the room and is replaced by cooler air that falls to the floor. Baseboard heaters should be placed on exterior walls under or next to windows or other openings. These units project a few inches into the room at floor level. Furniture arrangements should be a factor in locating any heating element, because furniture must be kept at a safe distance from the heater.

The recessed wall-mounted **fan heater** in Figure 13.10 is a zone heating option when the baseboard heater is undesirable. A fan heater has a heating element used to generate heat, and a fan circulates the heat into the room. These units should be placed to circulate the warmed air in each room adequately. Avoid placing the heaters on exterior walls, where a recessed unit reduces or eliminates insulation. The fan in these units causes some noise.

Split systems using zone heat in part of the building and central heat in the rest is an option. For example, placing a zone heater in a bathroom provides extra heat after a bath or shower or on a cold day when more heat is desired.

RADIANT HEAT

Radiant heat is heat that radiates from an electric or hot water element, warming objects rather than the air. Radiant elements are installed in the floor, walls, or ceiling. The elements heat the surrounding surface and radiates heat out into the rooms. Radiant heating and cooling systems provide a comfortable environment by controlling surface temperatures and minimizing excessive air motion within the space. Radiant heat systems can provide operating cost savings of 20 to 50 percent annually, compared with forced air systems. This saving is accomplished through lower thermostat settings. Three to four percent of the energy is saved for each degree the thermostat is lowered. Users of surface-mounted radiant panels take advantage of this fact in two ways. Daytime radiant heat comfort can be achieved at 60 to 64°F (16.6°C) as compared with forced air heating temperatures of 68 to 72°F (21°C). Night temperature can be reduced to 58°F (14.4°C) in areas used frequently, 55°F (12.8°C) in areas used occasionally, and 50°F (10°C) in areas used seldom.

THERMOSTATS

A **thermostat** is a mechanism for automatically controlling the amount of heating or cooling given by a central or zone heating or cooling system. The thermostat floor plan symbol and a typical thermostat is shown in Figure 13.11.

The thermostat location is important to the proper function of the system. For zone heating or cooling units, thermostats can be placed in each room, or a central thermostat panel that controls each room can be placed in a convenient location. For central heating and cooling systems, there can be one or more thermostats, depending on the layout of the HVAC system. For example, a very large building can have a split system that divides the building into two or more zones, with a thermostat for each zone.

Several factors contribute to the effective placement of the thermostat for a central forced-air system. A good common location is near the center of the building and close to a return air duct. The air entering the return air duct is usually temperate, thus causing little variation in temperature on the thermostat. A key to successful thermostat placement is to find a stable location where an average temperature reading can be achieved. There should be no drafts that adversely affect temperature settings. The thermostat should not be placed in a location where sunlight or a heat register causes an unreliable reading. The thermostat should not be placed close to an exterior door, where temperatures can change quickly when the door is opened and closed. Thermostats should be placed on inside partitions rather than on outside walls, where a false temperature reading can also be possible. Avoid placing

Figure 13.11   The thermostat floor plan symbol and a typical thermostat. (*Photo ©lucadp—Adobe Stock.*)

the thermostat near stairs or a similar traffic area, where bouncing or shaking can cause the mechanism to alter the actual reading. Energy consumption can be reduced by controlling thermostats in individual rooms. Central panels are available that make it easy to lower or raise temperatures in any room where each panel switch controls one remote thermostat.

Programmable microcomputer thermostats effectively help reduce the cost of heating or cooling, and automatically switch from heating to cooling as needed. These computers can be used to alter heating and cooling temperature settings automatically for different hours of the day, days of the week or different months of the year.

NOTE

A complete set of architectural drawings are provided in the image library found on the website for this textbook. Go to Downloads & Resources tab at **www.mhprofessional.com/ CommercialBuildingConstruction-Ancillaries** to access the complete set of plans for the Brookings South Main Fire Station. This set of plans includes the following drawings directly related to HVAC systems described in this chapter:

First Floor HVAC Ventilation Plan, Sheet M 202

First Floor HVAC Piping Plan, Sheet M 203

Mechanical Details, M 300 and M 301

Mechanical Schedules, M 400 and M 401

NOTE

HVAC is a specialty trade in the construction industry that requires postsecondary education in an HVAC program or on-the-job training with an HVAC contractor or both. Students learn how to read prints, HVAC equipment operation and maintenance, temperature control applications. Training also includes sheet metal layout and fabrication, along with completing HVAC installations. After completing school, candidates receive apprenticeship training through unions such as the Air Conditioning Contractors of America or Plumbing-Heating-Cooling Contractors Association. The apprenticeship is an alternative to a baccalaureate college education and generally takes 3 to 5 years to complete. Apprentices learn all aspects of the HVAC industry through on-the-job training and classroom instruction.

INTRODUCTION TO PLUMBING SYSTEMS

Building construction piping is called **plumbing.** A plumbing contractor, called a **plumber,** installs the plumbing system in two phases, which is the **rough-in** phase and the **finish** phase. Rough-in is the installation of the plumbing system that includes everything except putting **fixtures** in place. Finish plumbing is one of the last construction phases before the building is ready to occupy and includes installing all fixtures such as sinks, **vanities,** toilets, faucets, and final **valves.** A vanity is a bathroom lavatory fixture that is freestanding or in a cabinet. A valve is a fitting used to control the flow of fluid or gas.

The **plumbing system** includes all of the pipes, tanks, **fittings,** and **fixtures** required for the water supply, water heating, and sanitation in a building. The plumbing system provides these elements:

- **Water supply risers** and other associated pipes
- Fixtures and **fixture traps**

- **Soil pipe, soil stack, waste pipe, waste stack,** and **vent pipes**
- Drain and sewer
- Stormwater drainage

The pipe used in plumbing is made of copper, plastic, galvanized steel, or cast iron. A fitting is a standard pipe part such as a coupling, elbow, reducer, tee, and union used for joining two or more sections of pipe together. A fixture is a component used to supply and contain water, and discharge waste. Examples of fixtures are sinks, lavatories, showers, tubs, and water closets. A fixture trap is a U-shaped pipe below plumbing fixtures that holds water to prevent odor and sewer gas from entering the fixture. The water supply provides **potable water.** Potable water is drinking water that is free from impurities. A water supply riser is a pipe that extends vertically one story or more to carry water to fixtures. A soil pipe is a pipe that carries the discharge of water closets or other similar fixtures. A soil stack is a vertical pipe that extends one or more floors and carries discharge of water closets and other similar fixtures. A waste pipe is a pipe that carries only liquid waste free of fecal material. A waste stack is a vertical pipe that runs one or more floors and carries the discharge of fixtures other than water closets and similar fixtures. A vent pipe is the pipe installed to ventilate the building drainage system and to prevent drawing liquid out of traps and stopping **back pressure.** Back pressure is a resistant force applied to liquid or gas against the desired direction or flow in a pipe.

Copper pipes have **soldered** joints and fittings and are used for carrying hot or cold water. Solder is a low-melting alloy, especially one based on lead and tin, used for joining less fusible metals such as copper. Plastic pipes can have glued joints and fittings and are used for vents and for carrying fresh water or solid waste. Many contractors are replacing copper pipe with plastic piping for both hot and cold water. One example is a plastic pipe with the chemical name of **polybutylene (PB),** also known as poly pipe. Plastic **polyvinyl chloride (PVC)** pipe has been used effectively for cold water installations. Corrosion-resistant plastic piping is available in a thermoplastic with the chemical name **postchlorinated polyvinyl chloride (CPVC).** CPVC pipe has insulation value that retains heat and saves energy, when compared with metal pipe that loses heat. CPVC piping lasts longer than copper pipe because it is corrosion resistant, it maintains water purity even under severe conditions, and it does not cause condensation, as does copper. Plastic pipe is considered quieter than copper pipe, and it costs less to buy and install. Another plastic plumbing product, **cross-linked polyethylene (PEX)** pipe, provides a flexible and durable plumbing system that can withstand high pressures and temperatures. PEX is also freeze resistant because it expands and contracts to its original shape as needed to protect against freeze damage. Special fittings and tools are used for making connections when installing PEX product.

Steel pipe is used for large-distribution water piping and for natural gas installations. Steel pipe is joined by threaded joints and fittings or grooved joints. The steel pipe used for water is galvanized. **Galvanized pipe** is steel pipe that has been cleaned and dipped in a bath of molten zinc. The steel pipe used for natural gas applications is protected with a coat of varnish. This pipe is commonly referred to as **black pipe** because of its color. Steel pipe is strong, rugged, and fairly inexpensive. However, it is more expensive than plastic pipe, and labor costs for installation are generally higher because of the threaded joints. **Corrugated stainless steel tubing (CSST)** is also used for natural gas piping. CSST is a flexible piping system that is easier and less expensive to install than black pipe. This type of pipe comes in rolls that allow the plumbing contractor to easily run pipe through walls and under floors. The flexible piping system is easy to cut with traditional pipe cutters and has easy-to-assemble fittings and fixtures.

Cast-iron pipe is commonly used to carry solid and liquid waste as the sewer pipe that connects with a local or regional sewer system. Cast-iron pipe can also be used for the drain system throughout the building to help reduce water flow noise in the pipes. Cast-iron pipe is more expensive than plastic pipe but can be worth the price if a quiet plumbing system is desired.

Residential plans generally do not require a complete plumbing plan, but commercial and industrial buildings normally do require plumbing plans. The need for a complete plumbing plan should be verified with the local building code. In most cases, the plumbing requirements can be clearly provided on the floor plan in the form of symbols for fixtures and notes for specific applications or conditions. The plumbing fixtures are displayed in their proper locations on the floor plans. Floor **drains**, vent pipes, and sewer or water connections can be displayed on the floor plans. Floor drains are shown in their approximate location, with a note identifying size, type, and slope to drain. Sewer and water service lines can be located on the site plan in relationship to the position in which these utilities enter the building. A drain is any pipe that carries wastewater in a building drainage system. A drain pipe typically requires a **cleanout** be placed specific distances apart. A cleanout is a fitting with a removable plug that is placed in plumbing drainage pipe lines to allow access for cleaning out the pipe.

A very detailed plumbing layout is not normally provided, because the plumbing contractor is required to install plumbing of a quality and in a manner that meets local code requirements. Figure 13.12 shows typical floor plan plumbing fixture symbols.

SIZING OF PLUMBING PIPE

Plumbing must be sized properly to allow fixtures to operate properly and for proper draining and venting. The size of water supply piping is based on these conditions:

- Amount of water needed
- Supply pressure
- Pipe length
- Number of stories to be supplied
- Flow pressure needed at the farthest point from the source

The size of drainage pipe is based on standards established for the type of fixture and average amount of waste that can be discharged through the fixture in a given amount of time. The size of vent pipes is based on the number of fixture components that drain into the waste portion of the **vent stack**. A vent stack is a general term referring to any vertical pipe for soil waste or vent piping.

American National Standard taper pipe threads are the standard thread used on galvanized steel pipes and pipe fittings. These threads are designed to provide pressure-tight joints or not, depending on the intended function and materials used. American pipe threads are measured by the **nominal pipe size,** which is the inside pipe diameter. For example, a 1/2 in. pipe size has an outside diameter of 0.840 in. Pipe threads are identified using a thread note such as 3/4–14NPT, where the 3/4 is the nominal inside diameter of the pipe, 14 is the number of threads per inch, and NPT stands for National Pipe Thread.

The following table lists common minimum plumbing pipe sizes. Actual plumbing pipe sizes should be confirmed with local, state, and national codes that apply to your area.

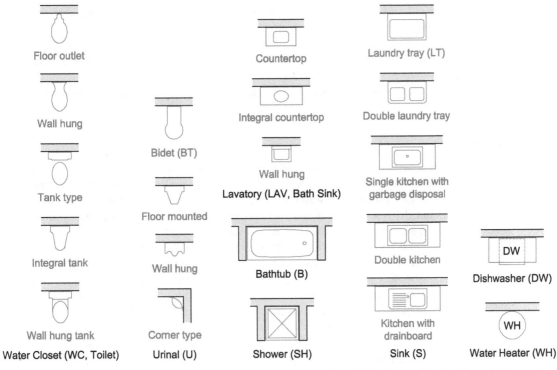

Figure 13.12 Typical floor plan plumbing fixture symbols. (*Previously published in Modern Residential Construction Practices, Routledge, 2017 and reproduced here with permission.*)

Fixture or Pipe Size	Water Size	Minimum Fixture Trap and Drain Size	Minimum Vent Size
Distributing pipe	3/4		
Bathroom group*	1/2		
Plus one or more fixtures	3/4		
Two fixtures	1/2		
Three fixtures	3/4		
Hose bibb	1/2		
Plus one or more fixtures	3/4		
Water closet	3/8	3	2
Bathtub	1/2	1-1/2	1-1/4
Shower	1/2	2	1-1/4
Lavatory	3/8	1-1/4	1-1/4
Kitchen sink	1/2	1-1/2	1-1/4
Laundry tray	1/2	1-1/2	1-1/4
Clothes washer	1/2	1-1/2	1-1/4
Dishwasher			
Values are in inches.	3/8	1-1/2	1-1/4

*The typical bathroom group consists of a lavatory, tub with shower, and toilet.

The following defines the new terms highlighted in the previous table:

A hose bibb is a faucet used to attach a hose.

A water closet water-flushing plumbing fixture, such as a toilet, that is designed to receive and discharge human excrement. This term is sometimes used to mean the compartment where the fixture is located.

A lavatory is a fixture designed for washing hands and face, usually found in a bathroom.

WATER SYSTEMS

The water supply starts at a **water meter** for public systems or at a water storage tank for private **water well** systems. A water meter is a device used to measure the amount of water that goes through the **water service.** A water well is structure in the ground created by digging, driving, boring, or drilling to access groundwater in underground **aquifers.** An aquifers is water-bearing porous soil or rock strata that yield significant amounts of water to wells. Water is pumped out of the well for distribution. The water service is a pipe from the **water main** or other supply to the **water-distribution pipes.** The water service can be 1 to 2 in. plastic or galvanized steel pipe for service to small buildings but can be much bigger for service to large commercial or industrial, or public facilities. This size can vary in relation to the service needed. A water distribution pipe carries water from the service to the point of use. The water main is a primary water supply pipe, generally located in the street or public utility easement for public water. The term **main** is used when referring to a water main or sewer main, depending on its purpose. The water service line joins a plastic or copper line within a few feet of the building. The water service supply often changes to 3/4 in. pipe, where a junction is made to distribute water to various specific locations. From the 3/4 in. lines, 1/2 in. pipe usually supplies water to specific fixtures, such as the kitchen sink. These pipe sizes are designed for the specific service requirements and can be larger for commercial, industrial, or public buildings. **Figure 13.13** shows a typical water service, connecting from the water main to the water meter and then to the building. The water meter location and water service are generally shown on the site plan. Verify local codes regarding the use of plastic pipe and the water meter location.

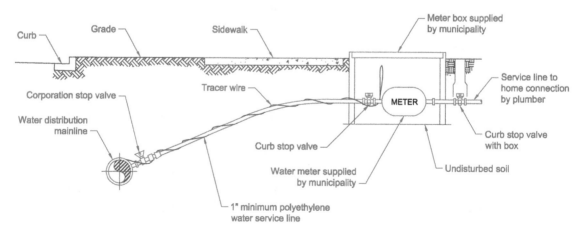

Figure 13.13 A typical water service, connecting from the water main to the water meter and then to the building. (*Previously published in Modern Residential Construction Practices, Routledge, 2017 and reproduced here with permission.*)

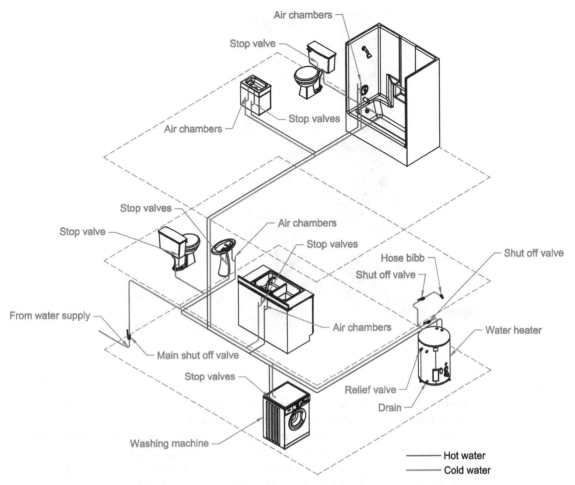

Figure 13.14 A detailed representation of a typical hot and cold water system. (*Previously published in Modern Residential Construction Practices, Routledge, 2017 and reproduced here with permission.*)

Figure 13.14 shows a detailed representation of a typical hot and cold water system. Cold water enters the building from the water service pipe and is plumbed to all fixtures where cold water is needed. The cold water supply pipe continues to the **water heater,** where water is heated and then distributed through hot water pipes to the fixtures where hot water is needed. A water heater is an appliance used for heating, storing, and distributing hot water.

Water Heaters

Water heaters are placed on a platform with an overflow tray, as shown in Figure 13.15. Water heaters with nonrigid water connections and over 4 ft (1200 mm) in height must be anchored or strapped to the building. If the water heater is located in the garage, it must be protected from impact by automobiles with a steel pipe embedded in concrete in front of the water heater. Fuel combustion water heaters cannot be installed in sleeping rooms, bathrooms, clothes closets, or closets or other confined spaces opening into a bedroom or bathroom.

Figure 13.15 Water heaters are placed on a platform with an overflow tray. (*©Alexey Stiop—Adobe Stock.*)

Confirm if local or national codes require the water heater and other mechanical equipment to be in a designated location inside the building and not in a garage.

Large commercial operations such as hotels often heat water with a gas or oil fired **boiler** located in the basement or utility area. A boiler is a fuel-burning container for heating water. The boiler generally has separate water distribution lines running to different areas of the facility, depending on the size and layout of the building. A large hotel can be divided into sections, with a separate boiler centrally located in each section. This system is more efficient than one large unit. A small hotel or motel can have a single boiler or each room can have its own demand unit described in the next section. Hot water lines running from the boiler to rooms usually lower water temperature to 140°F (60°C) using a mixing valve, while another non-mixed line runs to the commercial laundry to wash linens and other hotel items. The hot water for the laundry is usually 160°F (71°C) or more. The hot water going to each room and restrooms lavatory can be no more than 105°F (40.5°C) by code. This water temperature reduction is accomplished with a point of use mixing valve. The entire boiler hot water system uses a recirculation pump that pulls continuous hot water through a loop from the boiler and back to the boiler.

Tankless Hot Water Heater

Tankless water heaters, also called **instantaneous** or **demand water heaters,** provide hot water only when needed. A tankless water heater heats water directly without use of the storage tank found with traditional water heaters. Tankless water heaters avoid the **standby heat losses** associated with storage water heaters. Standby heat loss is the heat lost when water is heated and waiting for use. When a hot water faucet is turned on, cold water travels through a pipe to the fixture. In a tankless water heater, an electric element or a gas burner heats the water. As a result, tankless water heaters deliver a constant supply of hot water.

Typically, tankless water heaters provide hot water at a rate of 2 to 5 gal/min (7.6 to 15.2 L/min). Smaller tankless water heaters cannot supply enough hot water for multiple uses at the same time, such as taking a shower and running a clothes washer. To overcome this problem, you can use two or more tankless water heaters connected in parallel, or install separate tankless water heaters for appliances, such as a clothes washer or dishwater that uses extra hot water. Tankless water heaters can also be used for hot water supply at an outdoor sink, a pool shower, a remote bathroom, or a spa. Tankless water heaters are also used as a booster, eliminating long pipe runs, for solar water heating systems, dishwashers, and clothes washers.

Apartment complexes, hotels, motels, and other similar buildings often use tankless water heaters in each bathroom and service area such as kitchen and laundry to ensure guests, occupants, and works have adequate hot water instantly. Electric tankless waters have technology that allow the hot water system to maintain accurate water for correct and safe use. Code requires demand hot water heater to be no more than 2 ft from the use fixture. When in use, the average water faucet uses 2.5 gal/min, which is excessive for typical use. This is inefficient and wasteful for a business with many rooms. Water usage efficiency can be greatly improved by installing a low-flow faucet that operates automatically with the electric tankless water heater to save energy costs and save water by reducing average water flow to 0.5 to 1 gal/min.

Hot Water Circulation System

Depending on the distance from the hot water heater to a faucet or shower head, it can take up to a minute or more for hot water to reach the fixture. This time is spent running and wasting water. A **hot water circulation pump** can be used to circulate water through the main hot water line and the return-line back to the boiler or water heater to keep hot water in the main line and available to the fixtures at all times. Water consumption is reduced, because hot water is available instantly when using a hot water circulation system. Cold water can take longer to get to the fixture, because hot water in the line has to flow out first.

Solar Hot Water

Solar hot water collectors are available for producing hot water from the sun. Solar collectors can be located on a roof, on a wall, or on the ground near the building. Figure 13.16 shows a typical roof installation of solar hot water collectors. Solar systems vary in efficiency. The number of collectors needed to provide heat for hot water depends on the size and demand of the building and the volume of heat needed. The **flat-plate collector** is the heart of a hot water solar system. Its main parts are the transparent glass cover, absorber plate, flow tubes, and insulated enclosure. The flat-plate collector works by sun heating water that flows through the copper flow tubes inside the glass covered collector. The copper pipes are connected to a common manifold which is then connected to a slow flow circulation pump that pumps water to a storage tank located in a convenient place. Water is heated by the sun during the day and stored in an insulated tank for use as needed. Piping and wiring drawings for a complete solar hot water heating system are available through the manufacturer.

Hose Bibbs

Hose bibbs are faucets used to attach a hose, and normally placed at convenient locations around the outside of the building for watering plants, washing a car, or other uses. The hose bibb floor plan symbol and a typical hose bibb is shown in Figure 13.17. Outside hose bibbs require a separate valve that allows the owner to turn off water to the hose bibb during freezing weather. The separate valve must be located inside a heated area and is generally under a cabinet where a sink is located, such as in the kitchen or bathroom. This provides

Figure 13.16   A typical roof installation of a solar hot water collector. (*©vittavat—Adobe Stock.*)

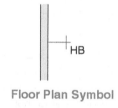

Floor Plan Symbol

Figure 13.17   A hose bibb floor plan symbol and a typical hose bibb. (*Photo ©didden—Adobe Stock.*)

easy access to the valve. Frost proof hose bibbs do not require a separate inside valve, but the stem must extend through the building insulation into an open heated or semi-conditioned space to avoid freezing.

DRAINAGE AND VENT SYSTEMS

The **drainage system** provides for the distribution of solid and liquid waste to the sewer line. Drainage pipes are required to have a minimum slope of 1/4 in./ft (6 mm per 300 mm). The **vent system** allows for a continuous flow of air through the system so gases and odors can dissipate and bacteria do not have an opportunity to develop. The vent system also protects from drawing liquid out of the traps and stops back pressure. Vent pipes are generally made of PVC plastic, although the pipe from the building to the concrete sewer pipe is commonly cast iron. Figure 13.18 shows a typical drainage and vent system.

PLUMBING PLACEMENT

There is an advantage to placing plumbing fixtures back-to-back in a **common plumbing wall** when possible. A common plumbing wall is any wall where plumbing pipes are installed. A common plumbing wall is a wall between rooms where plumbing fixtures are placed back-to-back. The common plumbing wall can require 2 × 6 or 2 × 8 stud framing for extra width to hold the plumbing pipes. This practice saves materials and labor costs. Placing plumbing fixtures one above the other in a multistory building is also an economical plumbing installation. If the functional design of the floor plan clearly does not allow for these practical applications, then good judgment should be used in the placement of plumbing fixtures so plumbing installation is physically possible by providing places for pipes to be installed. Figure 13.19 shows a back-to-back bathroom installation with a common plumbing wall. Another economical installation is a common plumbing wall between a bathroom and laundry room, as shown in Figure 13.20.

PLUMBING SCHEDULES

Plumbing schedules are similar to door, window, and lighting fixture schedules. Schedules provide specific information regarding plumbing equipment, fixtures, and supplies. The information is condensed in a chart so the floor plan is not unnecessarily crowded. A plumbing fixture schedule can provide fixture type, manufacturer name, model number, and color columns. Figure 13.21 shows a typical plumbing fixture schedule.

Other schedules can include specific information regarding floor drains, water heaters, pumps, boilers, or radiators. These schedules generally key specific items to the floor plan with complete information describing size, manufacturer, type, and other specifications as appropriate.

FIRE SPRINKLER SYSTEMS

Fire sprinkler systems are normally required in commercial buildings and are becoming required in residential construction by some local codes. The requirement is established by the local governing agency. The following describes fire sprinkler requirements administered by a local jurisdiction.

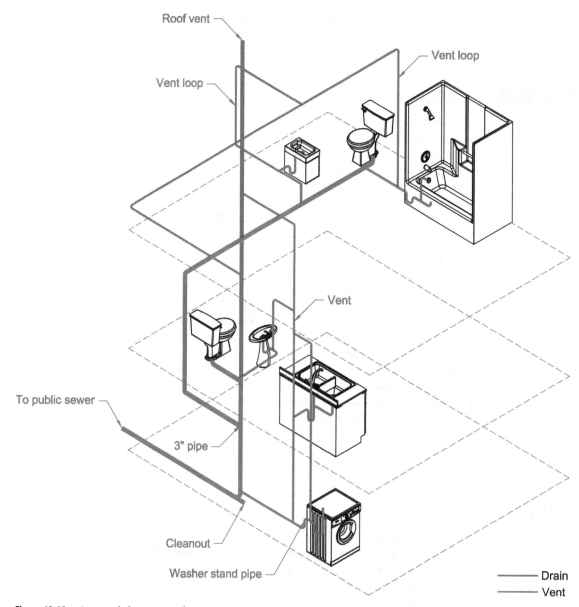

Figure 13.18 A typical drainage and vent system. (*Previously published in Modern Residential Construction Practices, Routledge, 2017 and reproduced here with permission.*)

Automatic sprinkler systems are required in newly constructed commercial buildings exceeding 3600 ft² of **fire area,** any remodel or addition that changes the footprint beyond 3600 ft². The term **fire area** refers to the combined floor area enclosed and bounded by fire walls, fire barriers, exterior walls, or fire-resistance-rated horizontal assemblies of a building.

An automatic sprinkler system is required in all townhouse occupancies with more than two dwelling units per building. The automatic sprinkler system is installed in accordance

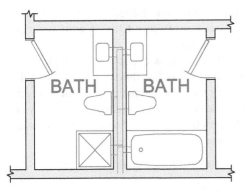

Figure 13.19 A back-to-back bathroom installation with a common plumbing wall. (*Previously published in Modern Residential Construction Practices, Routledge, 2017 and reproduced here with permission.*)

Figure 13.20 A common plumbing wall between a bathroom and laundry room. (*Previously published in Modern Residential Construction Practices, Routledge, 2017 and reproduced here with permission.*)

PLUMBING FIXTURE SCHEDULE			
LOCATION	**ITEM**	**MANUFACTURER**	**MODEL**
KITCHEN	33 X 22 X 9.625 4 HOLE WHITE	KOHLER	HEARTLAND SELF RIMMING CAST IRON
MASTER BATH	SOLID COPPER SELF RIMMING	ECOSINKS	HAMMERED ANTIQUE COPPER
GUEST BATH	SELF RIMMING WHITE	AMERICAN STANDARD	CADET
WATER HEATER	40,000 BTU NATURAL GAS	GE	50 GAL 12 YEAR
LAUNDRY	30 X 22 X 17.5 2 HOLE WALL MOUNTED	KOHLER	GILFORD APRON FRONT CHINA

Figure 13.21 A typical plumbing fixture schedule. (*Previously published in Modern Residential Construction Practices, Routledge, 2017 and reproduced here with permission.*)

with National Fire Protection Association (NFPA) to include garages, exterior balconies, and attached breezeways. Each building requires a minimum **Class A roof** covering. A Class A roof is effective against severe fire test exposure.

Some remote locations requiring fire sprinkler systems because of the logistics for fighting a fire. Even if a fire sprinkler system is not required, it should be a consideration for new construction. According to NFPA, properly installed and maintained automatic fire sprinkler systems help save lives, because fire sprinkler systems react so quickly, they can dramatically reduce the heat, flames, and smoke produced in a fire. Some of the advantages of an automated fire sprinkler system are as follows:

- Losses 90 percent lower than can be expected without a sprinkler system.
- Fire sprinkler systems provide added security.
- Each sprinkler is individually activated by heat. Ninety percent of all fires are controlled by a single sprinkler.
- A fire can easily spread throughout the building without a fire sprinkler system, and traditional firefighting generally takes over eight times the amount of water to fight a fire when compared to a building protected with a fire sprinkler system.

Fire sprinklers activate by a water plug that releases water when the heat reaches a certain temperature. Architects, fire sprinkler consultant, or a qualified plumbing contractor can create a fire sprinkler system design. Plumbing contractors, or a fire sprinkler specialist can install fire sprinkler systems.

SEWAGE DISPOSAL

A **sanitary sewer system** is a system of underground pipes designed for the collection and transfer of wastewater from domestic residences, businesses, and industries to a wastewater treatment plant, or private sewage treatment such as a **septic tank** or **cesspool**. The domestic sewage is from the bathroom, kitchen, and laundry drains. Sanitary sewers are usually not designed to handle stormwater. Strom water is carried in a separate **storm sewer** system. A storm sewer is used for carrying groundwater, rainwater, surface water, or other nonpolluting waste to locations where it can be safely dispersed. Septic tanks are used primarily for projects outside of public sewage districts. A cesspool is a cistern that receives untreated sewage.

Public Sewers

Public sewers are available in and near most cities and towns. Public sewers are generally located under the street or in a utility easement next to the construction site. The public sewer main line is often under the street, so new construction has a pipe line **run** from the building to the sewer. The term **run** refers to a portion of a pipe or fitting continuing in a straight line in the direction of flow in which it is connected. The pipe run from the building to the sewer main line is usually 4 or 6 in. diameter (100 or 150 mm) concrete pipe, but can be much larger for commercial and industry installations. The cost of this construction usually includes installation, street repair, sewer tap, and permit fees. Figure 13.22 shows a public sewer connection.

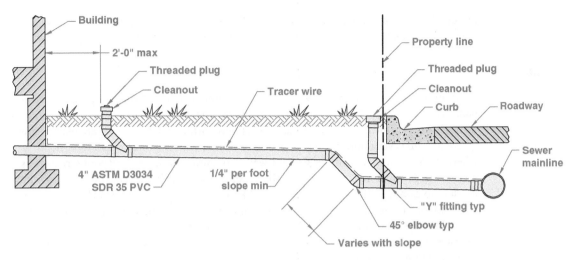

Figure 13.22   A public sewer connection. (*Previously published in Modern Residential Construction Practices, Routledge, 2017 and reproduced here with permission.*)

Septic System

A **septic tank** is an on-site treatment system for domestic sewage, in which the sewage is held to go through a process of liquefaction and decomposition by bacterial organisms. The flow continues to be safely disposed in a subsurface facility such as a tile field, leaching pools, or buried sand filter. A septic system consists of a storage tank and an absorption field and operates as solid and liquid waste enters the septic tank, where it is stored and begins to decompose into **sludge.** Sludge is a thick, soft, wet mixture of liquid and solid components. Liquid material, or effluent, flows from the tank outlet and is dispersed into a soil absorption field, or drain field, also called **leach lines.** When the solid waste has effectively decomposed, it also dissipates into the soil absorption field. The owner should use a recommended chemical to work as a catalyst for complete decomposition of solid waste. Septic tanks can become overloaded in a period of up to 10 years and can require pumping.

The characteristics of the soil must be verified for suitability for a septic system by a **soil feasibility test,** also known as a **percolation test.** This test, performed by a soil scientist or someone from the local government, determines if the soil can accommodate a septic system. The test should also identify certain specifications that should be followed for installation. The Veterans Administration (VA) and the Federal Housing Administration (FHA) require a minimum of 240 ft of drain field line, or more if the soil feasibility test shows it is necessary. Verify these dimensions with local building officials. When the soil characteristics do not allow a conventional system, there are some alternatives such as a **sand filter system,** which filters the effluent through a specially designed sand filter before it enters the soil absorption field. Check with your local code officials before calling for such a system. The serial system allows for one drain field line to fill before the next line is used. The drain field lines must be level and must follow the contour of the land perpendicular to the slope. The drain field should be at least 100 ft from a water well but verify the distance with local codes. There is usually no minimum distance to a public water supply. A septic system generally has gravity feed of solid and liquid waste to the septic tank. However, if gravity feed is not possible a pump system can be installed. The pump system has a chamber with an automatic pump that receives waste and sends it to the septic tank located at a higher elevation.

Cesspool System

A **cesspool** is a cistern that receives untreated sewage that goes through a process of liquefaction and decomposition by bacterial organisms. The decomposed sewage flow continues through an open bottom and perforated sides into porous soil. Cesspools are used in areas where the soil is very porous that has gravel or similar material to a considerable depth and where there is no possibility of ground water contamination. The cesspool structure is a series of large concrete cylinders with the lower cylinder having slots cut through to allow for fluid to escape to the surrounding gravel. Precast reinforced concrete cylinders are usually used for ease and convenient installation into an excavated hole. A gravel layer is placed around the cylinders to aid in fluid absorption. Cesspools can only be used in very specific locations where local building codes allow for this type of sewage system. Figure 13.22 shows a sample cesspool site plan and detail.

Storm Drains

Stormwater is typically drained off the roof into gutters and through down spouts to **rain drains** buried in the ground along the foundation walls. Rain drains are pipes buried along the foundation wall used to transfer weather-related waters to storm sewers or other designated locations. Rain drains are typically installed using **acrylonitrile-butadiene-styrene (ABS)** pipe. ABS pipe is black, which makes it easy to recognize when compared with white PVC pipe. Typically, 3 or 4 in. diameter ABS pipe is used to transfer rainwater. Each down spout enters the rain drain in a T, Y, or elbow fitting connected to the pipe run.

NOTE

A complete set of architectural drawings are provided in the image library found on the website for this textbook. Go to Downloads & Resources tab at **www.mhprofessional.com/ CommercialBuildingConstruction-Ancillaries** to access the complete set of plans for the Brookings South Main Fire Station. This set of plans includes the following drawings directly related to plumbing systems described in this chapter:

Under Floor Plumbing Plan, Sheet M 200

First Floor Plumbing Plan, Sheet M 201

METRIC VALUES IN PLUMBING

Pipe is made of a wide variety of materials identified by trade names, with nominal sizes related only loosely to actual dimensions. For example, a 2 in. galvanized pipe has an outside diameter of about 2-1/8 in. but is called 2 in. pipe for the inside diameter. Few pipe products have even inch dimensions that match their specifications, so there is no reason to establish even metric sizes. Metric values established by the International Organization for Standardization (ISO) relate nominal pipe sizes (NPS) in inches to metric equivalents, referred to as diameter nominal (DN). The following table shows equivalents that relate to all plumbing, natural gas, heating, oil, drainage, and miscellaneous piping used in buildings and civil engineering projects.

NPS (in.)	DN (mm)	NPS (in.)	DN (mm)
1/8	6	8	200
3/16	7	10	250
1/4	8	12	300
3/8	10	14	350
1/2	15	16	400
5/8	20	20	450
3/4	20	20	500
1	25	24	600
1-1/4	32	28	700
1-1/2	40	30	750
2	50	42	800
2-1/2	65	36	900
3	80	40	1000
3-1/2	90	44	1100
4	100	48	1200
4-1/2	115	52	1300
5	125	56	1400
6	150	60	1500

DN typically precede the metric value when you see metric pipe size specifications. For example, the conversion of a 2-1/2 in. pipe to metric is DN65.

The standard of threads for thread pipe is the National Standard Taper Pipe Threads (NPT). The thread on a 1/2 in. pipe reads 1/2–14NPT, where 14 is threads per inch. The metric conversion affects only the nominal pipe size, which is 1/2 in this case. The conversion of the 1/2–14NPT pipe thread to metric is DN15–14NPT.

NOTE

Plumbing is a specialty trade in the construction industry that requires completion of a plumbing apprenticeship program. Entry into plumbing apprenticeship programs is competitive, making it important for you to prepare with postsecondary education in a plumbing or vocational program. Apprenticeship training is typically operated through union organizations such as the United Association of Journeymen and Apprentices of the Plumbing and Pipefitting Industry of the United States and Canada or the National Association of Plumbing-Heating-Cooling Contractors. Apprenticeship programs take 4 to 5 years to complete and combine classroom instruction with on-the-job training. You can learn everything from print reading to the entire range of plumbing skills. Each region has different but similar requirements for plumbers, but most involve licensing. Typically, plumbers need to have from 2 to 5 years of experience in the trade and pass a test based on plumbing trade and code knowledge.

Related opportunities involve ground work such as sewer and water installations where training and experience in excavation and grading are unique skills allied to the plumbing industry.

INTRODUCTION TO ELECTRICAL SYSTEMS

The **electrical system** provides electrical power transmission to the building and contains all of the **circuits** and **fixtures** to be used by the electrical contractor during installation. Circuits are the various **conductors,** connections, and devices found in the path of electrical flow from the source through the components and back to the source. Electrical fixtures are all plugs, switches, cover plates, and lights. A conductor is material that permits the free flow of electricity. Copper is a common conductor in architectural wiring. Residential electrical wiring is always covered with sheath insulation to protect from electrical shock. Most residential wiring is **single phase** and 120 V, consisting of three wires, positive, negative, and neutral. Single phase is the distribution of alternating current electric power using a system with all the voltages of the supply fluctuates at the same time. For some more demanding appliances, such as an oven, air conditioner, furnace, refrigerators, and clothes dryer use a **two phase** circuit of 240 V. Two-phase electrical wiring is where two wires each provide the same voltage AC out of phase with each other to operate electric motors that require continuous current and higher voltage to start and run. Commercial electrical applications normally have wiring run through **conduits** or open construction spaces where the wiring is easily accessible for service. Conduit is a metal or fiber pipe or tube used to enclose one or more electrical conductors. Commercial wiring normally has increased insulation, called **thermoplastic, high-heat resistant, nylon coated (THHN).** The THHN shield helps to protect the electrical wiring from corrosive gases and liquids. In some cases, special outlets are installed for high electrical demand or specifically sensitive equipment. Commercial electrical wiring normally uses a **three-phase** design. Three-phase electrical systems use two smaller legs running 120 V each and one wider leg running 208 V. Three-phase electrical systems allow each wire to perform less workload, while creating a higher output when the wires are required to work together. The three-phase system provides greater efficiency and increases equipment operational life. Commercial higher voltage requirements are due to the increased power demands in a business or industry environment.

Electrical installation for new construction occurs in three phases described in the following.

Temporary

The installation of a **temporary** underground or overhead **electrical service** near the construction site and close to the final **meter** location provides electricity during construction. A temporary electric service is provided during construction and is used for construction electricity purposes. This service is installed on a temporary pole, which is placed near the permanent power service or **transformer.** The temporary service has a **meter base** and meter, and usually two 20 **A,** 120 **V (VAC), grounded,** duplex outlets and one 50 A, 4 wire, single-phase, 240 VAC (208 VAC at some locations) outlet mounted in a weatherproof enclosure. A meter is an instrument used to measure electrical quantities. The electrical meter for a building is where the power enters and is monitored for the electrical utility. A meter base is the mounting base on which the electrical meter is attached and contains all of the connections and clamps. A transformer is a device for reducing or increasing the voltage of an alternating current. Ampere is a measurement of electrical current flow. Referred to by its abbreviation amp or amps. Volt (V) is a unit of measure for electrical force. VAC is the abbreviation for volts **alternating current.** Alternating current is the type of electrical current used in buildings and is an electric current that reverses direction in a circuit at regular intervals. The term **ground** or **grounded** refers to an electrical connection to the earth by means of a rod, or a common return path for electric current.

> **NOTE**
>
> Typical electrical voltage is 110 or 120 and 220 or 240. North American power outlets provide 120 V at 60 Hz. Power comes in between 110 and 120 V or 220 and 240 V. One floor plan symbol is used for 110 or 120 V and a different symbol is used for 220 or 240 V.

Rough-In

Rough-in electrical, also typically called **rough-in,** is when the **electrical work** is done installing wiring and **boxes.** A box, also called an **electrical box,** is equipped with clamps, used to terminate a conduit. Connections are made in the box, and a variety of covers are available for finish electrical. A premanufactured box or casing is installed during electrical rough-in to house the switches, outlets, and fixture mounting. Types of boxes include an **outlet box, lighting outlet box, junction box,** and **switch box,** as shown in Figure 13.23. An outlet box is a box for an **outlet.** A lighting outlet box is an electrical box intended for the direct connection of a **light fixture** and can be ceiling or wall-mounted. A junction box is an electrical box that protects electrical wiring splices in conductors or joints in runs. The junction box has a removable cover for easy access. A switch box is an electrical box that houses a switch or group of switches.

Electrical Service Panel

An **electrical service panel,** also called a **distribution board, panel board, breaker panel,** or **electric panel,** is installed in a convenient location generally near where the electrical utility meter is located and in an accessible place such as in a garage or electrical service room in a commercial facility. The electrical service panel is a distribution box with **circuit breakers** connecting electrical wiring to circuits within the building. A circuit breaker, also called a **breaker,** is an electric safety switch that automatically opens a circuit when excessive **amperage** occurs. Amperage, referred to as **amps** is a measure of the amount of electricity used. **Arc fault circuit interrupter (AFCI)** is also required as a safety device during electrical overload conditions such as electrical storms. An AFCI is a duplex receptacle or circuit breaker that breaks the circuit when it detects a dangerous electrical arc, in order to prevent electrical fires.

There is normally one electrical service panel for the entire building, but there can be another subpanel installed to serve a specific area such as a kitchen, garage, or outdoor cooking and living area. Large commercial buildings can have several service panels that provide electrical service to specific areas. Circuit breakers are stacked in the electrical service panel and have a manual switch that can be set to an on or off position. A **main circuit breaker,** also called a **main,** is a large double circuit breaker at the top of the panel that controls power to all circuit breakers in the electrical service panel. The electrician generally places a label next to each breaker that identifies the breaker circuit, such as kitchen, water heater, furnace, bedroom, or living room outlets. The maximum amperage that an electrical service panel can distribute at one time is marked on the main breaker. A 100-amp (A) main is enough to handle all electrical needs for a small service. Although, most buildings typically install 150-A or 200-A services for enough capacity. Large commercial facilities can have larger capacity with several panels depending on the requirements. Each circuit breaker is rated for the type of wire and load required by the circuit. Typical capacities for lighting and outlet circuits are 15 and 20 A breakers. Standard breakers for 120-V circuits take one slot in the electrical service panel, and service typical electrical applications such as lighting and outlets. Typical 240-V circuits use two slots in the electrical service panel and service appliances such as electric

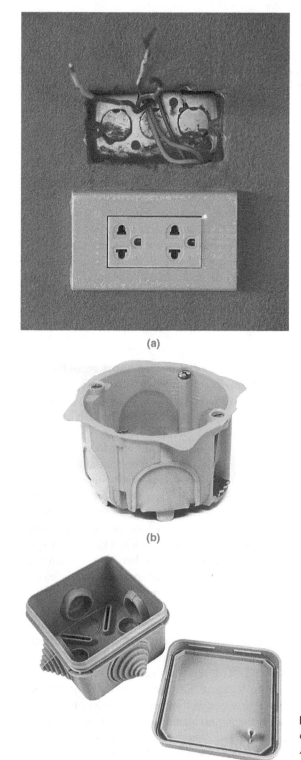

(a)

(b)

(c)

Figure 13.23 (*a*) Outlet box with duplex convenience outlet and cover. (©*geargodz—Adobe Stock.*) (*b*) Lighting outlet box. (©*Ponchy—Adobe Stock.*) (*c*) Junction box. (©*AVD—Adobe Stock.*)

range, water heater, and furnace. The term volt (V) is a measure of electrical pressure or force. Thin circuit breakers can be used, in some applications that occupy half the space of standard breakers and are often used when space is limited in the electrical service panel or when adding a breaker for a remodeling project.

Finish Electrical

Finish electrical is when the light fixtures, outlets and covers, and appliances are installed prior to moving in final construction phase. An outlet is an electrical connector used to plug in devices. A **duplex outlet,** also called a **duplex convenience outlet,** has two outlets and is the typical wall plug. A light fixture is any device that provides artificial light.

Tamper-resistant (TR) outlets are required for child safety. A TR outlet has a spring-loaded shutter that closes the contact openings.

READING ELECTRICAL SYMBOLS

Electrical symbols show the electrical arrangement. Electrical symbols are displayed on the floor plan or separate electrical plans. Figure 13.24 shows the typical electrical symbols that can be found on electrical plans. The electrical contractor follows the layout and symbol locations on the electrical plans to install the electrical system.

Figure 13.25 shows several typical electrical installations with switches to outlets. An **electrical circuit line,** also called a **switch leg,** is generally a curved dashed line that connects a switch to one or more light outlets. A switch leg is the electrical conductor from a switch to the electrical device being controlled. The following describes the electrical circuits shown in each Figure 13.25 example:

- A **single-pole switch** can be connected to one or more lights. A single-pole switch is a standard on and off wall switch that is the only switch controlling one or more light fixtures in a single electrical circuit.
- A **three-way switch** is two switches controlling one or more lights. You can turn the lights on or off at either switch.
- A **four-way switch** is three switches controlling one or more lights. You can turn the lights on or off at any switch.
- A single-pole switch controlling a wall-mounted light, such as a light outside of an entry door or porch.
- A single-pole switch with a circuit to a **split wired outlet**. A split wired outlet is a duplex outlet that has one outlet controlled by a switch. This is a common installation in a room where a table lamp is plugged into the outlet and the light is controlled by the wall switch.

When a fixture requires special characteristics, such as a specific size, location, or other specification, a note is placed next to the symbol, briefly describing the application, as shown in Figure 13.26. Figure 13.27 shows maximum spacing recommended for outlet installations. Figure 13.28 shows a typical electrical plan.

Commercial Electrical Plan Examples

Commercial electrical plans often follow much more detailed installation guidelines than residential applications. Separate electrical plan sheets are commonly used, so the only information provided is that of electrical installations.

ELECTRICAL SYMBOLS

THESE SYMBOLS COMPRISE A STANDARD LIST; NOT ALL SYMBOLS MAY APPEAR ON THIS PROJECT.

ALL MOUNTING HEIGHTS ARE TO CENTER OF DEVICE ABOVE FINISHED FLOOR, MOUNTING HEIGHTS INDICATED ON ARCH. WALL ELEVATIONS OR AS NOTED SPECIFICALLY ON THE DRAWINGS OR IN THE SPECIFICATIONS SHALL TAKE PRECEDENCE OVER MOUNTING HEIGHTS LISTED BELOW.

LIGHTING

CEILING SURFACE MOUNT FIXTURE. (Capital letter indicates fixture type. Small letter indicates switching. Typical for all fixture types).
EMERGENCY CEILING SURFACE MOUNT FIXTURE
WALL FIXTURE
EMERGENCY WALL FIXTURE
RECESSED FIXTURE
EMERGENCY RECESSED FIXTURE
EXTERIOR POLE LIGHT
BOLLARD LIGHT
SURFACE MOUNT FIXTURE
EMERGENCY SURFACE MOUNT FIXTURE

RECESSED FIXTURE
EMERGENCY RECESSED FIXTURE
WALL FIXTURE
FLOOD LIGHT
TRACK LIGHT
PC — PHOTO ELECTRIC CELL
LC — LIGHTING CONTACTOR (54"M.H.)
TC — TIME CLOCK (60" M.H.)
EMERGENCY LIGHTING W/BATTERY PACK
CEILING EXIT LIGHT (FACE(S) SHADED, ARROW INDICATES CHEVRON)
WALL EXIT LIGHT (FACE(S) SHADED, ARROW INDICATES CHEVRON)

OS — OCCUPANCY SENSOR
SINGLE POLE SWITCH (46" M.H.)
DOUBLE POLE SWITCH (46" M.H.)
THREE-WAY SWITCH (46" M.H.)
FOUR-WAY SWITCH (46" M.H.)
SWITCH WITH PILOT (46" M.H.)
KEY OPERATED SWITCH (46" M.H.)
MOMENTARY CONTACT SWITCH (60" M.H.)
DIMMER SWITCH (46" M.H.)
TIMER SWITCH (60" M.H.)
VARIABLE SPEED SWITCH
FUSED SWITCH

POWER

PUSH BUTTON STATION (62" M.H.)
DOUBLE PUSH BUTTON STATION
EMERGENCY SHUTDOWN PUSHBUTTON
ISOLATED GROUND RECEPTACLE (18" M.H.)
DUPLEX CONVENIENCE RECEPTACLE (18" M.H.)
SINGLE RECEPTACLE (18" M.H.)
DOUBLE DUPLEX CONVENIENCE RECEPTACLE (18" M.H.)
SPLIT WIRED DUPLEX RECEPTACLE (18" M.H)
SAFETY CONVENIENCE RECEPTACLE
POWER RECEPTACLE
EMERGENCY DUPLEX RECEPTACLE
TWIST LOCK RECEPTACLE
GFI DUPLEX CONVENIENCE RECEPTACLE
GFI DOUBLE DUPLEX CONVENIENCE RECEPTACLE
SPECIAL PURPOSE OUTLET OR CONNECTION
CORD/PLUG
CORD REEL
CEILING DUPLEX RECEPTACLE
FLUSH FLOOR DUPLEX RECEPTACLE
FLUSH FLOOR DOUBLE DUPLEX RECEPTACLE
FLUSH FLOOR MULTI-SERVICE OUTLET (WITH DEVICES INDICATED)
CP — MULTI-SERVICE POLE (WITH DEVICES INDICATED)

B — BLANK OUTLET
J — JUNCTION BOX
P — PULL BOX
MOTOR
DISCONNECT SWITCH
GAP — GENERATOR ANNUNCIATOR PANEL
ATS — AUTOMATIC TRANSFER SWITCH
VFD — VARIABLE FREQUENCY DRIVE
VFD — COMBINATION VARIABLE FREQUENCY DRIVE DISCONNECT
MAGNETIC STARTER
COMBINATION STARTER/DISCONNECT
MOTOR THERMAL SWITCH
TR — TRANSFORMER
M — ELECTRIC METER
SWITCHBOARD/DISTRIBUTION PANEL SECTION
PANELBOARD OR LOAD CENTER
PANELBOARD OR LOAD CENTER (EXISTING TO REMAIN)
TVSS — TRANSIENT VOLTAGE SURGE SUPPRESSER
CIRCUIT BREAKER
FUSE
H — HUMIDISTAT
T — THERMOSTAT

R — REMOTE HVAC SENSOR
RADIANT HEAT PANEL
BASEBOARD OR COVE ELEC. HEAT
ELECTRIC UNIT HEATER
ELECTRIC CABINET UNIT HEATER
M — MOTORIZED DAMPER
BUS DUCT
SURFACE MOUNT RACEWAY
CEILING PADDLE FAN
XX / # — TYPE OF EQUIPMENT / EQUIPMENT NUMBER — SEE SCHEDULES
ROOFTOP EQUIPMENT/CIRCUITING
EXISTING EQUIPMENT/CIRCUITING
GROUND
UG — CONDUIT IN FLOOR OR UNDERGROUND
PNL-1,2,3 — CONDUIT IN WALL OR CEILING SPACE, CROSS MARKS INDICATE NUMBER OF WIRES, NO MARKS INDICATE TWO WIRES. ARROWS INDICATE HOME RUNS TO PANEL. NUMBERS INDICATE PANEL AND CIRCUIT IN PANEL.
SWITCHLEG
TRAVELER
HOT
NEUTRAL WIRE
INDICATES SEPARATE GROUND WIRE TO BE INSTALLED IN RACEWAY

TELECOM

SPECIAL EQUIPMENT CABINET-AS NOTED
TERMINATION BOARD - AS NOTED
CABLE TRAY

INTERCOM
TELEPHONE/VOICE OUTLET (18" M.H.)
WALL PHONE (46" M.H.)
DATA OUTLET (18" M.H.)

CEILING MOUNT DATA OUTLET
COMBINATION VOICE/DATA OUTLET (18" M.H.)
TV — TELEVISION OUTLET (18" M.H.)
TV — CEILING MOUNT TELEVISION OUTLET

Figure 13.24 Typical electrical symbols found on electrical plans. (*Courtesy of JLG Architects, Alexandria, MN.*)

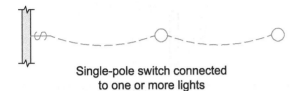

Single-pole switch connected
to one or more lights

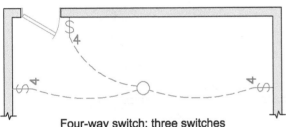

Three-way switch; two switches
control one or more lights

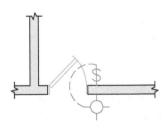

Four-way switch; three switches
control one or more lights

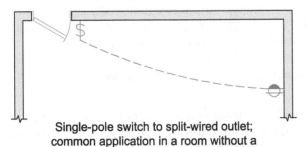

Single-pole switch to wall-mounted light;
typical installation at an entry or porch

Single-pole switch to split-wired outlet;
common application in a room without a
ceiling light; allows switching a table lamp

Figure 13.25 Typical electrical installations with switches to outlets. (*Previously published in Modern Residential Construction Practices, Routledge, 2017 and reproduced here with permission.*)

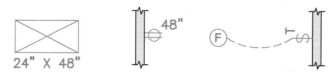

Figure 13.26 When a fixture requires special characteristics, such as a specific size, location, or other specification, a note is placed next to the symbol, briefly describing the application. (*Previously published in Modern Residential Construction Practices, Routledge, 2017 and reproduced here with permission.*)

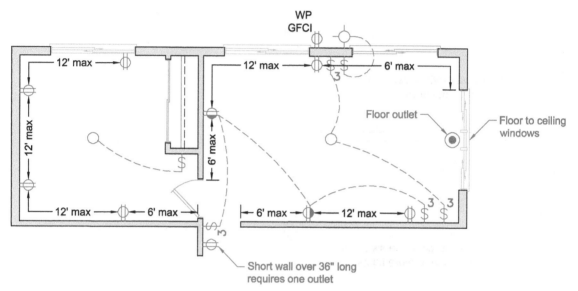

Figure 13.27 Maximum spacing recommended for outlet installations. (*Previously published in Modern Residential Construction Practices, Routledge, 2017 and reproduced here with permission.*)

The electrical circuit switch legs for commercial applications are generally solid lines rather than dashed lines as in residential electrical plans. The electrical circuit lines that continue from an installation to the service distribution panel are terminated next to the fixture and capped with an arrowhead, meaning that the circuit continues to the distribution panel. When multiple arrowheads are shown, this indicates the number of circuits in the electrical run as shown in Figure 13.29. For many installations in which a number of circuit wires are used, the number of wires is indicated by slash marks placed in the circuit run. The number of slash marks equals the number of wires, as shown in Figure 13.30.

There can be more than one commercial electrical sheet in a set of plans. For example, floor plan lighting, electrical plan power supplies, or reflected ceiling plan.

Floor Plan Lighting Layout

The floor plan lighting layout provides the location and identification of lighting fixtures and circuits. The floor plan lighting layout is usually coordinated with a **lighting fixture schedule** that provides a list of the light fixtures used in the building, as shown in Figure 13.31.

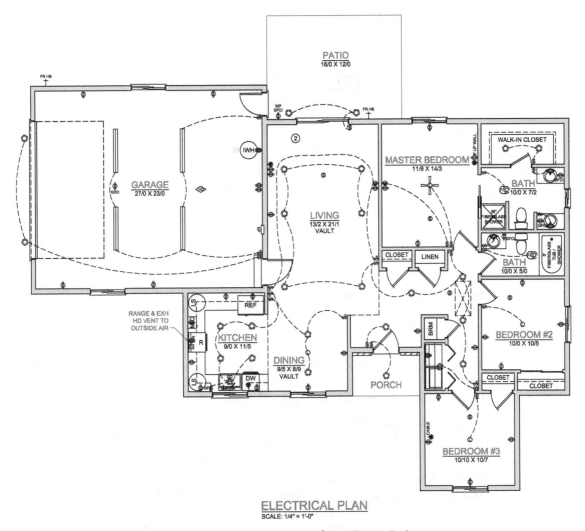

Figure 13.28 A typical electrical plan. (*Courtesy Engineering Drafting & Design, Inc.*)

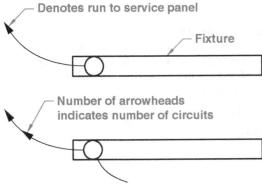

Figure 13.29 Electrical circuit switch legs for commercial applications are generally solid lines that continue from an installation to the service distribution panel and terminated next to the fixture and capped with an arrowhead, meaning that the circuit continues to the distribution panel. Multiple arrowheads indicate the number of circuits in the electrical run.

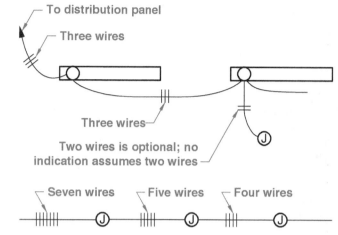

Figure 13.30   For installations in which a number of circuit wires are used, the number of wires is indicated by slash marks placed in the circuit run.

Power Supply Plan

In some applications, a **power-supply plan** is used to show all electrical outlets, junction boxes, and related circuits, as shown in Figure 13.32. A **power supply** provides components with electric power.

Reflected Ceiling Plan

The **reflected ceiling plan (RCP)** is used to show the layout for the suspended ceiling system, as shown in Figure 13.33. The RCP is called to as a reflected ceiling plan, because it is drawn to display a view of the ceiling as if it is reflected onto a mirror on the floor.

Electrical Equipment Plans

Electrical plans for equipment installations can also be needed to supplement the power-supply and lighting plans. Figure 13.34 shows the plan for roof installation of equipment.

Electrical Schematic Diagrams

Schematic diagrams are often provided for specific electrical installations, as shown in Figure 13.35. A schematic diagram is commonly associated with electrical circuits shown in a drawing representing the components of a process, device, or electrical system using lines and standardized symbols.

NOTE

A complete set of architectural drawings are provided in the image library found on the website for this textbook. Go to Downloads & Resources tab at **www.mhprofessional.com/CommercialBuildingConstruction-Ancillaries** to access the complete set of plans for the Brookings South Main Fire Station. This set of plans includes the following drawings directly related to electrical systems described in this chapter:

First Floor Reflected Ceiling Plan, Sheet A 701

First Floor Power and Communication Plan, Sheet E 200

First Floor Lighting Plan, Sheet E 201

Enlarged Electrical Plan, Sheet E 202

Electrical Details, Sheet 300

Electrical Schedules, Sheet E 400

LIGHTING FIXTURE SCHEDULE

FIXTURE MARK	FIXTURE TYPE	FIXTURE DIFFUSER	VOLTAGE	LAMP NUMBER AND WATTS	LAMP TYPE	MOUNTING TYPE	MOUNTING HEIGHT	MANUFACTURER	MODEL	COMMENTS
A	LED HIGH BAY		120	20,000 LUMENS, 174W	LED, 4000K	CHAIN	BOTTOM EVEN WITH JOIST - IN STRUCTURE	H.E. WILLIAMS OR EQUAL	GL	6' CORD AND NEMA TWISTLOCK, INTEGRAL OCC. SENSOR, WIREGUARD, L85 @ 50,000 HRS, WHITE FINISH
B	4' INDUSTRIAL		120	6300 LUMENS 52W	LED, 4000K	CHAIN	10' AFF	H.E. WILLIAMS OR EQUAL	80	WIRE GUARD, L70 @ 50,000 HRS
BP	INVERTER		120	100 VA	LED, 4000K	WALL		DUAL LITE OR EQUAL	LG	AC INVERTER, 3 YEAR WARRANTY, VERIFY CAPACITY IS ADEQUATE FOR ACTUAL FIXTURES PROVIDED.
C	2 X 4 LED ACRYLIC TROFFER	FROSTED PRISMATIC LENS	120	4500 LUMENS 47W	LED, 4000K	RECESS	CEILING	H.E. WILLIAMS OR EQUAL	50	L70 @ 50,000 HRS, WHITE FINISH
D	LED SURFACE DOWNLIGHT	ACRYLIC	120	650 LUMENS 9.5W	LED, 4000K	SURFACE	CEILING	LIGHTOLIER OR EQUAL	SR5	WET LOCATION, WHITE FINISH
EM	EMERGENCY WALL PACK	-	120	2-6W	LED	WALL	8' AFF	DUAL LITE OR EQUAL	EVHC	POLYCARBONATE WHITE HOUSING, BATTERY & CHARGER, SELF DIAGNOSTICS
F	LED TRACK HEAD	-	120	2500 LUMENS 45W	LED, 4000K	TRACK		PRESCOLITE OR EQUAL	AKTMLED	30 DEGREE FLOOD, COLOR BY ARCH AT SHOP DRAWINGS, L70@ 50,000 HRS
FT	TRACK	-	120	-	-	SURFACE	CEILING	PRESCOLITE OR EQUAL	AKT	COLOR BY ARCH AT SHOP DRAWINGS, CONTRACTOR TO PROVIDE ALL COMPONENTS FOR A COMPLETE AND FUNCTIONAL SYSTEM
H	8" LED DOWNLIGHT	CLEAR SPECULAR	120	6400 LUMENS 72W	LED, 4000K	RECESS	CEILING	H.E. WILLIAMS OR EQUAL	H85	L70 @60,000 HRS, WET LOCATION, MEDIUM DISTRIBUTION, WHITE FLANGE
W1	LED WALL PACK	-	120	6300 LUMENS 46W	LED, 4000K	WALL	12' AFG	CREE OR EQUAL	SEC-EDG	DIE-CAST ALUMINUM, BRONZE, WET LOCATION, BIRD SPIKES
W2	LED WALL PACK	-	120	2500 LUMENS 25W	LED, 4000K	WALL	10' AFG	CREE OR EQUAL	XSPW	DIE-CAST ALUMINUM, TYPE II MEDIUM DISTRIBUTION, BRONZE, WET LOCATION, BIRD SPIKES
X1	EXIT LIGHT	RED	UNIV.	LED	LED	UNIV.	PER PLANS	DUAL LITE OR EQUAL	LX	POLYCARBONATE WHITE HOUSING, BATTERY & CHARGER, SELF DIAGNOSTICS

UNV. =   UNIVERSAL BALLAST, SURF. = SURFACE

Figure 13.31 The floor plan lighting layout is often coordinated with a lighting fixture schedule that provides a list of the light fixtures used in the building. (*Courtesy of JLG Architects, Alexandria, MN.*)

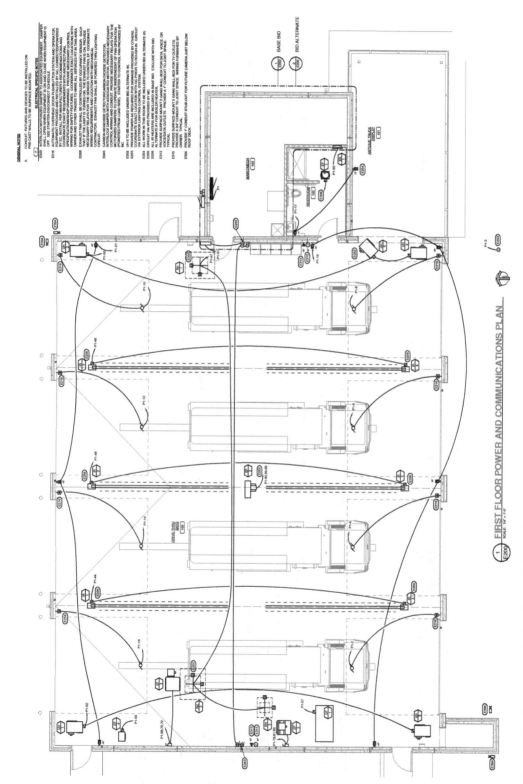

Figure 13.32 A power-supply plan shows all electrical outlets, junction boxes, and related circuits. (*Courtesy of JLG Architects, Alexandria, MN.*)

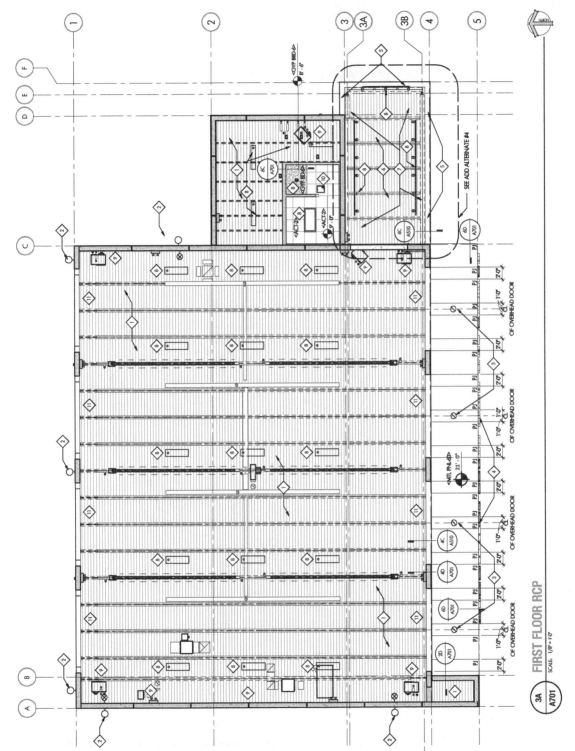

Figure 13.33 The reflected ceiling plan (RCP) shows the layout for the suspended ceiling system. (*Courtesy of JLG Architects, Alexandria, MN.*)

FIRST FLOOR RCP

SCALE: 1/8" = 1'-0"

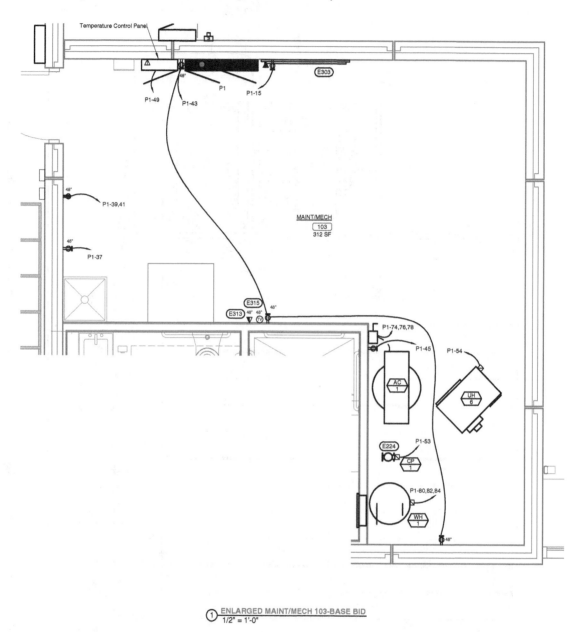

Temperature Control Panel

E303

P1-49

P1-43

P1

P1-15

P1-39,41

P1-37

MAINT/MECH
103
312 SF

E315

E313

P1-74,76,78

P1-45

P1-54

AC
1

UH
6

E224

P1-53

CP
1

P1-80,82,84

WH
1

1 ENLARGED MAINT/MECH 103-BASE BID
1/2" = 1'-0"

Figure 13.34 Electrical plans for equipment installations. (*Courtesy of JLG Architects, Alexandria, MN.*)

Smoke Detectors

Smoke detectors and alarms provide an opportunity for safe exit through early detection of fire and smoke. A smoke alarm that meets the requirements of **Underwriters Laboratories (UL)** UL 217 must be installed per National Fire Protection Association (NFPA) within 72 in. (1800 mm) from each sleeping room, and at a point centrally located in a corridor that provides access to the bedrooms. Confirm these requirements with local building code

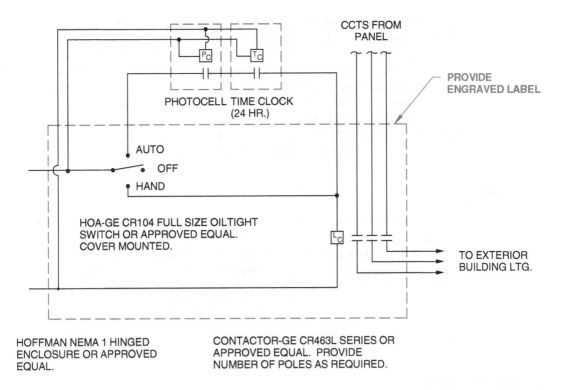

CCTS FROM
PANEL

PROVIDE
ENGRAVED LABEL

PHOTOCELL TIME CLOCK
(24 HR.)

AUTO

OFF

HAND

HOA-GE CR104 FULL SIZE OILTIGHT
SWITCH OR APPROVED EQUAL.
COVER MOUNTED.

TO EXTERIOR
BUILDING LTG.

HOFFMAN NEMA 1 HINGED
ENCLOSURE OR APPROVED
EQUAL.

CONTACTOR-GE CR463L SERIES OR
APPROVED EQUAL. PROVIDE
NUMBER OF POLES AS REQUIRED.

② EXTERIOR LIGHTING CONTROL DIAGRAM
NO SCALE

Figure 13.35 Schematic diagrams are often provided for specific electrical installations. (*Courtesy of JLG Architects, Alexandria, MN.*)

enforcement officials. Battery-operated smoke detectors are available, and the battery should be changed annually. New construction and remodeling projects have the smoke detectors **hard-wired** to the electrical system. Hard-wired means a fixed connection between electrical, electronic components, and electric devices by means of permanent wiring into the electrical system. Underwriters Laboratories is global safety science company that tests the latest products and technologies for safety before they are marketed to consumers.

Carbon Monoxide Alarms

A **carbon monoxide** alarm is required in all new residential construction that contains fuel-fired appliances or has an attached garage. Carbon monoxide is a tasteless, colorless, odorless gas found in the fumes of fuels that contain carbon, such as wood, coal, and gasoline. **Carbon monoxide poisoning** is a potentially fatal illness that occurs when people breathe in carbon monoxide. Carbon monoxide sources include motor vehicles, small gasoline engines, stoves, lanterns, furnaces, grills, gas ranges, water heaters, and clothes dryers. The risk of carbon monoxide poisoning is high when equipment is used in an enclosed place and ventilation is poor. Carbon monoxide poisoning symptoms vary depending on the concentration of carbon monoxide in the environment, the length of time you are exposed, and your health. Exposure to high levels of carbon monoxide gas can result in headache, shortness of breath, personality changes, unusually emotional behavior or extreme swings in emotions, fatigue, a generally sick feeling, dizziness, clumsiness or difficulty walking, vision problems, confusion and impaired judgment, nausea and vomiting, rapid breathing, chest pain, or rapid or irregular heartbeat. You can lose consciousness, have a seizure, enter a coma, and potentially die without immediate treatment. Death can result from only a few minutes of exposure to higher concentrations or from an hour of exposure to lower levels.

Alarms that meet UL 2034 must be placed outside of each separate sleeping area in the immediate vicinity of the bedrooms. Carbon monoxide alarms must be installed in accordance with the manufacturer's instructions and code requirements.

AUTOMATION

Many modern buildings are built with **automation systems.** This technology is rapidly changing and requires that you continuously research the available products and installations. An automation system is a method or process of controlling and operating mechanical devices electronically. Such operations include the computerized control of the heating, ventilating, air-conditioning, landscape sprinkling systems, lighting, security, intercom, audio, and visual systems. Automation system electrical symbols and specific notes are placed on the floor plan or on separate drawings and in specifications used for construction purposes. The following briefly describes some of the many automation system options and features available.

Entertainment Centers

Entertainment centers are available that provide an in-home or office meeting room theater and sound system. These are fully automatic and contain the best technology in picture and sound performance. For example, pressing one button can automatically dim the lights, open the curtain, and begin the presentation. Automated equipment can also provide sound throughout a designated area using a variety of sources, such as streaming, satellite radio, and computer access sources.

Computerized Programming

Computerized programming functions from a personal computer allows you to use a computer to set and monitor a variety of electrical circuits, including security, sound, landscape watering, cooking, and lighting. Automation systems often place the computer and monitors in locations such as an office where there is frequent activity. There are dedicated computer systems that can be installed, or a standard personal computer can be used with desired software and Internet access.

Security Systems

Security systems can be installed indoors at doors and windows, which sound an alarm when opened. Additional indoor motion sensors can detect inside movement when the system is armed. Video cameras can also be used for added security and surveillance. These security systems allow you to enter a special code at a control panel, allowing you to set the alarm when leaving, and to disarm the system when enter the building at the site or remotely. Most systems have a night setting that allows you to arm the doors and windows but keep the motion sensors off when the system active in order to provide exterior security for night employees or occupants. Outdoor security systems can also monitor the outside area. Outdoor systems can programmed to turn on flood lights when detecting movement, and typically have surveillance cameras.

Radio Frequency Systems

Radio frequency (RF) systems are available that allow you to control a variety of lighting and mechanical applications. RF is electromagnetic radiation waves that transmit audio, video, or data signals. The Federal Communication Commission (FCC) has defined RF ranges that can be used for wireless applications, such as those used in home or commercial systems. RF systems allow you to design options for specific lighting applications. For example, the RF system can control security lighting inside and outside by the press of a button within the home or business, or remotely. Other lighting schemes can also be designed that provide visibility or accent lighting to various areas. You can control the lighting of hallways at night, or specific light dimming applications for accent purposes.

Structured Wiring

As computer use and Internet access are common in nearly every home and business, **structured wiring systems** are part of the electrical work needed in the architectural design. Structured wiring systems are high-speed voice and data lines and video cables wired to a central service location. These wires and cables optimize the speed and quality of various communication signals. This kind of wiring is typically used in commercial construction projects and is gaining popularity in residential design. In a system of this type, each electrical outlet, telephone jack, or computer port has a dedicated line back to the central service location. The central service location allows the wires to be connected as needed for a network configuration or for dedicated wiring from the outside. High-quality structured wiring systems use network connectors and **parallel circuits,** because conventional outlets and **series circuits** degrade the communication signal. A parallel circuit is an electrical circuit that contains two or more paths for the electricity or signal to flow from a common source. A series circuit is a circuit that supplies electricity or a signal to a number of devices connected so the same current passes through each device in completing its path to the source. Structured wiring systems allow the use of fax, multiline telephone, and computers at the same time. Additional applications include digital satellite system (DSS), digital broadcast system (DBS), stereo audio, and closed-circuit security systems.

Most homes and businesses are equipped with **Wi-Fi.** Wi-Fi is the name of wireless networking technology that uses radio waves to provide wireless high-speed Internet and network connections. Homes and businesses with a wireless network have an **access point (AP).** The access point broadcasts a wireless signal used by computers and wireless devices. Computers and devices are equipped with wireless network adapters for connection to the access point to interact with the wireless network. The access point is generally located in a central place where it provides good service to the wireless devices.

ELECTRICAL SCHEDULES

Electrical fixtures can be identified in schedules, notes, description of materials, or in specifications. Light fixture schedules typically provide fixture symbol, manufacturers name, type, color, bulb, **wattage,** and remarks columns. Light fixtures are identified in a

LIGHT FIXTURE SCHEDULE

LOCATION	QTY	MANUFACTURER	DESCRIPTION	NOTES
A	25	CREE	LED LR6 120V EDISON BASE	6" LED RECESSED DOWNLIGHT
B	4	MAXIM	MX 92200 MANOR	MINI PENDANT
C	3	FEISS	MF VS10503 CLARIDGE	3 LIGHT 18" WIDE
D	2	MINKA LAVERY	ML1008 PL	FLOURESCENT

Figure 13.36 A typical light fixture schedule. (*Previously published in Modern Residential Construction Practices, Routledge, 2017 and reproduced here with permission.*)

schedule by placing a letter or number key next to the floor plan symbol and correlating the same letter or number to the schedule or fixtures by room name where the fixtures are located. The term **watt** or **wattage** refers to a unit measure of power corresponding to the power in an electric circuit in which the potential difference is one volt and the current one ampere.

Electrical circuit schedules can be used to provide electrical **distribution panel,** circuits, poles, and amperage requirements and specifications. A distribution panel, also called a **panel,** is where the conductor from the meter base is connected to individual circuit breakers, which are connected to separate circuits for distribution to various locations throughout the building. An electrical panel schedule can be used to provide panel size and type, main circuit size and type, panel manufacturer and model, and wiring specifications. Figure 13.36 shows a typical light fixture schedule.

WIND POWER ELECTRICITY GENERATION

Wind power electricity generators are available that can provide the entire energy needs for a home or business and also sell electricity back to the local public utility company. The wind power electricity generator converts wind into electricity. Modern wind generators have built-in controls and **inverter** designed specifically for **utility grid**-connected use. An inverter is an appliance used to convert **direct current (DC)** power, produced by the wind generator, into standard AC current. A utility grid is the transmission system for electricity that is a network of coordinated power providers and consumers that are connected by transmission and distribution lines and operated by one or more control centers. Direct current is a continuous electric current that flows in one direction only. Electricity consumers, in most areas, can take advantage of **net-metering,** which is the sale of unused energy back to the power grid.

A wind generator can also provide electricity for remote locations, telecommunications sites, water pumping, and other rural applications. The generator provides direct current to batteries, which store the energy until it is needed. Standard residential alternating current appliances can be used from the batteries when the power is run through an inverter before use.

PHOTOVOLTAIC MODULES

Photovoltaic cells turn sun light into electricity. The word photovoltaic comes from the Greek word photo, meaning light, and voltaic, meaning to produce electricity by chemical action. In the photovoltaic cells, photons strike the surface of a silicon wafer, which is a semiconductor diode that stimulates the release of electric charges that are guided into a circuit where they become a useful electric current. Photovoltaic modules produce direct current (DC) electricity. This type of power is useful for many applications and for charging storage batteries. An inverter is used to change DC current to alternating current (AC). AC is the type of current that powers electrical systems. Some photovoltaic systems use the DC current immediately to power DC motors as in hot water-pumping systems. The energy produced from photovoltaic cells must be stored when solar electric systems are not used immediately, or if an energy reserve is required for use when sunlight is not available. Batteries are the most common storage devices, allowing the stored electricity to be used when needed. DC and AC systems can be used to supply the electrical needs of the home, and AC systems can be connected to a utility grid where electricity can be shared. The system can draw on the grid for extra electricity if needed during times of peak power usage. The system can also return extra unused power to the grid where the electricity can be purchased by the utility, providing income or electricity credits for the owner. In most areas of the country, the utility (grid) is required to purchase excess power from private sources. Because battery storage is expensive and space consuming, a municipally connected solar electric system is the most popular, and least expensive way to take advantage of solar power.

The solar modules need to be located and positioned where they receive maximum exposure to direct sunlight for the longest period of time every day. It is also important to keep distances to electrical loads to a minimum. The electrical loads are the circuits and appliances that use electricity. It is also important to confirm that shade from buildings and trees do not block the sunlight. Photovoltaic solar collectors are constructed by placing individual photovoltaic cells in groups called modules, and the modules are combined in groups of six to create the photovoltaic collector. Figure 13.37 shows roof-mounted photovoltaic solar collectors.

Figure 13.37   Roof-mounted photovoltaic solar collectors.

METRIC VALUES IN ELECTRICAL INSTALLATIONS

Electrical conduit designations are expressed in millimeters. Electrical conduit is a metal or fiber pipe or tube used to enclose a single or several electrical conductors. Electrical conduit is produced in decimal inch dimensions and is identified in nominal inch sizes.

Nominal size is referred to as the conventional size; for example, a 1/2 in. diameter pipe has a 16 mm nominal size. The actual size of a conduit remains in inches but is labeled in metric. The following table shows inch and metric electrical applications:

Inch	Metric (mm)	Inch	Metric (mm)
1/2	16	2-1/2	63
3/4	21	3	78
1	27	3-1/2	91
1-1/4	35	4	103
1-1/2	41	5	129
2	53	6	155

Existing American **Wire Gage** (AWG) sizes remain the same without a metric conversion. The diameter of wires conform to various wire gauge systems. The AWG is one system for the designation of wire sizes. Wire gauge is a method of defining wire diameter by a number, with wire diameter increasing as the number gets smaller.

NOTE

Work as an electrician is a specialty trade in the construction industry that normally requires completion of an apprenticeship program. Preliminary skills needed to become an electrician apprentice include above average manual dexterity, good hand-eye coordination, physical fitness, an excellent sense of balance, and the ability to quickly and accurately solve math problems. An electrician apprentice candidate also needs good color vision to identify electrical wires by color. Entry into an electrician apprenticeship program requires secondary school graduation or equivalent, and pass an electrician apprentice aptitude test. Postsecondary education can be an asset for entry into an apprenticeship program. Several trade schools and community colleges offer electrician education that is allied with local electrician unions and contractor organizations. Entry employment as an electrician helper is an alternate way to gain basic experience before entering an electrician apprenticeship program.

Electrician apprenticeship programs are typically sponsored by local electrician unions of the International Brotherhood of Electrical Workers, and the National Electrical Contractors Association, or local chapters of the Associated Builders and Contractors and the Independent Electrical Contractors Association. Some electrical contracting companies also sponsor their own electrician apprenticeship programs. Apprenticeship programs take 4 to 5 years to complete and combine classroom instruction with on-the-job training. Electrician apprenticeship education can include mathematics, print reading, electrical codes, electrical safety, **soldering** skills, communication systems, alarm systems, and equipment operation. Soldering is the joining of metal parts, such as wires, with the use of heat on fusible **alloys,** usually tin and lead. An alloy is a metal made by combining two or more other metals to achieve a specific result. Completion of an electrician apprentice program, typically qualifies the graduate to do construction work and maintenance as an electrician.

Each region has different but similar requirements for electricians, but most involve licensing.

Related opportunities involve installation and maintenance of automation systems and related installations described in this chapter.

Go to Downloads & Resources tab at **www.mhprofessional.com/Commercial Building Construction-Ancillaries** to access the images provided in this chapter and correlate with the content of this chapter. Here you can look at the chapter figures on your computer screen to pan and zoom in and out for better observation, because full drawings reduced to fit on a textbook page are often difficult to read. The image bank also includes the complete set of plans for the Brookings South Main Fire Station used as examples throughout this textbook.

Go to **Downloads & Resources tab at www.mhprofessional.com/Commercial Building Construction-Ancillaries** to access the test and creative thinking problems for this chapter. The chapter test can be used for review or to evaluate content knowledge depending on your course objectives. Creative thinking problems are provided to help expand your knowledge by researching given subjects.

Print reading problems have questions that ask you to find information on the Brookings South Main Fire Station plans. The print reading problems can help you become familiar with the format of construction documents and reinforce your ability to seek specific information. This is an important skill for you to master as you prepare to enter the construction industry. Go to **Downloads & Resources tab at www.mhprofessional.com/Commercial Building Construction-Ancillaries** to access the complete set of plans for the Brookings South Main Fire Station.

Glossary

A-frame roof is an architectural home style with steep roof planes that also function as walls and meet at a central horizontal ridge.

Abbreviation is a shortened form of a word or phrase that takes the place of the whole word or phrase, such as mm for millimeter.

Access point broadcasts a wireless signal used by computers and wireless devices.

Accessibility see barrier-free.

Acrylonitrile-butadiene-styrene (ABS) rain drains are typically installed using this black pipe.

Accordion door is an interior door that folds back in small sections to open.

Active fire protection is manual or automatic fire detection and fire suppression, such as fire sprinkler systems.

Active solar systems, also called **mechanical solar systems** use mechanical devices to absorb, store, and use solar heat.

Actual size of brick, also referred to as specified size, is the physical dimension of each brick.

Actual size of lumber is the dimensional lumber after it is planned.

Addition adds a new portion of building to an existing structure.

Addendum to the contract, which is a written notification of the change or changes and is accompanied by a drawing that represents the change.

Advanced technology underlayment is a polyethylene product with a grid structure of square cavities, and an anchoring fleece laminated to the base for tile and stone is available in rolls for lightweight shipping and handling, and provides underlayment support and load distribution for tile flooring, with waterproofing and vapor-control.

Air exchanger see heat recovery and ventilation (HRV).

Air infiltration test shows how much air leakage there is in a house.

Air leakage (AL) is indicated by an air leakage rating expressed as the equivalent cubic feet of air passing through a square foot of window area (cfm/sq ft).

Air-lock entry, known as a **vestibule.** This is an entry that provides a chamber between an exterior and interior door to the building.

Air space between the masonry veneer and the housewrap on the exterior sheathing allows for ventilation, allows moisture from condensation and rain to drain down the weep holes, and aids in keeping the structure dry.

Air supply register is a grill with moving parts that can be opened, closed, and direct the air flow.

Air-to-air heat exchanger   see heat recovery and ventilation (HRV).

Aldehydes are reactive organic compounds that contribute to ozone production.

Algorithms are a process or set of rules to be followed in calculations or other problem-solving operations.

Alignment stakes are used to align a roadway.

All-wood roof system is anchored by long-span glued laminated timber framing that uses purlins attached to the primary glulam beams using pre-engineered metal hangers.

Alloy is a metal made by combining two or more other metals to achieve a specific result.

Alternating current (AC) is the type of electrical current used in homes and is an electric current that reverses direction in a circuit at regular intervals.

Americans with Disabilities Act of 1990 (ADA) prohibits discrimination and ensures equal opportunity for persons with disabilities in employment, State and local government services, public accommodations, commercial facilities, and transportation.

Ampere or amperage (amp) is a measurement of electrical current flow. Referred to by its abbreviation amp or amps.

Anchor bolts are steel L-shaped bolts that are imbedded in the top of the foundation wall and extend out far enough to fasten the sill with a washer and nut at each bolt.

APA refers to APA—The Engineered Wood Association.

APA rated refers to sheathing that is rated by the APA for use as subfloor, wall, roof, diaphragm and shear wall sheathing, and construction applications where strength and stiffness are required.

Aquifer is an underground bed or layer of earth, gravel, or porous stone that yields water.

American National Standard taper pipe threads are the standard thread used on galvanized steel pipes and pipe fittings.

Angle of repose is the slopes of cut and fill from the excavation site measured in feet of horizontal run to feet of vertical rise.

Arc Fault Circuit Interrupter (AFCI) is a duplex receptacle or circuit breaker that breaks the circuit when it detects a dangerous electrical arc, in order to prevent electrical fires.

Architect   see registered architect.

Architectural roofing, also called three-dimensional roofing, imitates the appearance of real cedar wood or slate shakes.

Architectural scales are based on each inch representing a specific increment of feet. Each foot is subdivided into 12 parts to represent inches and fractions of an inch, and the degree of precision depends on the specific scale.

Architectural shingles are multilayered asphalt shingles and heavier than regular asphalt shingles to add durability and weatherproofing to the building.

Architectural solar systems   see passive solar systems.

Architecture, engineering, and construction (AEC) used together and collectively referred to by AEC is a division of the construction industry that provides architectural design, engineering design, and building construction services.

Arrow side the welding symbol leader arrows connect the symbol reference line to one side of the joint.

Artificial intelligence is the theory and development of computer systems able to perform tasks that normally require human intelligence, such as visual perception, speech recognition, decision making, and translation between languages.

Asbestos board is a sheet of construction material made from asbestos cement.

Ash dump is a small opening with a cast iron door located in the hearth, used to dump the ashes down into an ash pit below the fire box.

Asphalt is a mixture of dark bituminous pitch and sand or gravel, used for purposes such as surfacing roads, flooring, and roofing.

Asphalt shingles are made with asphalt for weatherproofing and are the most common, cost-effective shingle on the market used for architectural applications.

Assemblies   see functional elements.

AutoCAD is a computer-aided design (CAD) program used for 2D and 3D design and drafting.

Autodesk Revit is BIM software for architects, structural engineers, engineers, and contractors.

Autodesk Seek (seek.autodesk.com) free Web service allows architects, engineers, and other design professionals to discover, preview, and download branded and generic building information modeling (BIM) files, models, drawings, and product specifications directly into active design sessions in **Autodesk Revit** or **AutoCAD** software.

Automatic level maintains a horizontal line of sight once leveled.

Automation system is a method or process of controlling and operating mechanical devices electronically.

Awning windows have a hinge at the top and single-sash that tilts outward.

Axial forces are forces working along the axis of a structure such as a column.

Back draft damper is a movable plate that regulates the draft or air flow in a chimney or vent pipe with blades that are activated by gravity, permitting air to pass through them in one direction only.

Back pressure is a resistant force applied to liquid or gas against the desired direction or flow in a pipe.

Back sight (BS) is the rod or target location in surveying from which a reference measurement is made.

Back-to-back double fireplace is used when it is desired to have a fireplace in two rooms or the home where the fireboxes are built into one structure.

Backer board or substrate is used as a base cabinet top upon which the surface finish material is installed.

Backer rod is a round open- or closed-cell polyethylene or polyurethane foam product used to fill joints between building materials.

Backfill is the earth or other material used in the process of backfilling, which closes the large space created by the excavation.

Backing material is a comparatively low-quality brick used behind **face brick** or other masonry.

Baffle is used to keep the insulation from plugging the screen vents.

Balloon is a circle placed on a drawing with an identification number or letter inside the circle, and the circle is connected to the view with a leader. The balloon number or letter correlate with the same balloon identifying the same location on other drawings.

Balloon framing uses continuous studs that rest on the mud sill at the foundation and run through each floor and to the top plate at the roof.

Baluster, also called a spindle or picket, is one of a series of vertical handrail supports placed equally spaced between newel posts.

Banister   see handrail.

Bar joist is a welded steel joist with an open web made with a single bent bar running in a zigzag pattern between horizontal upper and lower chords.

Barbecue is an appliance such as a grill, a spit, or a fireplace for cooking food over an open fire or heat source.

Bargeboard is attached to the ends of the lookout rafters and used trim the gable end overhang.

Barrier-free refers to the modification of buildings or facilities for use by people who are disabled or have physical impairments. The term is used primarily in non-English speaking countries, while in English-speaking countries, terms such as accessibility are common.

Base is the layer of compacted gravel, such as 3/4 minus, on top of the subbase and directly under the concrete slab.

Base metal sheet steel thickness excludes protective coatings.

Baseboard heaters are placed along the bottom of a wall and have an electric heating element, which causes a convection current as the air around the unit is heated.

Baseboards are placed at the intersection of walls and floors and are generally used for appearance and to protect the wall from damage.

Basement foundation uses a concrete footing and steel reinforced concrete or concrete block foundation wall that extends from the footing to the floor framing above, and a concrete slab that rests on the footing at the foundation wall.

Battens are strips of wood, plastic, or metal installed on the roof to support and fasten the roofing tiles.

Batts are insulation strips that are 93 in. (2350 mm) long to fit in the wall space between studs of a typical 8-ft (2400-mm) wall. See also blanket insulation.

Bay   see bay window.

Bay window is constructed by extending the exterior of a structure outward with short sides at a 45° or 30° angle and a wide outer wall where each side and the outer wall generally have windows.

Beam pocket is a recessed area to hold the end of a beam in a concrete or masonry wall, and is created by pouring concrete around a mold inserted in the concrete forms.

Beams are horizontal construction members that are used to support floor systems, wall or roof loads. Joist construction uses standard dimensional lumber or engineered wood products as joists that span between foundation walls and can be supported at mid span by post and beams or stem walls. Joists are dimensional lumber such as 2×8, 2×10, or 2×12, or engineered wood products, and spaced 12, 16, or 24 in. on center depending on the span and structural engineering.

Bearing locations on the floor framing plan are where bearing walls above the floor being framed rest on the joists or beams.

Bearing partitions carry structural weight from above and distribute the weight to the ground.

Bearing pressure is normally the number of pounds per square foot of pressure the soil is engineered to support.

Bearing wall   see load-bearing wall.

Bedding is specific material such as sand used under pipe and other ground work for uniform grade, protection, and support. Specific bedding requirements can be specified on the plans.

Bench or benching is a fairly level step excavated into the earth material on which fill is placed.

Bench mark (BM) is a survey mark intended to be permanent, which specifies the latitude, longitude, and elevation of a point. Measuring the elevation of points is an essential surveying and civil engineering design and construction process.

Berm is a mound or built-up area.

Besista steel tension rods is a patented system that consists of steel bars called tension rods with external threads that are connected to each other and to the structure by special connecting devices.

Best Management Practices (BMPs) are procedures that provide effective and practical means in achieving construction goals while making the best use of available resources.

Between pours means that it is often necessary to pour a portion of the slab either when the concrete truck is empty or at the end of the day, and then continue the slab with another truck at a later time.

Bevel groove weld is created when one piece is square and the other piece has a beveled surface.

Bevel siding is bevel-shaped wood, metal or plastic material applied with each horizontal row overlapping the previous row.

Bi-fold door has a hinge that holds two panels together so that one can swing relative to the other and provide a 100 percent opening in the doorway.

Bi-pass doors are two or more door panels that slide horizontally past each other for access to a closet or other opening.

Bid, also called **estimate,** is an estimate for the construction cost of a project or portion of a project.

Bidding requirements are used to attract bidders and provide the procedures to be used for submitting bids. Bidding requirements are the construction documents issued to bidders for the purpose of providing construction bids.

BIM model contains all the necessary information to describe the building.

BIM modeling software coordinates the 3D model, the 2D documentation and the attached BIM data.

Biofiltration is a pollution control technique using living material to capture and biologically degrade process pollutants.

Biofuel is biomass used as fuel.

Biomass means natural material, such as trees, plants, agricultural waste, and other organic material.

Biomass energy　see biopower.

Biopower, also referred to as **biomass energy,** uses biomass to generate electricity in a way that is cleaner and more efficient than most other electricity generation techniques. Biomass generating facilities typically use natural biofuels to produce steam, which drives a turbine that turns a generator to produce electricity.

Biotic Soil Amendment (BSA) is a generally hydraulically applied soil amendment used to improve poor soils by adding organic matter, water-holding capacity, and soil life.

Birdsmouth is a notch is cut out of the rafter at the location where the rafter meets the wall top plate.

Bituminous refers to a natural substance.

Black pipe is steel pipe used for natural gas applications is protected with a coat of varnish, referred to as black pipe because of its color.

Blade cut brick is brick cut with masonry saw blades designed to cut through brick, stone, ceramic, tile, and other types of masonry materials.

Blanket insulation is the most common and widely used type of insulation that comes in the form of batts or rolls.

Blocking (roof) is used between the ceiling joists at the top plate to stabilize the joists and enclose the space between rafters. See also solid blocking.

Blower-door system uses a fan to blow air into or out of the building, creating either a positive or negative pressure difference between the inside and outside.

Blue-line prints are made using the diazo process.

Blueprint is an old term sometimes used in the construction business when referring to prints.

Blueprinting is an old method that results in a print with a dark blue background and white lines.

Boiler is a fuel-burning container for heating water.

Bolt is a threaded fastener with a head on one end connected to a body called the shank. The end opposite of the head is threaded to fit a nut or to be fastened into a threaded feature. The bolt is generally used in construction to fasten two or more pieces together in combination with a washer under the head and a washer under the nut.

Bolt is a straight grained, knot free, section of log pre-cut to the desired length of the shingle or shake.

Bond beam is a horizontal reinforced concrete or concrete masonry beam used to strengthen and tie a masonry wall together at the top or in other locations where needed.

Bookshelves and display shelves can have simple construction with metal brackets and metal or wood shelving, or they can be built and detailed the same as fine furniture.

Bore piling   see replacement piling.

Borrow or borrow pit, also known as a sand box, is an area where material such as soil, gravel or sand has been dug for use at another location.

Borrow pit   see borrow.

Bottom rail is a short wall built up above a stringer to enclose the ends of the treads and risers and used to attach the balusters.

Bottom track used in steel wall framing replaces the sole plate used in wood framing.

Bow window is an arc-shaped bay window.

Box is equipped with clamps, used to terminate a conduit, also called an electrical box, outlet box, or junction box.

Box nail is a small diameter nail typically used in making boxes.

Braced panel   see shear panel.

Braced wall line   see shear wall.

Bracing is used in floor framing to provide stiffness to a floor system, to resist lateral loads, to prevent joist rotation, to provide a nailing surface for the bottom plate of a partition framed above, or used to support plumbing and heating equipment. The two types of floor joist bracing commonly used in construction are solid blocking and cross bracing.

Breaker, also called a circuit breaker, is an electric safety switch that automatically opens a circuit when excessive amperage occurs.

Breaking chain is when tape measurements must be made on a slope.

Brick mold is part of the exterior door or window assembly and used to trim and fasten a door installation around the exterior top and sides, or all around a window.

Bridging   see continuous steel wall frame bridging.

Broomed concrete has the surface textured with a broom while still wet. A broomed finish improves traction or to create a distinctive texture on the concrete surface.

Brownfields sites are parcels of land that have the presence or potential presence of a hazardous substance, pollutant, or contaminant.

Brush barriers are perimeter sediment control structures constructed of material such as small tree branches, root mats, stone, or other debris left over from site clearing.

Btu means British thermal unit, a measure of heat.

Btuh is heat loss calculated in Btu's per hour.

Bubble tube   see spirit level.

Buckling is bulging, bending, bowing, or kinking of the steel studs as a result of compression stress on the structure.

Builder Option Package (BOP) represents a set of construction specifications for a specific climate zone, based on performance levels for the thermal envelope, insulation, windows, orientation, heating, ventilating, and air-conditioning (HVAC) system, and water heating efficiency for the climate zone.

Building assembly is all the parts of the structure, including the floor, walls, and roof including all layers of material used in the construction.

Building bricks are typically used as **backing material** in nonstructural and structural applications where appearance is not a requirement.

Building codes are required laws that are intended to protect the public by establishing minimum design and construction standards.

Building envelope consists of the roof, exterior walls and floor of a structure, forming a barrier that separates the interior of the building from the outdoor environment.

Building information modeling (BIM) is a computer technology 3D modeling process that gives architecture, engineering, and construction (AEC) professionals the creativity and tools to efficiently plan, design, construct, and manage buildings and infrastructure throughout the entire process from concept to completion. BIM applications go beyond planning the project and proceeding throughout the building life cycle.

Building life cycle refers to the observation and examination of a building over the course of its entire life.

Building Officials Code Administration (BOCA) codes developed for the east coast.

Building permit is the permission granted by a jurisdiction to build a specific structure based on approved plans and specifications, and required in most jurisdictions for new construction, or adding on to preexisting structures, and in some cases for major renovations.

Building science is the collection of scientific knowledge that focuses on the analysis and control of a buildings physical environment.

Building section is a section that cuts through the entire house.

buildingSMART Data Model is the software-neutral Industry Foundation Classes (IFC) data model and represents a modern approach to interdisciplinary collaboration for all members of the AEC industry.

Built-up roofing systems, commonly referred to as tar and gravel roofs, are installed by alternating layers of asphalt or tar and supporting fabrics directly onto the roof.

Cabinet elevations are exterior views of the cabinets developed directly from the floor plan drawings.

Cabinets are generally a storage or display system with drawers, and doors with shelves behind.

California corner   see three-stud corner.

Callout is used in the construction industry to identify a given product, process or specification using words, numbers, abbreviations, and symbols to identifying an illustration or part of a drawing feature.

Camber is a slight arch in a beam that makes it better for supporting heavy loads.

Cantilever is a structure that is supported at one end and is self-supporting on the other end where it projects into space.

Cantilever beam means a rigid beam that is fixed to a support, such as a vertical column or wall at one end and again near the other end, and the other end of the beam is detached. The detached end is designed to carry vertical loads.

Cap sheds weather off the chimney and keep moisture from entering the masonry.

Carbon dioxide has the chemical formula CO_2, and is a colorless, odorless gas present in the atmosphere and formed when any fuel containing carbon is burned.

Carbon dioxide equivalent emissions path is a standard for measuring carbon footprints by express the impact of greenhouse gas in terms of the amount of CO_2 created using a ratio to convert the various gases into equivalent amounts of CO_2.

Carbon footprint is a measure of the impact human activity has on the environment by producing greenhouse gases.

Carbon monoxide is a colorless, odorless, poisonous gas, produced by incomplete burning of carbon-based fuels.

Carbon monoxide poisoning is a potentially fatal illness that occurs when people breathe in carbon monoxide.

Carbon neutral is a state in which the net amount of carbon dioxide or other carbon compounds emitted into the atmosphere is reduced to zero by carbon offsets.

Carbon offsets are an abstract tool used to reduce the impact of carbon footprints by lessening carbon emissions through the development of alternative projects such as solar or wind energy, or reforestation.

Carpet is a finish floor material made from woven fabric, such as nylon or wool.

Carpet pad is installed between the underlayment and the carpet to provide desired cushion, comfort, and prolong carpet life.

Carriage   see stringers.

Casting is the resulting concrete structure when describing poured-in-place concrete. **Casement windows** can be 100 percent operable and open from the outside vertical edge, generally using a crank mechanism.

Casing is the trim around doors and windows.

Cast-in-place concrete   see poured-in-place concrete.

Catalytic converter is a device that contains a catalyst for converting pollutant gases into less harmful emissions.

Catch basin is inlet structure for a drain or drain system and is designed to drain excess rain and ground water from the adjacent area. The grade is sloped toward the catch basin so that water naturally flows into the structure.

Category II asbestos containing material (ACM) is any material containing more than 1 percent asbestos as determined using the methods specified by the EPA.

Cathedral ceiling has equal sloping sides that meet at a ridge, or stop at a flat ceiling part way to the ridge.

Caulking is the use of a soft, waterproof product to reduce air infiltration by filling gaps, openings, seams, cracks, and voids to make the structure air and water tight.

Cavity refers to the blank area or void between construction members such as studs in walls or joists in ceilings where insulation is placed.

Ceiling joist the horizontal member of the roof system that is used to resist the outward spread of the rafters and to provide a surface on which to mount the finished ceiling.

Cement board, also called backer board, is a combination of cement and reinforcing fibers formed into 1/4 to 1/2 in. thick 4 × 8 ft (1200 × 2400 mm) sheets, or 3 × 5 ft sheets 900 × 1500 mm), that are typically used as a tile backing board.

Centerline stakes represent the centerline of a construction feature such as a driveway, road, or pipe.

Central forced air systems provide heating and air-conditioning for a home that circulates the air from living spaces through or around heating or cooling devices. A fan forces the air into ducts, which connect to openings called diffusers, or air supply registers. Warm air or cold air passes through the ducts and registers to enter the rooms, heating or cooling as needed. Air then flows from the room through another opening into the return duct, or return air register.

Cesspool is a cistern that receives untreated sewage that goes through a process of liquefaction and decomposition by bacterial organisms.

CFM is cubic feet per minute, which is the rate of airflow.

Chaining is a term for measuring in surveying leftover from the use of the old 66-ft-long Gunter chain.

Chair rail is traditionally placed horizontally on the wall at a height where chair backs would otherwise damage the wall.

Chairs rebar can be positioned with these prefabricated supports.

Chase is a continuous recessed area built to carry or conceal ducts, pipes, or other construction products.

Check dams used as runoff control are small, temporary structures constructed across a swale or channel.

Chemical stabilizers or soil binder provide temporary soil stabilization. These are typically vinyl, asphalt, or rubber sprayed on the ground to hold soil in place and minimize erosion from runoff and wind.

Chimney is a structure containing a passage through which smoke and gases escape from a fireplace, furnace, vent or other application.

Chimney cap see cap.

Chimney draft see draft.

Chord is a principal member of a truss extending across the span, and can be parallel or with a bottom chord connected to a top chord forming the roof slope. The top and bottom chords are connected by a web of members.

Circuit breaker see breaker.

Circuits are the various conductors, connections, and devices found in the path of electrical flow from the source through the components and back to the source.

Circular stairs see spiral stairs.

Circulator is a pump used in a hot water systems.

Civil engineer scales are commonly used to draw large projects such as site plans, roads, utilities, and bridges.

Civil engineering is a professional engineering discipline that deals with the design, construction, and maintenance of the physical and naturally built environment, including projects like roads, bridges, canals, dams, and buildings.

Cladding is a wide range of materials as a general term used to describe the application of one material over another to provide usually an exterior layer over construction.

Class A fire resistance is the most stringent rating available for building materials by the National Fire Protection Association (NFPA).

Class A roof is effective against severe fire test exposure.

Class 4 hail-rated product is designed to withstand high winds and hail damage.

Clay tile is made by baking molded clay shapes in a kiln.

Clean gravel contains no fines or small particles of ground rock, which is normally not used for compacting but is used for good drainage.

Cleanout is a fitting with a removable plug that is placed in plumbing drainage pipe lines to allow access for cleaning out the pipe.

Clear distance can have many applications that basically mean the minimum clear spacing between features or objects. For example, the minimum clear space between the edge of concrete to the rebar or welded wire reinforcement, the minimum clear spacing between reinforcement, or the minimum clear spacing between any specific construction members.

Clerestory windows can be used to provide light and direct solar gain to a second-floor living area and increase the total solar heating capacity of a home. Clerestory windows are a row of windows set along the upper part of a wall.

Climate zones are divisions of the earth's climates into general areas according to average temperatures and rainfall.

Coefficient of thermal expansion is the amount of expansion per unit length of a material resulting from one degree change in temperature.

Cold-formed steel (CFS) is the common term for formed by cold processes of rolling, pressing, and stamping thin gauges of sheet steel into products.

Cold rolled refers to a steel forming process when the cold metal is rolled into sheets or other shapes such as steel framing members.

Cold rolled channel system that uses a continuous cold rolled steel channel as bridging that runs continuously through aligned **knockouts** in the studs.

Collar ties are used depending on structural engineering requirements to help tie the rafters together and support the ridge.

Column is a vertical structural member designed to transmit a compressive load, and are typically made of reinforced concrete, but can be constructed from wood, steel, fiber-reinforced polymer, cellular PVC, or aluminum.

Combination film and combustion foam is an underlayment with a sheet of moisture barrier attached, and is used on projects where moisture can rise from concrete.

Combustion air is outside air supplied in sufficient quantity for fuel combustion.

Comfort zone is the range of temperatures, humidity and air velocities at which people generally feel comfortable.

Commercial construction is the project used for business, commercial, or industry applications with documents that are often more complex and comprehensive than those for residential construction.

Commercial doors are specifically manufactured for use in business and industry-related activities.

Common areas are not part of individual private properties and are used by all property owners.

Common nail has a smooth uncoated shank less than one-third the diameter of the nail head and is most commonly used framing.

Common plumbing wall is a wall between rooms where plumbing fixtures are placed back-to-back.

Common rafters are typically dimensional lumber such as 2 × 6 or 2 × 8 spaced 16 or 24 in. (400 or 600 mm) on center that run at the desired roof pitch from the wall top plate at the bottom to the ridge at the top.

Common wall, also called a **party wall,** is a structural fire-resistant-rated wall and floors dividing partition between two adjoining living units or buildings shared by the separate occupants of each residence or business.

Compaction refers to increasing the soil density.

Composite construction refers to two different materials securely bonded together so the two or more individual materials act together as a single structural unit.

Composite pile is a pile made of two or more different types and shapes of material fastened together, end to end, to form a single pile. Also see fiberglass pilings.

Composition roofing is a durable long-lasting roofing product made from a variety of materials such as asphalt, fiberglass, and natural materials depending on the manufacturer.

Compost blankets are a layer of compost material placed on the soil in disturbed areas to reduce storm water runoff and erosion.

Compost filter berm is a long raised bed of compost placed along the site contours to slow runoff and control erosion.

Compost filter sock is compost in a round or oval-shaped mesh tube. Compost filter socks are placed along the site contours to slow runoff and control erosion.

Compression refers to pressing together, or forcing something into less space.

Compressive strength is the force applied by weight.

Compressor (HVAC) maintains adequate pressure to cause refrigerant to condense and flow in sufficient quantities to meet the cooling requirements of the system.

Computer numerical control (CNC) is the automation of machine tools through the use of software in a microcomputer attached to the machinery.

Concrete is a construction material made from cement, sand, gravel, and water mixed together and set in a form to make a solid structure when cured.

Concrete blocks are prefabricated construction blocks made from concrete. Standard size concrete blocks are $4 \times 8 \times 16$ and $8 \times 8 \times 16$ in. A concrete block is also called a concrete masonry unit (CMU), cement block, or foundation block.

Concrete board is concrete pressed into sheets.

Concrete forms are temporary or permanent structures or molds into which concrete is poured.

Concrete masonry unit (CMU)   see concrete blocks.

Concrete piers are placed below the slab within the perimeter at locations where needed to support structural loads distributed to the ground below.

Concrete slab details are drawn to provide information of the intersections of the concrete slabs.

Concrete slab foundation plan is drawn to outline the concrete used to construct the concrete floor or floors. Items often found in concrete slab plans include floor slabs, slab reinforcing, expansion joints, pedestal footings, metal connectors, anchor bolts, and any foundation cuts for doors or other openings.

Concrete slab foundation system uses minimum 3-1/2 in. (90 mm) reinforced concrete that is referred to as a slab that is a flat concrete pad poured directly on the ground.

Concrete tiles are made using Portland cement, sand, and water that are mixed and extruded in molds under high pressure.

Condensation is water that collects when humid air comes in contact with a cold surface.

Condensation resistance (CR) measures the ability of a product to resist the formation of condensation on the interior surface of the product.

Condenser (HVAC) is the portion of a refrigeration system where the compression and condensation of refrigerant is accomplished.

Conditioned air refers to a heated or cooled space within the home.

Conduit is a metal or fiber pipe or tube used to enclose one or more electrical conductors.

Conductor is material that permits the free motion of electricity. Copper is a common conductor in architectural wiring.

Coniferous trees are cone-bearing trees that are normally evergreen.

Construction documents are drawings and written specifications prepared and assembled by architects and engineers for communicating the design of the project and administering the construction contract.

Construction drawings are a principal part of the set of construction documents. The individual drawings needed depend on the specific requirements of the construction project.

Construction entrance stabilization provides gravel on the entrance driveway and road to stabilize the ground and keep mud and dirt off vehicle tires.

Construction fabric   see geotextiles.

Construction loan is commonly used to fund the building of a new home.

Construction sequencing is a specified work schedule that coordinates the coordination and timing of construction activities and stages from the start of land development to final occupancy.

Consulting engineer is a professional expert engineer in a specific discipline who works with the architect on a certain phase of a project.

Contemporary architecture is the modern architecture of today.

Continuous lateral bracing allows the trusses to be accurately spaced and stabilizes the roof system.

Continuous perimeter footings are the lowest member of the foundation system used to spread the loads of the structure on the supporting ground, and the continuous perimeter footings form the boundary of a building.

Continuous steel wall frame bridging is used in bearing walls to provide lateral bracing called bridging.

Contour interval the vertical distance between contour lines.

Contour lines, also called contours, are lines that join points of equal elevation (height) above a given level, such as mean sea level, and they help demonstrate the general lay of the land.

Contract documents are the legal requirements that become part of the construction contract. Contract documents are where the construction drawings and specifications are found.

Control joints (concrete block) allow the concrete masonry wall structures to shrink independently between joints, and to transfer lateral wind loads from one side of the joint to the other.

Control joints (concrete slab) placed periodically across the concrete slab to help restrict the cracking to the locations where the control joints are placed. Control joints can be a cut made in the slab, or a prefabricated joint that separates the slab at specific locations.

Convection air is the movement of heated air that rises and is replaced by cooler air causing air circulation.

Convection current is a natural condition where warm air rises and cool air falls. The heated air rises into the room and is replaced by cooler air that falls to the floor.

Conventional construction used in this textbook is the typical wood-frame construction practices used in most residential construction.

Copper waterstop is a premanufactured copper product that fits into a masonry joint to control expansion, keep out moisture, and keep mortar or debris from entering the joint.

Corbel is a projection of stone, brick, timber, or metal projecting out from a wall to support a structure above.

Cork underlayment, also referred to as acoustic underlayment, is used under wood and laminate floors when noise reduction is desired between floors.

Corner blocks are square decorative prefabricated blocks that are traditionally used at the top corners of doors for ornamental appearance, and to simplify casing installation.

Corner cube prism is a corner cut off of a cube of glass that reflects the beam of light back onto the incoming path, and can be slightly out of alignment with the EDM device and still provide an accurate measurement.

Corner fireplace is a single fireplace built at the intersection of two walls.

Corner staking is placing survey stakes at each corner of the house.

Cornice is generally any horizontal decorative molding that is commonly found over a door or window, or along the corner where a wall meets the ceiling, and is a projection out from the building to protect the home from weather, and generally has decorative moldings used in the finish construction. See also eave.

Corrugated galvanized steel is a building material made in sheets of hot-dip galvanized mild steel, cold-rolled to produce a linear corrugated pattern.

Corrugated stainless steel tubing (CSST) is also used for natural gas piping.

Cost estimate developed by the architect or the general contractor, or both based on materials and labor costs necessary to complete the project.

Counterbore is a flat-bottomed cylindrical enlargement of the mouth of a hole with enough depth to hide the bolt head or washer and nut below the surface.

Course is a row of bricks.

Cove   see cornice.

Covenants, Conditions, and Restrictions (CC&R) refers to any document, contract, or agreement that restricts the usage and enjoyment of real property.

Craftsman style architecture is influenced by the use of natural materials and colors, heavy posts and wood beams, trim boards, masonry veneer accents, and wide porches.

Crawl space refers to a space that is at least 18 in. (450 mm) between the bottom of the wood structure and the ground that provides access to install and service plumbing, and heating and cooling systems where appropriate.

Crawl space system has a perimeter footing and foundation wall with interior piers and stem walls to support the mid span structure above. There are two basic types of crawl space construction, which are the post and beam and joist construction.

Crawl space ventilation is used to prevent moisture buildup and possible moisture damage to floor construction materials and other structural elements exposed to the crawlspace.

Cricket is generally a small sloped surface or roof built to divert water over an area where water would otherwise collect, such as behind a chimney or a place where a slope roof meets a vertical wall.

Cripple is a short stud framed above a door rough opening, or above and below a window rough opening between the header and a top plate, or between the sill and sole plate.

Cross-laminated timber (CLT) is an engineered wood product that is prefabricated using several layers of kiln-dried lumber, laid flat with each layer placed alternating 45° or 90° with each other, and glued together on their wide faces.

Cross-linked polyethylene (PEX) pipe provides a flexible and durable plumbing system that can withstand high pressures and temperatures.

Cross section   see section.

Crown   see cornice.

Cul-de-sac is a street that is closed at one end and usually has a large radius turning area.

Curb, used in roof construction, can have one of several different purposes, but are basically a raised member used to support roof penetrations above the level of the roof surface, or a raised roof perimeter or projection relatively low in height.

Curb, used in road and street construction, is a raised concrete edging to a road, street, or path.

Curb, used in skylight construction, is a 2 × 4 or 2 × 6 frame built on the roof, or a premanufactured curb provided by the supplier, upon which a skylight is mounted.

Curtain wall is any non-load-bearing exterior wall building facade that hangs like a curtain from the floor slabs.

Custom cabinets are usually designed by the architect, architectural designer, or custom cabinet shop, and are generally built at a local cabinet shop and delivered to the project for installation.

Custom home is a home designed specifically for a client who wants to have a new home built.

Cut and fill is the excavation process involving the removal of earth, which is the cut, and moving earth to another location, which is the fill.

Cut stakes indicate a lowering of the ground or elevation.

Cutout template is a paper outline of the exact substrate cutout required for the specific fixture.

Cutting-plane line is commonly used to correlate the section drawing to the location where the cut is made through the building.

d in a nail specification refers to penny, which is an old term still used to specify nail sizes today. The nail size gets larger as the number gets larger. For example, the 8d nail is 2-1/2 in. long and the 16d nail is 3-1/2 in. long.

Datum is theoretically exact reference point, axis, or plane.

Daylighting is the use of various design techniques to enhance the use of natural light in a building.

Dead loads is the weight of the structure, including its walls, floors, finishes, and mechanical systems.

Deadbolt is a locking mechanism that adds extra security to an entry door in addition to the regular lockset, and has a solid cylinder that engaged by turning a knob or key, rather than by spring action.

Decibel is a unit used to measure the intensity of a sound.

Deciduous trees lose their leaves in the winter.

Deed is a written document used to transfer land or other real property from one person to another.

Deep foundation   see piling foundation.

Deflection is the movement of a structure as a result of stress applied to the structure.

Deflection track is the top track in a steel frame wall designed and installed to allow vertical movement of the structure without damage to the wall system and wall finish material.

Deformed steel rebar is rod-shaped steel used for reinforcing concrete, having surface irregularities such as ridges to improve the bond with concrete.

Delta note   see flag note.

Dense-pack is installing higher-than-normal density of loose-fill insulation to reduce settling.

Description of materials includes specific information about construction and finish materials, products, appliances, and equipment.

Design drawings contain all the details required to prepare structural drawings.

Design for all (dfa) is used to describe a design philosophy targeting the use of products, services, and systems by as many people as possible without the need for adaptation.

Design notes are notes that relate to the design characteristics of the plan where they are located.

Details are enlarged sections used to show exact construction requirements at a specific location, such as a footing, connection, or building application, and range in scale from 3/4″ = 1′-0″, 1″ = 1′-0″, 1-1/2″ = 1′-0″, or 3″ = 1′-0″.

Detention basins are created to temporarily detain storm water and release it slowly; and prevent sediment-laden water from leaving the site, or both.

Diaphragm is a structural element that transmits lateral loads to the vertical resisting elements of a structure, such as **shear** walls.

Diazo (see blue-line prints) process uses a chemically coated sheet placed under an original drawing and both exposed to light that removes the chemical except under images. The sheet is then exposed to ammonia vapor that turns the remaining chemical to a blue color.

Differential leveling refers to the method used to determine the difference in elevation between two points.

Dig-out   see excavation.

Digital level can be used to process elevations and distances electronically when used with a special rod that has a face graduated with a bar code.

Dimension is shown on a drawing using *extension lines*, a *dimension line*, and a numerical value giving the measurement.

Dimensional lumber is lumber that is cut and planed to standard width and depth specified in inches, and lengths specified in feet.

Dimensional tolerance (brick) refers to how tightly the manufactured brick is held to its nominal size.

Direct current is a continuous electric current that flows in one direction only.

Direct leveling, also known as **spirit leveling,** is the theory of measuring vertical distances relative to a horizontal line, and using the measurements to calculate the differences between the elevations of points.

Direct-vent fireplaces are generally sealed-combustion heaters, having exhaust gases exit through the center, and combustion air is drawn into the fireplace through the outer chamber.

Displacement piles are columns installed without removing material from the ground.

Distribution panel, also called a panel, is where the conductor from the meter base is connected to individual circuit breakers, which are connected to separate circuits for distribution to various locations throughout the home.

Division in the *MasterFormat: Master List of Numbers and Titles for the Construction Industry*, 2004 edition, is made up of level 2, level 3, and occasionally level 4 numbers and titles assigned by MasterFormat, each of which defines a gradually more detailed area of work results to be specified.

Documents are the drawings and written information related to a project.

Dolly is a solid steel bar used to hold the end of a rivet head in place while a head is formed on the other end.

Dome skylight has generally plastic that rounds up over the frame.

Dormer is a small structure extending out from the slope of the main roof of a home.

Double-acting doors swings both ways for entering and leaving a room.

Double faced see through fireplace has a firebox that is open to two adjacent rooms.

Double-hung window has two sashes with one above the other. Both sashes open to any desired extent by sliding vertically.

Double wall construction uses structural sheathing over the exterior studs, followed by a vapor barrier, followed by finish siding material.

Dowel-laminated timber (DLT) is a structural and economic mass timber panel that is used for floor, wall, and roof construction that has mating boards drilled with a pilot hole and hardwood dowels are press-fit into the holes holding the boards tightly together.

Down-draft range exhaust system is generally directly behind the range and has an exhaust fan that ventilates the range to the outside.

Downspout is a vertical pipe that is connected to the gutters for the purpose of moving rainwater from the gutter to the ground or to a rain drain pipe in the ground.

Draft, also referred to as chimney draft, is a current of air and gases through the fireplace and chimney, which is the process of air being pulled from the area inside the home and through the fireplace opening, allowing hot air to escape through the flue and out of the chimney.

Drain is any pipe that carries wastewater in a building drainage system.

Drainage system provides for the distribution of solid and liquid waste to the sewer line.

Drawing revisions are common in the architectural, structural, and construction industry. Revisions can be caused for a number of reasons. For example, changes requested by the owner, job-site corrections, correcting errors, or code changes.

Drawing scale   see scale.

Drill is a rotating cutting tool used for making holes.

Drip cap is a formed aluminum or vinyl projection installed at the top of windows and doors that allows water to run off.

Drip edge is an L-shaped galvanized steel or aluminum flashing that keeps water from penetrating the edge of the sheathing at the bottom of the roof.

Drip line is the area directly located under the outer circumference of the tree branches.

Driven piles   see displacement piles.

Drive types refer to the type of driver tool used to drive or turn the screw into the material.

Dropped beams are beams that have the joists running over the top of the beams and the beams are exposed in the room below.

Dry-in is a term used to describe placing a weatherproof layer of material over the roof sheathing in an effort to keep the home dry.

Drywall is also called gypsum board, gyprock, plasterboard, sheetrock, or wallboard, is a plaster panel made of **gypsum** pressed between two thick sheets of paper.

Drywall clips or other nailing brackets are used to secure the other drywall material edge at a corner that has no framing support.

Drywall knife is a small hand tool with a blade used for applying joint compound when finishing drywall.

Drywall tape is a paper, vinyl, or mesh product used to cover drywall joints and corners.

Duct system is square, rectangular, and round sheet metal or plastic pipes used to conduct hot or cold air of the HVAC system.

Ductless heat pumps, also called mini-split-system heat pumps, are heat pump systems that heat or cool directly into the room without the use of ductwork.

Ducts are the pipes used to move hot or cold air in a heating, air-conditioning, and ventilation system.

Dummy knob looks just like other door knobs or leavers, but they do not operate a lock bolt, and are often mounted with a magnetic latch or other latching mechanism on closet, pantry, and utility doors.

Duplex outlet, also called a **duplex convenience outlet,** has two outlets and is the typical wall plug.

Dutch doors are used when it is desirable to have a door that can be half open and half closed.

Dutch hip roof is a combination hip and gable roof, with the gable portion at the top.

Earth anchors   see helical pilings.

Earthen floors are a green building technology that combines the mixture of sand, clay soil, and fibers that are poured in place and sealed with linseed oils and waxes to make durable washable surfaces.

Earthquake engineering, also called **seismic engineering,** is the study of the behavior of buildings and structures that can be affected by earthquakes.

Easement is a right given to a person or entity to access a portion of land that owned by another person or entity.

Eave is the lowest part of the roof that projects from the exterior wall, also referred to as a cornice or overhang.

Edging trowel is used to make concrete edges and joints formed with a radius along the slab edge that helps resist chipping and damage after the forms are removed.

Egress refers to the way for people to exit a building in an emergency.

Electric fireplaces require 110 or 220 V electrical wiring depending on the manufacturer specifications, and produce very little heat, allowing them to be installed directly against wood framing with no ventilation required, so electric fireplaces can be used in locations where no venting source is available and other types of fireplaces are not suitable.

Electric system is a network of wiring, outlets, and fixtures used to transmit and supply the electrical needs for a home.

Electrical box   see box.

Electrical circuit line   see switch leg.

Electrical conduit is electrical piping system used to protect and rout electrical wiring.

Electrical legend on the plan correlates fixture symbols to descriptive text.

Electrical legend on the plan correlates electrical fixture and wiring symbols with their name or description.

Electrical loads are the circuits and appliances that use electricity.

Electrical plan shows the location and details of electrical wiring and fixtures.

Electrical service panel is a distribution box with circuit breakers connecting electrical wiring to circuits within the home.

Electrician is the subcontractor who performs the electrical work for the project.

Electrochromic glass, also called **smart glass** or **electronically switchable glass,** is a technologically advanced building glass that can be used to create partitions, windows, and skylights that change opacity from clear to a desired shade using electricity.

Electrochromism allows certain materials to change color or opacity when an electrical charge is applied.

Electronic distance measurement (EDM) is a device used to measures distance using electromagnetic or microwave radio waves, or infrared or laser light waves.

Electronic or **laser level** is commonly used for construction work such as leveling to set elevation grade stakes for road, parking lot, and driveway grades, and elevations for excavations, fills, concrete work, plumbing, and floor level verification.

Electrostatic air filters clean the air flowing through the HVAC system by using static electricity in a unit placed in the return air duct directly next to the furnace or air handler of the system. The electrostatic air filter works by an electrostatic charge generated by air flowing through a network of static fibers.

Elevation (surveying) is the vertical difference between two points.

Elevation is the two-dimensional (2D) exterior view of a building. An elevation is drawn at each side of the building to show the relationship of the building to the final grade, the location of openings, wall heights, roof slopes, exterior building materials, and other exterior features, and interior elevations show interior features such as cabinets, and other interior characteristics.

Elevation is related to surveying, site planning, and disciplines such as plumbing and HVAC. The height of a feature from a known base, which is usually given as 0 (zero elevation).

Embodied energy is the combination of all energy required to produce a product or service, as if the energy is in the product.

End bearing piles are columns driven into the ground and the bottom end of the pile rests on a layer of strong soil or rock, where the load of the building is transferred through the pile onto the ground layer at the bottom.

End dams, also called **side dams,** are made by bending the end of flashing up to stop water from flowing across the flashing and into the adjacent construction.

End-result specifications provide final characteristics of the products and methods used in construction, and the contractor can use a desired method for meeting the requirements.

Energy positive home is when the home energy system creates more electricity than needed for the home, and the system allows the home owner to sell electricity back to the utility company.

Energy Rating Index (ERI) is a performance path that gives builders options for meeting target energy efficiency score through a wide range of performance options to demonstrate compliance.

Energy recovery ventilators (ERV) system is closely related to a HRV, except the ERV also transfers the humidity level of the exhaust air to the intake air.

Energy Star is a joint program of the U.S. Environmental Protection Agency (EPA) and the U.S. Department of Energy created to help save money and protect the environment through energy-efficient products and practices, and through superior product energy efficiency.

Engineered wood products are a combination of smaller components to used make structural products that have been engineered for specific applications and fabricated in a manufacturing facility and delivered to the construction site.

Entry door is a door that allows access to the home from the front, back or other location through an exterior wall.

Envelope, related to home design and construction, is the entire exterior of the home that includes exterior walls, the roof, the foundation, and doors and windows.

Environmentally friendly and **environmentally sound** refer to design and construction using renewable materials.

EPDM is the abbreviation and commercial name for ethylene propylene diene terpolymer, which is an extremely durable synthetic rubber roofing membrane widely used on low-slope roofs.

Erosion is the process that causes soil and rock to be removed from the earth's surface and deposited in other locations by natural processes such as wind or water flow.

Estimate   see bid.

Ethylene propylene diene terpolymer (EPDM) is an extremely durable synthetic rubber roofing membrane () widely used in low-slope buildings.

Evaporator (HVAC) is a device in which a liquid refrigerant is vaporized.

Excavation as any man-made cut, cavity, trench, or depression in the earth surface formed by earth removal. This includes the excavation for a house or other building that is referred to as a dig-out in the construction industry.

Expanding foam insulation is a spray-applied insulating foam in a pressurized container that is installed from a nozzle as a liquid and then expands many times its original size until it hardens in its final form upon contact with the air.

Expansion joints are used to control horizontal movement by separating masonry into sections to prevent cracking, and stop water penetration and air infiltration in the masonry wall.

Expansion valve (HVAC) controls the amount of refrigerant flow into the evaporator.

Exposed aggregate concrete is a method of finishing concrete by washing the cement and sand away from the top surface of the concrete, exposing the aggregate immediately under the surface.

Extensible Markup Language (XML) is a general-purpose description language that can describe different kinds of data.

Exterior door is any door in an exterior wall or in a wall between the home and garage.

Exterior elevations are drawings that show the external appearance of the building.

Exterior envelope is made up of elements of a building that enclose conditioned spaces through which thermal energy transfers to or from the exterior.

Extrusion is a manufacturing process used to create long objects of a fixed cross-sectional profile, where material is pressed or pushed under pressure through a desired shaped opening.

Face brick is used for visual quality on the exposed surface of a building wall or other structure.

Face plate is a rectangular plate on the edge of the door around the latch bolt.

Face stringer is an exposed stringer on the open side of stairs.

Fan heater has a heating element used to generate heat, and a fan circulates the heat into the room.

Fascia is usually 1× dimensional lumber or plastic material installed over the ends of rafters or at the rake as a finish or trim.

Fenestration is any opening in a building envelope including windows, doors and skylights.

Fiber carpet pad is made from a nylon that provides a high-density product.

Fiber rolls are tube-shaped erosion-control devices filled with straw, flax, rice, coconut fiber material, or composted material.

Fiberglass is a reinforced plastic material composed of glass fibers embedded in a resin matrix.

Fiberglass pilings are considered a composite piling, because of the material from which they are made. Also see composite pilings.

Field is the internal area of the plywood between the edges.

Field data based on leveling contain location and elevation information.

Field weld is a weld that is performed on the job site, which is referred to as the field, rather than in a fabrication shop.

Fill is moving earth to another location.

Fill stakes indicate raising the ground or elevation.

Filler are pieces of wood or other material used to fill in the space between modular cabinet units or between the cabinet and a wall.

Fillet weld is formed in the internal corner of the angle formed by two pieces of metal.

Filter fabric is a water-filtering porous material and soil stabilization product made to allow water to pass through, but keep dirt and other materials from passing through.

Filter fabric sock is a filter fabric product made to fit over the drain tile to keep material from plugging the drain pipe.

Fines describes small miscellaneous gravel, dirt, and debris sizes found in any gravel and crushed rock. All gravel has fines, but the amount of fines allowed depends on the gravel specifications.

Finish refers to preparation and final process used to improve the appearance and protect surfaces in the home.

Finish grade means the grade after all required work is performed to cut or fill the area.

Finish wood is the best quality wood that has been milled and sanded to provide the best possible material for final wood projects.

Fink truss provides the typical pitch roof and flat ceiling.

Fire area refers to the combined floor area enclosed and bounded by fire walls, fire barriers, exterior walls, or fire-resistance-rated horizontal assemblies of a building.

Fire barrier wall is a nonstructural wall that provides fire-rated protection.

Fire blocking fills and seals the construction cavity to prevent or slow down the spread of fire.

Fire brick is brick made of refractory ceramic material which resists high temperatures.

Firebox is the chamber built or manufactured to hold the fire in a fireplace.

Fireplace footing is a solid reinforced concrete structure engineered to carry the weight of the fireplace and effectively distribute the load to the ground.

Firewall is a wall or partition designed to deter or prevent the spread of fire in the home.

Fit home is designed to fit on lots where traditional plans do not fit.

Fitting (plumbing) is a standard pipe part such as a coupling, elbow, reducer, tee, and union; used for joining two or more sections of pipe together.

Fixed window has a single non-operable glass pane, or a group of large windows each separated by a vertical support.

Fixture (plumbing) is a component used to supply and contain water, and discharge waste.

Fixtures (electrical) are all plugs, switches, cover plates, and lights.

Flag note or **delta note** is a note keyed to a triangle with a revision number inside that is placed next to the revision cloud or along the revision cloud line. The number in the flag note triangle is then correlated to a revision note placed somewhere on the drawing or in the title block.

Flame cutting process uses a high-temperature gas flame to preheat the metal to a kindling temperature, at which time a stream of pure oxygen is injected to cause the cutting action.

Flange is a metal or vinyl strip that goes all around the window frame and is used to seal and fasten the window to the home framing.

Flare-bevel groove weld is commonly used to join flat parts to rounded or curved parts.

Flare-V groove weld is commonly used to join two rounded or curved parts.

Flashing (masonry) between the concrete or concrete block ledge and the masonry veneer acts to collect and divert moisture from the wall.

Flashing (roof) components used to weatherproof or seal the roof system edges at perimeters, penetrations, walls, expansion joints, valley, drains, and other places where the roof covering is interrupted or terminated. Types of flashing include **base flashing** that is metal or composition flashing at the joint between a roofing surface and a vertical surface, such as a wall or parapet. **Cap flashing** is used to cover the top of various building components. **Counter flashing** a strip of sheet metal in the form of an inverted L built into a vertical wall of masonry and bent down over the flashing to make it watertight.

Flat-plate collector is the heart of a hot water solar system, with the transparent glass cover, absorber plate, flow tubes, and insulated enclosure that works by sun heating water that flows through the copper flow tubes inside the glass covered collector.

Flat roof for a home is constructed almost horizontal or level with a slight slope that allows for rain water drainage.

Flight is a set of steps between one floor or **landing** and the next.

Floor joists are structural members that are normally made from wood or other engineered products used to span between supports such as foundation walls, frame walls, or other construction members to construct the floor system.

Floor plan is a scale drawing showing the arrangement, sizes, and location of rooms in one story of a building. A floor plan is a representation provided by an imaginary horizontal cut made through the home at approximately 4 ft (1220 mm) above the floor line.

Flow rate is calculated in gallons per minute (GPM).

Flue is any vent or chimney that connects a combustion device with the outside, and the passageway in a chimney for conveying gases to the outdoors.

Flush beams are beams that have the joists intersecting the beams and the bottom of the beams are flush with the bottom of the joists. When the beam height is greater than the joists, then the beam can be partly exposed above or below the joists depending on the desired appearance and construction requirements.

Flush contour weld is when the weld surface is the same level as the surrounding metal surface.

Flush-cut saw has no **saw set** on one side, so that the saw can be laid flat on a surface and cut without scratching the surface.

Fly ash is a fine powder that is a by-product of burning pulverized coal in electric generation power plants.

Fly-through is similar, but the camera is like a helicopter flying over the area.

Foam carpet pad is made from urethane that is not designed for use in high traffic areas.

Foam underlayment is commonly used under wood and laminate floors.

Footings are the lowest member of the foundation system used to spread the loads of the structure on the supporting ground.

Footplate used with some level rods provides a stable, well-defined foundation for the rod, and is a cone-shaped device that has pointed feet anchored in the ground.

Footprint is used to identify the foundation dimensions upon which the home is constructed.

For sight (FS) is the rod or target location in surveying to which an elevation and/or location reading is taken.

Formaldehyde is a chemical found in disinfectant, preservative, carpets, furniture, and the glue used in construction materials, such as plywood and particle board, as well as some insulation products.

Foundation is the construction system used to support the structure loads and distribute the loads to the ground.

Foundation details provide information on the concrete foundation at the perimeter walls, and the foundation and pedestals at the center support columns.

Foundation plan is a scale drawing used to display construction features and dimensions for the **foundation** of the building. The foundation plan scale is normally 1/4″ = 1′-0″.

Foundation vents are closable with built-in doors, louvers, or insulating foam inserts. This allows the foundation vents to be closed during the winter.

Foundation walls are the vertical walls of the foundation system than connect between the footings and the structure above.

4D modeling is a term often used in the construction industry as a time dimension to a 3D CAD model that allows teams to analyze the sequence of events on a timeline and visualize the time it takes to complete tasks within the construction process.

Four-way switch is three switches controlling one or more lights.

Foyer is an entrance hall or other open area in a home used for people to gather generally after entering.

Frame (steel construction) structural system that supports the walls and roof.

Frame is a fixed part of a window that encloses the sash and consists of the head jamb, sill, side jambs, jamb extension, and casing.

Framer is the person who does framing, also known as a rough carpenter.

Framing refers to the construction of the structural parts of a house.

Framing plan shows the construction members used in floor framing, roof framing, and the framing of construction features such as decks.

Freestanding fireplace is a manufactured metal appliance made to hold a fire.

Freestanding stairs, also called open stairs, has stringers or carriages that are unsupported by walls, and attached only to the floor at the bottom and the structural framing at the top.

French doors normally refers to double exterior or interior doors that have glass panels and swing into a room.

French drains can be used to capture and drain water away from the house and direct it to local storm water channels. French drains are common drainage systems, primarily used to prevent ground and surface water from penetrating or damaging building foundations.

Freon is a trademark used for a variety of nonflammable gaseous or liquid fluorinated hydrocarbons employed primarily as working fluids in refrigeration and air-conditioning and as aerosol propellants.

Frothed foam carpet pad is made of densely packed urethane, and designed for use under any carpet, and wears under heavy traffic and weight.

Functional elements, also referred to as systems or assemblies, are common major components in buildings that perform a known function regardless of the design specification, construction method, or materials used.

Furnaces for residential heating produce heat by burning fuel oil or natural gas, or by using electric heating coils or heat pumps.

Furring is pressure-treated 2×4 studs fastened to a concrete wall.

Gable is the vertical triangular wall built on each end of a home with a gable roof.

Gable roof is a roof system that slopes downward both ways from a central horizontal ridge so as to leave a **gable** at each end.

Galvanize is a zinc coating.

Galvanized pipe is steel pipe that has been cleaned and dipped in a bath of molten zinc.

Gambrel roof has a double slope on each side, where the lower slope is normally steeper than the upper slope.

Gas logs are an open flame appliance with manufactured ceramic or ceramic fiber logs placed over a burner to provide dramatic realism of a traditional flame.

Gas metal arc welding process can be used to weld thin material or heavy plate. It was originally used for welding aluminum using a metal inert gas shield, a process that was referred to as MIG. The application now employs a current-carrying wire that is fed into a joint between pieces to form the weld.

Gas tungsten arc welding process is sometimes referred to as **tungsten inert gas welding (TIG),** or as **Heliarc®,** which is a trademark of the Union Carbide Corporation.

Gauge is a range of numbers from 10 to 26 specifying thickness for sheet metal or wire. The sheet metal gets thinner as the gauge number gets larger.

General contractor is the person or business who contracts for the construction of an entire building or project, rather than just a portion of the work.

General notes relate to the entire plan.

Genuine Stone® is the Natural Stone Council trademarked **Genuine Stone®** as a defining name for natural stone.

Geodetic vertical datum is the principle datum used to express land elevations and to provide vertical survey control.

Geotechnical engineer is a civil engineering involved in engineering the performance of construction taking place on the surface or in the ground, using principles of soil and rock mechanics to investigate and monitor site conditions, earthwork and foundation construction.

Geotextiles are porous fabrics also known as filter fabrics, road rugs, or construction fabrics.

Geothermal heating and cooling closed-loop system has water or water-antifreeze fluid pumped through polyethylene tubes.

Geothermal heating and cooling open-loop system has water pumped from a well or reservoir through a heat exchanger, and then discharged into a drainage ditch, a field tile, a reservoir, or another well.

Geothermal heating and cooling refrigerant-based system is also known as direct exchange where refrigerant flows in copper tubing around a heat exchanger.

Geothermal heating and cooling systems are designed to use the constant, moderate temperature of the ground to provide space heating and cooling, or domestic hot water, by placing heat exchangers in the ground, or in water wells, lakes, rivers, or streams.

Girder is a horizontal structural member made of wood, laminated wood, engineered wood, or steel that spans between two or more supports at the foundation level, or above any floor level.

Girder truss is a truss used to hold trusses that are perpendicular to the girder truss.

Girt is a horizontal structural member in a framed wall that provides lateral support to the wall panel to resist wind loads.

Glass sliding doors have two or more glass panels that slide horizontally on a track, where one panels slides next to another panel to open, and are made with wood or metal frames and tempered glass for safety.

Glaze is a layer or coating that has been fused to a ceramic object through firing to high temperatures in a kiln.

Glazed bricks have a glaze.

Global navigation satellite system (GNSS) is a generic term for any system of satellites that provide global coverage for geographic positioning.

Global Positioning System (GPS) is the United States global navigation satellite system (GNSS) which uses a network of satellites in combination with Earth-based receivers to determine locations on the earth.

Glued-laminated beam (glulam) is a structural member made up of layers of lumber that are glued together.

Glulam truss made with glulam chords and webs interconnected with heavy metal gusset nail plates.

Glycol is ethylene glycol or ethylene alcohol that is a colorless, sweet liquid used as antifreeze and solvent.

Grade of land or constructed site, refers to the amount of incline of the surface using earth removal practices.

Grade control station provides grade and offset information to the specific grade at construction.

Grading plan shows the existing and proposed contours, elevations and grades of the construction site before and after excavation.

Grade stakes provide elevation information.

Grass-lined channel can be an attractive part of the landscaping and used to carry storm water runoff.

Grate is a framework of metal bars in the form of a grille set into a catch basin, wall, pavement, or other location, serving as a cover or guard allowing air, light and water to pass through.

Gravel is a mixture of three sizes or types of material: stone, sand, and fines.

Gravel filter berm is a temporary ridge made up of loose gravel, stone, or crushed rock used as an efficient form of sediment control on gentle slopes.

Gravity loads include **dead loads,** which is the weight of the structure, including its walls, floors, finishes, and mechanical systems, and **live loads,** which is the weight of contents, occupants, and weight of snow.

Green building refers to a structure and construction processes that are environmentally responsible and resource-efficient throughout a building life cycle.

Green home uses less energy, water, and other natural resources; creates less waste; and is healthier and more comfortable for the occupants.

Green lumber is air-dried lumber.

Green roof, also known as a **rooftop garden,** has plants over the roof structure to help reduce building temperatures, filter pollution, and lessen water runoff.

Green Seal is a nonprofit organization that uses science-based programs to empower consumers, purchasers, and companies to create a more sustainable world.

Green space is an open space area that can be used for a park, a playing field, a natural area with or without walking trails. The green space can be well landscaped or totally natural, and can be available for use only by the adjoining landowners or open to the public.

Greenhouse effect is atmospheric heating caused by solar radiation being transmitted inward through the earth's atmosphere with less radiation transmitted outward due to absorption by gases in the atmosphere.

Greenhouse gases are the gases that absorb global radiation and contribute to the greenhouse effect. The main greenhouse gases are water vapor, methane, carbon dioxide, and ozone.

Groove welds are commonly used to make edge-to-edge joints, and used in corner joints, T joints, and joints between curved and flat pieces.

Ground or grounded refers to an electrical connection to the earth by means of a rod.

Ground fault circuit interrupter (GFCI or GFI) trips a circuit breaker when there is any unbalance in the circuit current.

Grout   see mortar.

Grout joint   see mortar joint.

Guard stake is usually a 48 in. lath with paint and flagging used as visibility marker and protection for the guard stake.

Gunite see shotcrete.

Gusset is a steel or other material plate used to strengthen a joint in the structure.

Gutter is a channel at the edge of the eave for moving rainwater from the roof to the downspout.

Gypsum is a soft white or gray mineral consisting of hydrated calcium sulfate and is used to make plaster, which is mixed with water and allowed to harden for various applications such as making drywall.

Gyprock   see drywall.

Gypsum board   see drywall.

Half bathroom is a bathroom without a shower or bathtub.

Hand is door swing direction.

Handrail is generally a single rail attached to a wall or baluster following the rise of the stairs and used for hand support and safety.

Hang is slang used in the construction industry to mean something that is installed by placing and fastening in a desired location, such as hang a door, or hang drywall. Hung is the past tense of hang.

Hard conversion means that the typical inch units are converted directly to metric.

Hard-wired means a fixed connection between electrical, electronic components, and electric devices by means of permanent wiring into the electrical system.

Hardwood is wood from a broadleaf tree, but more specifically one of the fine hardwoods typically used in finish architectural applications, such as cherry, oak, and walnut.

Hardwood plywood is manufactured using the same veneer system as softwood plywood, except hardwood is used as the facing surface veneer for quality wood appearance.

Head is the vertical distance that water falls, measured in feet.

Head jamb is the horizontal part of the door frame above the door.

Header is a horizontal structural member that supports the load over an opening such as a door or window, or around an opening.

Headers are used to break up the typical stretchers by providing units oriented perpendicular to the face of the wall with the end of each brick exposed.

Hearth is a reinforced concrete slab that rests on the structure below and upon which the fireplace is constructed.

Heat exchanger (fireplace) transfers the heat created in the firebox to the **convection air** and into the room.

Heat exchanger see heat recovery and ventilation (HRV) a device for transferring heat from one medium to another.

Heat pump is a forced-air central heating and cooling system that operates using a compressor and a circulating refrigerant system.

Heat recovery and ventilation (HRV) system is also referred to as a heat exchanger, air exchanger, or air-to-air heat exchanger. A heat recovery ventilation system is a ventilation system that uses a counter-flow heat exchanger between the inbound and outbound air flow.

Heating, ventilating, and air-conditioning (HVAC) is the terminology used to refer to the industry that deals with the heating and air-conditioning equipment and systems found in a building.

Height of the instrument (HI) is the back sight reading added to the elevation of the known point.

Helical piles, also called screw pilings, are steel tubes with helical blades attached.

Herringbone brick placement provides an interesting pattern a pattern by placing bricks in angular rows of parallel groups with any two adjacent rows angled in opposite directions.

Hidden lines are short equally spaced dashed lines that represent a feature that is not visible in the current view.

High-efficiency particulate air (HEPA) filters are designed to be 99.97 percent effective in capturing particles as small as 0.3 μm.

Hinge jamb is part of the door frame where the hinges are attached.

Hip is the corner of an external angle formed by two intersecting exterior sides.

Hip jack rafter is a jack rafter that connects to the hip.

Hip rafter is the rafter that forms the hip and runs from the ridge to the top plate.

Hip roof has the ends and sides sloping with the same pitch.

Hold-down anchors refers to fasteners that are embedded in concrete foundation walls and attach to wood structural walls above, or attach wood structural walls between floors, and especially at shear walls.

Hollow bricks are used in the manufacture of building bricks and facing bricks, with cavities used to place wall anchoring and for steel reinforced masonry where the cavities have steel reinforcing and are filled with grout.

Holz-Stahl-Komposit (HSK)™ plates are high-capacity holddowns for mass timber construction.

Home Energy Rating Systems (HERS) rating is an evaluation of the energy efficiency of a home, compared to a computer-simulated reference house of identical size and shape.

Hopper gun has a container with a spray nozzle that is operated by an air compressor. The container or hopper is filled with joint compound that is sprayed out through the nozzle in a consistent pattern on the drywall.

Hopper windows hinge at the bottom and swing inward.

Hose bibb is a faucet used to attach a hose.

Hot water circulation pump can be used to circulate water through the main hot water line and the return-line back to the water heater to keep hot water in the main line and available to the fixtures at all times.

Hot water system, the water is heated in an oil- or gas-fired boiler and then circulated through pipes to radiators or convectors in the rooms.

Hotlinking is a process of linking directly to other project files or predefined typical project components.

House wrap is the term that describes a variety of synthetic products that have replaced tar paper for use as a vapor barrier.

Housed stringer, also called a routed stringer, is a closed stringer that has the ends of the treads and risers recessed into channels cut into the stringer.

Hub stakes are normally driven flush with the ground and a survey tack is set in the top of the hub to mark the exact survey point.

Hubs are important survey starting points or they mark important survey points, work points, or reference points, which are to remain in place for future use.

Humidistat device that automatically regulates the humidity of the air in a room or building.

Hybrid concrete construction is a commercial construction system that combines precast concrete and cast-in-place concrete to take advantage the different qualities of each practice.

Hybrid panelized roof system combines panelized wood components connected to open web steel joists.

Hydroelectric generators convert the energy from falling water into electricity.

I-joist is an engineered wood product with a web between a top and bottom flange, creating the I shape.

Index contours contour lines that show the elevation values. Generally every fifth line is used as an index contour line.

Industry Foundation Classes (IFC) data model is intended to describe architectural, building and construction industry data. It is a platform neutral, open file format specification that is not controlled by a single vendor or group of vendors.

Infill is where extra ground material is brought onto site to build up a sloping area, cavity, or depression in the existing terrain.

Inspections are required to insure that the construction is proceeding according to the approved plans and specifications, and that all current building code requirements are met.

Instantaneous or **demand water heaters**   see tankless water heater.

Insulated concrete forms (ICF)   see insulating concrete forms.

Insulated structural sheathing combines engineered wood structural sheathing, added insulation, air-infiltration barrier, and moisture-resistant barrier in exterior wall and roof sheathing panels.

Insulating concrete forms (ICF) are rigid foam forms that hold concrete in place during pouring and curing.

Insulation is material used to restrict the flow of heat, cold, or sound, saves energy costs and makes the home comfortable; properly insulated walls, ceilings, and floors stay warmer in winter and cooler in summer.

Insulation cover (IC) means that you are allowed to insulate around and over the recessed light to help avoid heat loss.

Insurance coverage for construction is specifically for general contractors and subcontractors, and includes coverage for general liability, physical loss damage to owned property, and business interruption, including loss of income.

Integrated Project Delivery (IPD) is a new approach to the design and construction of buildings that is based on a cooperative working relationship, shared risk and reward, and open exchange of data.

Interior door is any door on the inside of the home used to enter rooms, closets, or other areas.

Interior elevations or details such as the interior finish elevation Interior elevations are drawings that show the inside appearance of specific characteristics such as cabinets, architectural details, and other inside features that need to be represented for construction.

Intermediate contours contour lines other than the index contour lines.

Intermediate foresight (IFS) is a separate elevation taken in addition to the run of levels for information purposes.

International Conference of Building Officials (ICBO) codes developed for the west.

Inverter is an appliance used to convert direct current (DC) power, produced by the wind generator, into standard household AC current.

Isolation joints are used in concrete block and masonry veneer construction to create a joint between the masonry and other material that allows both structures to contract and expand independently.

Jack rafter is any rafter than is shorter than a common rafter and connecting to a hip rafter or a valley rafter.

J groove weld is necessary when one piece is a square cut and the other piece is in a J-shaped groove.

Jack stud is a partial stud nailed next to full studs to support the header at door or window openings.

Jalousie window has overlapping narrow glass, metal, plastic, or wooden louvers, operated with a crank handle for adjusting the louver angle outward.

Jamb is a vertical framing member that forms the sides of a door opening, window opening, or other opening.

Jamb extensions are used over the framing between the window frame and the edge of the drywall or other wall finish material.

Joint compound, commonly called mud, is used to finish drywall joints, corners, screw locations, and is used as a skim coating.

Joint trowel creates a groove without extending the opening through the concrete.

Joints (concrete) to provide a clean surface to bond between **pours,** to control expansion, and to isolate stress in the concrete.

Joist construction uses standard dimensional lumber or **engineered wood products** as joists that span between foundation walls and can be supported at mid span by post and beams or stem walls.

Joist dimensional lumber such as 2×8, 2×10, or 2×12, or engineered wood products that are spaced 12, 16, or 24 in. on center depending on the span and structural engineering.

Joist hanger is a manufactured metal angle, bracket, or strap used to support and attach the ends of floor joists to beams.

Junction box is an electrical box that protects electrical wiring splices in conductors or joints in runs.

Jurisdiction refers to the city, county, or state location where the official authority exists to make legal decisions and judgments.

Key is a slot in the footing that is runs down the center of the footing and along the entire length of the footing created when the footing is poured.

Key brick is the brick at the top center of a brick arch and has an equal number of masonry units laid on each side down to the supporting structure.

Keyed means that the footing is keyed to the foundation wall to keep the wall connected to the footing. The key is a slot in the footing that is created when the footing is poured.

Keyhole saw is a small handsaw with a long narrow blade used for cutting holes, short radius curves, and other small features.

Keynote is a note found on the drawings. Each keynote has a letter or number, or combination of letters and numbers or symbols next to or pointing at a specific feature on the drawing that correspond to the description of the keynote in a legend or in a general note.

Keystones are wedge-shaped blocks used in the center of curved and straight arches for a decorative appearance or to match Italian architecture.

Kiln is a furnace or oven used to heat products to a desired temperature.

Kilowatt (kW) is a measure of power.

King studs are studs used to support and trim both ends of a header, and run from the sole plate to the top plate.

Knee wall   see stem wall or pony wall.

Knockdown texturing is created by spraying texturing material on the drywall and then using a trowel to flatten the high places while still wet.

Knockouts are prepunched holes at regular intervals to allow rapid installation of electrical conduit, mechanical, piping, and structural applications.

Kyoto Protocol, also known as the Kyoto Accord, is an international treaty among industrialized nations that sets mandatory limits on greenhouse gas emissions.

L-shape stairs makes a 90° turn at a landing or with **winders** between flights.

Lag bolt, also called a **lag screw,** is a heavy wood screw with a tapered point and a square or hex head.

Laid means placing one masonry unit at a time to build the masonry structure.

Laminate, also referred to as a laminated structure or material, especially one made of layers fixed together to form a hard, flat, or flexible material.

Laminate flooring, also called floating wood floor, is a multilayer synthetic flooring product fused together with a lamination process.

Lamstock is quality high-grade dimensional lumber with low-moisture content and minimal defects, designed for specific strength requirements for use in mass timber products.

Land surveying is the technique, profession, and science of accurately determining the three-dimensional location of points and the distances and angles between points on the earth.

Lap siding has a notch on one side that allows each piece to lap over the previous.

Lap splice is the most common method of creating a single structural object from splicing together two rebar or welded wire reinforcement segments. A lap splice is made by overlapping and wiring together two lengths of reinforcement. The most important part of a lap splice is the overlap length between the reinforcement.

Laser detector, also known as a **laser receiver,** is used to detect the beam from a laser level when the beam cannot be seen, such as over far distances or in bright light.

Laser level   see electronic level.

Laser receiver   see laser detector.

Lath is a thin strip of wood, typically available in $1/4'' \times 1\text{-}1/2'' \times 48''$.

Latch bolt bore is a hole cut in from the latching side of the door that intersects the center of the lockset bore, and is where the latch bolt slides.

Lateral load is a force working on a structure applied parallel to the ground, and diagonally to the structure. Wind and seismic loads are the most common lateral loads.

Latitude is an angle measured from the point at the center of the earth, and lines of latitude, also called **parallels,** are imaginary lines running east to west around the earth.

Lavatory is a fixture designed for washing hands and face, usually found in a bathroom.

Laydown space, also called **laydown area,** is an area used for the receipt, temporary storage, construction equipment use, and for assembly of building components.

Leach lines, absorption field, or drain field where liquid material, or effluent, flows from the septic tank outlet and is dispersed into a soil.

Ledger fastened to the siding directly across from the bottom of the rafter tails used to support lookout joists that connect to each rafter tail.

LEED for homes is a rating system that promotes the design and construction of high-performance green homes.

Left-hand (LH) inswing, when the door swings in away from you and to the left.

Left-hand outswing, when the door swings out toward you and to the left. This is also called left hand reverse (LHR), because you have to backup or move in reverse when opening the door.

Legal description is established for every piece of property in the United States for legal identification making each piece of property unique.

Legend is a feature on a drawing that shows and names symbols used on the drawing.

Legs are the sides of the right triangle forming a fillet weld.

Lentil is the term used in masonry construction to describe the horizontal support over a door or window opening.

Level is the term used to describe true horizontal.

Level one is the first two numbers, representing the divisions in the MasterFormat numbers.

Level pole   see level rod.

Level rod, also known as a **level pole** or **level staff,** is a rod graduated to fractions of feet or meters used to measure elevations and distances when viewed through a level or transit.

Level staff   see level rod.

Level 3 is the third pair of numbers representing a sub-subcategory of the divisions in the MasterFormat numbers.

Level 2 is the second pair of numbers representing a subcategory of the divisions in the MasterFormat numbers.

Leveling is a process of determining the height of one level relative to another. Leveling is used in surveying and in construction to establish the elevation of a point relative to a **datum,** or to establish a point at a given elevation relative to a datum.

Lien is a legal term meaning a form of security interest or charge granted over real or personal property to secure the payment of a debt or performance of some other obligation.

Light-emitting diode (LED) is a highly efficient semiconductor device that emits light.

Light fixture is any device that provides artificial light.

Light-framing construction   see platform framing.

Light wood frame is construction that uses small and closely spaced, such as 16 and 24 in., 400 and 600 mm, wood members that are generally assembled by nailing.

Lighting outlet box is an electrical outlet that is intended for the direct connection of a lighting fixture.

Line laser level projects a laser beam or beam in front of the instrument that appears on a surface as a line or crosshairs.

Linkages, related to LEED certification, means that the specific home site is environmentally linked to the community.

Linoleum is a material consisting of a canvas backing thickly coated with a preparation of linseed oil and powdered cork, used especially as a floor covering.

Lintel is a horizontal support made of timber, masonry, concrete, or steel across the top of a door, window, or other opening.

Lip is the part of a C shape that extends from the flange at the open end. The lip increases the strength characteristics of the member and acts as a stiffener to the flange.

Lite, also called light, is a framed opening in a door, sidelight or transom containing a pane of glass.

Live loads is the weight of contents, occupants, and weight of snow.

Load-bearing wall, also called a bearing wall, is a wall that supports the weight of the structure resting on it from above, and by transferring the weight to the foundation structure.

Local control points are called temporary bench marks (TBM) because they are temporary, but serve the same purpose as a BM.

Local magnetic influence is also called local magnetic disturbance and local attraction. Local magnetic influence is an abnormality of the magnetic field of the earth, extending over a relatively small area, due to local magnetic influences.

Local note   see specific note.

Location, as related to surveying, specifies a geographic point or area.

Lockset is complete locking system, including knobs, plates, latch bolt, and a locking mechanism.

Lockset bore is a hole cut through the door where the lockset is installed.

Long break line is a graphic symbol used to break away a portion of a drawing that is not shown.

Longitudes, also called **meridians,** are imaginary lines on the earth running north and south.

Lookout joists connect ledger to each rafter tail for enclosed soffit construction. Also see tail joists.

Lookout rafters that extend out over the gable top plate the desired distance of the overhang.

Loose-fill insulation is fibers or granules made from cellulose, fiberglass, rock wool, cotton, or other materials.

Lot or plot is a parcel of land that can be an individual piece of real estate but is often one of several lots in a plat.

Lot and block legal description system can be established from either the metes and bounds or the rectangular system when a portion of land is subdivided into individual building lots, the subdivision is established as a legal plat and recorded by name in the local county records. The subdivision is given a name and broken into blocks of lots.

Low-e glass   see low-emissive glass.

Low-emissive glass is a technology that improves window energy efficiency. Low-e glass has a transparent coating that acts as a thermal mirror, which increases insulating value, blocks heat from the sun, and reduces fading of objects inside the building.

Low-VOC materials refers to paints, sealants, adhesives, and cleaners and other products that have a very low or zero VOC (volatile organic compounds) that are not harmful to the environment and humans.

Magnetic declination the difference between true north and magnetic north.

Magnetic north is the direction of the earth's magnetic pole, and is the direction where the north-seeking pole of a compass points when free from local magnetic influence.

Main   see main circuit breaker.

Main circuit breaker, also called a main, is a large double circuit breaker at the top of the panel that controls power to all circuit breakers in the electrical service panel.

Major appliance is usually a large product that is usually built-in as a permanent part of the cabinets and is used for daily home activities, such as cooking, or food storage, cleaning, and disposal.

Mansard roof has a double-hip roof on each face with a steeper lower part portion, and the upper story of a home is generally under the lower slope of the roof.

Mantel is generally a shelf above the fireplace opening that can decorative with ornate craftsmanship, a wood timber mounted to the wall, or a masonry projection depending on the architectural style.

Manufactured stone, also called **artificial stone, cast stone, engineered stone, reconstructed stone,** or **simulated stone** is a concrete masonry product that looks like natural stone for use in architectural applications in the building construction industry.

Marker stake, also called a *reference stake*, is usually next to a hub stake, and is a short lath stake with a marker card attached that provides survey information for a hub.

Mask means to cover, usually with special tape, plastic, or paper to keep specific areas from getting painted.

Mason is a person skilled in masonry construction.

Masonry is one of the most durable, long lasting, and maintenance-free sidings available. The material is referred to as masonry units, which are laid next to each other and bound together by mortar. Common masonry materials of masonry construction are brick, stone, marble, granite, travertine, limestone, cast stone, concrete block, glass block, and tile.

Masonry units are masonry laid next to each other and bound together by mortar.

Masonry veneer walls are made of a single nonstructural exterior layer of masonry that takes the place of traditional siding over wood-frame construction.

Mass timber is a term used to describe a variety of large engineered wood products that commonly manufactured by compressing multiple layers of wood together to create solid panels of wood or a combination of materials such as concrete and wood.

MasterFormat™: Master List of Numbers and Titles for the Construction Industry is a master list of numbers and subject titles for organizing information about construction work results, requirements, products, and activities that are divided into a standard sequence.

Mean sea level (MSL) is the average of high and low tides taken by a tide gage over an extended period of time.

Measuring wheel is a device that measures and records distance as a wheel is rolled along the ground.

Mechanical phase of construction involves the heating, ventilating, and air-conditioning (HVAC).

Mechanical describes the HVAC system and the subcontractor who performs the installation.

Mechanical drive-fit splicers are premanufactured fasteners used to connect a variety of steel piling shapes together.

Mechanical solar systems   see active solar systems.

Mechanical system of the home, provides heat, fresh or **conditioned air,** and ventilation.

Melamine is a plastic-based product used to cover cabinet surfaces, drawers, and shelves.

Melt through is a term that refers to the weld melting through the bottom of the weld or the opposite side of where the weld is being applied.

Membranes are generally thin, flexible materials that come in rolls or as parts of building materials.

Meridians   see longitude.

Meter is an instrument used to measure electrical quantities. The electrical meter for a building is where the power enters and is monitored for the electrical utility.

Meter base is the mounting base on which the electrical meter is attached. It contains all of the connections and clamps.

Metes and bounds legal description is used to identify the perimeters of any property using metes that are measurements, and bounds that are boundaries.

Method specifications outline material selection and construction operation process to be followed in providing construction materials and practices.

Metric scales are based on a ratio that is the relationship of one measurement to another. A common metric scale is 1:50, where 1 m on paper equals 50 m on the drawing.

Micrometer is a unit of length equal to one millionth of a meter.

Mid span is structural support provided at or near the center of the joists or other construction members.

Mil 1 mil equals 1/1000 of an inch.

Milling is the use of cylindrical or rotational cutting tools used to cut millwork in any desired contour or shape.

Milling plant is an industry that manufactures millwork.

Millwork is finished woodwork that has been manufactured in a milling plant, and is anything that is considered finish trim or finish woodwork, including cabinets.

Mini-split-system heat pumps   see ductless heat pump.

Minimum Efficiency Reporting Value (MERV) refers to the filtration efficiency of an air filter.

Minus in a gravel note, refers to gravel containing fines, which is good for compacting.

Miter is a joint made by cutting each of two surfaces to be typically joined at a 45° angle, usually forming a 90° angle corner.

Model Green Home Building Guidelines established by the National Association of Home Builders (NAHB) are designed as a tool kit for builders who want to use green building practices, and for local home builder associations that want to start their own green building programs.

Modern architecture refers to a unique architectural style that is different from other styles and can be seen in any time period.

Modular cabinets are designed and built at a manufacturing company and delivered to the home for installation during construction, and used by placing a group of modules side by side in a given space.

Modules are a selected unit of measure used as a basis for the planning and standardization of building materials.

Moisture barrier, also called a vapor barrier, is waterproof material used to protect the finish flooring from moisture coming up from the subfloor.

Mold is a superficial growth produced especially on damp or decaying organic matter or on living organisms by a fungus.

Moments are a measure of resistance to changes in the rotation of an object, also referred to as **moment of inertia.**

Monolithic means formed as a single unit.

Monument, known as the **point of beginning (POB)** is a fixed location that is generally an iron rod driven into the ground to start a metes and bounds survey, also a fixed point such as a section corner, a rock, a tree, or an intersection of streets.

Mortar is a mixture of lime with cement, sand, and water, used in construction to bond bricks, concrete blocks, or rockwork.

Mortar joint is a mortar-filled space between concrete blocks, bricks, and other masonry materials.

Mortise (door) for the strike plate is a square hole, pocket or relief cut into the jamb to accept the latch bolt and to keep the door closed.

Mortise and tenon joint is commonly used when adjoining construction members connect at a 90° angle. The mortise is a rectangular hole and the tenon is cut to fit the mortise hole exactly.

Movement joints are used to separate masonry construction into segments in an effort to prevent wall damage such as buckling and cracking.

Mud (drywall)   see joint compound.

Mud sill is a continuous **pressure-treated wood** member that provides a barrier between the foundation wall and the framing above.

Mulching is an erosion control practice that uses materials such as grass, hay, wood chips, wood fibers, straw, or gravel to stabilize exposed or recently planted soil surfaces.

Mullions are horizontal or vertical divides between sections of a window.

Muntins are horizontal or vertical dividers within a section of a window.

Nail head is the normally flat round enlarged top of a nail, but can be a slightly enlarged and rounded end as on a finish nail.

Nail-laminated timber (NLT) is made using dimensional lumber placed on edge and fastened together with nails.

Nail set is a tool used for driving a nail head set below or flush with a surface.

Nailer is a wood member fastened to the structure and used for attaching other wood members or finish materials.

Nanometer is one billionth of a meter.

Narrow resawn glulam refers to cutting large glulams into smaller sizes.

National Energy Conservation Code regulates the design and construction of the **exterior envelope** and selection of heating, ventilating, and air-conditioning (HVAC); service water heating; electrical distribution and illuminating systems; and equipment required for effective use of energy in buildings for human occupancy.

National Fenestration Rating Council (NFRC) is a nonprofit organization that administers the only uniform, independent rating and labeling system for the energy performance of windows, doors, skylights, and attachment products.

Natural area is a geographical area that has a physical and cultural originality developed through natural growth rather than design or planning.

Natural stone refers to products that are quarried or mined from the earth and used as building materials and other applications.

Neoprene is synthetic material similar to rubber that is resistant to heat, and weathering.

Net-metering   see net-zero energy.

Net-zero energy means that the home generates as much or more energy than the occupants consume through the course of a year.

Newel post is the large vertical support for the handrail at the ends of each flight of stairs, and often placed at regular intervals when additional support is needed.

Nominal pipe size is the inside pipe diameter.

Nominal size brick is the actual size plus the width of the mortar joint.

Nominal size lumber, also referred to as rough lumber or rough sawn lumber, is the size before it is plained.

Non-bearing partitions do not carry a structural load.

Non-load-bearing wall, also called a non-load-bearing wall, is a wall that supports only its own weight and does not support structural weight from above.

Nosing is the edge part of the tread that projects from the face of the riser below.

Notes are written information that describe features on the plan.

Nut is square or hex and has the same thread specification as the bolt so they can fit together.

Occupational Safety and Health Administration (OSHA) is a government agency under the U.S. Department of Labor that helps employers reduce injuries, illnesses, and deaths in the workplace.

Odometer is similar to a measuring wheel and is also used to measure and record distance.

Offset stakes are placed a desired distance from the actual building corner stakes to prevent the loss of reference information during construction activity.

On-grade stakes indicate the ground is at the desired grade and does not need a cut or fill.

One and a half story building has one level of living space on the main level with one full-height wall, and with a second floor or living space under the rafters.

One pipe hot water system has one pipe that leaves the boiler and runs through the rooms of the building and back to the boiler.

One-story home has one level of living space with no stairs. The roof over a one-story home is generally over an unusable attic space, and the rafters are typically fairly low pitch.

Opaque glass is made translucent, which is not clear, instead of transparent, which is clear.

Open corner   see three-stud corner.

Open stairs   see freestanding stairs.

Optical-micrometer level uses a special lens that can be rotated to vertically deflect the incoming light ray from the line of sight, where level instrument can then subdivide the level rod graduations to achieve an accuracy of ±0.02 of the level rod graduation.

Orange peel texturing makes a surface finish that looks similar to an orange and provides a delicate uniform appearance that is easy to paint and maintain.

Organic material is any material that originated as a living organism. Biomass refers here to plant matter grown to generate electricity or produce biofuel, but it also includes plant or animal matter used for production of fibers, chemicals, or heat.

Oriel window is a bay window that projects from an upper story and is self-supported, or supported by corbels.

Oriented strand board (OSB) is structurally engineered board manufactured from cross-oriented layers of thin, rectangular wooden strips compressed and bonded together with wax and resin adhesives.

Other side the side opposite the location of the arrow on a welding symbol.

Outer hearth is the part that extends past the fireplace opening into the room to provide protection from fireplace heat and from sparks that can result when burning firewood.

Outlet is an electrical connector used to plug in devices.

Outlet box is a box for an outlet.

Overflow drain (OD) a backup in case the roof drains fail.

Overhang   see eave.

Overhead garage door is generally a series of horizontal panels hinged together that fold up vertically on side rails that continue overhead upon which the door panels ride to open.

Oxygen gas welding, commonly known as **oxyfuel welding** or **oxyacetylene welding,** also can be performed with fuels such as natural gas, propane, and propylene. Oxyfuel welding is most typically used to fabricate thin materials, such as sheet metal and thin-wall pipe or tubing.

Pacing is a way to measure distance by walking and counting the number of steps taken. Pacing is inaccurate compared to other forms of distance measurement, but is useful for quickly taking rough measurements without using equipment.

Paint grade refers to a wood product that is more suitable for painting than for a clear finish where the actual wood grain is visible.

Painting is the process of using a solid, semi-transparent, or transparent surface covering as a decoration and protective coating.

Pan is the term used to describe the flat part of the metal roofing that comes in contact with the roof.

Pane is a single sheet of glass in a window or door.

Panel   see distribution panel.

Paneling is material used to cover an interior wall, and usually manufactured in 4 × 8 ft (1200 × 2400 mm) sheets.

Panelization, or assembling the components of the building into panels, including walls, floors, and roofs.

Panelized wall system see panelization and structural insulated panel.

Parallel circuit is an electrical circuit that contains two or more paths for the electricity or signal to flow from a common source.

Parallels   see latitude.

Parapet is a protective wall along the edge of a roof, or other structure.

Particle board is wood fibers that are glued and pressed into 4 × 8 ft sheets.

Passive fire protection (PFP) is the combination of all construction components of structural fire protection and fire safety in a building, designed to contain fires or slow the spread of fire.

Passive solar systems, also called **architectural solar systems,** use no mechanical devices to retain, store, or radiate solar heat.

Paving bricks are used as the wearing surface of patios, walks, and roads for people and vehicle traffic.

Pedestal footing is a concrete structure provided to carry the loads from supported elements like columns to a footing below the ground, and the pedestal width is usually greater than the height.

Pedometer is an instrument that estimates the distance traveled by walking and records the number of steps taken.

Pegs are cylindrical wooden fasteners driven into a hole that connects between two or more construction members.

Penny see d.

Percolation test is where the characteristics of the soil must be verified for suitability for a septic system by a soil feasibility test.

Performance bond is issued by an insurance company or a bank to guarantee satisfactory completion of a project by a contractor.

Pitch is the distance from one point on a weld length to the same corresponding point on the next weld.

Pitch is a polymer that can be natural or manufactured, derived from petroleum, coal tar, or plants.

Purlins are attached horizontally to the roof framing system and are used to attach the metal sheets of roofing material in steel construction or sheathing in penalized wood construction.

Permeability is the ability to let water vapor pass through. Permeability is expressed as perms. The acceptable amount of permeability depends on the application, such as example 0.3 perms is normally recommended for residential construction.

Perspective refers to a drawing that shows height, width, depth, and position of objects when viewed from a specific point.

Philadelphia rod see level rod.

Photosynthesis is the process by which green plants and some other organisms use sunlight to synthesize foods from carbon dioxide and water.

Photovoltaic comes from the Greek word *photo,* meaning light, and *voltaic,* meaning to produce electricity by chemical action.

Photovoltaic (PV) cells turn sun light into electricity. In the photovoltaic cells, photons strike the surface of a silicon wafer, which is a semiconductor diode that stimulates the release of electric charges that are guided into a circuit where they become a useful electric current. Photovoltaic modules produce direct current (DC) electricity.

Picket see baluster.

Pier is a cylindrical, square, or rectangular-shaped cube made of concrete or concrete block and used to support individual foundation members.

Pilaster is a reinforcing column built into or against a masonry or other wall structure.

Piling foundation, also called a **deep foundation**, is a foundation that transfers building loads to the earth farther down from the surface than a conventional foundation, with pilings are set to the depth of a subsurface layer based on below grade material and structural engineering requirements.

Pillar is a vertical structure of stone, wood, or metal, used as a support or decoration.

Pilot hole is a small hole drilled with a diameter designed to for a specific nail or screw size. See predrill.

Pilot hole, also called a **tap hole,** is a hole drilled into material to a specified depth or through the material prior to a bolt, screw or other fastener being inserted.

Pipe laser level projects a laser beam in front of the instrument that appears on a surface as a dot.

Pipe slope drain   see temporary slope drain.

Plain wire is smooth and is designated with a W in welded wire reinforcement.

Plan and profile contour map and its related profile.

Plan view is a two-dimensional (2D) view looking down on the feature being represented.

Pitch   see slope.

Planed means to plane lumber by using a planer, which is a machine with cutters that remove material from the surface of the lumber to desired smoothness and finished dimensions.

Planned use development or **planned unit development (PUD)** has been used in zoning. A PUD allows zoning with mixed uses, such as single- and multifamily homes, parks, offices, small businesses, restaurants, and grocery stores.

Plans refers to the complete set of drawings needed to build a structure.

Plasterboard   see drywall.

Plat is a map of part of a city or township showing some specific area, such as a subdivision made up of several individual lots.

Platform (computer) refers to a major piece of software, such as an operating system, operating environment, or database, under which smaller applications can run.

Platform is any intermediate landing in a stairway, and is also an extension of the floor landing, which is often used as the top tread of a spiral stairs.

Platform framing, also called light-frame construction, is the most popular residential framing practice used in the United States and Canada, and is like building a box where the floor joists and rim joist form the sides of the box and the subfloor is attached as the top of the box.

Plenum is a chamber which can serve as a distribution area for heating or cooling systems, generally between a false ceiling and the actual ceiling, or between construction members.

Plot   see lot.

Plug weld is made in a hole in one piece of metal that is lapped over another piece of metal.

Plumb is the term used to describe true vertical.

Plumb bob is a weight hanging from a string used to establish a vertical line to a point.

Plumb laser level projects a beam in front of the instrument that appears on a surface as a dot.

Plumber is a plumbing contractor.

Plumbing is residential piping.

Plumbing schedules provide specific information regarding plumbing equipment, fixtures, and supplies.

Plumbing system includes all of the pipes, tanks, fittings, and fixtures required for the water supply, water heating, and sanitation in a home.

Plumbing wall is any walls in a home where plumbing pipes are installed.

Plywood is sheets of material generally 4 × 8 ft (1200 × 2400 mm) made of thin layers of wood called veneer. The veneer is glued together with the grain of adjoining layers at right angles to each other.

Pneumatically projected refers to the use of a pneumatic mechanism operated by air or gas under pressure.

Pocket door slides horizontally into a frame that is enclosed in the wall.

Pocket door kit is a complete package that contains the door, split studs, a header assembly that has a track on which the door runs slides, and door hardware.

Podium construction, also called a **podium building,** is the construction of multiple stories of **wood-frame construction** over one or more levels of concrete structure.

Point of beginning (POB)   see monument.

Pole buildings are similar to timber-frame buildings except the vertical supports are used directly from the round tree. Also see post and beam.

Pole construction   see post and beam.

Polybutylene (PB) plastic piping for both hot and cold water.

Polystyrene is a synthetic resin which is a polymer of styrene, used chiefly as lightweight rigid foams and films.

Polyurethane is a synthetic resin used mainly as ingredients of paints, varnishes, adhesives, and foams.

Polyvinyl chloride (PVC) is a widely used building material found in products such as window frames, flooring, and shower units.

Pond-loop system geothermal exchanger is inserted into a lake, river, or other natural body of water to extract heat by pumping the water up to the heat exchanger.

Pony wall is a general term used to describe any short height wall, or partial wall, also called a knee wall.

Post and beam or **pole construction** is also referred to as **timber-frame construction** or **pole buildings,** generally using large posts and beams or timbers for the horizontal and vertical members. Pole buildings are constructed using a combination of post and beam, timber framing, and conventional stud framing depending on the requirements of the specific building. The vertical poles or posts are used as the building supports to which the horizontal framing beams are fastened.

Post and beam construction uses wood **posts** supported by concrete piers or footings with beams above. The beams are generally spaced 48 in. (1200 mm) apart and run between the foundation walls. Wood decking is placed above and perpendicular to the beams.

Post bases are used to provide a rigid connection between the footing.

Post caps are used to provide a rigid connection between the post and beam above.

Postchlorinated polyvinyl chloride (CPVC) pipe has insulation value that retains heat and saves energy, when compared with metal pipe that loses heat.

Posts are vertical wood members that connect between a pier or footing and support the beam above. Posts can be any size depending on structural engineering, but they are commonly 4×4 or 4×6.

Potable water is water of a quality that is suitable for drinking.

Pour is the process of flowing the concrete into the forms.

Poured-in-place concrete is the concrete construction method where concrete is poured into forms.

Powder coating is a polyester or epoxy powder placed on the surface and then heated to fuse with the metal forming a protective layer.

Power supply provides components with electric power.

Power-supply plan is used to show all electrical outlets, junction boxes, and related circuits.

Pre-hung door is a premanufactured door assembly with jambs and door mounted and ready for installation in the home.

Precast concrete is concrete structures made by casting concrete with steel reinforcing in a form or mold.

Predictive design technology works by analyzing historical and current data and generating a model to help predict future outcomes.

Predrill means to drill a hole through the jam using a drill diameter slightly smaller than the nail or screw used for fastening.

Prefabricated   see premanufactured.

Prefinished cabinets and woodwork is finished at the manufacturer and then delivered to the project for installation.

Premanufactured, also referred to as pre-fabricated, are products that are built away from the job site by a manufacturing company and delivered to the construction project for assembly into the building.

Prequalified welds are most of the common welded joints used in steel construction and are exempt from tests and qualification.

Pressure-treated means that wood has had a liquid preservative forced inside to protect against deterioration due to rot or insect damage.

Prestressed concrete beam has steel reinforcing stretched from both ends in the concrete form or mold, which is released after the concrete cures. The finished concrete beam has a slight arch that makes it better for supporting heavy loads than precast concrete.

Prevailing winds refers to the direction from which the wind most frequently blows in a given area of the country.

Prints are generally made on a printer that is a mechanical process involving the transfer of text, images, or designs from a computer file to paper.

Procurement and Contracting Requirements are referred to as series zero in the MasterFormat, because they begin with a 00 level one numbering system prefix, and are not specifications, and establish relationships, processes, and responsibilities for projects.

Professional Engineer (PE) an engineer who is registered or licensed within a specific state or states where they pass rigorous qualifications needed to offer professional services directly to the public, and is generally licensed in one or more specific disciplines, such as a structural engineer who performs the engineering on the building structure, or civil engineer who does earth-related engineering.

Professional home designer is normally not a registered architect, but can design homes for construction in some jurisdictions, and usually work with a PE who performs the required engineering and stamps the plans with a PE certification.

Profile is a section view cut through the construction site that is a vertical section of the surface of the ground, and underlying earth that is taken along any desired fixed line.

Project life cycle is the sequence of phases that a project goes through from its initiation to its closure.

Property line is the legal boundary line of a piece of land.

Proprietary product specification provides specific product names and models for desired applications.

PSDE stands for Public or Private Storm Drain Easement.

Public land states are states in an area of the United States starting with the western boundary of Ohio to the Pacific Ocean, and including some southeastern states and Alaska, established by the U.S. Bureau of Land Management to create the rectangular survey system.

Puddle weld is a type of plug weld for joining two sheets of light-gauge material by burning a small hole in the upper sheet and then filling the hole with a puddle of weld metal to fuse the upper sheet to the lower sheet.

Punched window refers to a construction practice where a hole is punched in the exterior wall of the building and filled with a window.

Punchout or **web opening** is a hole or opening in the web of a steel-framing member allowing for the installation of plumbing, electrical, and other utility installation. A punchout can be made during the manufacturing processor in the field with a hand punch, hole saw, or other suitable tool.

Purlins are horizontal lateral members that extend the length of a roof, used to support and tie rafters together and can be used to attach roof decking depending on the installation practice.

R-2000 Standard is a specific Canadian home building technology that has earned a worldwide reputation for quality, comfort, and environmental responsibility.

R410a is a hydro-fluorocarbon (HFC) which does not contribute to ozone depletion and has been approved for use in new residential air conditioners.

R-value of a material is a measure of thermal resistance to heat flow.

RA is return air.

Radiant heat is heat that radiates from an electric or hot water element, warming objects rather than the air.

Radio frequency (RF) is electromagnetic radiation waves that transmit audio, video, or data signals.

Radon is a naturally occurring radioactive gas that breaks down into compounds that are can cause cancer when large quantities are inhaled over a long period of time.

Rafters are the sloped structural members of a roof system used to support the roof loads and connect to the ridge.

Rafting is a system where trusses are erected, braced, and sheathed on the ground then an entire section of truss framing is hoisted in place.

Railings are used for safety at stairs, landings, decks, and open balconies where people can fall.

Rain drains are pipes buried along the foundation wall used to transfer weather-related waters to storm sewers or other designated locations.

Raised hearth is built above the floor level by continuing the lower structure to support the hearth to the desired height.

Rake board   see fascia.

Rake molding   see fascia.

Ranch style home has one-story.

Range hood is a metal hood over a range that has lights and an exhaust fan that vents to the outside.

Rebar the term used to identify steel reinforcing bars in the construction industry.

Rebond carpet pad is made from high-density foam that is bonded together to make a variety of densities and thicknesses.

Rectangular survey system   see public land states.

Reference datum used for a survey can be any solid object with a known elevation from the vertical datum, such as a bench mark (BM) or a temporary control point established by a surveyor.

Reference stake   see marker stake.

Reflected ceiling plan (RCP) is used to show the layout for the suspended ceiling system.

Reflected light is light bounced off a ceiling that has a uniform quality to help reduce the need for artificial lighting.

Reflective glass has a coating that provides privacy and heat reduction by reflecting light.

Reflective insulation includes radiant barriers that are typically highly reflective aluminum foil backing on one or both sides of the insulation.

Refrigeration system includes the cooling coil called an evaporator, the compressor, the condenser, and the expansion valve.

Refusal point is the location where a pile cannot be driven any further.

Register   see air supply register.

Registered Architect (RA) a person trained in the planning, design, and oversight of the construction of buildings, and is licensed to practice architecture in a specific state or states where they pass rigorous qualifications.

Reinforced concrete is concrete poured around steel bars placed in the forms.

Relative humidity is a percentage ratio of the amount of water vapor in the air at a specific temperature to the maximum amount that the air could hold at that temperature.

Relief   see topography.

Remodeling is changing an existing structure, either internally or externally.

Renovation is improving by renewing and restoring.

Resilient flooring, also called vinyl flooring, is either sheet or tile vinyl material that is firm and flexible.

Retaining walls are concrete, masonry, or wood structures designed to restrain soil on a slope between two different elevations.

Return duct directs the air from the rooms over the heating or cooling device.

Revision cloud is a cloudlike circle around a change made to a print after the drawing has been released for construction.

Revision history block, also called the revision block, is used to record changes to the drawing and is generally located in or next to the title block.

Revolving doors provide an entry system that is always open, an always closed at the same time, and eliminate drafts, while reducing noise and air pollution.

Ribbed concrete floor system refers to a pattern of raised features serving to strengthen or support the structure.

Ridge is the top horizontal member connecting two intersecting roofs.

Right-hand (RH) inswing, when the door swings in away from you and to the right. This is also called right-hand reverse (RHR).

Right-hand outswing, when the door swings out toward you and to the right.

Rigid fiber board insulation contains fiberglass or mineral wool material primarily used for insulating ducts, and for insulation that can withstand high-temperature applications.

Rim joist, also called a **rim board, band joist,** or **header** is attached perpendicular to the joists, and provides lateral support for the ends of the joists while capping off the end of the floor or deck system.

Ring shank nails have ridges or grooves along the shank.

Riprap is a layer of large rocks used to protect soil from erosion on steep slopes or slopes that are unstable because of seepage problems.

Riser (plumbing) is a pipe that extends vertically one story or more to carry water to fixtures.

Riser (stairs) is the vertical part of a stair, forming the space between each step.

Rivet is a metal pin with a head used to fasten two or more materials together. The rivet is placed through holes in mating parts and the end without a head is extends through the parts to be headed-over.

Road rugs   see geotextiles.

Roll insulation is the same as batts except it comes in approximately 40 ft (12,000 mm) rolls rather than strips. See also blanket insulation.

Roof is the supporting structure and exterior surface on top of a building.

Roof drain means a fitting or device that is installed in the roof to allow storm water to discharge into a downspout.

Roof drainage plan shows the elevations of the roof and provide for adequate water drainage required on low slope and flat roofs.

Roof framing plan shows the major structural components in plan view that occur at the roof level.

Roof joists are flat roof rafters because they serve as rafters and ceiling joist for the rooms below.

Roof plan displays the outline of the roof, roof construction members, roof vents, and roofing materials. The outline of the walls below can be shown with hidden lines and dimensions, such as eave overhang are provided.

Roof pitch is the slope of the roof in rise over run, such as 1/12. This means that the roof has a slope of 12 in. of *rise* in 12 in. of *run*. Rise is the vertical distance and run is the horizontal distance. Roof pitch can also be referred to as roof slope.

Roof pond is usually constructed of containers filled with antifreeze and water on a flat roof. The water is heated during the winter days, and then at night the structure is covered with insulation, which allows the absorbed heat to radiate into the living space. This process functions in reverse during the summer, when the water-filled units are covered with insulation during the day and uncovered at night to allow stored heat to escape.

Roof slope   see roof pitch.

Roof truss is a manufactured structural support for the roof system.

Roof vents are screened venting devices that are attached to the roof for providing attic ventilation.

Roofer, or roofing contractor, is the person who installs the roofing material.

Roofing contractor   see roofer.

Roofing felt, also commonly called tar paper, is a heavy-duty material made by impregnating paper with tar to make a waterproof material used as an underlayment for roofing material.

Roofing underlayment that is a weatherproof layer of material over the roof sheathing in an effort to keep the home dry before the final roofing material is installed.

Rooftop garden   see green roof.

Root opening in a weld joint, is the two pieces of metal are spaced apart a given distance.

Rotating beam laser level, or **rotating laser level** projects a 360° rotating laser beam in a plane that appears on a surface as a line that can be detected anywhere around the instrument, and used for a variety of purposes from basic surveying and site layout to establishing grades.

Rotation is an action of studs rotating around their axis caused by tension applied to the wall system by wind or seismic activity.

Rough carpenter   see framer.

Rough-in electrical, also known as **rough-in** is when the electrical boxes and wiring are installed.

Rough lumber   see nominal.

Rough opening (RO) is any unfinished opening that is framed to specific measurements to accommodate the finish product.

Rough sawn   see nominal.

Routed stringer   see housed stringer.

Rowlocks are bricks oriented perpendicular to the face of the wall similar to headers, except they are used with the end and face exposed at sills and at the top of walls.

Run (plumbing) refers to a portion of a pipe or fitting continuing in a straight line in the direction of flow in which it is connected.

Run (stairs) is the horizontal dimension one tread, measured from the face of one riser to the face of the next riser, or from the edge of one nosing to the edge of the next nosing.

Safety glass is glass that has been toughened or laminated so that it is less likely to splinter when broken.

16d nail is commonly used in framing and is 3-1/2 in. long, and the d in the nail specification refers to penny, which is an old term still used to specify nail sizes today. The nail size gets larger as the number gets larger.

Safety glass is a general term that refers to glass that has been toughened or laminated so that it is less likely to splinter when broken.

Sand box   see borrow.

Sand fences, also called **wind fences** are barriers made of small, evenly spaced wooden slats or fabric used to reduce wind velocity and to trap blowing sand.

Sand filter system, which filters the effluent through a specially designed sand filter before it enters the soil absorption field.

Sanitary sewer system is a system of underground pipes designed for the collection and transfer of waste water from domestic residences, businesses, and industries to a wastewater treatment plant, or private sewage treatment such as a septic tank or cesspool.

Sash, plural sashes, is a framework that holds the panes of a window in the window frame.

Sash block is a concrete block unit manufactured with a vertical groove where the shear lug is placed.

Saw set is the distance a saw tooth is bent away from the saw blade. A saw typically has each tooth set alternating on opposite sides.

Scale is a measurement unit representing a proportional relationship between a reduced-size drawing and the actual full-sized feature.

Schedule is a grouping of related items that formats information into rows and columns in order to more easily present design information.

Schematic diagram is commonly associated with electrical circuits shown in a drawing representing the components of a process, device, or electrical system using lines and standardized symbols.

Scissors truss is a truss that has the bottom chord at a slope, creating a sloped ceiling.

Screed is a flat board, or specially made aluminum tool as shown here, used to smooth a concrete slab after it has been poured in the forms.

Screed is the process of using a screed to level and remove excess wet concrete to the top surface of a slab and to the accurate grade.

Screened soffit vents are part of a roof ventilation system that circulates cool air in from the soffit vents into the attic space and moves hot air out through vents high in the roof and in gable ends in some cases.

Screened vents are provided in the blocking to help ventilate the attic space.

Screw pilings   see helical piles.

Scupper is an outlet in the side of a building or from a roof for draining water.

Section is a type of drawing that shows a cut through the building to display the construction practices being used along with construction materials and principal dimensions.

Sectional view   see section.

Sediment is earth particles and other material that drop out of the storm water runoff.

Sediment basins and rock dams are used to confine sediment from storm water runoff in an excavated pool or natural depression.

Sediment control is basically any practice used to keep soil on a construction site, so that it does not flow on to other properties or cause water pollution in a stream, river, lake, or ocean.

Sediment traps are small temporary excavated pools that allow sediment from construction runoff to collect.

Seeding is used to control and reduce erosion and sediment loss by providing permanent stabilization disturbed areas by establishing perennial vegetative cover from seed.

Seismic engineering   see earthquake engineering.

Seismic loads are caused by earthquakes.

Self-reading rods are rods that can be read by the person operating the instrument.

Self-tapping is the ability of a screw to creating its own thread without the need of a pilot hole.

Self-venting fireplaces can be vented directly out the wall behind the fireplace.

Septic tank is an on-site treatment system for domestic sewage, in which the sewage is held to go through a process of liquefaction and decomposition by bacterial organisms.

Series circuit is a circuit that supplies electricity or a signal to a number of devices connected so that the same current passes through each device in completing its path to the source.

Set refers to the time it takes concrete to harden after it has been poured.

Set means to sink the nail head below or flush with the surface.

Set of working drawings is a complete set of construction documents contains drawings and specifications. See also plans.

Setback is the minimum distance required between the structure and the property line.

Shear forces are forces caused when two construction pieces move over each other.

Shear lugs are made with hard rubber or polyvinyl chloride (PVC) plastic is placed in the sash block groove running the entire height of the concrete block wall to transmit externally applied loads to the structure.

Shear panel, also called a braced panel, is typically part of a wood frame stud wall that is covered with structural sheathing such as plywood, but other materials such as steel and bracing systems can be used.

Shear plate is a special round plate inserted in the face of a timber to improve shear resistance in the wood-to-wood joint.

Shear wall, also called a braced wall line, is a braced wall made of a shear panel to oppose the effects of lateral load acting on a structure.

Sheathing is fastened to the floor joists, wall studs, or roof rafters to reinforce the structure and provide a backing for finish materials.

Shed roof has a single slope that is often used in modern and contemporary architecture.

Sheet blocks are a group of informational areas normally surrounded by boarder lines and grouped in one consistent location on the drawings.

Sheet metal screws have deep spiral threads along the entire body length, and have a pointed end for easy start when threading.

Sheetrock   see drywall.

Shelf angle is a horizontal steel angle that provides a break in the veneer for the masonry to expand and the connected structure to shrink.

SHGC is the fraction of incident solar radiation admitted through a window, both directly transmitted and absorbed and later released inward. SHGC is expressed as a number between 0 and 1. The lower a window's solar heat gain coefficient, the less solar heat it transmits.

Shielded metal arc or **stick electrode welding** is the most traditionally used welding method that uses a flux-covered metal electrode to carry an electrical current forming an arc that melts the work and the electrode. The molten metal from the electrode mixes with the melting base material, forming the weld.

Shims are a thin strip of wedge-shaped or rectangular wood used for leveling, making plumb, and positioning wood members, especially door frames.

Shingle siding is installed using wood shingles, which are thin, tapered pieces of wood placed side by side in rows with each row overlapping the row below.

Shoe is a blocking used to reinforce and stabilize the ends of stringers at the floor.

Shoe rail is used to receive the square bottom end of balusters when they are not connected directly to the treads.

Shop drawing   see shop ticket.

Shop tickets, also called shop drawings are drawings or sets of drawings produced by the contractor, supplier, manufacturer, subcontractor, or fabricator for use in a fabrication shop to manufacture a construction item.

Shotcrete, also called gunite or sprayed concrete is concrete or mortar transported through a hose and pneumatically projected at high velocity onto a surface.

Side dams   see end dams.

Siding can be wood, metal, plastic, composite materials, stucco, or masonry. The type of siding can be used to define or enhance the architectural style.

Silicon wafer is a semiconductor diode that stimulates the release of electric charges that are guided into a circuit where they become a useful electric current.

Sill is a continuous pressure treated wood member that provides a barrier between the foundation wall and the framing above.

Sill is the framing member that forms the bottom edge of an exterior door or window opening.

Sill pan   see through-wall flashing.

Sill plate   see sole plate.

Silt fence is a temporary sediment barrier made of porous fabric normally available in 3 ft tall × 100 ft long rolls.

Simple span refers to a construction member such as a beam, girder, or truss that spans from one support to another.

Single-faced fireplace   see single fireplace.

Single fireplace has one opening on one side of the structure.

Single-hung window has two sashes of glass, with the top sash stationary and the bottom sash movable.

Single phase is the distribution of alternating current electric power using a system with all the voltages of the supply fluctuates at the same time.

Single-ply membranes are sheets of rubber and other synthetics that can be ballasted, mechanically fastened or chemically adhered to insulation creating a layer of protection on a commercial facility.

Single-pole switch is a standard on and off wall switch that is the only switch controlling one or more light fixtures in a single electrical circuit.

Sinker nail is used for framing and is thinner than a common nail with a funnel-shaped head, a grid stamped on the top of the head, and coated with adhesive for smooth driving and to improve holding.

Site identifies property corners, border lines, elevations, and can include the location of construction corners, building outlines, and corner elevations.

Site benching is where each adjacent site or lot is graded separately toward the adjoining property line.

Site grading is the construction process of changing the elevation and slope of the land to civil engineering specifications and to site survey requirements at and near the proposed construction site.

Site orientation is the placement of a structure on the property with certain environmental and physical factors taken into consideration.

Site plan is a drawing that describes how a parcel of land is to be improved.

Site section   see profile.

Site survey identifies property corners, border lines, elevations, and can include the location of construction corners, building outlines, and corner elevations.

Skim coating is the application of a thin layer of joint compound over drywall to smooth out walls and ceilings.

Skin friction piles develop most of the pile-bearing capacity by shear stresses along the sides of the pile, by friction between the surface of the pile and the earth where the pile is driven.

Skirt board is a nonstructural fascia used to trim the sides of stairs to which the treads and risers are fitted.

Skylight is a window in a roof used to allow sunlight to enter.

Skylight flashing is galvanized sheet metal that is bent in an L shape and used around the skylight to direct water away from the curb and over the roofing material.

Slab that is a flat concrete pad poured directly on the ground or on compacted gravel over the ground.

Slab door is a rectangular slab of wood that is not drilled or processed for hinges or a locking mechanism.

Slab rubber carpet pad is a single thickness of rubber material that stands up under pressure for a long period of time, and is used in high traffic areas.

Slate is a natural occurring material composed of clay or volcanic ash.

Sliding door   see bi-pass door.

Sliding window is a popular 50 percent operable window when there are two panes in the window.

Slope of land or constructed site, refers to the amount of incline of the surface using earth removal practices. Slope is described by the ratio of the rise divided by the run between two points on a line or plane.

Slope (roof), also called pitch, is the amount of rise a roof has compared to a horizontal measurement called the run.

Slope diversions are constructed by creating channels laterally across slopes to intercept the down-slope flow of runoff reduce the possibility of erosion.

Slope stakes are used to determine the point at which the proposed slope intersects the existing ground.

Sloped glazing is a glass and framing assembly that is sloped more than 15° from the vertical, forms the entire roof of the structure, and is generally a single-slope construction.

Slot weld is a weld applied to a slot.

Sludge is a thick, soft, wet mixture of liquid and solid components.

Smoke chamber is a tapered area above the fireplace and below the flue, used to allow smoke to mix and rise into the flue.

Smoke shelf, also called a **wind shelf,** is a ledge at the bottom of a smoke chamber, used to deflect or break downdrafts from the chimney, and collect rain that enters the chimney.

Sod, also called turf grass, is grass that has already been planted in sections that are grass and soil held together by roots or other materials, and available in pieces 16 to 24 × 48 in., or 12 to 18 in. wide × 40 to 60 in. long rolls.

Sodding is an immediate and permanent erosion control practice where grass sod is installed on exposed soils.

Soffit is the term used to describe the underside of any architectural feature, such as a beam, arch, ceiling, overhang, or vault.

Soft metric conversion the actual metric conversion of 4 × 8 ft = 1219 × 2438 mm is rounded to the nearest 100 mm modules, resulting in 1200 × 2400 mm.

Software is the program or instructions that run the computer.

Softwood is wood that comes from a coniferous tree, such as fir, hemlock, and pine.

Softwood plywood is structural material made of layers of softwood veneer glued together, under heat and pressure, with the grains of adjoining layers placed at right angles.

Soil feasibility test   see percolation test.

Soil pipe is a pipe that carries the discharge of water closets or other similar fixtures.

Soil retention structures are used to hold soil in place on a slope site.

Soil stack is a vertical pipe that extends one or more floors and carries discharge of water closets and other similar fixtures.

Soil support system can be made up of a subgrade, subbase and base gravel layers.

Solar access refers to the availability of direct sunlight to a structure or construction site.

Solar architectural products absorb and collect heat from the sun and outside air and transfer the heat into water, glycol, or another heat-transferring fluid passing through embedded tubes.

Solar energy is produced by sunlight that can be captured when it is transferred to something that has the ability to store heat or energy.

Solar heat gain coefficient (SHGC) measures how well a product blocks heat caused by sunlight.

Solar orientation allows for excellent exposure to the sun.

Solar water heating (SWH) or **solar hot water (SHW)** systems use solar radiation to heat water or air in buildings.

Solarium is a room with walls of glass that go beyond the eaves. Solariums often have roofs made entirely out of glass.

Solder is a low-melting alloy, especially one based on lead and tin, used for joining less fusible metals such as copper.

Soldering is the joining of metal parts, such as wires, with the use of heat on fusible alloys, usually tin and lead.

Soldiers add accent style to the brick wall by placing units oriented vertically with the full face exposed.

Solid blocking is placed between floor joists to help resist lateral loads, to prevent joist rotation, and to provide nailing surface for the bottom plate of a wall or partition framed above, to provide stiffness to a floor system, to provide fire blocking, and to support plumbing and heating equipment.

Solid fuel-burning appliances are products such as airtight stoves, freestanding fireplaces, fireplace stoves, room heaters, zero-clearance fireplaces, antique stoves, and fireplace inserts for existing masonry fireplaces.

Sone is a sound rating, where the lower the number the quieter the sound.

Soot is finely divided carbon deposited from flames during the incomplete combustion of organic substances such as coal.

Sound transmission class (STC) is a rating of how well a building partition reduces airborne sound, which is any sound transmitted by the air.

Southern Building Code Congress (SBCC) codes developed for the south.

Span refers to the horizontal distance between two supporting members.

Span (roof) is the horizontal dimension across the building measure from the outside of top plates.

Spandrel panels is a preassembled structural panel used as a wall or roof panel that can have a variety of textures and colors, and used to replace the appearance of a masonry wall.

Spec refers to speculation, where the contractor is building the home based on the assumption that it will sell for a profit. Also a slang term used when referring to a specification. See specification.

Spec home is a home designed for a contractor to build and offer for sale.

Specific heat is the amount of energy required to raise the temperature of any substance 1°F.

Specific notes describe individual features.

Specifications provide detailed written information placed on the drawing or in separate documents and provide exact statements describing the characteristics of particular aspects of the project.

Specifications group in the MasterFormat contains the construction specifications subgroups and their related divisions.

Spectrally selective glazing is high-performance glazing that allows as much daylight as possible, while preventing transmission of as much solar heat as possible.

Spindle   see baluster.

Spiral stairs, also called a circular stairs, has treads winding around a center newel.

Spirit level, also called a **bubble tube**, is a glass tube containing liquid and a bubble that is level when the bubble is adjusted between two marks on the tube.

Spirit leveling   see direct leveling.

Split studs are narrower than normal studs and used to frame each side of the pocket part of a pocket door kit, providing space between the split studs for the door to slide into the wall.

Split-wired outlet is a duplex outlet that has one outlet controlled by a switch.

Spotface is a flat-bottomed cylindrical enlargement of the mouth of a hole with slight depth to insert a washer or shear plate below the surface.

Spray polyurethane foam, commonly referred to as SPF, is a material that is sprayed as a liquid so it can expand into a foam, creating a solid layer across an existing roof.

Sprayed concrete   see shotcrete.

Square, used in construction, is any four-sided shape with four straight sides, four right angles, and equal diagonal measurements.

Square groove weld is applied to a butt joint between two pieces of metal.

Stack is a general term referring to any vertical pipe for soil waste or vent piping.

Staggered means to alternate the nail placement so they are not in a straight row.

Stain is a penetrating liquid used to preserve and tint woodwork for a desired appearance.

Stainless steel is steel containing chromium, used to resistant tarnishing and rust.

Stair, commonly referred to as stairs, is a set of steps leading from one floor of a building to another.

Stair horse   see stringers.

Stairbuilder is a specially skilled finish carpenter who designs and builds stairways, and guardrail systems.

Staircase is the entire stair construction including framing the stairwell opening, adjacent walls, and the railing systems.

Stairway   see stair.

Stairwell is the opening in the floor where the stairs are located.

Stamped concrete is a process of using rubber molds pressed into the concrete to create patterns resembling brick, slate, cobblestone, flagstone, or tile. When used with concrete colors, the stamped surface has a decorative appearance without the cost of natural stone.

Standing seam roofing is a raised seam between each parallel metal sheet where one sheet is joined to the next, creating an interlocking, watertight seam.

Static slicing is the insertion of a narrow custom-shaped blade at least 10 in. into the ground, and at the same time pulling silt fence fabric into the opening created, as the blade is pulled through the ground. **Stem walls** are used at the interior of the foundation system to support structural members at mid span.

Steel construction refers to buildings built using steel.

Steel hanger is a premanufactured steel bracket used to hold and support a construction member such as a beam, joist, post, or rafter.

Steel knife plate is flat plate steel timber post connector with rebar embedded in the concrete foundation, and a post base is attached to the timber post through the knife plate and held together with a steel pin.

Stem wall is a general term used to describe any short height wall, or partial wall, also called a pony wall or knee wall.

Step flashing is small pieces of flashing material used to overlay each other where a vertical surface meets a sloping roof.

Stick framing one board or *stick* is used at a time to assemble the structure.

Stock house plan has been preemptively designed and made available for sale by a professional home designer.

Stockpile refers to the earth material that is piled and stored during excavation for later use on the site, such as backfill, and the on-site storage of other construction material, such as lumber, or the piling of construction waste for later removal.

Stop molding is a small rectangular-shaped molding positioned on the jamb, slightly past the width of the door, and is used to stop the door in the jamb at a specific location.

Storefront is defined as a nonresidential, non-load-bearing assembly of commercial entrance systems and windows, usually spanning between the floor and the structure above, designed for high use and strength.

Storm sewer is used for carrying groundwater, rainwater, surface water, or other nonpolluting waste to locations where it can be safely dispersed.

Stormwater easement is a low area or swale between properties that allows storm water to drain away from homes.

Stormwater Pollution Prevention Plans (SWPPPs) are a requirement of the National Pollutant Discharge Elimination System (NPDES) that regulates water quality when associated with construction or industrial activities. The SWPPP addresses all pollutants and their sources, including sources of sediment associated with construction, construction site erosion, and all other activities associated with construction activity and controlled through the implementation of Best Management Practices (BMPs).

Straight flight of stairs is a flight between landings without any turns.

Stretchers are the most common brick pattern with units oriented horizontally with the full face exposed.

Strike jamb is the vertical side of the door frame on the lock side of the door.

Strike plate is a metal plate attached to the door frame that the lock engages.

Stringers, also called carriage or stair horse, are the supporting member running the length of a stair incline on which treads and risers are mounted.

Strip windows are a row of horizontal windows placed next to each other at each floor of a building.

Stripped concrete forms are removed after the pour has hardened.

Structural brick is used here to describe economic brick used to build a structure and not selected for appearance.

Structural engineer works with architects to engineer the structural components of a building.

Structural engineering is a branch of engineering that deals with the design and construction of structures to withstand physical forces or displacements without danger of collapse, or without loss of serviceability or function.

Structural insulated panels (SIP) are a high energy-efficient building system for residential and light commercial construction that is custom manufactured for each home and delivered to the project site for installation.

Structural masonry is the primary structural system for home or commercial building construction that can be combined with wood, steel, and reinforced concrete applications.

Structural masonry walls are typically constructed using concrete masonry units.

Structural panels are the structural wood panels attached to 2 × 4 or 2 × 6 dimensional lumber subpurlins or stiffeners.

Structured wiring systems are high-speed voice and data lines and video cables wired to a central service location.

Stucco siding is composed of cement, sand, and lime, and is applied while wet. The stucco dries to a hard durable material for exterior and interior walls.

Stud wall the wall constructed using studs.

Studs are vertical framing members used to construct walls and partitions, and are usually 2 × 4 or 2 × 6 and spaced 16 or 24 in. on center.

Subbase is a layer of compacted gravel, such as crushed rock with 10 to 20 percent fines, on top of the subgrade.

Subcontractor is a person or business who carries out work for a general contractor as part of a complete project, and there can be a number of subcontractors on a project, including electrical, heating, ventilating, air-conditioning, plumbing, framing, concrete work, and roofing just to name a few.

Subdivision is parcel of land that has been divided into two or more pieces and developed with roads and utilities for easier sale.

Subfloor is a layer of structural material fastened above the floor joists to tie the floor joists together and to provide support for the finished floor.

Subgrade is a preliminary grade prior to doing work to finish grade.

Subgrade is the native soil that meets engineering requirements or soil improved by compacting to engineering specifications. Subbase is a layer of gravel on top of the subgrade.

Subpurlins or **stiffeners,** typically spaced 24 in. on center. Subpurlins or stiffeners are light construction framing member resting on or between **purlins** and usually running at right angles to purlins.

Substrate   see backer board.

Sunroom is a room with walls of glass that stop at the eaves and normally have traditional looking roofs.

S4S surfaced on four sides.

Survey is the result of establishing the exact corners, boundaries, and elevations of a piece of land using **surveying** techniques.

Surveyor level is an instrument with a telescope and spirit level or compensator used to establish a horizontal line for measuring elevations.

Survey stakes are used to control alignment and grade of building corners, roads, and other features during construction.

Survey tack is a small, sharp, broad-headed nail.

Surveying is a branch of applied mathematics that is a science concerned with determining the area of any portion of the earth surface by measuring distances, angles, and directions of characteristics such as bounding lines and the contour of the surface by accurately defining the features in notes and on drawings.

Surveying for construction is the measurement of dimensional relationships, horizontal distances, elevations, directions, and angles, on the earth's surface especially for use in locating property boundaries, construction layout, and site plan drafting.

Surveyor is a person who uses current technology surveying equipment and practices for surveying as a specialist who provides a map file of the job site from which civil designers and drafters develop drawings.

Sustainable describes anything that is capable of being continued or maintained with minimum or no long-term effect on the environment.

Sustainable buildings are buildings capable of maintaining their desired function into the future.

Sustainable sites offer enrichment with trees and plants that provides shade, aesthetic value, habitat for native species, and a way to absorb carbon and enrich the soil.

Sweep 45° to 90° turns in the conduit with large radius that makes it easy to pull wire through the conduit during the electrical work.

Switch box is an electrical box that houses a switch or group of switches.

Switch leg is the electrical conductor from a switch to the electrical device being controlled.

Symbol is an image that represents a feature on a drawing or in specifications, for example, the symbol for feet is ′, the symbol for inch is ″, and the symbol for diameter is Ø.

System development charges (SDCs) are charged by local jurisdictions to fund the development of parks, schools, transit systems, and other municipal facilities associated with an increase in population due to the construction of new homes.

Systems   see functional elements.

T-intersection occurs when an interior partition intersects an exterior wall, and the number of studs used needs to be kept to a minimum for the maximum amount of insulation to be used.

T1-11 siding is the most common structural plywood siding that has a textured finish exterior veneer that provides the appearance of traditional solid-wood siding.

Tack means to drive the nail partially into place to hold the work before driving the nail all the way.

Tackifiers are a tacking agent material, of various chemical compositions, used as an additive in hydraulically applied erosion control or soil amendments to increase product to soil adhesion.

Tail joists are used to tie the outside wall to the adjacent joists when the main joists run perpendicular to the rafters.

Tail is added to the welding symbol when it is necessary to designate the welding specification, procedures, or other supplementary information needed to fabricate the weld.

Tails are the bottom of the rafters that generally extend past the exterior wall a desired distance to create an **eave** or **overhang.**

Tamper-resistant (TR) outlet has a spring-loaded shutter that closes the contact openings for child safety.

Tankless water heaters, also called instantaneous or demand water heaters, heats water directly without use of the storage tank found with traditional water heaters.

Taping is the process of using a tape measure, or tape, to measure distances.

Tar paper is a thick product that is manufactured in rolls and is made by impregnating paper with tar, producing a waterproof material.

Target vernier is a secondary scale that has ten divisions equal in distance to nine divisions on the main scale that allows measurements of 0.001′.

Teamwork is based on a client-server architecture and is designed to ensure maximum flexibility, speed and data safety to enable teams—even those spread out around the world—to collaborate on large projects.

Tek screw is s self-tapping sheet metal screw that eliminates the need to pre-drill a pilot hole.

Tempered glass is widely used in commercial windows and required by code for specific installations. Tempered glass, also called toughened glass, is a type of safety glass made by controlled thermal or chemical treatments to increase the glass strength.

Tempered glass is a safety glass that is four to five times stronger than standard glass and shatters into small oval-shaped pebbles when broken.

Temporary (electrical) the installation of a temporary underground or overhead electrical service near the construction site and close to the final meter location provides electricity during construction.

Temporary bench marks (TBM)   see local control points.

Temporary electric service is provided during construction and is used for construction electricity purposes. This service is installed on a temporary pole, which is placed near the permanent power pole or transformer. The temporary service has a meter base and meter, and usually two 20 A, 120 VAC, grounded, duplex outlets and one 50 A, 4 wire, single phase, 240 VAC (208 VAC at some locations) outlet mounted in a weatherproof enclosure.

Temporary slope drain can be used to divert storm water from one elevation to another with a corrugated metal, plastic, or concrete pipe extending from the upper to the lower elevation

Tensile strength is the strength the material has against pulling forces.

Tensile stress or tensile strength is a measurement of the force required to pull something such as rope, wire, or a structural beam to the point where it breaks.

Tension is caused by stretching.

Tension system   see continuous steel wall frame bridging.

Termination bar is used to secure the top edge of the flashing to the concrete block wythe.

Terrain is the characteristics of land, especially as considered with reference to its natural features such as flat or sloping.

Texturing is applied to drywall to make the surface look even and cover any surface irregularities.

Thermal break system is a longitudinal channel, longitudinal flange, or side walls that create a hollow center in which a thermal barrier is integrally formed.

Thermal mass is a dense material that can effectively absorb and store heat, and release the heat as the home cools at night.

Thermal radiation is the heat that comes from the sun to the earth or emitted by other heat sources.

Thermal resistance is the ability of materials to slow heat transfer.

Thermal storage wall receives and stores energy from the sun during the day and releases the heat slowly at night.

Thermochromic technology is the property of substances to change color due to a change in temperature.

Thermoplastic, high-heat resistant, nylon coated (TTHT) is commercial wiring with increased insulation.

Thermostat is an automatic mechanism for controlling the amount of heating or cooling given by a central or zone heating or cooling system.

Thin-set mortar is used to adhere tile to the floor or wall, with a very thin layer of cement often containing other additives, such as acrylic for strength.

Thin veneer bricks have normal face dimensions with reduced thickness for application to surfaces with adhesive.

36/7 pattern is a common deck material attachment pattern is used to place deck fasteners every 36 in. on center length and 7 in. on center apart in a grid pattern that results in the fasteners being placed in the deck flat channel.

3D model is a mathematical representation of any three-dimensional surface of an object developed using specialized software.

3/4 inch minus refers to the size of the rock, where 3/4 in. (19 mm) is the largest piece that fits through a 3/4 in. screen, and the approximate amount of fines in a product. Minus material can have 60 to 70 percent fines.

Three-dimensional roofing   see architectural roofing.

Three faced fireplace is open on three sides.

Three phase electrical systems use two smaller legs running 120 V each and one wider leg running 208 V.

Three-stud corner uses three studs to secure and strengthen the exterior corner and provide a nailing surface for sheathing, siding, and interior drywall or other interior finish materials.

Three-tab roofing is the most popular type of asphalt shingle usually 12×36 in. with three profile sections along the bottom of the shingle called tabs.

Three-way switch is used when two switches control a single light or group of lights.

Threshold is the metal, plastic, or wood sill on an exterior door.

Throat is an opening between a fire box and the chimney where the **damper** is located.

Through-wall flashing, also called a sill pan flashing at the bottom, is a special fabricated flashing with sealed seams to keep water from entering the structure and allow rainwater to flow out.

Tie-beams connect horizontally across the structure between posts.

Ties are wrapped around vertical steel in a column or placed horizontally in a wall or slab to help keep the structure from separating, and they keep the rebar in place while the concrete is being poured into the forms.

Timber-concrete-composite construction uses the combination of timber beams or panels connected to a reinforced concrete slab using connectors that secure the concrete to the wood.

Timber-frame construction uses generally large posts and beams or timbers for the horizontal and vertical members of the structural system. See also post and beam.

Title block provides a variety of information about the company, client, and the drawing, such as company and client name, the title of the drawing, sheet size, predominate scale, and sheet page number.

Ton is the sizing of a heat pump where each ton of rating removes approximately 12,000 Btu per hour (Btuh) of heat.

Tongue and groove (T&G) edges where one joining edge has a tongue that fits into the groove of the other joining edge

Top plate is a framing member on top of a stud wall on which joists rest to support an additional floor or to form a ceiling, or upon which rafters rest to form a roof.

Top soil is the upper part of the soil, which is usually rich in nutrients and most favorable for landscaping lawn and plant growth.

Top track replaces the top plate used in wood framing.

Topography is shown as lines representing given heights of the geographical landscape.

Total rise is the vertical dimension from the top of the plate to the top of the ridge.

Total run (rafter) is half the span measured from the outside of a top plate to the center of the ridge.

Total run (stairs) of the stairs is the horizontal distance from the face of the first riser to the face of the last riser.

Total station is a tripod-mounted instrument that contains an EDM device. Horizontal and vertical angles between the total station and the measured points, when combined with a distance measurement, can provide location and elevation (XYZ) data for any point.

TPO is the abbreviation and commercial name for thermoplastic polyolefin, which is a single-ply reflective roofing membrane made from polypropylene and ethylene-propylene rubber polymerized together.

Track steel framing member used for applications, such as band or rim joists, for flooring systems. A track has a web and two flanges, but no lips. Track web depth measurements are taken to the inside of the flanges.

Tract homes are built in a housing development where homes are similar and economical by keeping costs to a minimum by using standardized designs, materials, products, and labor.

Traditional architecture used here refers to architectural styles that evolved from the early American influence and from regions around the world.

Transformer is a device for reducing or increasing the voltage of an alternating current.

Transom is a horizontal crosspiece over a door or between a door and a window above the door, or a normally small horizontal operable window above a door or another window.

Trap is a U-shaped pipe below plumbing fixtures that holds water to prevent odor and sewer gas from entering the fixture.

Tread is the horizontal portion of a stair where you step when going up or down the stairs.

Trenching is the digging of a ditch to a desired width and depth with a trenching machine or excavation machine such as a backhoe.

Trimmer is any construction member that runs parallel to other framing members and used as support or to strengthen the perimeter of an opening.

Trombe wall is a thermal storage wall constructed as a massive dark-painted masonry or concrete wall situated a few inches inside and next to south-facing glass. The sun heats the air between the wall and the glass. The heated air rises and enters the room through vents at the top of the wall. At the same time, cool air from the floor level of adjacent rooms is pulled in through vents at the bottom of the wall.

Trowel (concrete) is a flat-bladed tool used for leveling, spreading, shaping, and smoothing concrete and mortar. This is the final concrete finishing process, unless the concrete shall be broomed, stamped, or has an exposed aggregate finish.

Trowel (drywall) is a flat-bladed hand tool used for spreading, leveling, or shaping joint compound in drywall finishing applications.

True firewall is a structurally stable wall that is certified to prevent fire from spreading from one side of a building to the other.

True north is determined by a line from the North Pole to the South Pole. True north is the same as geographic north, which is the North Pole.

Truss a prefabricated or job-built construction member formed of triangular shapes used to support roof or floor loads over long spans.

Turf grass   see sod.

Turnbuckle is a device for adjusting the tension or length of ropes, cables, or rods, and generally made of two threaded fasteners, one with a left-hand thread and the other with a right-hand thread.

Turning point (TP) is The most temporary TBM used when level measurements must be made over long distances or large changes in elevation, and can be a large rock, fire hydrant (FH), concrete curb or foundation, wooden stake, or metal spike, on which the rod is rested and used as a pivot for the rod.

24 in. (600 mm) modules, which means that every dimension is a 24-in. increment, such as 30 × 48 ft (9000 × 14,600 mm).

Two-faced fireplace is open on two sides.

Two-pipe hot water system, has two pipes running throughout the home. One pipe supplies heated water to all of the outlets. The other is a return pipe, which carries the water back to the boiler for reheating.

Two phase electrical wiring is where two wires each provide the same voltage AC out of phase with each other.

Two-story home has two floors with two full-height walls built with the second framed above the first, and there is a stairs for access between floors.

Two-stud corner has the least number of studs and allows for the maximum amount of insulation in the corner.

Type X gypsum is 5/8 in. thick drywall, manufactured for use in locations where building codes require a fire resistance rating in home construction.

Typical cross sections shows the general arrangement of construction through the entire building and often have details correlated to them.

Typical wall section provides more detail than is normally found in a building section, and is drawn at a larger scale such as 1/2″″ = 1′-0″, 3/4″ = 1′-0″, or 1″ = 1′-0″.

U groove weld is created when the groove between two parts is in the form of a U.

U-shaped stairs makes a 180° turn at a landing.

U-value is the coefficient of heat transfer expressed as Btuh sq ft/°F of surface area.

UV-C light has a wavelength of 253.7 nm, which is called the germicidal bandwidth.

Ultraviolet (UV) light can be beneficial for improving indoor air quality.

Unbalanced beams is when the quality of lumber used on the tension side of the beam is higher than the lumber used on the corresponding compression side, allowing a more efficient use of the timber resource.

Underlayment is construction material used over subflooring and under the finish floor material to provide a base for finish floor material.

Underwriters Laboratories (UL) is global safety science company that tests the latest products and technologies for safety before they are marketed to consumers.

UniFormat is a uniform classification system for organizing preliminary construction information into a standard order or sequence on the basis of functional elements.

Unit skylight is a complete factory-assembled glass- or plastic-glazed opening consisting of not more than one panel of glass or plastic installed in a sloped or horizontal orientation primarily for natural daylighting.

Universal design refers to broad range concepts meant to produce buildings, products, and environments that are characteristically accessible to older people, people without disabilities, and people with disabilities.

Urban heat island effect and potentially tempers heating and cooling loads in the building. The urban heat island effect means that city areas are warmer than suburbs or rural areas due to less vegetation, more land coverage, and other infrastructure.

Utility grid, also referred to as **grid,** is the transmission system for electricity that is a network of coordinated power providers and consumers that are connected by transmission and distribution lines and operated by one or more control centers.

V groove weld is formed between two adjacent parts when the side of each part is beveled to form a groove between the parts in the shape of a V.

VAC is the abbreviation for volts alternating current.

Valley is the corner of an interior angle formed by two intersecting interior roof surfaces.

Valley jack rafter is the jack rafter that connects to a valley.

Valley rafter is the rafter that forms the valley and runs from the ridge to the top plate.

Valve is a fitting used to control the flow of fluid or gas.

Vanity is a bathroom lavatory fixture that is freestanding or in a cabinet.

Vapor barrier or vapor diffusion retarder is a material that reduces the rate at which water vapor can move through a material.

Vapor diffusion retarder   see vapor barrier.

Variance is a request to depart from zoning requirements.

Vaulted ceiling can have unequal sloping sides, a single sloping side, an arch shape. Vaulted ceilings can have the same pitch or a different pitch than the roof.

Veneer is thin sheets of wood glued together to form plywood or glued to a wood base material.

Vent is an opening designed to convey air, heat, water vapor, or other gas from inside a building or a building component to the atmosphere.

Vent-free fireplace is installed without a flue, does not draw outside air to fuel the fire, and has no exhaust gases to the outside.

Vent pipe is the pipe installed to ventilate the building drainage system and to prevent drawing liquid out of traps and stopping back pressure.

Vent stack is a general term referring to any vertical pipe for soil waste or vent piping.

Vent system allows for a continuous flow of air through the system so that gases and odors can dissipate and bacteria do not have an opportunity to develop.

Verdyol Biotic Earth™ is the original biotic soil amendment used to achieve sustainable vegetation on less than ideal soils.

Vertical datum provides the zero elevation reference for elevations of surfaces and features on the earth.

Vestibule   see air-lock entry.

View orientation provides optimum exposure to a view and can be a major factor in the purchase of property for home construction.

Vinyl flooring   see resilient flooring.

Virtual refers to something that appears to have the properties of a real or actual object or experience.

Virtual reality (VR) refers to a world that appears to be a real or actual world, having many of the properties of a real world.

Visible transmittance (VT) measures how much light comes through a product.

Volatile organic compounds (VOC) are chemicals contained in the items used in home construction, which can emit pollutants throughout the lifespan of the product.

Volt (V) is a unit of measure for electrical force.

WA means warm air.

Waffle concrete floor system refers to a pattern of indentations on each side, formed by the grid-like design on each side or one side.

Waffle rubber carpet pad is rubber material pressed into a waffle shape that provides a soft cushion that is not designed for use in high traffic areas.

Wainscot is any wall finish where the material on the bottom portion of the wall is different from the upper portion.

Walk-through can be described as a camera in a computer program that is set up like a person walking through a building, around a building, or through a landscape.

Wall ties are used to transfer lateral forces, such as wind loads, on the masonry veneer wall back to the wood frame structure.

Wallpaper is a paper or vinyl product that usually has printed decorative patterns and colors, used for pasting in vertical strips over the walls and ceilings to provide a decorative or textured surface.

Walls are typically vertical structures made from wood, steel, concrete, or masonry used to enclose, or divide the floor area based on the design provided in the plans.

Waste pipe is a pipe that carries only liquid waste free of fecal material.

Waste stack is a vertical pipe that runs one or more floors and carries the discharge of fixtures other than water closets and similar fixtures.

Water-based geothermal system has closed-loop and open-loop options. The closed-loop system has water or water-antifreeze fluid pumped through polyethylene tubes. The open-loop system has water pumped from a well or reservoir through a heat exchanger, and then discharged into a drainage ditch, a field tile, a reservoir, or another well.

Water closet water-flushing plumbing fixture, such as a toilet, that is designed to receive and discharge human excrement. This term is sometimes used to mean the compartment where the fixture is located.

Water distribution pipe carries water from the service to the point of use.

Water heater is an appliance used for heating, storing and distributing hot water.

Water main is a primary water supply pipe, generally located in the street or public utility easement for public water.

Water meter is a device used to measure the amount of water that goes through the water service.

Water service is generally 1 inch plastic or galvanized steel pipe. This size may vary in relation to the service needed.

Water well is structure in the ground created by digging, driving, boring or drilling to access groundwater in underground aquifers.

Waterstop is a component of a concrete or masonry structure, intended to prevent the passages of water running continuously through the joints.

Watt or wattage refers to a unit measure of power corresponding to the power in an electric circuit in which the potential difference is one volt and the current one ampere.

Web is the part of a C-Shape or track that connects the two flanges.

Web opening   see punchout.

Web stiffener is additional material attached to the web to strengthen the members.

Weep holes (masonry) are openings in the first course of masonry that allow water to drain out through the bottom of the wall, and help dry the structure by provide air circulation behind the masonry veneer.

Weep holes (windows) are fabricated through the frame at the outside sill of aluminum and vinyl frame windows to allow moisture and rain to escape from the window.

Weld-all-around is when a welded connection must be performed all around a feature, the weld-all-around symbol is attached to the reference line at the junction of the leader.

Weld length is given when a weld is not continuous along the length of a part.

Weld symbol indicates the type of weld and is part of the welding symbol.

Welded wire reinforcement (WWR) is steel wires spaced a specified distance apart in a square grid, and the wires are welded together.

Welding is a process of joining two or more pieces of like metals by heating the material to a temperature high enough to cause softening or melting.

Welding specification is a detailed statement of the legal requirements for a specific classification or type of product.

Welding symbols identify the location of the weld, the welding process, the size and length of the weld, and other weld information.

Weldment is an assembly of parts welded together.

Welds are created using welding processes, and are classified according to the type of joint on which they are used.

Whole building refers to the building assembly designed and built to maximize sustainable and economic function through the use of energy and other resources, building materials, site preservation, and indoor air quality for a structure to run at its maximum efficiency, provide a comfortable and healthy environment, and have the minimum impact on the environment.

Wi-Fi is the name of wireless networking technology that uses radio waves to provide wireless high-speed Internet and network connections.

Wind farm is a large-scale application using several technologically advanced wind turbines grouped together.

Wind fence   see sand fence.

Wind shear refers to the variation of wind over either horizontal or vertical surfaces.

Wind shelf   see smoke shelf.

Winder is a special type of **tread** used for making a turn in a staircase at a mid-staircase landing.

Window is a fixed or operable framework enclosing a pane of glass or more than one pane of glass separated by mullions and muntons and is typically framed in a wall or roof and functions to let in light and air to the area or room where the window is mounted in the home.

Window jambs are the sides, top and bottom members that continue from the window frame inside to cover the framing and extend flush with the drywall or other wall finis material.

Window tint is special film or coating applied to windows for a variety of reasons, most commonly to prevent a specific spectrum or amount of sunlight from passing through the glass.

Window walls are framed window panels mounted in between the concrete floor slabs from the top of the floor slab at the bottom of the window wall to the bottom of the next floor slab above attaching the top of the window wall.

Wire gauge is a method of defining wire diameter by a number, with wire diameter increasing as the number gets smaller.

Wood flooring is any finish flooring product manufactured from wood.

Wood-frame construction is the use of wood to build all exterior and interior walls, floor, and roof constructions.

Wood shakes are split from slabs of wood, making them rougher and less uniform than wood shingles.

Wood shingles are sawn from slabs of wood with a taper, making one end thicker than the other.

Work results are traditional construction practices that typically result from an application of skills to construction products or resources.

Working drawings are the complete set of drawing for a project.

Wythe is a continuous vertical section of masonry one unit in thickness.

Zero clearance fireplace is a manufactured steel fireplace that can be placed safely near combustible material.

Zero-lot-line refers to real estate property where the structure is allowed, by zoning, to be on or very near to the **property line.**

Zipper trusses use compression and tension members to transfer gravity loads to the timber structural system.

Zone of influence is a zone surrounding a piling or group of pilings, in which the stresses caused by the piling construction is at a desired engineered level for supporting the structural loads.

Zoned heating system requires one heater and one thermostat per room.

Zoning (land use planning) used by local governments for land use planning based on mapped zones which separate one set of land uses from another.

Zoning (drawing sheets) is a system of numbers along the top and bottom margins and letters along the left and right margins of a sheet. Zoning allows the drawing to be read like a road map.

Index

Note: Figures are indicated by *f*